Phytochemical Drug Discovery for Central Nervous System Disorders

Phytochemical Drug Discovery for Central Nervous System Disorders

Biochemistry and Therapeutic Effects

Edited by

Chukwuebuka Egbuna
Africa Centre of Excellence for Public Health and Toxicological Research (ACE-PUTOR)
University of Port Harcourt, Choba, Nigeria

Mithun Rudrapal
Vignan's Foundation for Science, Technology & Research (Deemed to be University)
Guntur, Andhra Pradesh, India

WILEY

Published by John Wiley & Sons, Inc., Hoboken, New Jersey.
Published simultaneously in Canada.

Library of Congress Cataloging-in-Publication Data
Names: Egbuna, Chukwuebuka, editor. | Rudrapal, Mithun, editor.
Title: Phytochemical drug discovery for central nervous system disorders : biochemistry and therapeutic effects / edited by Chukwuebuka Egbuna, Mithun Rudrapal.
Description: Hoboken, NJ : Wiley, 2023. | Includes bibliographical references and index.
Identifiers: LCCN 2023002458 (print) | LCCN 2023002459 (ebook) | ISBN 9781119794097 (cloth) | ISBN 9781119794103 (adobe pdf) | ISBN 9781119794110 (epub)
Subjects: MESH: Central Nervous System Diseases–drug therapy | Phytochemicals–therapeutic use | Phytotherapy–methods | Drug Discovery
Classification: LCC RC350.C54 (print) | LCC RC350.C54 (ebook) | NLM WL 301 | DDC 616.8/0461–dc23/eng/20230429
LC record available at https://lccn.loc.gov/2023002458
LC ebook record available at https://lccn.loc.gov/2023002459

Cover Design: Wiley
Cover Images: © ARTFULLY PHOTOGRAPHER/Shutterstock; picmedical/Shutterstock

Set in 9.5/12.5pt STIXTwoText by Straive, Pondicherry, India

Contents

Contributors *xv*
Preface *xxi*

1 **Central Nervous System Disorders and Food and Drug Administration–Approved Drugs** *1*
Estella U. Odoh, Chukwuebuka Egbuna, Chukwuma M. Onyegbulam, Diovu E. Obioma, Linda A. Onugwu, Obinna S. Onugwu, and Mithun Rudrapal
1.1 Incidence and Prevalence of Major Neurologic Disorders *2*
1.2 Etiology *2*
1.3 Pathogenesis *3*
1.4 Central Nervous System Disorders and Drugs Approved by the Food and Drug Administration *3*
1.4.1 Attention-Deficit/Hyperactivity Disorder *4*
1.4.2 Migraine *8*
1.4.3 Parkinson's Disease *9*
1.4.4 Multiple Sclerosis *11*
1.4.5 Alzheimer's Disease *11*
1.4.6 Muscular Dystrophy *12*
1.4.7 Epilepsy/Seizure *12*
1.5 Conclusion *13*
References *13*

2 **Drug Discovery from Medicinal Plants against Parkinson's Disease** *17*
Dunya Al-Duhaidahawi
2.1 Pathogenesis of Parkinson's Disease *17*
2.1.1 Misfolding and Aggregation of Proteins *17*
2.1.2 Mitochondrial Dysfunction *18*
2.1.3 Neuroinflammation *18*
2.1.4 Metal Ion Imbalance in the Brain *18*
2.1.5 Protease-Mediated Degradation Inhibition *18*
2.1.6 Oxidative Stress *19*
2.2 Natural Dopaminergic Neuroprotective Compounds *19*
2.2.1 Polyphenols *19*
2.2.1.1 Resveratrol *19*
2.2.1.2 Sesamin *19*
2.2.1.3 Curcumin *19*
2.2.1.4 6-Shogaol *20*
2.2.1.5 Fustin *20*
2.2.1.6 Biochanin A *20*

2.2.1.7 Acacetin *20*
2.2.1.8 Baicalein *20*
2.3 Nitrogenated Phytochemicals *21*
2.3.1 Alkaloids *21*
2.3.1.1 Tetrahydro Berberine *21*
2.3.1.2 Berberine *21*
2.3.1.3 Celastrol *21*
2.3.2 Saponins *21*
2.3.2.1 Astragaloside IV *21*
2.3.2.2 Ginseng *21*
2.4 Chinese Herbal Medications and Parkinson's Disease *22*
2.4.1 *Acanthopanax* *22*
2.4.2 *Alpinia* *22*
2.4.3 *Camellia* *22*
2.4.4 *Cassia* *23*
2.4.5 *Chrysanthemum* *23*
2.4.6 *Cistanche* *23*
2.4.7 *Gastrodia* *23*
2.4.8 *Ginkgo* *24*
2.4.9 *Gynostemma* *24*
2.4.10 *Paeonia* *24*
2.4.11 *Panax* *25*
2.4.12 *Polygala* *25*
2.4.13 *Polygonum* *25*
2.4.14 *Psoralea* *26*
2.5 Herbal Medicines from India and Parkinson's Disease *26*
2.5.1 *Withania somnifera* *26*
2.5.2 *Tinospora cordifolia* *26*
2.5.3 *Mucuna pruriens* *26*
2.6 European Plants *27*
2.7 α-Synuclein as a Potential Therapeutic Target *27*
2.7.1 Phytochemicals Targeting the α-Synuclein Cascade *27*
2.8 Conclusion *28*
References *39*

3 Drug Discovery from Medicinal Plants against Alzheimer's Disease *43*
Mona M. Marzouk, Mai M. Farid, Nesrine M. Hegazi, and Shahira M. Ezzat
3.1 Pathogenesis *44*
3.1.1 Amyloid Plaques *44*
3.1.2 Neurofibrillary Tangles *45*
3.1.3 Synaptic Damage *45*
3.1.4 Cholinergic Functions *45*
3.1.5 Oxidative Stress *45*
3.1.6 Neuroinflammation *45*
3.2 Treatment Strategies for Alzheimer's Disease *46*
3.2.1 Amyloid Hypothesis *46*
3.2.2 Tau Proteins Hypothesis *46*
3.2.3 Dendritic Hypothesis *54*
3.2.4 Cholinergic Hypothesis *55*
3.2.5 Strategies for 5-HT_6 Obstruction *55*
3.2.6 Metabolic Hypothesis *55*

3.3 Medicinal Plants Having Effects against Alzheimer's Disease *56*
3.3.1 Medicinal Plants Targeting the Cholinergic System *56*
3.3.2 Medicinal Plants Targeting Amyloid Beta *61*
3.3.3 Medicinal Plants Targeting Tau-Related Pathways *65*
3.4 Natural Products with Proven Anti-Alzheimer's Activity *66*
3.5 Conclusion *66*
References *70*

4 Effects of Medicinal Plants and Phytochemicals on Schizophrenia *81*
Nithya Rani Raju, S.V. Rashmitha, S. Pavithra, Erika Amparo Torres, M. Kishor, and Raghu Ram Achar
4.1 Mechanisms of Action Related to Schizophrenia *81*
4.1.1 Vesicular Monoamine Transporters *82*
4.1.2 Dopamine Receptors/Transporters *82*
4.1.3 Serotonin Receptors/Transporters *83*
4.1.4 Glutamate Transporters *84*
4.1.5 Gamma-Aminobutyric Acid Level/Receptors *84*
4.1.6 Genes Related to Schizophrenia *87*
4.2 Ayurvedic Plants Used as Treatment for Schizophrenia and Related Psychoses *88*
4.2.1 *Allium cepa* *88*
4.2.2 *Acorus calamus.* Linn *88*
4.2.3 *Bacopa monnieri* *89*
4.2.4 *Carum carvi* *89*
4.2.5 *Celastrus panicutalus* *90*
4.2.6 *Centella asiatica* *90*
4.2.7 *Convolvulus pluricaulis* *91*
4.2.8 *Coriandrum sativum* *91*
4.2.9 *Cuminum cyminum* L. *91*
4.2.10 *Cyperus Rotundus* Linn. *92*
4.2.11 *Eclipta alba* *92*
4.2.12 *Ficus religiosa* *92*
4.2.13 *Glycyrrhiza glabra* *93*
4.2.14 *Moringa oleifera* *93*
4.2.15 *Nigella sativa* *93*
4.2.16 *Piper longum* *94*
4.2.17 *Rauwolfia serpentina* *94*
4.2.18 *Sesbania grandiflora* *95*
4.2.19 *Sphaeranthus indicus* *95*
4.2.20 *Tinospora cordifolia* *96*
4.2.21 *Valeriana wallichii* *96*
4.2.22 *Withania somnifera* *97*
4.2.23 *Ziziphus mauritiana* *97*
4.3 Conclusion *97*
References *122*

5 Drug Discovery from Medicinal Plants and Phytochemicals against Neuropathic Pain *137*
Santwana Palai, Shyam S. Kesh, and Mithun Rudrapal
5.1 Mechanisms of Neuropathic Pain *138*
5.2 Animal Models for Studying Neuropathic Pain *139*
5.2.1 Streptozotocin-Induced Diabetes *139*

5.2.2 High-Fat Diet *139*
5.2.3 Sciatic Nerve Injury *139*
5.2.4 Chemotherapy-Induced Peripheral Neuropathy *139*
5.3 Medicinal Plants and Phytochemicals against Neuropathic Pain *139*
5.4 Role of Plants and Phytochemicals in Different Neuropathic Pain Models *140*
5.4.1 Diabetic Neuropathy *140*
5.4.2 Chemotherapy-Induced Peripheral Neuropathy *153*
5.4.3 Sciatic Nerve Chronic Constriction Injury *154*
5.4.4 Other Neuropathic Pain Signaling Pathways *154*
5.5 Future Perspectives *155*
5.6 Conclusion *155*
References *155*

6 Brain Function, Stroke, and Medicinal Herbs *161*
Shahira M. Ezzat, Rana Marghany, Nehal El Mahdi, and Mohamed A. Salem
6.1 Brain Function and Stroke *161*
6.2 Strategies for Treatment of Ischemic Stroke *162*
6.2.1 Neuroprotective Strategy *162*
6.2.1.1 Oxidative Stress Targeting *162*
6.2.1.2 Excitotoxicity Targeting *162*
6.2.1.3 Apoptosis Targeting *162*
6.2.2 Neurorestorative Strategy *162*
6.2.2.1 Angiogenesis Targeting *162*
6.2.2.2 Neurogenesis Targeting *163*
6.3 Medicinal Plants for the Treatment of Stroke *163*
6.4 Natural Products for the Treatment of Stroke *165*
6.5 Recent Applications of Nanomedicine for Treatment of Stroke *165*
6.6 Conclusion *174*
References *174*

7 Plant-Based Analgesics *181*
Gabriel O. Anyanwu and Dorathy O. Anzaku
7.1 Current Analgesic Drugs and Their Mechanisms of Action *182*
7.2 Plant-Derived Lead Compounds with Analgesic Activities *182*
7.2.1 Saponins and Terpenoids *184*
7.2.2 Flavonoids *184*
7.2.3 Alkaloids *184*
7.2.4 Glycosides *185*
7.3 Analgesic Effects of Medicinal Plants Found in Nigeria *185*
7.4 Limitations of Plant-Based Analgesics *194*
7.5 Future Directions and Perspective for Plant-Based Analgesics *194*
7.6 Conclusion *194*
References *195*

8 Medicinal Plants and Phytochemicals against Depression *203*
Neelma Munir, Ayesha Qamar, Maria Hasnain, Huma Waqif, Maria Hanif, Zirwa Sarwar, and Shagufta Naz
8.1 Causes of Depression *203*
8.1.1 Genetic Factors *204*
8.1.2 Environmental Factors *204*

8.1.3 Risk Factors for Depression *204*
8.2 Symptoms of Depression *204*
8.2.1 Symptoms of Depression in Men *204*
8.2.2 Symptoms of Depression in Women *205*
8.2.3 Symptoms of Depression in Children *205*
8.3 Diagnosis of Depression *205*
8.4 Types of Depression *205*
8.4.1 Major Depressive Disorder *206*
8.4.2 Persistent Depressive Disorder *206*
8.4.3 Depression with Psychosis *206*
8.4.4 Depression in Pregnancy *207*
8.5 Treatment of Depression *207*
8.5.1 Natural Treatment for Depression *208*
8.5.2 Medicinal Plants as Antidepressants *208*
8.5.2.1 *Lavandula officinalis* *209*
8.5.2.2 *Kaempferia parviflora* *210*
8.5.2.3 *Asparagus racemosus* *211*
8.5.2.4 *Passiflora foetida* *211*
8.5.2.5 *Momordica charantia* *212*
8.5.2.6 *Bacopa monniera* *212*
8.5.2.7 *Valeriana officinalis* *212*
8.5.2.8 *Rhodiola rosea* *212*
8.5.2.9 *Withania somnifera* *212*
8.5.2.10 *Matricaria recutita* *213*
8.5.2.11 *Ginkgo biloba* *213*
8.6 Conclusion *213*
References *214*

9 Anti-inflammatory Agents from Medicinal Plants *219*
Bui T. Tung, Tran V. Linh, Trinh P. Thao, and Nguyen D. Thuan
9.1 Role of Neuroinflammation in Neurodegenerative Diseases *220*
9.2 Neuroinflammatory Drugs *222*
9.3 Medicinal Plants as Sources of Anti-inflammatory Agents *231*
9.3.1 *Allium sativum* *232*
9.3.2 *Azadirachta indica* *233*
9.3.3 *Cassia tora* *233*
9.3.4 *Euphorbia hirta* *233*
9.3.5 *Garcinia mangostana* *234*
9.3.6 *Punica granatum* *235*
9.3.7 *Ricinus communis* *235*
9.3.8 *Scutellaria baicalensis* *236*
9.3.9 *Solanum melongena* *237*
9.4 Bioactive Compounds as Anti-inflammatory Agents *237*
9.4.1 Curcumin *238*
9.4.2 Eugenol *238*
9.4.3 Epigallocatechin 3-Gallate *239*
9.4.4 Guggulsterone *240*
9.4.5 Resveratrol *240*
9.4.6 Thymoquinone *241*
9.5 Conclusion *241*
References *242*

10 Plant-Based Products and Phytochemicals against Viral Infections of the Central Nervous System *251*
Santwana Palai, Shyam S. Kesh, and Mithun Rudrapal
10.1 Viral Infections of the Central Nervous System *252*
10.2 Plant and Phytochemicals as Antiviral Agents for Central Nervous System Viral Infections *252*
10.2.1 SARS Cov-2 Virus *252*
10.2.2 Japanese Encephalitis Virus *262*
10.2.3 West Nile Virus *263*
10.2.4 Tick-Borne Encephalitis Virus *264*
10.2.5 Herpes Simplex Virus *264*
10.2.6 Rabies Virus *265*
10.2.7 Varicella Zoster Virus *266*
10.2.8 Poliomyelitis Virus *266*
10.2.9 Human Immunodeficiency Virus *267*
10.3 Controlling Vectors of Viral Diseases of the Central Nervous System *268*
10.4 Future Perspectives *269*
10.5 Conclusion *269*
References *270*

11 Fruits and Nutraceuticals for the Prevention and Treatment of Central Nervous System Disorders *273*
Abeer M. A. El Sayed and Chukwuebuka Egbuna
11.1 Fruits for Cognition and Brain Health *274*
11.1.1 *Aegle marmelos* *274*
11.1.2 *Citrullus lanatus* *274*
11.1.3 *Citrus grandis* *274*
11.1.4 *Phoenix sylvestris* *274*
11.1.5 *Phyllanthus emblica* *274*
11.1.6 *Emblica officinalis* *275*
11.1.7 *Solanum torvum* *275*
11.1.8 *Terminalia chebula* *275*
11.1.9 Blackberries *275*
11.1.10 Blueberries *276*
11.1.11 Strawberries *276*
11.1.12 Raspberries *276*
11.1.13 Cherries *276*
11.1.14 Oranges *277*
11.1.15 Plums *277*
11.1.16 Prunes *277*
11.1.17 Red Grapes *277*
11.1.18 Pomegranates *278*
11.2 Nutraceuticals in Ameliorating Neurodegeneration *278*
11.2.1 Quercetin and Kaempferol *278*
11.2.2 Withanine *278*
11.2.3 Asiatic Acid *279*
11.2.4 Bhilavanol A and B *279*
11.3 Nutraceuticals in Alzheimer's Disease *279*
11.3.1 Flavonoids *279*
11.3.2 Apigenin *280*
11.3.3 Genistein, Daidzein, Glycitin *280*

11.3.4 Resveratrol *280*
11.3.5 Curcumin *280*
11.3.6 Carotenoids *280*
11.3.7 Crocin *281*
11.3.8 Carnosic Acid and Rosmarinic Acid *281*
11.3.9 Alkaloids *281*
11.4 Nutraceuticals in Parkinson's Disease *281*
11.4.1 Vitamins: Folate, Cobalamin, Pyridoxin *281*
11.5 Nutraceuticals in Depression *283*
11.6 Nutraceuticals in Psychotic Disorders *283*
11.7 Conclusion *283*
References *284*

12 Neurorestorative Potential of Medicinal Plants and Their hytochemicals *291*
Babatunde O. Adetuyi, Kehinde A. Odelade, Grace O. Odine, Oluwatosin A. Adetuyi, Semiloore O. Omowumi, Olubanke O. Ogunlana, and Chukwuebuka Egbuna
12.1 Therapeutic Value of Some Medicinal Plants and their Importance *292*
12.2 Types of Medicinal Plants and Their Uses *293*
12.3 Phytochemicals *293*
12.4 Phytochemical Constituents in Some Medicinal Plants *295*
12.4.1 Onions *295*
12.4.1.1 Phytochemical Constituents in Onions *295*
12.4.2 Turmeric *296*
12.4.2.1 Phytochemical Constituents in Turmeric *296*
12.4.3 Ginger *297*
12.4.3.1 Phytochemical Constituents in Ginger *297*
12.4.4 Garlic *297*
12.4.4.1 Phytochemical Constituents in Garlic *298*
12.5 The Brain *298*
12.5.1 Brain Physiology *299*
12.5.1.1 Neurotransmitters and Receptors *299*
12.5.1.2 Electrical Activity *300*
12.5.1.3 Metabolism *300*
12.5.2 Functions of the Brain *300*
12.5.2.1 Perception *300*
12.5.3 Motor Control *300*
12.5.4 Homeostasis *301*
12.5.5 Motivation *301*
12.5.6 Learning and Memory *302*
12.6 Brain Conditions *302*
12.7 Protective Effects of Medicinal Plants on the Brain *303*
12.7.1 *Crocus sativus* *303*
12.7.1.1 Medicinal Properties of *Crocus sativus* *303*
12.7.2 *Nigella sativa* *304*
12.7.2.1 Medicinal Properties of *Nigella sativa* *304*
12.7.3 *Coriandrum sativum* *304*
12.7.3.1 Medicinal Properties of *Coriandrum sativum* *304*
12.7.4 *Ferula assafoetida* *304*
12.7.4.1 Medicinal Properties of *Ferula assafoetida* *304*
12.7.5 *Thymus vulgaris* *305*
12.7.5.1 Medicinal Properties of *Thymus vulgaris* *305*

12.7.6 *Curcuma longa* *305*
12.7.6.1 Medicinal Properties of *Curcuma longa* *305*
12.8 Conclusion *305*
References *306*

13 Neurotransmitter Modulation by Phytochemicals *311*
Prachee Dubey and Kanti Bhooshan Pandey
13.1 Sources, Structures, and Classifications of Phytochemicals *311*
13.2 Neurotransmitters and Their Functions *316*
13.3 Modulation of Cholinergic Signaling by Phytochemicals *317*
13.3.1 Effect of Phytochemicals on Acetylcholinesterase *318*
13.4 Effect of Phytochemicals on GABAergic Signaling *318*
13.5 Effect of Phytochemicals on Glutamatergic Signaling *319*
13.6 Modulation of Serotonergic and Dopaminergic Signaling by Phytochemicals *320*
13.7 Conclusion *321*
Acknowledgments *321*
References *321*

14 Antipyretic Agents from Plant Origins *327*
Kamoru A. Adedokun, Sikiru O. Imodoye, Akeem O. Busari, Malik A. Sanusi, Abdullah Olawuyi, and Maroof G. Oyeniyi
14.1 Pyrexia Development, Its Mechanisms, and the Roles of Plant Metabolites as Antipyretics *328*
14.1.1 Fever Development *328*
14.1.2 Mechanisms of Fever Development *328*
14.1.2.1 Humoral Pathway *328*
14.1.2.2 Neural Pathway *330*
14.1.3 Roles of Plant Metabolites as Antipyretics *331*
14.2 Antipyretic Agents of Plant Origin *341*
14.2.1 *Arbutus andrachne* (Family Ericaceae) *341*
14.2.2 *Berberis* spp. (Family Berberidaceae) *343*
14.2.3 *Cassia fistula* Linn. (Family Caesalpiniaceae) *343*
14.2.4 *Crataeva magma* (Family Capparidaceae) *344*
14.2.5 *Echinops kebericho* M. (Family Asteraceae) *345*
14.2.6 *Enicostema littorale* (Family Gentianaceae) *346*
14.2.7 *Piper nigrum* (Family Piperaceae) *346*
14.2.8 *Viola betonicifolia* (Family Violaceace) *347*
14.3 Conclusion and Future Perspectives *348*
References *349*

15 Medicinal Herbs against Central Nervous System Disorders *359*
Ahmed Olatunde, Neelma Munir, Godwin Anywar, Maria Hanif, Huma Waqif, Habibu Tijjani, Barbara Sawicka, and Akram Muhammad
15.1 Medicinal Plants as Interventions for Central Nervous System Disorders *360*
15.2 Some Medicinal Plants with Neuroprotective Action on Central Nervous System Disorders *362*
15.2.1 *Desmodium adscendens, Cleome rutidosperma, Withania somnifera,* and *Panax ginseng* *362*
15.2.2 *Baccopa monnieri* and *Rauvolfia serpentina* *363*
15.2.3 *Avena sativa* and *Annona squamosa* *364*
15.2.4 *Acorus calamus, Emblica officinalis,* and *Strychnos nux-vomica* *365*
15.3 Some Central Nervous System Disorders and Medicinal Plant Interventions *366*

15.3.1 Depression *366*
15.3.1.1 Family Amaryllidaceae – *Allium cepa* *366*
15.3.1.2 Family Plantaginaceae – *Bacopa monnieri* *367*
15.3.1.3 Family Fabaceae – *Glycyrrhiza glabra* *368*
15.3.1.4 Family Lamiaceae – *Rosmarinus officinalis* *368*
15.3.1.5 Family Zingiberaceae – *Zingiber officinale* *369*
15.3.2 Epilepsy *369*
15.3.2.1 Family Hypoxidaceae – *Hypoxis hemerocallidea* *370*
15.3.2.2 Family Piperaceae – *Piper methysticum* *371*
15.3.2.3 Family Caprifoliaceae – *Valeriana officinalis* *372*
15.3.2.4 Family Phyllanthaceae – *Bridelia micrantha* *372*
15.3.2.5 Family Rubiaceae – *Sarcocephalus latifolius* *372*
15.3.3 Huntington's Disease *372*
15.3.3.1 Family Ginkgoaceae – *Ginkgo biloba* *373*
15.3.3.2 Family Araliaceae – *Panax ginseng* *373*
15.3.3.3 Family Asteraceae – *Calendula officinalis* *373*
15.3.3.4 Family Primulaceae – *Embelia ribes* *374*
15.3.3.5 Family Theaceae – *Camellia sinensis* *374*
15.4 Some Mechanistic Actions of Medicinal Herbs against Central Nervous System Disorders *374*
15.4.1 In Vitro Studies *375*
15.4.2 In Vivo Studies *375*
15.5 Conclusion *376*
References *376*

16 Important Antihistaminic Plants and Their Potential Role in Health *385*
Salwa Bouabdallah, Hagar A. Sobhy, Babatunde O. Adetuyi, Omayma A. Eldahshan, and Chukwuebuka Egbuna
16.1 Antihistaminic Plants *386*
16.1.1 Family Acanthaceae *386*
16.1.2 Family Amaranthaceae *386*
16.1.3 Family Amaryllidaceae *387*
16.1.4 Family Asclepiadaceae *387*
16.1.5 Family Asteraceae *387*
16.1.6 Family Caesalpiniaceae *387*
16.1.7 Family Casuarinaceae *387*
16.1.8 Family Cruciferae *388*
16.1.9 Family Cucurbitacea *388*
16.1.10 Family Euphorbiaceae *388*
16.1.11 Family Fabaceae *388*
16.1.12 Family Lamiaceae *388*
16.1.13 Family Moraceae *388*
16.1.14 Family Myricaceae *388*
16.1.15 Family Myrtaceae *389*
16.1.16 Family Olacaceae *389*
16.1.17 Family Piperaceae *389*
16.1.18 Family Poaceae *389*
16.1.19 Family Polygalaceae *390*
16.1.20 Family Scrophulariaceae *390*
16.1.21 Family Verbenaceae *390*
16.1.22 Family Zygophylaceae *390*

16.2 Bioactive Compounds with Antihistaminic Activities *390*
16.3 Conclusion *394*
References *394*

17 Effect of Plant-Based Anticonvulsant Products and Phytochemicals *397*
Muhammad Akram, Sadia Zafar, Hassan Shah, Zerfishan Riaz, Khawaja S. Ahmad, Muhammad Riaz, Naveed Munir, Muhammad Jahangeer, Imtiaz M. Tahir, Michael P. Okoh, Muhammad A. Ishfaq, David Pérez-Jorge, Vanessa de Andrade Royo, Muhammad M. Aslam, Chukwuebuka Egbuna, and Chukwunonso O. Igboekwe
17.1 Types of Epileptic Seizures *398*
17.2 Basic Mechanisms of Epilepsy *398*
17.3 Epilepsy and Oxidative Stress *400*
17.4 Epilepsy and Inflammation *401*
17.5 Tests for Seizure Induction *402*
17.6 Medicinal Plants Used to Treat Epilepsy *402*
17.7 Conclusion *403*
References *407*

18 Application of Nanophytomedicine for the Treatment of Central Nervous System Disorders *413*
Nilesh Rarokar, Nilambari Gurav, and Shailendra Gurav
18.1 Neurodegenerative Disease and the Blood–Brain Barrier *414*
18.1.1 Problems Associated with Treatment of Central Nervous System Disorders *414*
18.1.2 Role of the Blood–Brain Barrier *415*
18.1.3 Blood–Brain Barrier Crossing Mechanism *415*
18.1.4 Phytomedicine/Neuroprotective Drugs Reported for Central Nervous System Disorders *416*
18.2 Nano Approaches to Central Nervous System Drug Delivery *418*
18.2.1 Types of Nanocarriers *418*
18.2.1.1 Liposomes *418*
18.2.1.2 Micelles *418*
18.2.1.3 Solid Lipid Nanoparticles *418*
18.2.1.4 Phytosomes *419*
18.2.1.5 Nanosponges *419*
18.2.1.6 Nanoemulsions *419*
18.2.1.7 Dendrimers *420*
18.2.1.8 Nanoparticles *420*
18.2.2 Techniques/Preparation/Methods *420*
18.2.3 Mechanism of Action of Drug Release *421*
18.3 Nanophytomedicine for Treatment of Central Nervous System Disorders *422*
18.3.1 Alzheimer's Disease *422*
18.3.2 Parkinson's Disease *422*
18.3.3 Epilepsy *423*
18.3.4 Stroke *423*
18.3.5 Huntington's Disease *424*
18.3.6 Multiple Necrosis *425*
18.3.7 Tumors/Gliomas/Glioblastomas *425*
18.4 Challenges in Nanophytomedicine *425*
18.5 Conclusion *426*
References *426*

Index *431*

Contributors

Raghu Ram Achar
Division of Biochemistry, School of Life Sciences
JSS Academy of Higher Education and Research
Mysuru, Karnataka, India

Kamoru A. Adedokun
Department of Immunology
Roswell Park Comprehensive Cancer Center
Buffalo, NY, USA

Babatunde O. Adetuyi
Department of Natural Sciences
Faculty of Pure and Applied Sciences
Precious Cornerstone University
Ibadan, Oyo State, Nigeria

Oluwatosin A. Adetuyi
Department of Biochemistry
Osun State University
Osogbo, Osun State, Nigeria

Khawaja S. Ahmad
Department of Botany
University of Poonch
Rawalakot, Azad Jammu & Kashmir
Pakistan

Muhammad Akram
Department of Eastern Medicine and Surgery
Government College University Faisalabad
Faisalabad, Punjab, Pakistan

Dunya Al-Duhaidahawi
Department of Pharmacognosy
College of Pharmacy
University of Kufa
Al-Najaf, Iraq

Gabriel O. Anyanwu
Department of Biochemistry
Faculty of Science and Technology
Bingham University
Karu, Nasarawa State, Nigeria

Godwin Anywar
Department of Plant Sciences
Microbiology & Biotechnology
College of Natural Sciences
Makerere University
Kampala, Uganda

Dorathy O. Anzaku
Department of Biochemistry
College of Science and Technology
Covenant University
Ota Ogun State, Nigeria

Muhammad M. Aslam
Nuclear Institute of Agriculture (NIA)
Tando Jam, Sindh, Pakistan

Salwa Bouabdallah
Environmental Biomonitoring Laboratory
LBE (LR01/ES14)
Faculty of Sciences Bizerta
Carthage University
Zarzouna, Tunisia
Faculty of Sciences Tunis
University of Tunis El-Manar
Tunis, Tunisia
Pharmacognosy Department
Faculty of Pharmacy
Ain Shams University
Cairo, Egypt

Akeem O. Busari
Department of Medical Laboratory Science
Ladoke Akintola University of Technology
Ogbomosho, Oyo State, Nigeria

Prachee Dubey
Government G.I. College
Malwan, Fatehpur, India

Chukwuebuka Egbuna
Africa Centre of Excellence for Public
Health and Toxicological Research
(ACE-PUTOR)
University of Port Harcourt
Choba, Rivers State, Nigeria
Department of Biochemistry
Chukwuemeka Odumegwu
Ojukwu University
Uli, Anambra State, Nigeria

Omayma A. Eldahshan
Department of Natural Sciences
Faculty of Pure and Applied Sciences
Precious Cornerstone University
Ibadan, Oyo State, Nigeria

Nehal El Mahdi
Department of Pharmaceutics and
Industrial Pharmacy, Faculty of Pharmacy
October University for Modern Sciences
and Arts (MSA)
Giza, Egypt

Abeer M.A. El Sayed
Department of Pharmacognosy
Faculty of Pharmacy
Cairo University
Cairo, Egypt

Shahira M. Ezzat
Department of Pharmacognosy
Faculty of Pharmacy
Cairo University
Cairo, Egypt
Department of Pharmacognosy
Faculty of Pharmacy
October University for Modern Sciences
and Arts (MSA)
Giza, Egypt

Mai M. Farid
Department of Phytochemistry and Plant
Systematics
National Research Centre
Giza, Egypt

Nilambari Gurav
PES's Rajaram and Tarabai Bandekar
College of Pharmacy, Ponda
Goa University
Panaji, Goa, India

Shailendra Gurav
Department of Pharmacognosy
Goa College of Pharmacy
Goa University
Panaji, Goa, India

Maria Hanif
Department of Biotechnology
Lahore College for Women University
Lahore, Punjab, Pakistan

Maria Hasnain
Department of Biotechnology
Lahore College for Women University
Lahore, Punjab, Pakistan

Nesrine M. Hegazi
Department of Phytochemistry and Plant
Systematics
National Research Centre
Giza, Egypt

Chukwunonso O. Igboekwe
Africa Centre of Excellence in Public Health and Toxicological Research (ACE-PUTOR)
University of Port-Harcourt
Choba, Rivers State, Nigeria

Sikiru O. Imodoye
Department of Oncological Sciences
Huntsman Cancer Institute
University of Utah
Salt Lake City, UT, USA

Muhammad A. Ishfaq
Department of Chemistry
University of Poonch
Rawalakot, Azad Jammu & Kashmir Pakistan

Muhammad Jahangeer
Department of Biochemistry
Government College University Faisalabad
Faisalabad, Punjab, Pakistan

Shyam S. Kesh
Department of Veterinary Clinical Complex (Veterinary Biochemistry)
Faculty of Veterinary and Animal Sciences
West Bengal University of Animal and Fishery Sciences
Kolkata, West Bengal, India

M. Kishor
Department of Psychiatry
JSS Medical College & Hospital
JSS Academy of Higher Education & Research
Mysuru, Karnataka, India

Tran V. Linh
VNU University of Medicine and Pharmacy
Vietnam National University
Ha Noi, Vietnam

Rana Marghany
Department of Pharmacognosy
National Research Centre
Giza, Egypt

Mona M. Marzouk
Department of Phytochemistry and Plant Systematics
National Research Centre
Giza, Egypt

Akram Muhammad
Department of Eastern Medicine
Government College University Faisalabad
Faisalabad, Punjab, Pakistan

Naveed Munir
Department of Biochemistry
Government College University Faisalabad
Faisalabad, Punjab, Pakistan

Neelma Munir
Department of Biotechnology
Lahore College for Women University
Lahore, Punjab, Pakistan

Shagufta Naz
Department of Biotechnology
Lahore College for Women University
Lahore, Punjab, Pakistan

Diovu E. Obioma
Department of Pharmacognosy and Environmental Medicine
University of Nigeria
Nsukka, Enugu State, Nigeria

Kehinde A. Odelade
Department of Natural Sciences
Faculty of Pure and Applied Sciences
Precious Cornerstone University
Ibadan, Oyo State, Nigeria

Grace O. Odine
Department of Natural Sciences
Faculty of Pure and Applied Sciences
Precious Cornerstone University
Ibadan, Oyo State, Nigeria

Estella U. Odoh
Department of Pharmacognosy and Environmental Medicine
University of Nigeria
Nsukka, Enugu State, Nigeria

Olubanke O. Ogunlana
Department of Biochemistry
Covenant University
Ota, Ogun State, Nigeria

Michael P. Okoh
Department of Medical Biochemistry
Faculty of Basic Medical Sciences
College of Health Sciences, University of Abuja
Abuja, Federal Capital Territory, Nigeria

Ahmed Olatunde
Department of Medical Biochemistry
Abubakar Tafawa Balewa University
Bauchi, Bauchi State, Nigeria

Abdullah Olawuyi
Department of Medical Laboratory Science
General Hospital
Offa, Kwara State, Nigeria

Semiloore O. Omowumi
Department of Natural Sciences
Faculty of Pure and Applied Sciences
Precious Cornerstone University
Ibadan, Oyo State, Nigeria

Linda A. Onugwu
Department of Pharmaceutics
University of Nigeria
Nsukka, Enugu State, Nigeria

Obinna S. Onugwu
Department of Pharmacognosy
Enugu State University of Science and Technology
Agbani, Enugu State, Nigeria

Chukwuma M. Onyegbulam
Department of Pharmacognosy and Environmental Medicine
University of Nigeria
Nsukka, Enugu State, Nigeria

Maroof G. Oyeniyi
Department of Medical Laboratory Science
General Hospital
Offa, Kwara State, Nigeria

Santwana Palai
Department of Veterinary Pharmacology & Toxicology, College of Veterinary Science and Animal Husbandry
Odisha University of Agriculture and Technology
Bhubaneswar, Odisha, India

Kanti Bhooshan Pandey
CSIR – Central Salt & Marine Chemicals Research Institute
Bhavnagar, Gujarat, India

S. Pavithra
Division of Biochemistry
School of Life Sciences
JSS Academy of Higher Education and Research
Mysuru, Karnataka, India

David Pérez-Jorge
Department of Didactics and Educational Research
University of La Laguna
San Cristóbal de La Laguna
Tenerife, Spain

Ayesha Qamar
Department of Biotechnology
Lahore College for Women University
Lahore, Punjab, Pakistan

Nithya Rani Raju
Division of Biochemistry
School of Life Sciences
JSS Academy of Higher Education and Research
Mysuru, Karnataka, India

Nilesh Rarokar
Department of Pharmaceutical Sciences
Rashtrasant Tukadoji Maharaj
Nagpur University
Nagpur, Maharashtra, India

S.V. Rashmitha
Division of Biochemistry
School of Life Sciences
JSS Academy of Higher Education and Research
Mysuru, Karnataka, India

Vanessa de Andrade Royo
Department of General Biology
Laboratory of Natural Products
State University of Montes Claros
Montes Claros, Minas Gerais, Brazil

Muhammad Riaz
Department of Allied Health Sciences
Government College University Faisalabad
Faisalabad, Punjab, Pakistan

Zerfishan Riaz
Department of Eastern Medicine and Surgery
Government College University Faisalabad
Faisalabad, Punjab, Pakistan

Mithun Rudrapal
Department of Pharmaceutical Sciences
School of Biotechnology and Pharmaceutical Sciences
Vignan's Foundation for Science Technology & Research (Deemed to be University)
Guntur, Andhra Pradesh, India

Mohamed A. Salem
Department of Pharmacognosy and Natural Products, Faculty of Pharmacy
Menoufia University
Menoufia, Egypt

Malik A. Sanusi
Department of Biochemistry and Molecular Biology
Obafemi Awolowo University
Ile-Ife, Osun State, Nigeria

Zlrwa Sarwar
Department of Biotechnology
Lahore College for Women University
Lahore, Punjab, Pakistan

Barbara Sawicka
University of Life Sciences in Lublin
Lublin, Poland

Hassan Shah
Department of Eastern Medicine and Surgery
Government College University Faisalabad
Faisalabad, Punjab, Pakistan

Hagar A. Sobhy
Department of Natural Sciences
Faculty of Pure and Applied Sciences
Precious Cornerstone University
Ibadan, Oyo State, Nigeria

Imtiaz M. Tahir
College of Allied Health Professionals
Government College University Faisalabad
Faisalabad, Punjab, Pakistan

Trinh P. Thao
VNU University of Medicine and Pharmacy
Vietnam National University
Ha Noi, Vietnam

Nguyen D. Thuan
VNU University of Medicine and Pharmacy
Vietnam National University
Ha Noi, Vietnam

Habibu Tijjani
Department of Biochemistry
Bauchi State University
Gadau, Bauchi State, Nigeria

Erika Amparo Torres
Department d'Enginyeria Quimica
Universitat Rovira i Virgili
Av Paisos
Catalans, Tarragona
Spain

Bui T. Tung
VNU University of Medicine and Pharmacy
Vietnam National University
Ha Noi, Vietnam

Huma Waqif
Department of Biotechnology
Lahore College for
Women University
Lahore, Punjab, Pakistan

Sadia Zafar
Department of Botany
Division of Science and Technology
University of Education Lahore
Lahore, Punjab, Pakistan

Preface

The growing incidence of central nervous system (CNS) disorders has become a serious public health problem worldwide. This is particularly because of the increased number of people aged over 65, which has led to growing market demand for safe, effective, and affordable medicines for CNS disorders. Though a number of synthetic drugs are available for use, their clinical efficacy, ability to survive the blood–brain barrier, safety, and affordability have been the major bottlenecks.

The effectiveness of phytomedicines against a wide variety of chronic CNS disorders has been documented. This book, *Phytochemical Drug Discovery for Central Nervous System Disorders: Biochemistry and Therapeutic Effects*, focuses on the use of plants and herbal medicines/products for effectively targeting and treating chronic CNS disorders, including Alzheimer's disease, Parkinson's disease, schizophrenia, convulsions, depression, and viral CNS infections. The chapters offer valuable information on various phytochemicals, their sources, biochemical effects, and therapeutic efficacy, which could help them further develop as lead molecules/candidates for drug discovery against CNS disorders. The exceptional structure and rich contents of this book are timely and the need of the hour.

The book discusses CNS-active plant products, neuroprotective phytochemicals, plant-based nutraceutical products, drug discoveries from phytomedicines/medicinal plants, and nanophytomedicines used for the treatment/management of CNS disorders. Some of the key features of the book are as follows:

Presents up-to-date information on Food and Drug Administration-approved drugs, plant-based products, and phytochemicals used in CNS disorders.

Details drug discovery opportunities from medicinal plants and phytochemicals for CNS disorders.

Highlights the nutraceutical and nanophytomedicine-based therapeutic management of CNS disorders.

Written by a global team of experts, this book will be useful to drug discovery scientists, drug developers, medicinal chemists, phytochemists, pharmacologists, biochemists, healthcare professionals, researchers, faculty, and students.

1

Central Nervous System Disorders and Food and Drug Administration–Approved Drugs

Estella U. Odoh[1], Chukwuebuka Egbuna[2,3], Chukwuma M. Onyegbulam[1], Diovu E. Obioma[1], Linda A. Onugwu[4], Obinna S. Onugwu[5], and Mithun Rudrapal[6]

[1] *Department of Pharmacognosy and Environmental Medicine, University of Nigeria, Nsukka, Enugu State, Nigeria*
[2] *Africa Centre of Excellence for Public Health and Toxicological Research (ACE-PUTOR), University of Port Harcourt, Choba, Rivers State, Nigeria*
[3] *Department of Biochemistry, Chukwuemeka Odumegwu Ojukwu University, Uli, Anambra State, Nigeria*
[4] *Department of Pharmaceutics, University of Nigeria, Nsukka, Enugu State, Nigeria*
[5] *Department of Pharmacognosy, Enugu State University of Science and Technology, Agbani, Enugu State, Nigeria*
[6] *Department of Pharmaceutical Sciences, School of Biotechnology and Pharmaceutical Sciences, Vignan's Foundation for Science, Technology & Research (Deemed to be University), Guntur, Andhra Pradesh, India*

The central nervous system (CNS), which is a sophisticated system that regulates and coordinates body activity, consists of nerves in the brain and spinal cord. The brain is a part of the CNS housed inside the skull. The brain controls movement, sleep, hunger, thirst, and practically every other essential activity required for living. The brain logically governs animals' actions. A centralized brain allows groups of muscles to be co-activated in complicated patterns, allowing stimuli impinging on one region of the body to elicit reactions in another, preventing various body parts from acting in opposition [1].

A CNS disorder is a disease condition in which the brain functions abnormally, with consequent health limitations. CNS disorder could be a result of an inherited metabolic disorder; harmful effects of microbial infection, a degenerative condition, stroke, brain tumor, or other problem; or might arise from unknown or multiple factors. Disorders of the CNS encompass a broad array of maladies, ranging from restless leg syndrome to traumatic brain injury [2].

CNS disorder is indicative of dysfunction of the brain or the spinal cord, or of faulty genes [3]. Stroke, infections such as meningitis, carpal tunnel syndrome, and functional issues such as headache are all causes of nervous system illnesses. A transient ischemic attack (TIA), subarachnoid hemorrhage, subdural hemorrhage, hematoma, and extradural hemorrhage are all vascular disorders that cause a stoppage of the blood flow to the brain. Approximately every 60 seconds in Nigeria, someone has a stroke due to a blocked blood vessel resulting in leakage of blood to the brain. When there is insufficient oxygen to the blood and brain, the brain cell can die and result in permanent damage. In the case of an autoimmune disorder, the immune system attacks and destroys healthy body tissues, which is caused by protein intolerance in the body, making immune cells recognize these as foreign and directing the immune system against them. Infections such as meningitis, encephalitis,

Phytochemical Drug Discovery for Central Nervous System Disorders: Biochemistry and Therapeutic Effects, First Edition. Edited by Chukwuebuka Egbuna and Mithun Rudrapal.

polio, and an epidural abscess are often caused by the invasion of a microorganism or virus. Structural disorders include brain or spinal cord damage, Bell's palsy, cervical spondylosis, carpal tunnel syndrome, brain or spinal cord tumors, peripheral neuropathy, and Guillain–Barré syndrome. The functional conditions are headache, epilepsy, dizziness, and neuralgia. A degenerative condition such as Parkinson's disease, a promoting illness of CNS, results in the death of dopamine-producing brain cells affecting motor skills and speech, multiple sclerosis (MS), damage to the myelin sheath of the neuron, amyotrophic lateral sclerosis (ALS), Huntington's chorea, and Alzheimer's disease [4]. Children born with a structural disorder may have malformed limbs and heart and facial abnormalities.

1.1 Incidence and Prevalence of Major Neurologic Disorders

CNS disorder causes significant morbidity and mortality and often requires neurosurgical intervention for proper diagnosis and treatment. The incidence of CNS disorder is consistently highest in low-income countries, followed by middle- and then high-income countries. About a million persons are suffering from chronic pain and various forms of sleep disturbance, and approximately 70 million individuals are experiencing vestibular and balance-associated problems. These CNS disorders constitute five out of the top ten neurologic diseases. In addition, chronic tinnitus, substance-abuse disorders, blindness, and visual impairment affect about 20 million adults. Of the 1.2 million diagnosed adult-onset brain disorders, 51.3% and 21% are due to stroke and Alzheimer's disease, respectively. The sum of annual incidences of Parkinson's disease and traumatic brain injury is the same as the number of epilepsy episodes (135 million). In addition, nearly 500 000 cases of brain tumor, MS, and ALS are diagnosed each year. For example, 6.5–7.9% of adults in the United States are affected by a brain impairment [3].

An analysis of FDA-approved new drugs shows a slow but steady increase in the number of FDA approvals of CNS drugs, with the exception of two bursts of inactivity in the 1950s and 1990s. The average rate of neurologic diseases has almost doubled relative to its baseline since the beginning of the 2000s [2]. Between 2015 and 2021, the FDA approved 323 new drugs (in 2015, 45 novel drugs; in 2016, 22; in 2017, 46; in 2018, 59; in 2019, 48; in 2020, 53; in 2021, 50), of which 31 were for CNS disorders. This listing excludes vaccines, allergenic products, blood and blood products, plasma derivatives, cellular and gene therapy products, and other products approved by the Center for Biologics Evaluation and Research [4–6].

1.2 Etiology

CNS disorders are caused mainly by traumatic brain injury (TBI) or injury to the spinal cord causing a wide range of disabilities in a person [7]. Infectious diseases are one of the causative agents of CNS disease, some of which may directly affect the brain and the spinal cord [8]. In addition, CNS diseases could result from degenerative spinal disorders associated with loss of function in the spine. Brain degeneration also causes the CNS diseases Alzheimer's, Parkinson's, and Huntington's diseases [9]. One of the leading causes of CNS disease is the blockage of a blood vessel by a blood clot or rupture of blood vessels causing hemorrhage in the brain [10]. This condition is medically known as stroke.

1.3 Pathogenesis

The most common pathway for the spread of pathogenic organisms to access the CNS is predominantly hematogenic spread. Viruses usually gain access by colonizing the mucosal surfaces throughout the body and then entering the blood. Therefore, they multiply at different neural sites and then cross the blood–brain barrier (BBB) before destroying the CNS [10]. Although the majority of viruses penetrate the CNS directly through the cerebral capillary endothelial cells, some infect cerebral microvascular endothelial cells, some gain access through the choroid plexus, and some are carried through the barrier by infected leukocytes. Some viruses can also enter through the olfactory nerve [11]. In bacterial infections of the CNS, the first spread is usually mucosal colonization in the nasopharynx to the bloodstream. Once in the bloodstream, bacteria have to survive the host defense mechanism. They achieve that through a polysaccharide capsule that resists phagocytosis by neutrophils and classic complement-mediated bactericidal activity [12]. The case of bacterial meningitis is cerebral edema caused by a vasogenic, cytotoxic mechanism. It contributes to elevated pressure in the brain and may result in cerebral herniation-induced deaths. Cerebral blood flow is tampered with and tends to decrease, and relative anoxia follows suit, thus contributing to neuronal damage. The localized purulent spread of bacteria within the CNS originates from adjacent foci of the disease, such as otitis, sinusitis, and septic phlebitis, which arrive through septic emboli from distant sites of infection [13].

1.4 Central Nervous System Disorders and Drugs Approved by the Food and Drug Administration

All medications, including vaccines, have to go through a quality control process to ensure they are safe and effective before being approved for public use. The FDA oversees the process from start to finish – and even after approval. Emergency use authorizations (EUAs) may be granted in some instances. The FDA may also approve drugs before a Phase III clinical trial. This process is called accelerated approval. No medication is entirely safe with 100% effectiveness. Therefore, the drug's side effects and long-term efficacy will be monitored by the FDA – even after the drug has been approved [14].

The emergence of new drugs for neurologic diseases has almost doubled since the beginning of the year 2000 [2]. Most FDA-approved CNS drugs are palliatives that provide only symptomatic relief rather than cure the disorders. So this area of drug therapy presents a challenge and an opportunity for the development of new and effective medicines. Unfortunately, the failure rate for the development of new medicinal entities for CNS disorders is higher than for other disorders, and this has been attributed to a lack of understanding of the mechanism of CNS diseases, the possibility of CNS side effects, difficulty identifying and measuring appropriate clinical endpoints, a lack of relevant animal models, and the penetration obstacle posed by the BBB [15–17]. According to a report by the Tufts Center for the Study of Drug Development, the time taken to develop and approve CNS drugs is 20% and 38% longer, respectively, than for non-CNS drugs. Consequently, many pharmaceutical companies such as Pfizer, GSK, and Eli Lilly have stopped the development of drugs for CNS disorders [18].

Nevertheless, the FDA has approved many drugs for these disorders over the last few decades, and researchers are optimistic about meeting the health needs in this area. In 2021, 17 new drugs and devices were approved for CNS disorders. The approvals were for the management of migraine and pain (5), Parkinson's disease (3), epilepsy/seizure (3), neuromuscular diseases (3), Alzheimer's disease (1), MS (1), and sleep disorder (1). In addition, researchers are looking at drug repurposing as a faster and cheaper way of providing effective medicine. Some of the repurposed drugs are verapamil, amantadine, propranolol, sunitinib, fenfluramine, and minocycline [19]. New disease biomarkers are also being developed to effectively measure the pathophysiologic processes of CNS disorders. New FDA-approved drugs for some significant CNS disorders – including attention-deficit/hyperactivity disorder (ADHD), Alzheimer's disease, migraine, MS, Duchenne muscular dystrophy (DMD), and seizure – are shown in Table 1.1.

1.4.1 Attention-Deficit/Hyperactivity Disorder

ADHD is a chronic neuropsychiatric disorder characterized by attention difficulty, hyperactivity, and impulsiveness [20]. ADHD usually starts in childhood and symptoms can persist into adulthood. It is the most prevalent mental disorder affecting children. Common comorbidities may include anxiety disorder, autism spectrum disorder, specific communication and learning or motor disorders (e.g. reading disability, developmental coordination disorder), and intellectual disability. The specific cause of ADHD is not known, although there is evidence of genes playing an important role in the disorder. Other risk factors implicated in ADHD are brain injury, premature birth and low birth weight, use of tobacco and/or alcohol during pregnancy, and exposure to other harmful substances during pregnancy. Drug therapy and lifestyle changes are employed in the symptomatic management of ADHD as there is no cure yet. Drugs used in the treatment of ADHD are grouped into stimulant and non-stimulant medications.

Stimulant medications are the first-line drugs for ADHD and act by elevating the amount of dopamine and norepinephrine in the brain [21]. Dopamine and norepinephrine are neurotransmitters that regulate attention, thinking, pleasure, and movement. Stimulants are primarily of two classes – amphetamine-based and methylphenidate-based medications. FDA-approved amphetamine-based medications include:

- Amphetamine (Adzenys XR-ODT®, Aytu BioPharma, Englewood, CO, USA; Dyanavel XR®, Tris Pharma, Monmouth Junction, NJ, USA).
- Amphetamine sulfate (Evekeo®, Arbor Pharmaceuticals, Woburn, MA, USA).
- Dextroamphetamine sulfate (Dexedrine®, GSK, Research Triangle Park, NC, USA; ProCentra®, Outlook Pharmaceuticals, Cincinnati, OH, USA; Zenzedi®, Arbor Pharmaceuticals).
- Dextroamphetamine/amphetamine (Adderall®, Shire Pharmaceuticals, Wayne, PA, USA; Mydayis®, Takeda Pharmaceuticals, Lexington, MA, USA).
- Lisdexamfetamine dimesylate (Vyvanse®, Takeda Pharmaceuticals) [22–24].

Methylphenidate hydrochloride, an example of a methylphenidate-based medication, has been approved for treating ADHD and is available in different dosage forms, including:

- Tablets (Ritalin®, Novartis, Basel, Switzerland; Methylin®, SpecGX, Webster Groves, MO, USA; Metadate® ER, UCB, Smyrna, GA, USA; Cotempla XR-ODT®, Aytu BioPharma; Quillichew ER®, Tris Pharma; Concerta®, Janssen Pharmaceuticals, Beerse, Belgium) [25].
- Capsules (Adhansia XR®, Purdue Pharma, Stamford, CT, USA; Ritalin LA; Aptensio XR®, Rhodes Pharmaceuticals, Coventry, RI, USA; Metadate CD; Jornay PM®, Ironshore Pharmaceuticals & Development, Morrisville, NC, USA).
- Oral solution (Quillivant XR®, Tris Pharma; Methylin).
- Skin patch (Daytrana®, Noven Pharmaceuticals, Miami, FL, USA).

Table 1.1 Recent Food and Drug Administration–approved drugs for central nervous system disorders.

Drug/dosage form	Indication	Mechanism of action	Approval date	Company
Vyepti (eptinezumab) IV infusion	Prevention of migraine	Humanized IgG1 monoclonal antibody specific for binding to CGRP ligand	February 2020	Lundbeck Seattle BioPharmaceuticals
Nurtec ODT (rimegepant) Oral disintegrating tablet	Treatment of acute migraine with or without aura	CGRP antagonist	February 2020	Biohaven Pharmaceuticals
Zeposia (ozanimod) Capsule	Relapsing forms of MS, CIS, RRMS, and active secondary progressive MS	S1P receptor modulator	March 2020	Bristol-Myers Squibb
Ongentys (opicapone) Capsule	As an adjunct to levodopa/carbidopa to reduce "off" episodes in patients with Parkinson disease	COMT inhibitor	April 2020	Neurocrine Biosciences
Koselugo (selumetinib) Capsule	Neurofibromatosis type 1 (NF1) in pediatric patients aged 2 years or older who have symptomatic, inoperable plexiform neurofibroma	Inhibitor of mitogen-activated protein kinases 1 and 2 (MEK1/2)	April 2020	AstraZeneca Pharmaceuticals
Tauvid (flortaucipir F18) IV injection	Diagnostic agent for patients with Alzheimer's disease	Complexes with aggregated tau protein to form neurofibrillary tangles (NFTs), a component required for neuropathologic diagnosis of Alzheimer's disease	May 2020	Avid Radiopharmaceuticals
Bafiertam (monomethylfumarate) Oral capsule	Relapsing forms of MS, CIS, RRMS, and active secondary progressive MS	Not clear. May have antioxidant properties that could be protective against damage to the brain and spinal cord	April 2020	Banner Life Sciences

(Continued)

Table 1.1 (Continued)

Drug/dosage form	Indication	Mechanism of action	Approval date	Company
Fintepla (fenfluramine) Oral solution	For seizures associated with Dravet syndrome in patients aged 2 years or older	Amphetamine derivative	June 2020	Zogenix
Qutenza (capsaicin) Transdermal patch	Treatment of neuropathic pain associated with diabetic peripheral neuropathy of the feet in adults	Reversible inhibition of the TRPV1 (transient receptor potential vanilloid 1) receptor	July 2020	Averitas Pharma
Viltepso (viltolarsen) IV injection	DMD in patients with confirmed alteration of the *DMD* gene responsive to exon 53 skipping	Antisense oligonucleotide that promotes the activity of dystrophin	August 2020	NS Pharma
Kesimpta (ofatumumab) Subcutaneous injection	Relapsing MS, CIA, RRMS, and active secondary progressive disease in adults	B-cell therapy, anti-CD20 antibody	August 2020	Novartis
Amondys 45 (casimersen) IV injection	DMD in patients with a confirmed mutation responsive to exon 45 skipping	Antisense oligonucleotide of the phosphorodiamidate morpholino oligomer (PMO) subclass	Feb 2021	Sarepta Therapeutics
Nulibry (fosdenopterin) IV injection	Indicated to decrease mortality rate in patients with molybdenum cofactor deficiency type A (MoCD-A), a rare disease	Exogenous source of cyclic pyranopterin monophosphate (cPMP)	February 2021	Origin Biosciences
Azstarys (dexmethylphenidate and serdexmethylphenidate) Capsules	ADHD	CNS stimulant	March 2021	KemPharm
Ponvory (ponesimod) Oral tablet	Relapsing forms of MS, CIS, RRMS, and active secondary progressive disease	S1P receptor 1 modulator	March 2021	Janssen Pharmaceuticals

Aduhelm (aducanumab) IV infusion	For patients with mild cognitive impairment or a mild dementia stage of Alzheimer's disease	Monoclonal antibody that targets beta-amyloid and binds to aggregated forms of beta-amyloid	June 2021	Biogen
Xywav (oxybate mixed salts) Oral solution	Treatment of idiopathic hypersomnia in adults	CNS depressant	August 2021	Jazz Pharmaceuticals
Qulipta (atogepant) Tablet	Preventive treatment of episodic migraine	CGRP receptor antagonist	September 2021	AbbVie
Vyvgart (efgartigimodalfa) IV injection	Generalized myasthenia gravis in AChR antibody-positive adult patients	Fc receptor (FcRn) antagonist preventing FcRn from recycling IgG back into the blood	December 2021	Argenx BV
Ztalmy (ganaxolone) Oral suspension	Seizures in cyclin-dependent kinase-like 5 (CDKL5) deficiency disorder patients	GABAA receptor-positive modulator	March 2022	Marinus Pharmaceuticals

AChR, acetylcholine receptor; ADHD, attention-deficit/hyperactivity disorder; CGRP, calcitonin gene-related peptide; CIS, clinically isolated syndrome; CNS, central nervous system; COMT, catechol-O-methyl transferase; DMD, Duchenne muscular dystrophy; IgG, immunoglobulin G; IV, intravenous; MS, multiple sclerosis; RRMS, relapsing–remitting multiple sclerosis.

Dexmethylphenidate hydrochloride (Focalin XR®, Novartis) is another example in this class of ADHD medications [25]. Azstarys® (KemPharm, Celebration, FL, USA), a combination of serdexmethylphenidate and dexmethylphenidate, was recently approved by the FDA for the management of ADHD in both children and adults [26]. Stimulants have a potential risk for abuse and dependence. Addiction and withdrawal symptoms are possible with amphetamine-based ADHD medications, which are also associated with increased blood pressure and heart rate.

Non-stimulant medications may be used if stimulants are not effective or cause unacceptable side effects. Non-stimulants employed in treating ADHD consist of selective norepinephrine reuptake inhibitors (SNRIs), α-2 agonists, and antidepressants. SNRIs increase the level of norepinephrine in the brain by preventing its breakdown. The SNRIs approved by the FDA for ADHD are atomoxetine hydrochloride (Strattera®, Eli Lilly, Indianapolis, IN, USA) and viloxazine (Quelbree®, Supernus Pharmaceuticals, Rockville, MD, USA) [27]. Strattera is the first non-simulant FDA-approved drug for ADHD. The α-2 agonists inhibit the release of norepinephrine in the brain, which has sedative and hypotensive effects. They were initially approved for treatment of hypertension, but later approved as monotherapy and as adjunctive therapy to stimulant medications in the management of ADHD. FDA-approved α-2 agonists for ADHD include clonidine hydrochloride (Kapvay®, Concordia Pharmaceuticals, Fort Lauderdale, FL, USA; Catapres®, Catapres-TTS skin patch, Boehringer Ingelheim, Ingelheim am Rhein, Germany) and guanfacine hydrochloride (Intuniv®, Takeda Pharmaceuticals), which is approved for children aged 6–17 years as adjunctive therapy to stimulants [28].

Antidepressants are medications to treat depression that are repurposed to treat ADHD. Most antidepressants for ADHD are long-acting, with duration of action lasting up to 24 hours, and are typically taken once a day. Tricyclic antidepressants increase the level of serotonin, a hormone that regulates mood, sleep, and digestion. Tricyclic antidepressants utilized in the therapy of ADHD include imipramine hydrochloride (Tofranil®, Mallinckrodt Pharmaceuticals, Staines-upon-Thames, UK), desipramine hydrochloride (Norpramin®, Validus Pharmaceuticals, Parsippany, NJ, USA), and nortriptyline hydrochloride (Pamelor®, Mallinckrodt Pharmaceuticals) [29]. Atypical antidepressants employed in treating ADHD include bupropion hydrochloride (Wellbutrin, Bausch Health, Bridgewater, NJ, USA) and venlafaxine hydrochloride (Effexor®, Pfizer, New York, USA) [30, 31]. They elevate dopamine, serotonin, and norepinephrine levels in the brain.

1.4.2 Migraine

Migraine is a complex disorder mostly on one side of the head, characterized by recurrent episodes of headache. It is usually associated with an aura (visual or sensory symptoms) that often occurs before the headache, but can occur during or afterward. Migraine has a strong genetic component. It occurs predominantly in women. Typical migraine symptoms include throbbing or pulsatile headache, nausea and vomiting, and sensitivity to light and sound.

Migraine treatment can be abortive (stop the acute phase) or prophylactic (preventive). Abortive treatment aims to reverse or stop the progression of a headache. It is most effective when given as soon as the headache begins. Abortive medications approved by the FDA include non-steroidal anti-inflammatory drugs (NSAIDs), selective serotonin receptor (5-hydroxytryptamine–1, or 5-HT1) agonists (triptans), serotonin 5-HT1F agonists (titans), calcitonin gene-related peptide (CGRP) receptor antagonists, and ergot alkaloids. FDA-approved NSAIDs for managing migraine include ibuprofen, diclofenac potassium (Cambia®, Assertio Therapeutics, Lake Forest, IL, USA), and naproxen sodium [32]. They are the most widely used drugs for migraine attacks and first-line treatment for mild to moderate migraine. Triptans are first-line acute therapy for moderate to severe migraine attacks. FDA-approved triptans are sumatriptan (Tosymra®, Upsher-Smith, Morristown, NJ, USA; Zembrace® SymTouch, Upsher-Smith; Onzetra® Xsail®, Currax Pharmaceuticals, Brentwood, TN, USA; Imitrex, GSK), eletriptan hydrobromide (Relpax®, Pfizer), frovatriptan (Frova®, Endo Pharmaceuticals, Malvern, PA, USA), and rizatriptan benzoate (Maxalt®, Merck, Rahway, NJ, USA) [33]. Other examples are zolmitriptan, naratriptan, and almotriptan. Titans have fewer vasoconstriction side effects compared to triptans. They have selective activity on the receptor subtype 5-HT1F. Lasmiditan Reyvow® (Eli Lilly) was the first ditan to be approved for migraine treatment, in 2019.

Other FDA-approved drugs for acute therapy of migraine are rimegepant (Nurtec® ODT, Biohaven Pharmaceuticals, New Haven, CT, USA), ubrogepant (Ubrelvy®, AbbVie, Chicago, IL, USA), and dihydroergotamine mesylate (Trudhesa®, Impel Pharmaceuticals, Seattle, WA, USA; Migranal®, Bausch Health) [34].

Prophylactic medications for migraine include repurposed drugs initially approved for other diseases such as hypertension, depression, and epilepsy. Antiepileptic drugs acts as channel blockers, thereby preventing the delivery of electrical impulses to nerve, muscle, and brain cells. Additionally, antiepileptic medications potentiate the actions of a neurotransmitter (gamma-aminobutyric acid, GABA) that modulates motor skills, vision, and anxiety. The mechanism of action of antiepileptic drugs in the management of migraine is not precise. Still, they act on multiple target sites in the brain, modifying the transmission of information through their inhibitory actions on ion channels and metabolism of GABA, as well as their activity on neurotransmitter receptors. The FDA-approved antiseizure medications for the prophylactic treatment of migraine are divalproex sodium (Depakote®, AbbVie), topiramate (Topamax®, Janssen Pharmaceuticals; Qudexy® XR, Upsher-Smith; and Trokendi XR®, Supernus Pharmaceuticals), and valproic acid (Stavzor®, Bionpharma, Princeton, NJ, USA) [35]. The FDA approves only two β-blockers for migraine. These are propranolol (InnoPran XL®, ANI Pharmaceuticals, Baudette, MN, USA) and timolol [36]. β-Blockers are drugs that were initially developed to treat hypertension (high blood pressure) since they block adrenaline receptors, causing the vessels to relax.

Another FDA-approved drug for prophylactic management of chronic migraine is botulinum toxin A [37] (Botox®, AbbVie; or Onabotulinum A), a toxin produced by the naturally occurring *Clostridium botulinum* that causes botulism. Botox decreases the number of migraine headaches in people prone to them who use the drug and was approved in 2010 to treat chronic migraine.

CGRP inhibitors (gepants) are another class of drugs used in migraine prophylaxis. They are monoclonal antibodies and small-molecule CGRP receptor antagonists that block the activities of CGRP. Six CGRP inhibitors have gained FDA approval for migraine prevention: erenumab (Aimovig®, Amgen, Thousand Oaks, CA, USA), fremanezumab (Ajovy®, Teva Pharmaceuticals, Parsippany, NJ, USA), galcanezumab (Emgality®, Eli Lilly), eptinezumab (Vyepti®, Lundbeck Seattle BioPharmaceuticals, Bothell, WA, USA), rimegepant (Nurtec ODT), and atogepant (Qulipta®, AbbVie) [38].

1.4.3 Parkinson's Disease

Parkinson's disease is among the most common neurologic disorders. It affects approximately 1% of individuals older than 60 years and causes progressive disability. The two significant neuropathologic findings in Parkinson's disease are the depletion of pigmented dopaminergic neurons of the substantia nigra pars compacta and the presence of Lewy bodies and Lewy neuritis [39].

Parkinson's disease is a situation when nerve cells (neurons) in the substantia nigra of the brain become impaired or die, leading to low dopamine production. Dopamine is essential for the operation of basal ganglia, a part of the brain responsible for coordinating body movements. Deficiency of dopamine causes the movement symptoms seen in people with Parkinson's disease. Loss of norepinephrine is also implicated in Parkinson's disease and causes some of the non-movement-related symptoms. Common symptoms of this disorder are tremors, slowness of movement, stiff muscles, unsteady walking, balance, and coordination problems. Other symptoms are a subtle decrease in dexterity, soft voice, decreased facial expression, sleep disturbances, decreased sense of smell, constipation, sweating abnormalities, sexual dysfunction, and seborrhea dermatitis. There is no cure for the disease. However, most patients can maintain a good quality of life with medications.

Medical management of Parkinson's disease is to provide control of signs and symptoms for as long as possible while minimizing adverse effects. The gold standard for symptomatic therapy of Parkinson's disease is levodopa, a dopamine precursor. Levodopa is commonly taken with a dopamine decarboxylase inhibitor, carbidopa (Sinemet®, Merck; Parcopa®, UCB; Rytary®, Amneal Pharmaceuticals, Bridgewater, NJ, USA; Duopa®, AbbVie) [40]. Inbrija® (Acorda Therapeutics, New York, USA) is an FDA-approved levodopa inhalation powder [41]. A combination of carbidopa, levodopa, and entacapone (Stalevo®, Novartis) is also approved [42].

The dopamine agonists approved for treating Parkinson's disease are ropinirole (ReQuip®, GSK), rotigotine (Neupro®, UCB), apomorphine (Apokyn®, Supernus Pharmaceuticals), and pramipexole (Mirapex®, Boehringer Ingelheim) [43]. Neupro is given as a transdermal patch, Apokyn is a short-acting injectable medication, while Kynmobi®, Sunovion Pharmaceuticals, (Marlborough, MA, USA) is a sublingual film.

The catechol O-methyltransferase (COMT) inhibitors block the breakdown of dopamine in the brain by inhibiting the activity of COMT. Entacapone (Comtan®, Novartis), tolcapone (Tasmar®, Bausch Health), and opicapone (Ongentys®, Neurocrine Biosciences, San Diego, CA, USA) are the three FDA-approved COMT inhibitors for Parkinson's disease. Opicapone is the most recent medication in this class, which received FDA approval in 2020 [44].

Monoamine oxidase (MAO) B inhibitors block the monoamine oxidase B (MAO B), another enzyme involved in the breakdown of dopamine in the brain. This allows dopamine to have longer-lasting effects on the brain. Examples of MAO B inhibitors are selegiline (Eldepryl®, Orion Pharma, Espoo, Finland; Zelapar®, Bausch Health), rasagiline (Azilect®, Teva Pharmaceuticals), and safinamide (Xadago®, Supernus Pharmaceuticals) [45]. Anticholinergic agents, such as benztropine (Cogentin®, Merck) and trihexyphenidyl (Artane®, Teva Pharmaceuticals), help reduce tremor and muscle stiffness. They are the oldest class of drugs to treat Parkinson's disease. Amantadine (Symmetrel®, Endo Pharmaceuticals), an antiviral agent, helps reduce the involuntary movements (dyskinesia) resulting from levodopa medication. There are two extended-release types of the drug, Gocovri® (Supernus Pharmaceuticals) and Osmolex ER® (Supernus Pharmaceuticals) [46]. Side effects include confusion and memory problems. Another medication, istradefylline (Nourianz®, Kyowa Kirin, Tokyo, Japan), is an adenosine A2A receptor antagonist approved for use in patients taking carbidopa-levodopa but experiencing off symptoms. Istradefylline is the first non-dopaminergic drug approved by the FDA for Parkinson's disease in the last two decades. These drugs act to increase the effectiveness of levodopa and also increase its side effects, including involuntary movements (dyskinesia) and hallucinations [47].

Pimavanserin (Nuplazid®, Acadia Pharmaceuticals, San Diego, CA, USA) is an atypical antipsychotic agent and the only drug approved to treat hallucinations and delusions related to Parkinson's disease. In contrast, rivastigmine transdermal system (Exelon® Patch, Novartis) was approved to treat non-severe dementia associated with Parkinson's disease.

1.4.4 Multiple Sclerosis

MS is an immune-mediated inflammatory disease. It destroys the myelinated axons in the CNS, producing significant physical disability within 20–25 years in more than one-third of patients. MS is the most prevalent cause of non-traumatic disability in young adults. Generally, MS is detected between 20 and 40 years of age, but less than 1% can occur in childhood and approximately 2–10% after 50 years of age. MS affects women more than men. The cause of MS is not yet precise. However, it is considered to be caused by multiple factors, including genetic and environment-related factors [48, 49].

The treatment of MS is divided into disease-modifying therapy and symptomatic therapy. The majority of disease-modifying agents for MS (DMAMS) that have been approved are indicated for treating relapsing forms of MS only. However, siponimod (Mayzent®, Novartis), ocrelizumab (Ocrevus®, Genentech, South San Francisco, CA, USA), ozanimod (Zeposia®, Bristol-Myers Squibb, New York, USA), ofatumumab (Kesimpta®, Novartis), and cladribine (Mavenclad®, Merck) are also approved for active secondary progressive disease. The FDA-approved DMAMS include the following:

- Interferons (IFNs), e.g. IFN β-1a (Avonex®, Biogen, Cambridge, MA, USA; Rebif®, Merck), IFN β-1b (Extavia®, Novartis), and peginterferon β-1a (Plegridy®, Novartis) [49].
- Sphingosine 1-phosphate (S1P) receptor modulators, e.g. siponimod (Mayzent), fingolimod (Gilenya®, Novartis), and ozanimod (Zeposia) [50].
- Monoclonal antibodies, e.g. natalizumab (Tysabri®, Biogen), alemtuzumab (Lemtrada®, Genzyme, Cambridge, MA, USA), ocrelizumab (Ocrevus), and ofatumumab (Kesimpta) [51, 52]. Ofatumumab was approved in 2020 for treating relapsing forms of MS in adults. It is the first B-cell therapy for MS and can be self-administered once monthly using an autoinjector pen.
- Other immunomodulators employed in the therapy of MS are glatiramer (Copaxone®, Teva Pharmaceuticals), mitoxantrone, teriflunomide (Aubagio®, Genzyme), dimethyl fumarate (Tecfidera®, Biogen), and cladribine (Mavenclad) [53].
- Diroximel fumarate (Vumerity®, Biogen), a fumarate therapy, is also approved for relapsing forms of MS. In addition, diroximel fumarate is indicated for treating clinically isolated syndrome (CIS) and relapsing–remitting MS (RRMS), as well as for the treatment of active secondary progressive disease in adults [54].

Several symptomatic therapies have been licensed specifically for MS. These include nabiximols (Sativex®, GW Pharma, Cambridge, UK) for neuropathic pain and fampridine (Ampyra®, Acorda Therapeutics, Waltham, MA, USA) for walking difficulties.

1.4.5 Alzheimer's Disease

Alzheimer's disease is among the types of dementia that develop gradually with the expression of neuritic plaques and neurofibrillary tangles caused by the buildup of amyloid beta (Aβ) peptide in the brain's most affected area, the medial temporal lobe and neocortical structures [55]. Alzheimer's disease has no cure, although available treatments improve the symptoms [56]. Several physiologic processes in Alzheimer's disease destroy acetylcholine-producing cells, reducing cholinergic transmission through the brain. Donepezil hydrochloride (Aricept®, Eisai R&D Management, Nutley, NJ, USA), galantamine hydrochloride (Reminyl®, Janssen Pharmaceuticals), and rivastigmine tartrate (Exelon®, Novartis) are acetylcholinesterase inhibitors (AChEIs) and act by blocking cholinesterase enzymes (AChE and butyrylcholinesterase [BChE]) from metabolizing acetylcholine, which results in increasing acetylcholine levels in the synaptic cleft. On the other hand, overactivation of N-methyl-D-aspartate receptor (NMDAR) leads to rising levels of Ca2+, which promotes cell death and synaptic dysfunction. NMDAR antagonists such as memantine HCl (Namenda®, Merz Pharma, Frankfurt, Germany) prevent the overactivation of the NMDAR glutamate receptor, and hence the Ca2+ influx, restoring its regular activity.

Namzaric® (Merz Pharma) combines memantine hydrochloride and donepezil hydrochloride in a fixed dose. It is a medication for moderate to severe dementia of the Alzheimer's type in patients already stabilized on memantine hydrochloride and donepezil hydrochloride.

Aduhelm® (aducanumab-avwa; Biogen) is a human immunoglobulin gamma 1 (IgG1) monoclonal Aβ-directed antibody. It was approved in 2021 for treating Alzheimer's disease. Aduhelm reduces Aβ plaques, a pathophysiological feature of Alzheimer's disease [57].

1.4.6 Muscular Dystrophy

Muscular dystrophy is a progressive muscle disorder without a central or peripheral nerve abnormality. The disease causes muscle weakness and degeneration and affects breathing and heart function, which can be life-threatening. DMD is among the types of muscular dystrophy. Patients with these symptoms are typically diagnosed between 2 and 3 years of age.

DMD is caused by mutations in the *DMD* gene (encoding dystrophin) that prevents the production of the muscle isoform of dystrophin, resulting from non-sense or frame-shifting mutations in the dystrophin gene. Most of these mutations could be resolved by removing an extra exon to generate shorter but in-frame transcripts and establish partly functional proteins.

Amondys 45® (Sarepta Therapeutics, Cambridge, MA, USA) was approved in 2021 for treating confirmed mutation of the *DMD* gene responsive to exon 45 skipping. Also, Exondys 51® (Sarepta Therapeutics) and Viltepso® (NS Pharma, Paramus, NJ, USA) are specifically indicated for treating patients with confirmed alterations in the *DMD* gene responsive to exon 51 and 53 skipping, respectively [58].

Deflazacort (Emflaza®, PTC Therapeutics, South Plainfield, NJ, USA), a corticosteroid that exerts anti-inflammatory and immunosuppressive effects, is specifically indicated for treating DMD in patients not less than 2 years old.

1.4.7 Epilepsy/Seizure

Epilepsy is a chronic medical disorder, usually resulting in unpredictable, unprovoked recurrent seizures affecting various mental and physical functions. Seizures can be partial, generalized, or unclassified. A shift in the average balance of excitation and inhibition within the CNS and abnormal brain function cause an attack. Causes of epileptic seizures include a genetic predisposition, head trauma, stroke, brain tumors, alcohol or drug withdrawal, repeated episodes of metabolic issues, hypoglycemia, and other conditions.

Many medications have been approved for the management of epilepsy, and the choice of drugs depends on the diagnosis of the epileptic syndrome. Although some anticonvulsants (e.g. lamotrigine, topiramate, valproic acid, zonisamide) have various means of eliciting anticonvulsant action, some (e.g. phenytoin, carbamazepine, ethosuximide) have only one known mechanism of action. FDA-approved anticonvulsant agents are classified into large groups based on their mechanisms, as follows:

- Blockers of repetitive activation of the sodium channel: phenytoin, carbamazepine (Carbatrol®, Takeda Pharmaceuticals), oxcarbazepine, lamotrigine, topiramate.
- Enhancers of slow inactivation of the sodium channel: lacosamide (Vimpat®, UCB), rufinamide (Banzel®, Novartis).
- GABA(A) receptor enhancers: clobazam (Onfi®, Lundbeck Seattle BioPharmaceuticals), diazepam (Valium®, Roche, Basel, Switzerland), lorazepam (Ativan®, Bausch), clorazepate (Tranxene®, Recordati Rare Diseases, Lebanon, NJ, USA), alprazolam (Xanax®, Pfizer), gabapentin (Neurontin®, Pfizer), phenobarbital (Luminal®, Bayer, Leverkusen, Germany), pregabalin (Lyrica®, Pfizer), vigabatrin (Sabril®, Lundbeck Seattle BioPharmaceuticals), ganaxolone (Ztalmy®, Marinus Pharmaceuticals, Radnor, PA, USA).
- Glutamate blockers: felbamate, perampanel, topiramate.
- Calcium channel blockers: ethosuximide, lamotrigine, topiramate (Trokendi XR; Topamax), zonisamide, valproate (Stavzor).
- H-current modulators: gabapentin (Neurontin®, Pfizer), lamotrigine (Lamictal®, GSK).
- Blockers of unique binding sites: gabapentin, levetiracetam (Keppra®, UCB).
- Carbonic anhydrase inhibitors: topiramate (Qudexy), zonisamide.
- Neuronal potassium channel (KCNQ [Kv7]) opener: ezogabine.

The FDA approved ganaxolone in 2022 for seizures that are caused by cyclin-dependent kinase-like 5 (CDKL5) deficiency disorder (CDD). This is the first FDA-approved treatment that is specific to CDD-related seizures.

1.5 Conclusion

Research and development of drugs for CNS disorders is a complex area in the medical world. Many of the drug applications for approval have failed to achieve the primary endpoints and show the medications' safety. An insight into the cellular and molecular processes of the disorders will go a long way to resolving this problem. There is a need to identify and develop new biomarkers in CNS drug development, which will provide researchers with the tools and techniques to better understand the diseases' biological mechanisms and help facilitate the development of effective drugs for CNS disorders.

References

1 Jao, T., Tsai, T.H., and Jeng, J.S. (2009). Aggrenox (Asasantin retard)-induced Stevens-Johnson syndrome. *Br. J. Clin. Pharmacol.* 67 (2): 264–265.

2 Klinch, M.S. (2015). An analysis of FDA-approved drugs for neurological disorders. *Drug Discov. Today* 20 (9).

3 Pal, S. (2018). Incidence and prevalence of major neurologic disorders. *US Pharm.* 43 (1): 24.

4 US Food and Drug Administration (2021). Novel drug approvals for 2021. https://www.fda.gov/drugs/new-drugs-fda-cders-new-molecular-entities-and-new-therapeutic-biological-products/novel-drug-approvals-2021 (accessed 8 February 2023).

5 US Food and Drug Administration (2016). Novel drug approvals for 2016. https://www.fda.gov/drugs/new-drugs-fda-cders-new-molecular-entities-and-new-therapeutic-biological-products/novel-drug-approvals-2016 (accessed 8 February 2023).

6 US Food and Drug Administration (2017). Novel drug approvals for 2017. https://www.fda.gov/drugs/new-drugs-fda-cders-new-molecular-entities-and-new-therapeutic-biological-products/novel-drug-approvals-2017 (accessed 8 February 2023).

7 Alve, O.L. and Bullock, R. (2001). Exatotoxic damage in traumatic brain injury. In: *Brain Injury* (ed. R.S. Clark and P. Kochanek), 1–36. Boston, MA: Kluwer Academic.

8 Wilbur, C., Bitnum, A., Kronenberge, S. et al. (2019). PANDAS/PANS in childhood: Controversies and evidence. *Paediatr. Child Health* 24 (2): 85–91.

9 Jauch, E.C. (n.d.). Stroke. *healthywomen*. https://www.healthywomen.org/condition/stroke (accessed 8 February 2023).

10 Mandell, G.L., Bennett, J.E., and Dolin, R.C. (2000). *Principles of Infectious Diseases*, 5e. Philadelphia, PA: Churchill Livingstone.

11 Sheld, W.M., Witley, R.J., and Durack, D.I. *Infections of the Central Nervous System*, 2e. Philadelphia, PA: Lippincott-Raven.

12 Whitley, R.J. and Gnann, J.W. (2002). Viral encephalitis: familiar infections and emerging pathogens. *Lancet* 359: 507–514.

13 Saez-llorens, X.M. and Ccracker, G.H. Jr. (2003). Bacterial meningitis in children. *Lancet* 361: 2139–2148.

14 Adams, M. (2021). 5 things to know about the FDA approval process. Cancerwise. https://www.mdanderson.org/cancerwise/5-things-to-know-about-the-fda-approval-process-.h00-159463212.html (accessed 8 February 2023).

15 US Food and Drug Administration (2019). Novel drug approvals for 2019. https://www.fda.gov/drugs/new-drugs-fda-cders-new-molecular-entities-and-new-therapeutic-biological-products/novel-drug-approvals-2019 (accessed 8 February 2023).

16 US Food and Drug Administration (2020). Novel drug approvals for 2020. https://www.fda.gov/drugs/new-drugs-fda-cders-new-molecular-entities-and-new-therapeutic-biological-products/novel-drug-approvals-2020 (accessed 8 February 2023).

17 Gribkoff, V.K. and Kaczmarek, K.L. (2017). The need for new approaches in CNS drug discovery: why drugs have failed, and what can be done to improve outcomes. *Neuropharmacology* 120: 11–19.

18 Howes, O.D. and Mehta, M.A. (2021). Challenges in CNS drug development and the role of imaging. *Psychopharmacology* 238: 1229–1230.

19 Mao, X.-Y. (2020). Drug repurposing in neurological diseases: opportunities and challenges. In: *Drug Repurposing: Hypothesis, Molecular Aspects and Therapeutic Applications* (ed. F.A. Badria). London: IntechOpen https://doi.org/10.5772/intechopen.93093.

20 Thapara, A. and Cooper, M. (2016). Attention deficit hyperactivity disorder. *Lancet* 387: 1240–1250.

21 Verkuijl, N., Perkins, M., and Fazel, M. (2015). Childhood attention-deficit/hyperactivity disorder. *BMJ* 350: h2168–h2168.

22 Biederman, J.F.S., Mick, E., Fried, R. et al. (2011). Are stimulants effective in the treatment of executive function deficits? Results from a randomized double blind study of OROS-methylphenidate in adults with ADHD. *Eur. Neuropsychopharmacol.* 27: 508–515.

23 Childress, A.C., Kando, J.C., King, T.R. et al. (2019). Early-onset efficacy and safety pilot study of amphetamine extended-release oral suspension in the treatment of children with attention-deficit/hyperactivity disorder. *J. Child Adolesc. Psychopharmacol.* 29: 2–8.

24 Cutler, A.J., Suzuki, K., Starling, B. et al. (2022). Efficacy and safety of dextroamphetamine transdermal system for the treatment of attention-deficit/hyperactivity disorder in children and adolescents: results from a pivotal phase 2 study. *J. Child Adolesc. Psychopharmacol.* 32: 89–97.

25 Wigal, L., Nordbrock, S.B., Adjei, E. et al. (2015). Efficacy of methylphenidate hydrochloride extended-release capsules (Aptensio XRTM) in children and adolescents with attention-deficit/hyperactivity disorder: A phase III, randomized, double-blind study. *CNS Drugs* 29: 331–340.

26 Childress, A.C., Spencer, T., Lopez, F. et al. (2009). Efficacy and safety of dexmethylphenidate extended-release capsules administered once daily to children with attention-deficit/hyperactivity disorder. *J. Child Adolesc. Psychopharmacol.* 19: 351–361.

27 Kollins, S.H., Braeckman, R., Guenther, S. et al. (2021). A randomized, controlled laboratory classroom study of serdexmethylphenidate and d-methylphenidate capsules in children with attention-deficit/hyperactivity disorder. *J. Child Adolesc. Psychopharmacol.* 31 (9): 597–609.

28 Faraone, S.V., Gomeni, R., Hull, J.T. et al. (2021). Executive function outcome of treatment with viloxazine extended-release capsules in children and adolescents with attention-deficit/hyperactivity disorder: a post-hoc analysis of four randomized clinical trials. *Pediatr. Drugs* 23: 583–589.

29 Naya, N., Sakai, C., Okutsu, D. et al. (2021). Efficacy and safety of guanfacine extended-release in Japanese adults with attention-deficit/hyperactivity disorder: exploratory post hoc subgroup analyses of a randomized, double-blind, placebo-controlled study. *Neuropsychopharmacol. Rep.* 41: 26–39.

30 Prince, J.B., Wilens, T.E., Biederman, J. et al. (2000). A controlled study of nortriptyline in children and adolescents with attention deficit hyperactivity disorder. *J. Child Adolesc. Psychopharmacol.* 10: 193–204.

31 Reimherr, F.W., Hedges, D.W., Strong, R.E. et al. (2005). In adults with ADHD: a short-term, placebo-controlled trial. *Neuropsychiatr. Dis. Treat.* 1: 245–251.

32 Becker, W.J. (2015). Acute migraine treatment in adults. *Headache.* 55: 778–793.

33 Thorlund, K., Mills, E.J., Wu, P. et al. (2014). Comparative efficacy of triptans for the abortive treatment of migraine: a multiple treatment comparison meta-analysis. *Cephalalgia* 34: 258–267.

34 Lipton, R.B., Dodick, D.W., Ailani, J. et al. (2019). Effect of ubrogepant vs placebo on pain and the most bothersome associated symptom in the acute treatment of migraine. *J. Am. Med. Assoc.* 322: 1887.

35 Krymchantowski, A.V. and Jevoux, C.C. (2011). Topiramate vs divalproex sodium in the preventive treatment of migraine: a prospective "real-world" study. *Headache.* 51: 554–558.

36 Kurian, A., Reghunadhan, I., Thilak, P. et al. (2020). Short-term efficacy and safety of topical β-blockers (timolol maleate ophthalmic solution, 0.5%) in acute migraine. *JAMA Ophthalmol.* 138: 1160.

37 Rothrock, J.F., Adams, A.M., Lipton, R.B. et al. (2019). FORWARD study: evaluating the comparative effectiveness of onabotulinum toxin A and topiramate for headache prevention in adults with chronic migraine. *Headache* 59: 1700–1713.

38 Tao, X., Yan, Z., Meng, J. et al. (2022). The efficacy and safety of atogepant for the prophylactic treatment of migraine: evidence from randomized controlled trials. *J. Headache Pain* 23: 19.

39 Schwedt, T.J., Lipton, R.B., Ailani, J. et al. (2022). Time course of efficacy of atogepant for the preventive treatment of migraine: results from the randomized, double-blind ADVANCE trial. *Cephalalgia.* 42: 3–11.

40 Hauser, D.N. and Hastings, T.G. (2013). Mitochondrial dysfunction and oxidative stress in Parkinson's disease and monogenic parkinsonism. *Neurobiol. Dis.* 51: 35–42.

41 DeMaagd, G. and Philip, A. (2015). Parkinson's disease and its management: part 4: treatment of motor complications. *P T.* 40: 747–773.

42 Farbman, E.S., Waters, C.H., LeWitt, P.A. et al. (2020). A 12-month, dose-level blinded safety and efficacy study of levodopa inhalation powder (CVT-301, Inbrija) in patients with Parkinson's disease. *Parkinsonism Relat. Disord.* 81: 144–150.

43 Koller, W., Guarnieri, M., Hubble, J. et al. (2005). An open-label evaluation of the tolerability and safety of Stalevo (carbidopa, levodopa and entacapone) in Parkinson's disease patients experiencing wearing-off. *J. Neural Transm.* 112: 221–230.

44 Hauser, R.A., Schapira, A.V.H., Rascol, O. et al. (2010). Randomized, double-blind, multicenter evaluation of pramipexole extended release once daily in early Parkinson's disease. *Mov. Disord.* 25: 2542–2549.

45 Reichmann, H., Lees, A., Rocha, J.-F. et al. (2020). Effectiveness and safety of opicapone in Parkinson's disease patients with motor fluctuations: the OPTIPARK open-label study. *Transl. Neurodegener.* 9: 9.

46 Hattori, N., Tsuboi, Y., Yamamoto, A. et al. (2020). Efficacy and safety of safinamide as an add-on therapy to L-DOPA for patients with Parkinson's disease: a randomized, double-blind, placebo-controlled, phase II/III study. *Parkinsonism Relat. Disord.* 75: 17–23.

47 Hauser, R.A., Lytle, J., Formella, A.E., and Tanner, C.M. (2020). Amantadine delayed release/extended release capsules significantly reduce OFF time in Parkinson's disease. *NPJ Park. Dis.* 8: 29.

48 Chen, J.-F. and Cunha, R.A. (2020). The belated US FDA approval of the adenosine A2A receptor antagonist istradefylline for treatment of Parkinson's disease. *Purinergic Signal* 16: 167–174.

49 Kamm, V.P., Uitdehaag, B.M., and Polman, C.H. (2014). Multiple sclerosis: current knowledge and future outlook. *Eur. Neurol.* 72: 132–141.

50 Cohan, S.L., Hendin, B.A., Reder, A.T. et al. (2021). Interferons and multiple sclerosis: lessons from 25 years of clinical and real-world experience with intramuscular interferon beta-1a (Avonex). *CNS Drugs* 35: 743–767.

51 Cohen, J.A., Comi, G., Selmaj, K.W. et al. (2019). Safety and efficacy of ozanimod versus interferon beta-1a in relapsing multiple sclerosis (RADIANCE): a multicentre, randomised, 24-month, phase 3 trial. *Lancet Neurol.* 18: 1021–1033.

52 Montalban, X., Hauser, S.L., Kappos, L. et al. (2017). Ocrelizumab versus placebo in primary progressive multiple sclerosis. *N. Engl. J. Med.* 376: 209–220.

53 Giovannoni, G. (2018). Disease-modifying treatments for early and advanced multiple sclerosis: a new treatment paradigm. *Curr. Opin. Neurol.* 31: 233–243.

54 Naismith, R.T., Wolinsky, J.S., Wundes, A. et al. (2020). Diroximel fumarate (DRF) in patients with relapsing-remitting multiple sclerosis: interim safety and efficacy results from the phase 3 EVOLVE-MS-1 study. *Mult. Scler.* 26: 1729–1739.

55 De-Paula, V.J., Radanovic, M., Diniz, B.S., and Forlenza, O.V. (2012). Alzheimer's disease., subcell. *Biochemist* 65: 329–352.

56 Yiannopoulou, K.G. and Papageorgiou, S.G. (2020). Current and future treatments in Alzheimer disease: an update. *J. Cent. Nerv. Syst. Dis.* 12: 1179573520907397.

57 Salloway, S., Chalkias, S., Barkhof, F. et al. (2022). Amyloid-related imaging abnormalities in 2 phase 3 studies evaluating aducanumab in patients with early Alzheimer disease. *JAMA Neurol.* 79: 13.

58 Dzierlega, K. and Yokota, T. (2020). Optimization of antisense-mediated exon skipping for Duchenne muscular dystrophy. *Gene Ther.* 27: 407–416.

2

Drug Discovery from Medicinal Plants against Parkinson's Disease

Dunya Al-Duhaidahawi

Department of Pharmacognosy, College of Pharmacy, University of Kufa, Al-Najaf, Iraq

Parkinson's disease (PD) affects 1–2% of persons over 60 years, or 5–6 million worldwide. The loss of substantia nigra compacta dopaminergic neurons and striatal nerve fibers reduces dopamine levels, a critical neurotransmitter for cerebral control. Non-motor symptoms include sleep problems, sadness, and cognitive decline. Oncogenic pathways such as mitochondrial failure, cellular damage, metal buildup, oxidative stress, inflammation, and protein misfolding and aggregation may promote dopamine neuron injury. Recent research links oxidative stress to an increase in reactive oxygen species (ROS), which promotes apoptotic neuronal damage [1]. Dopamine supplementation therapy is often used to address PD symptoms of pain, agitation, and depression. Toxic ROS as well as other toxic intermediates produced by dopamine metabolism accumulate after 10 years of use. They also cannot stop or slow disease progression.

2.1 Pathogenesis of Parkinson's Disease

2.1.1 Misfolding and Aggregation of Proteins

Misfolding and aggregation of proteins are the most associated molecular manifestations and pathogenic attributes for the pathogenesis of PD. For instance, genetic changes in the genes *SNCA*, *PARK2*, *PINK1*, *DJ-1*, and *LRRK2* cause protein misfolding in the SNpc of the brainstem. Lewy bodies (LBs) are a type of inclusion of neurons that occurs in only specific areas of the brain and is a main pathological defining feature of PD and memory loss. Although synuclein is the key part of LBs, it is also discovered in other LBs [2]. Thus, a design for synuclein fibrillation has been suggested, for which the unfolded native protein is conformationally converted into a partly folded aggregation-competent intermediate. Several accumulated synuclein forms have been observed. Herbicides, pesticides, and trace metals have been proven to accelerate synuclein aggregation. Furthermore, synuclein-encoding gene mutations (*A30P*, *G51D*, *E46K*, *A53T*, and *A30P*) have been discovered in PD family members from Spanish, Italian, American, and German populations.

Phytochemical Drug Discovery for Central Nervous System Disorders: Biochemistry and Therapeutic Effects, First Edition. Edited by Chukwuebuka Egbuna and Mithun Rudrapal.

2.1.2 Mitochondrial Dysfunction

Multiple studies in patients with PD and associated animal studies have linked mitochondrial damage (especially ROS) to pathogenesis. Inhibition of the mitochondrial electron transport chain (ETC) causes 1-methyl-4-phenylpyridinium (MPP+) to be established by monoamine oxidase B. A rise in calcium ion concentration and excitotoxicity-mediated neuronal injury are caused by MPP+ in neurons. Notably, 1-methyl-4-phenyl-1,2,3,6-tetrahydropyridine (MPTP) consumption causes oxidative stress and irreversible PD clinical signs in different testing models. MPTP substances were frequently found in PD patients' substantia nigra, inactivating ETC elements (i.e. complex I) [3]. Toxicants like rotenone (lipophilic pesticides) end up causing mitochondrial dysfunctions that induce PD-related morphologic, behavioral, neurochemical, and neuropathologic disorders in humans.

2.1.3 Neuroinflammation

The pathophysiology of PD has been linked to innate and adaptive immune responses. For example, PD patients' dopaminergic neurons had higher levels of nuclear translocated NF-B (nuclear factor kappa-light chain stimulator of activated B cells) interpretation. Inflammation-induced processes increase cytokine levels, including T-cell signaling pathway cytokine (interleukin [IL]-2), inflammatory cytokines (tumor necrosis factor [TNF]-α, IL-1β, and IL-6), anti-inflammatory interleukins (IL-4), and many growth factors (epidermal growth factor [EGF] and transforming growth factor [TGF]-1). Brain inflammation caused by Tlr4 is perhaps one of the main causes of neurodegeneration in PD [4]. Enhancement of CD8+ T-cytotoxic and CD4+ T-helper cells has been discovered in the substantia nigra system of MPTP-injected mice.

2.1.4 Metal Ion Imbalance in the Brain

Biologically, ions (especially Ca and Fe) are clearly involved in DNA formation, myelin sheath and transmitters, mitochondrial respiration, brain growth, and metabolism. Iron buildup in the SNpc and reticulata of PD patients is common and rises with symptom severity. Lithium therapy raises nigral and cortical iron levels, while it decreases tau levels in mice, causing cognitive decline and parkinsonian symptoms [5]. DMT1 (divalent metal transporter 1, which involves iron transfer) single-nucleotide polymorphisms (SNPs) or mutations were found in dopaminergic neurons of PD patients. An iron-dependent type of cellular damage defined by oxidation to phospholipids is involved in the etiology of Parkinson's disease in human induced pluripotent stem cells (iPSC)-derived neurons. Ferroptosis produces a buildup of 15-hydroperoxy-eicasotetraenoyl-phosphatidylethanolam ines (15-HpETE-PE), a death signal.

2.1.5 Protease-Mediated Degradation Inhibition

Proteasome is a molecular machine that destroys undesired or misfolding polypeptide with ploy-ubiquitin changes. An abnormal ubiquitin proteasome system (UPS) is linked to PD symptoms. Brain-derived neurotrophic factor (BDNF) and tyrosine kinase B (TRKB) receptors were shown to support the production of important synaptic proteins in relation to neurotransmission, which plays a major role in PD pathogenesis [6]. Many sorts of mutations have been reported in *PARK2*, which codes for the key ubiquitin ligase protein of UPS. It was found that therapeutic F-box protein (FBXO7) was associated with early-onset PD.

2.1.6 Oxidative Stress

The role of oxidative stress in the pathophysiology of PD has been found in multiple experimental studies on dopamine metabolism, lipid peroxide (LPO), and glutathione (GSH) depletion. These byproducts can alter cellular nucleophiles, such as low molecular weight (LMW) sulfhydryl (like GSH) and protein cysteinyl residues. LPO can remove hydrogens from methyl group bridges (CH_2) to generate H_2O_2. Parkinson's substantia nigra, an intermediate basal malondialdehyde in LPO production, is considerably higher than in other brain areas, implying that LPO may contribute to PD. In 2021, Demmings et al. investigated the impacts of ATF4 in neuronal damage in MPP+- and 6-OHDA-induced PD mice and showed that α-synuclein aggregation could cause a considerable increase of ATF4 expression in animal cortex and mesencephalic dopaminergic neurons [7].

2.2 Natural Dopaminergic Neuroprotective Compounds

Numerous plants produce phytochemicals, which are bioactive compounds with physiologic activities (antioxidant, anti-inflammatory, antitumor, antimutagenic, etc.) [8].

2.2.1 Polyphenols

Polyphenols are chemicals with antioxidant, anti-inflammatory, and immunomodulatory properties that may transfer electrons and hydrogens. They have medicinal effects in the treatment of diabetes, malignancy, heart disease, and neurodegenerative disorders. Grapes, red wine, and other foodstuffs contain resveratrol, a kind of polyphenol.

2.2.1.1 Resveratrol

Resveratrol preserves GSH levels and reduce ROS and oxidative stress triggered by 20 mM of the neurotoxic MPP+. Resveratrol reduces p53 acetylation, which inhibits apoptotic cell death caused by the DNA-damaging drug N-methyl-N′-nitrosoguanidine. Resveratrol at 0.01, 0.1, and 1 μM protected rat midbrain neurons, reducing dead cell count. For example, resveratrol promotes neurite growth and reduces superoxide radical ($O2^-$) creation by 68% [9]. In SH-SY5Y dopaminergic neurons, resveratrol (20 μM) increases HO-1 expression, and reduces ROS and oxidative stress, autophagy, and rotenone-induced induce apoptosis. This chemical promotes HO-1 expression by Nfr-2 in a cerebral ischemia model. The combining of resveratrol and quercetin reduces the expression of proinflammatory mediators such IL-1 and TNF-α, preventing neurodegeneration.

2.2.1.2 Sesamin

Sesamin's antiapoptotic and antioxidant properties have been shown to reduce genetic expression and IL-6, IL-1β, and TNF-α levels in MPP+-activated microglial N9 cells, and lowered inducible nitric oxide (iNOS) expression and $O2^-$ generation in co-cultured pheochromocytoma (PC12) neurons, saving dopaminergic neurons from death. Sesamin protects dopaminergic neurons against PC12 cells. After MPP+ treatment, a three-hour sesamin therapy protects dopaminergic neurons by boosting enzymatic activity and reducing oxidative damage [10].

2.2.1.3 Curcumin

Curcumin, a polyphenol from *Curcuma longa*, is anti-inflammatory and antioxidant. Curcumin protects SH-SY5Y neuronal cells by inhibiting rotenone-induced caspase-3 activation at 5 μM. By reducing phosphorylated p38 and toxic quinoprotein production, curcumin protects against oxidative damage, and restores tyrosine hydroxylase and cell viability. Further, 10 μM curcumin protects midbrain dopaminergic neurons from oxidative stress by raising dopamine uptake, reducing inflammatory cytokines such as nitric oxide (NO), prostaglandin E2 (PGE2), IL-1β, and TNF-α, and reducing nuclear factor B (NK-B) and activator protein-1 transcription. However, the researchers discovered stronger neuroprotective effects in microglial cells, indicating that microglia protected from lipopolysaccharide (LPS)-induced neurotoxicity. In order to protect SH-SY5Y dopaminergic neuron from MPP+-mediated neurotoxicity (3 mM MPP+), curcumin therapy was proposed [11].

2.2.1.4 6-Shogaol

An anti-inflammatory chemical derived from ginger, 6-shogaol prevented tyrosine hydroxylase-immunoreactive cellular damage and reduced NO and TNF-α concentrations in MPP+-treated rat midbrain cells [12].

2.2.1.5 Fustin

Fustin is an anti-inflammatory and antimutagenic flavonoid derived from *Rhus verniciflua*. Pretreatment of SK-N-SH dopamine neurons with fustin at concentrations over 50 μM for 30 minutes prior to exposure to 6-OHDA at 125 μM for 24 hours resulted in lower ROS levels and intracellular Ca2+. Fustin also inhibits Bax/Bcl-2, caspase-3, and p38 phosphorylation [13].

2.2.1.6 Biochanin A

Biochanin A, an anti-inflammatory isoflavone that resides in *Trifolium pratense*, protects dopaminergic neurons in the rat midbrain. It has been demonstrated to adequately shield neuronal cells against LPS-induced neurotoxic effects at concentration of 10 ng/ml, which results in a 36.7% drop in dopamine uptake and a 52% reduction in cell population. Biochanin A enhances dopamine uptake by 55.9%, 77.9%, and 88.7% at concentrations of 0.25, 1, and 2.5 μM, and inhibits LPS-induced damage at concentrations of 62.5%, 81.9%, and 89.4%, respectively. It decreases the generation of TNF-α, NO, and $O2^-$ in a dose-dependent manner [14].

2.2.1.7 Acacetin

Acacetin is an anticancer and anti-inflammatory flavone that can also be identified in chrysanthemums and safflowers, particularly the roots of *Scutellaria baicalensis*. Dopaminergic neurons in primary cultures treated with 10 M MPP+ and 50–200 nM acacetin generated more tyrosine hydroxylase-immunoreactive cells, as well as conserved neuronal morphology, notably dendrite shortening. Acacetin decreases the synthesis of proinflammatory factors such as NO, PGE2, and TNF-α [15].

2.2.1.8 Baicalein

Baicalein at concentrations of 0.05, 0.5, and 5 g/ml was tested in SH-SY5Y dopaminergic cells and revealed in vitro resistance to 6-OHDA-induced toxicity at a concentration of 100 nM. At doses of 0.5 g/ml or over, baicalein was noticed to have protective characteristics, with 64% survival relative to controls. In addition, 5 g/ml baicalein treatment greatly reduces apoptotic cell fraction [16].

2.3 Nitrogenated Phytochemicals

2.3.1 Alkaloids

2.3.1.1 Tetrahydro Berberine

Tetrahydro berberine protects the brain by inhibiting adenosine triphosphate (ATP)-sensitive K+ channels in *Rhizoma corydalis.* Dopamine neurons' membranes depolarize from 61.7 to 46.7 mV when exposed to tetrahydro berberine. Antagonizing D2 receptors restored hyperpolarization and increased action potential firing in tetrahydro berberine–treated cells.

2.3.1.2 Berberine

Berberine, an antidiarrheal agent derived from *Coptis xhinensis*, is anti-inflammatory, anticarcinogenic, and cardioprotective. Berberine protects SH-SY5Y cells from 6-OHDA-induced mortality. Three hours of berberine dosing lowered caspase-3 activity and ROS levels. Berberine enhances heme oxygenase-1 (HO-1) mRNA expression by promoting Nrf2 translocation [17].

2.3.1.3 Celastrol

Celastrol has anti-inflammatory and anticancer activities of *Tripterygium wilfordii* root extracts. The chemical has been found to reduce ROS levels, and to reduce cytochrome C cytosolic release and Bax expression at 10 nM. Celastrol may also protect against membrane potential loss and raise anti-apoptotic protein Bcl-22 levels [18].

2.3.2 Saponins

2.3.2.1 Astragaloside IV

Astragaloside IV, a component of *Astragalus membranaceus*, is used in traditional Chinese medicine to cure neurologic illnesses. In 6-OHDA-exposed midbrain dopaminergic neurons in primary cultures, 100 μM astragaloside IV protects neurons, enhances cell viability, and inhibits neurite loss and shortening. The same chemical enhances cell viability by reducing nuclear size, DNA fragmentation, and chromatin condensation, as well as reducing ROS, proapoptotic BAX/Bcl2 ratio, and caspase-3 activation [19].

2.3.2.2 Ginseng

Ginsenosides Rb1, Rg1, and Rd are among 30 kinds of ginsenosides derived from ginseng root. They act as radical scavengers and boost neural energy utilization. In neuronal cell cultures, ginsenosides Rb1 and Rg1 promote the preservation of mitochondrial transmembrane potential and reduce lactate dehydrogenase synthesis. They also lengthen and upsurge the quantity of neurites in glutamate-resistant SH-SY5Y human neuronal cells. The anti-inflammatory effect of ginsenoside Rd (50 μM) increases cell viability and reduces NO and PGE2 production [20]. At concentrations of 1–10 μM, ginsenoside Rd protects SH-SY5Y cells from MPP+-induced apoptosis. It has been shown that elevating antioxidant enzyme activity such as superoxide dismutase (SOD) and glutathione peroxidase (GPx) reduces lactate dehydrogenase (LD), ROS, and Bax levels. Gensinoside Rd also increased intracellular ATP levels and mitochondrial membrane potential (MMP) stabilization. Gensinoside Rd-mediated neuroprotection involves the PI3K/Akt survival signaling pathway.

2.4 Chinese Herbal Medications and Parkinson's Disease

2.4.1 *Acanthopanax*

Traditional Chinese medicine uses *Acanthopanax senticosus* roots and stems (ASRS), also called *wujipi*. In MPTP-induced PD mice, a root extract (45.5 mg/kg daily) was neuroprotective. The extract significantly lowered dopaminergic D1/2-positive cells and apoptotic caspase-3 total proteins in the striatum. The plant extract boosts SOD activity and lowers catalase and NO synthase protein expression in MPP+-induced neuronal PC12 cells. ASRS's eleutheroside B can assist with memory and cognition. Eleutheroside B shields PC12 cells against MPP+ cytotoxicity, enhances extracellular signal-regulated kinase (ERK)1/2 phosphorylation, and decreases c-Fos and c-Jun expression. LncRNA microarrays have been used to study ASRS pathology and physiology. However, under physiologic conditions, there was no significant change in the expression of lncRNA/mRNA, which might also be neurotoxic comparable to synuclein. These results suggest that *Acanthopanax*'s bioactivities for healthy and diseased neurons may be bidirectional [21].

2.4.2 *Alpinia*

The dried, ripe seeds of *Alpinia oxyphylla Miq*. (AOF) are often used in clinics to treat a variety of ailments. In a zebrafish model of PD, AOF ethanol extract reversed 6-OHDA-induced neurodegeneration in dopaminergic neurons and improved locomotor activity by reducing inflammation (IL-1 and TNF-α expression) and oxidative stress (NO production). AOF also has neuroprotective bioactivities via the PI3K-Akt pathway. The AOF polyphenols protocatechuic acid and chrysin improve 6-OHDA-treated PC12 cell viability and prevent neuronal cell death in zebrafish and animal Parkinson's models. Protocatechuic acid and chrysin (i) increase nuclear factor-erythroid 2-related factor (NRF2) expression; (ii) alter cellular redox state; and (iii) reduce malondialdehyde levels. Oxyphylla A, an AOF bioactive compound, exhibits neuroprotective effects: it prevents primary neuron damage and behavioral impairment in vivo. Oxyphylla A induces synuclein degradation in a cellular PD model by increasing PSMB8 expression through protein kinase (PK)A-Akt–mammalian target of rapamycin (mTOR) stimulation. It protects A53T-synuclein transgenic mice from synuclein-induced neurotoxicity by lowering triton-soluble and -insoluble buildup [22].

2.4.3 *Camellia*

An infusion of the dried leaves of *Camellia sinensis* L., from the plant family Theaceae, was traditionally used to treat gastrointestinal issues, headaches, and nervous tension. Epicatechin, epigallocatechin gallate, and catechin are anti-inflammatory, antioxidant, and neuroprotective polyphenols found in green tea. These extracts improve cognitive function and reduce depression in the 6-OHDA mouse model of PD. The toxicant paraquat (PQ) reduces lifespan and locomotor activity while promoting lipid peroxidation and neurotoxicity in *Drosophila melanogaster* flies [23]. On the other hand, saponin reduces inflammation, improves cognition, and improves behavior in PD mice treated with MPTP. Theacrine in animal models of PD shows dopaminergic neuron loss and behavioral abnormalities. To reduce ROS accumulation and restore mitochondrial function, theacrine directly promotes SIRT3 [24].

2.4.4 *Cassia*

The dried full-grown seed of *Cassia obtusifolia* L. or *Cassia tora* L. is *Cassiae* semen (*juemingzi* in Chinese) from the plant family Leguminosae. Cassia extract inhibited ROS generation, GSH deprivation, mitochondrial depolarization, and caspase-3 activity in 6-OHDA-treated PC12 cells. Acute MPTP treatment reduced dopaminergic mouse substantia nigra and striatum neuron degeneration [25]. One of the most important oxidants in the body, peroxynitrite ($ONOO^-$), has been related to the cause of PD. In vitro alaternin, an antioxidant compound found in the *C. tora* L. plant, effectively reduces $ONOO^-$-mediated tyrosine nitration by donating electrons. Cinnamaldehyde (at doses of 5 and 10 μM for 24 hours) increased vitality and decreased ROS in 6-OHDA-treated PC12 cells.

2.4.5 *Chrysanthemum*

The dried flower head of *Chrysanthemum morifolium* R. is known as *Chrysanthemi Flos* (*jvhua* in Chinese). The crude extract of *Chrysanthemi Flos* efficiently increased cell survival and decreased the raised ROS level, Bax/Bcl-2 ratio, and caspase-3 cleavage in MPP+-pretreated SH-SY5Y human cells. ROS generation was reduced, the Bcl-2/Bax ratio was increased, and poly-ADP ribose polymerase (PARP) proteolysis was inhibited by *Chrysanthemum indicum* L. crude extract. It also reduced inflammation in BV-2 microglial cells by inhibiting IkB breakdown and NF-kB (p65) activation. Acacetin (5,7-dihydroxy-4-methoxy flavone), a flavonoid found in chrysanthemums, has prevented 6-OHDA-induced neurodegeneration by inhibiting the mitochondrial-mediated cascade of apoptosis and ROS production [15]. Furthermore, acacetin inhibits c-Jun N-terminal kinase (JNK) phosphorylation (p38 MAPK), PI3K, and glycogen synthase kinase-3 (GSK-3).

2.4.6 *Cistanche*

Echinacoside is a phenylethanoid glycoside compound isolated from *Cistanche deserticola* stems. In 6-OHDA-induced PC12 it was found that echinacoside drastically reduced cell cytotoxicity, oxidative damage, and mitochondria-mediated cell turnover while maintaining MMP. *Cistanche tubulosa* nano powder improves neuronal survival and tyrosine hydroxylase expression in MPTP-induced rats and lowers apoptotic cells. *Ci. tubulosa* nano powder promotes GDNF, GFR1, and Ret protein expression in mouse substantia nigra neurons. Echinacoside protects neurons from 6-OHDA-promoted endoplasmic reticulum and oxidative damage. Echinacoside also lowers seizure activity by increasing seipin breakdown via the Grp94/Bip-ATF4-CHOP signaling cascade [26]. By increasing FoxO expression in neurons, echinacoside promotes autophagy and also reduces NLRP3/CASP-1/IL-1 signaling in microglia, and thereby increases dopaminergic neuron lifespan in MPTP-induced PD mice.

2.4.7 *Gastrodia*

Gastrodia elata Blume is a traditional Oriental plant rich in gastrodin and bibenzyl compounds. It helps protect dopaminergic neurons by removing ROS and reducing cell death in both in vitro and in vivo studies, in contrast to the bibenzyl component that lowers synuclein buildup, endoplasmic reticulum stress, and intracellular ROS generation. Levodopa (L-DOPA) is the mainstay of therapy for PD; however, long-term use induces dyskinesia [27].

G. elata aqueous extract decreases dyskinesia produced by L-DOPA via decreasing pERK1/2 and FosB. Gastrodin promotes HO-1 expression by activating p38 MAPK/NRF2 signaling in MPP+-induced SH-SY5Y cells. Also, bibenzyl compounds inhibit rotenone-induced cell death by activating the NRF2/ARE/HO-1 signaling cascade. Ferroptosis is a type of necrosis produced by an accumulation of Fe and driven by GPx enzyme activities. In H_2O_2-treated C6 cells, gastrodin increased NRF2, GPX4, ferroportin-1, and HO-1 protein expression.

2.4.8 *Ginkgo*

Brain-boosting *Ginkgo biloba* extract (EGb761) is used in dementia patients. EGb761 also stimulates neuronal stem cell proliferation in the subventricular zone of PD rats. A natural flavonoid in *Gi. biloba* leaves reduces intracellular ROS levels while retaining MMP in MPP+-induced rats. Ginkgetin reduces apoptosis induced by MPP+ through the caspase-3 and Bcl2/Bax networks, intercalates ferrous ion to decrease L-ferritin, and enhances transferrin receptor 1. In vitro, synuclein (monomeric or aggregated) reduces apoptosis in SY5Y cells [28]. *Gi. biloba* extract boosts locomotion, but lowers malondialdehyde (MDA) levels, and promotes tyrosine hydroxylase and dopamine transporter interpretation in A53T synuclein transgenic mice. It reduces oxidative and inflammatory stress when used orally in PD rats. To reduce intracytoplasmic synuclein aggregation and SUMO-1 levels, ginkgolic acid (from *Gi. biloba* leaves) increased autophagosome numbers. Protocatechuic acid, a *Gi. biloba* leaf component, may boost ginkgolide B's efficacy in treating PD.

2.4.9 *Gynostemma*

The ethanol extract of *Gynostemma pentaphyllum* decreases cytotoxicity and mortality while boosting cell survival in PD cell culture and MPTP-induced animal studies. In 2020, the neuroprotective effects of *Gynostemma pentaphyllum* on an A53T-synuclein recombinant mouse model of PD were reevaluated. It restores dephosphorylated tyrosine hydroxylase, ERK1/2, Bad (Bcl-2-linked death promoter), Ser112, and JNK1/2 in disease. On the other hand, a saponin extract from *Gynostemma pentaphyllum* alleviated anxiety in PD-induced mice. Dyskinesia and spatial memory are improved in L-DOPA–induced PD rats. Gypenoside alters ERK1/2 phosphorylation in hippocampal tissues [29].

2.4.10 *Paeonia*

Paeoniflorin, a monoterpenoid glycoside from *Paeoniae Alba Radix* aqueous leaf extract, promotes autophagic breakdown of α-synuclein and thereby protects cells from cytotoxicity. Another study demonstrated that paeoniflorin protects neurons from glutamate or MPP+ damage through modulating MMP and the Bcl-2/Bax signaling pathways. Paeoniflorin reduces apoptosis in 6-OHDA-induced PC12 cells via suppressing the ROS/PKC/NF-B signaling pathway. Paeoniflorin therapy improves behavioral impairments and lowers dopaminergic cell death in MPTP-treated PD mice [30]. Moreover, paeoniflorin stimulates dopamine depletion, which is primarily due to PD mice's striatum and substantia nigra losing dopaminergic transporter and tyrosine hydroxylase protein levels. The root cortex of *Paeonia suffruticosa Andrews* is termed *Moutan Cortex Radicis*. It enhances locomotor function and reduces bradykinesia in an MPTP-induced animal model, and provides protection against dopaminergic degeneration and locomotor impairment in PD zebrafish models. Cell death in PC12 cells is reduced by reducing intracellular ROS buildup and restoring total GSH. Gold nanoparticles from PM root extract have been shown to decrease inflammation in vitro and enhance motor function in PD rats.

2.4.11 *Panax*

In vivo, Panax notoginseng saponins provide neuroprotection against dopaminergic neuron loss and behavioral impairment produced by MPTP therapy. Osmotic treatment of *Panax ginseng* extract inhibits the development of locomotor impairments in rats exposed to β-sitosterol-D-glucoside-induced progressive PD. In H_2O_2-treated PC12 cells, ginsenoside Rg1, a natural substance derived from *P. ginseng*, inhibited NF-B/p65 phosphorylation and nuclear translocation while activating Akt and ERK1/2. Moreover, ginsenoside Rd, a primary active monomer component in *P. ginseng*, restored the depletion of tyrosine hydroxylase–positive neurons in MPTP-treated rats. Ginsenosides increase SOD and aconitase enzyme activity, reduce MMP depolarization, and restore calcium levels in rotenone-induced SH-SY5Y cells. Caspase-3-mediated cell death and the unfolded protein response have been shown to be affected by the antioxidant and anti-inflammatory properties of Korean red ginseng in PD animal models [30].

2.4.12 *Polygala*

In MPP+-induced toxicity, a water extract of *Polygala tenuifolia Radix Polygalae* was tested in PC12 cells. *Po. tenuifolia*'s principal active ingredient, tenuigenin, increases tyrosine hydroxylase-immunoreactive neuron survival, decreases dopamine concentration in the substantia nigra, and inhibits TNF-α and IL-1 production in a LPS-induced PD model. In 6-OHDA-damaged SH-SY5Y cells, tenuigenin protects MMP and enhances GSH and SOD expression [31]. Tenuigenin improves motor function and boosts striatal dopamine levels in MPTP-induced rats by blocking NLRP3 inflammasome activation and ROS generation. A53T-synuclein and mutated huntingtin are removed faster by onjisaponin B from *Radix Polygalae* in PC12 cells. MPTP-treated mice show improved motor function, inflammatory markers, and reduced microglia overactivation with onjisaponin B. In PD mice, onjisaponin B lowers the expression of the NF-B complex p65 subunit, RhoA, and ROCK2 proteins.

2.4.13 *Polygonum*

The active ingredient of *Polygonum multiflorum Thunb* is 2,3,5,4′-tetrahydroxystilbene-2-O-D-glucoside (TSG). TSG improved cell viability while reducing mortality and ROS production in vitro by regulating JNK, p38, and PI3K-Akt signal transduction. Dopaminergic neurons are protected for two weeks against neurotoxic 6-hydroxydopamine (6-OHDA) after TSG is injected intraperitoneally daily, and it also decreases microglial activation in 6-OHDA-induced rats. TSG improves behavioral function and MPTP-induced dopamine neuron loss in mice, reestablishing the fibroblast growth factor-2 (FGF2)–Akt and BDNF–TRKB signal transduction pathways in the striatum and substantia nigra. In 6-OHDA-induced PD animals, resveratrol from *Polygonum cuspidatum* reduces aberrant rotation, nigral cell loss and apoptosis, and total ROS. Juglanin, a natural chemical derived from *Polygonum aviculare*, has anti-inflammatory, antioxidant, and anticancer properties. It has been revealed that juglanin reduces amyloid and p-Tau levels in the brain. Juglanin suppresses the Toll-like receptor-4 (TLR4)/NF-B pathways to reduce LPS-induced inflammation [32].

2.4.14 *Psoralea*

The flavanone bavachinin from *Psoralea corylifolia* L. seeds significantly lowers monoamine oxidase B (MAO-B) activity. *Ps. corylifolia*'s isobavachalcone efficiently decreases neuron necrosis in MPTP-induced PD mice. Isobavachalcone reduces inflammation caused by microglia through regulating NF-B signaling. Some *Ps. corylifolia* prenylchalcones such as isobavachalcone, bavachromene, and kanzonol B have been demonstrated to reduce iNOS and cyclooxygenase-2 (COX-2) expression in LPS-induced inflammation by inhibiting 1β depletion and reducing NF-B levels [33].

2.5 Herbal Medicines from India and Parkinson's Disease

2.5.1 *Withania somnifera*

Withania somnifera (WS), often known as winter cherry or poison gooseberry, is a medicinal plant from the Solanaceae family. Triterpene lactones, alkaloids, tropine, and withanolides are among the bioactive chemicals found in WS. It is structurally related to ginseng. For this reason, WS is called "Indian ginseng." WS has been utilized medicinally in India for 5000 years to treat dementia, depression, epilepsy, and neurodegeneration. Studies on WS root extract previously revealed antiaging, antioxidant, and antitumor effects. In 2017, Nayak et al. investigated WS's neuroprotective potential in the MB/PQ mouse model [34]. The WS root ethanol extract boosted dopamine secretion and improved locomotor activity in PD rats. In PD mice, WS root ethanol extract lowered iNOS and GFAP (a proinflammatory biomarkers of astrocyte stimulation).

2.5.2 *Tinospora cordifolia*

Tinospora cordifolia (TC) is a tropical plant from the Indian subcontinent from the Menispermaceae family. TC has long been used in traditional Chinese medicine as an immunomodulator and antidiabetic medication. In 6-OHDA-induced PD mice, TC ethanol extract preserved neurons and restored locomotor function. In sleep-deprived rats, TC ethanol extract enhances behavioral performance, lowers stress-induced brain injury, and reduces inflammation. TC aqueous extract showed anti-inflammatory efficacy in PD mice. TNF-α and IL-12/1β levels were successfully recovered by the aqueous extract of TC in MPTP-intoxicated rats. The anti-inflammatory effects of aqueous extract of TC are neuroprotective, as prolonged inflammation can kill dopaminergic neurons. TC has neuroprotective bioactivity in PD animal models, similar to Chinese herbal anti-PD medications [35].

2.5.3 *Mucuna pruriens*

Mucuna pruriens (MP) is a tropical leguminous plant found in Africa and Asia, including China and India. MP's seed pods contain serotonin and mucunain, which cause itching when touched. MP seeds are utilized in traditional Indian medicine to treat aging, arthritis, diabetes, and neurologic diseases. The MP plant is a major source of L-DOPA, which it naturally contains at 5% dry weight [36]. MP seed powder water extract had less dyskinesia than regular levodopa treatment. The role of MP ethanol extract was investigated on lipid peroxidation in a PQ-induced PD animal model. In PQ-induced PD mice, ethanol extract of MP preserves substantia nigra dopaminergic neurons. Crude extract of MP (100 mg/kg body weight) was examined in PD rats provoked by MPTP. To reverse the behavioral abnormalities caused by MPTP, the MP aqueous extracts lowered lipid oxidation and nitrite concentrations while boosting pAKT1 activity. MP's neuroprotective actions are anti-inflammatory and antioxidant.

2.6 European Plants

Narcotic and toxic effects are acknowledged rather than neurodegenerative disorders. *Atropa belladonna* and *Hyoscyamus niger* are listed as plants to treat PD and other neurologic disorders (such as Alzheimer's disease and dementia) using extreme care and competent guidance. *A. belladonna* and *H. niger* both contain atropine, hyoscyamine, and scopolamine. Some PD medications contain these active components. Traditional PD treatments include *A. belladonna* powder and *H. niger* leaf tincture. Their anti-parkinsonian, anti-Alzheimer's, and anti-dementia activities are neuroprotective.

- *Atropa belladonna* L. (nightshade) is believed to be native to Europe, Asia, and North Africa. *A. belladonna*-based medications are used to block acetylcholine (Ach) in three ways. In the central nervous system, methacholine (MCh) receptors are inhibited, as are ACh receptors in cholinergic neurons and smooth muscles. High doses of scopolamine cause learning and memory problems [37].
- *Hyoscyamus niger* L. (black henbane) has been used in Ayurvedic medicine to treat PD, but as a tiny component of herbal preparations. Reduced MAO activity and intestinal hydroxyl radical protection are putative PD therapy strategies [38].

2.7 α-Synuclein as a Potential Therapeutic Target

The 140-residue presynaptic protein α-syn has 1 μM in the brain and 70 pM in the cerebrospinal fluid. An α-helix-like secondary structure is formed by unfolded monomer. As a result, synaptosomal stability is reduced. The α-syn monomer is a disordered protein with many conformations. It is implicated in metabolic reactions and an increasing variety of misfolding diseases, including PD. Synphilin-1 and ubiquitin intracytoplasmic inclusions form LBs in dopaminergic neurons. Small oligomers/protofibrils and finally sheet-rich syn fibrils are formed. Many theories exist on how α-syn contributes to dopaminergic neurotoxicity in PD. Conformationally, defective α-syn can adapt many forms, varying from condensed to fully expanded. The N- and C-termini of α-syn engage to stabilize it into a non-toxic monomeric shape. It is hypothesized that neuroprotective drugs establish a loop structure between the N- and C-termini of α-syn. Those that cause more compacted structures are deemed neurotoxic [39].

2.7.1 Phytochemicals Targeting the α-Synuclein Cascade

Many phytoconstituents are used in PD and have been shown to reduce oligomerization and fibrillization of syn. Plants like *A. senticosus* (*Eleutherococcus senticosus*, *Bacopa monnieri*), precipitate of cinnamon extract *Cinnamomum verum*, *Centella asiatica*, *P. ginseng*, *Po. tenuifolia*, *Rehmannia glutinosa*, and *Padina pavonica* (*Sagina japonica*, *Carthamus tinctorius* L., and *Crocus sativus* L.) protect against PD by oligomerizing, fibrillating, and disrupting the α-syn fibrils produced. Many plant extracts target α-syn in PD models (typically in vitro). *Caenorhabditis elegans* A53T-syn-induced PD is prevented by *B. monnieri*. Asiatic pennywort, *Ce. asiatica* (L.) Urb., inhibits α-syn aggregation in vitro [40]. In A53T-syn-induced PD

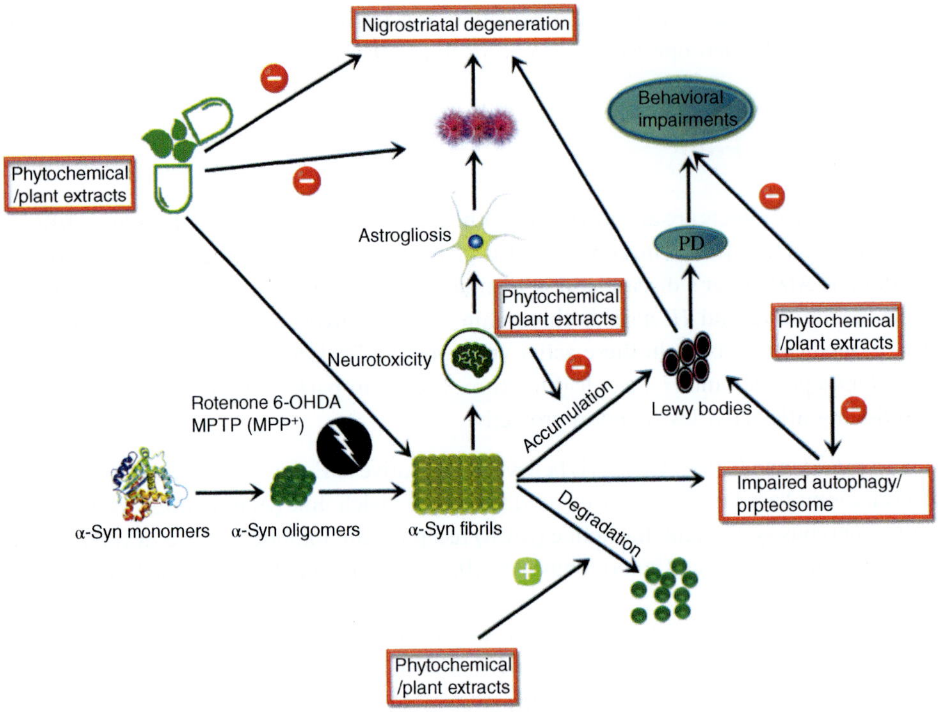

Figure 2.1 The effect of plants on α-syn. 6-OHDA, 6-hydroxydopamine; MPP+, 1-methyl-4-phenylpyridinium; MPTP, 1-methyl-4-phenyl-1,2,3,6-tetrahydropyridine; PD, Parkinson's disease.

in flies, precipitate of cinnamon extract reduces aggregation and stabilizes oligomers. *Eucalyptus citriodora* improves cell viability and minimizes cell damage in human–syn transgenic drosophila. The effects of *Eu. citriodora*'s bioactive compounds citronellol, linalool, and isopulegol on α-syn remain unknown. Components of saffron, *Cr. sativus* L., such as safranal, hexadecanedioic acid, norbixin, and transmuconic acid, have been shown to influence syn fibrillation and aggregate formation. The same was true for some crocetin analogues. In a *Ca. elegans* model of PD, the Korean mountain ash, *Sorbus alnifolia*, increased longevity, improved food sensing, and diminished dopaminergic neurotoxicity [41]. Figure 2.1 explains how phytochemicals affect α-syn aggregation.

2.8 Conclusion

Numerous in vitro studies have found that certain phytochemicals protect against neuronal cell damage in PD models. Several mechanisms are involved, including suppression of NF-B, caspase-3, and Bax expression, lower ROS levels, enhanced expression of endogenous antioxidative enzymes, and mitochondrial activity conservation. The goal of this chapter was to look into the potential of a variety of natural substances, such as polyphenols, alkaloids, and saponins, for the treatment of PD, and these are summarized in Tables 2.1 and 2.2. Due to their molecular actions, polyphenols may offer a new therapy option by slowing down the degeneration of dopaminergic neurons.

Table 2.1 Medicinal plant components, sources, structural model, and effect on Parkinson's disease.

Name of compound	Model	Dose	Structure	Effect
(−)-Epigallocatechin gallate	Male C57BL/6J mice MPTP	25 mg/kg	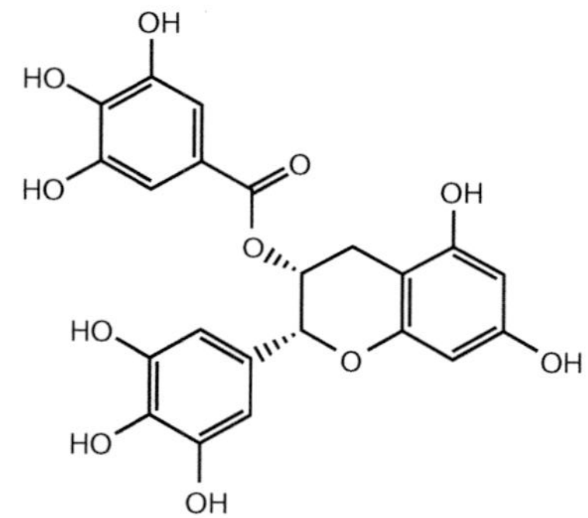	↑ Movement style ↑ TH-positive cells in the SN area ↑ CD3+ CD4+ to CD3+ CD8+ T-cell lymphocyte ratio in blood ↓ IL-6 and TNF-α expression in serum [42]
Acacetin	*Caenorhabditis elegans* model system	25 μM	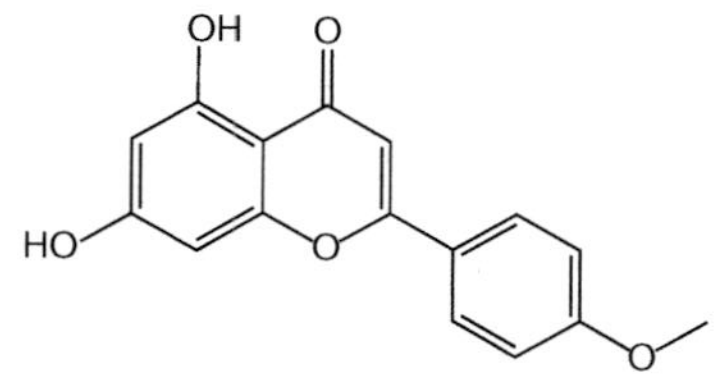	↑ Dopamine and treating behavioral issues ↓ NO, PGE2, iNOS, COX-2, TNF-α, and IL-1 [15]
Acteoside	Rotenone-induced PD in rats	50–200 μM	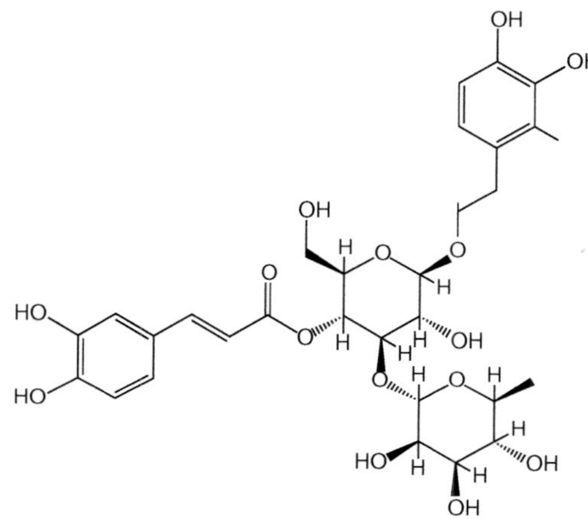	Suppresses α-syn, caspase-3 activation, and MAP2 downregulation [24]

(*Continued*)

Table 2.1 (Continued)

Name of compound	Model	Dose	Structure	Effect
Apigenin (flavone found in fruits and vegetables)	6-OHDA-induced rat	150 and 300 mg/kg	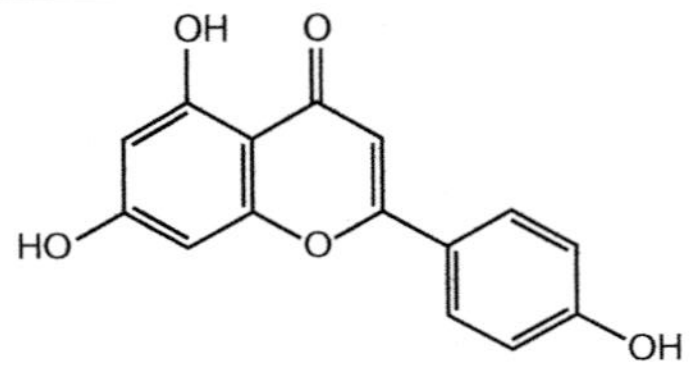	↓ Rotational behavior ↑ GSH, GPx, glutathione reductase, SOD, catalase, protein carbonyl level, MDA level [24]
Asiaticoside	MPTP/p, MPP+	100 mg/kg	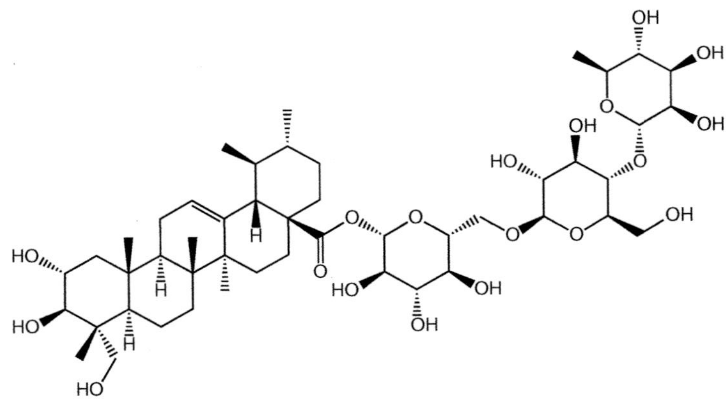	↑ Phosphorylated PI3K, Akt, GSK-3β, and mTOR ↑ TRKB ↓ NLRP3 protein expression [24]
Astragaloside IV	Primary nigral cell cultures (PNCs)	50, 100, 200 μM	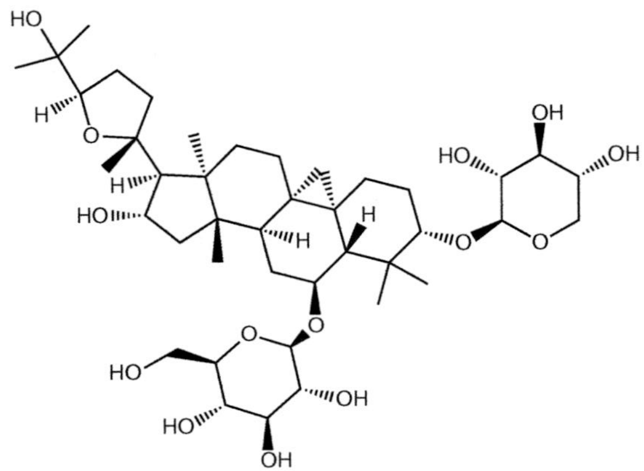	↑ TH and NOS immune reaction. Antidegeneration in DN [19]

Baicalein (flavonoid from *Scutellaria baicalensis*, Lamiaceae)	6-OHDA-induced rat; α-syn aggregation assay; dopaminergic cell lines (SN4741)	0.5–5 μg/ml	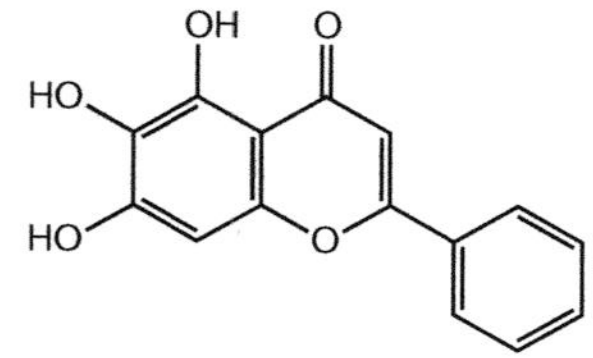	Reduce SH-SY5Y apoptotic cell death Increase PC12 cell neurite growth Reduce tremors Boost TH-positive neurons [16]
Berberine	6-OHDA-induced PC12 cells; rats with bilateral 6-OHDA lesion	5, 10, 30 μM	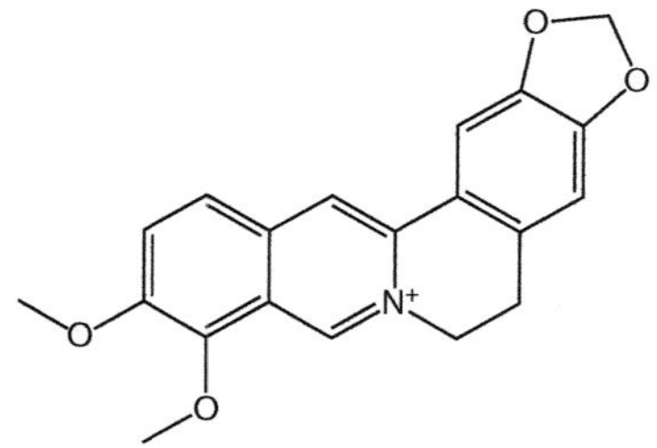	Substantia nigra immunodepleted TH cells Reduced striatal dopamine and norepinephrine levels [17]
Caffeic acid phenethyl ester	In vitro model Cerebral granule neuron cells	10 μl	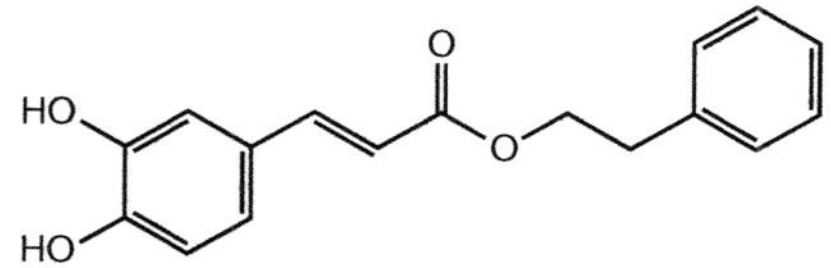	Modulate Ca2+ promoted cytochrome c discharge Suppress caspase-3 activation Halt cell apoptosis [24]
Carnosic acid	6-OHDA-induced rat	20 mg/kg	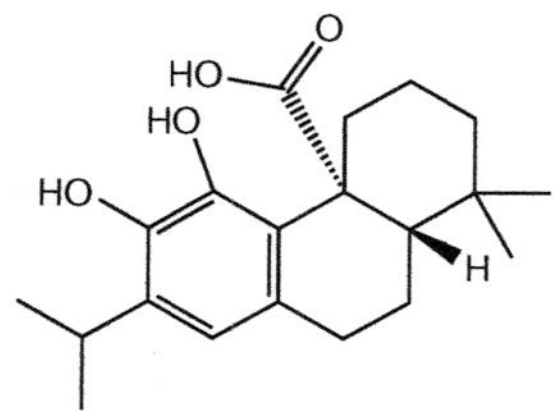	↑ Motion function ↓ Apomorphine-induced spin ↓ Lipid peroxidation ↓ GSH ↑ Expression of glutamate-cysteine ligase SOD and GR proteins ↓ Bcl2/Bax ratio ↓ Caspase-3 cleavage ↓ PARP cleavage induction [43]

(*Continued*)

Table 2.1 (Continued)

Name of compound	Model	Dose	Structure	Effect
Curcumin	6-OHDA-induced rat α-Syn aggregation, cytotoxicity, and A53T-syn PC12 induction in cells	80 mg/kg	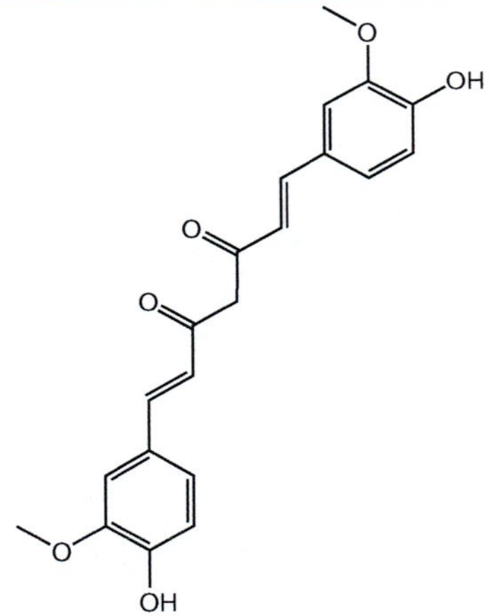	↓ MDA ↑ GSH, GPx, GR, SOD, catalase, TH, and D2 receptors [11]
Echinacoside	MPTP-induced C57BL/6 mice	40 mg/kg	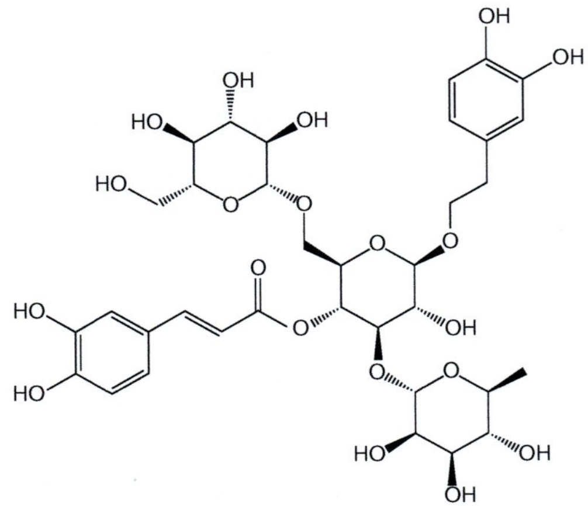	↓ Dopamine neuron loss; caspase-3–8 activation ↓ DA and DOPAC depletion [27]

Ellagic acid	6-OHDA-induced rat	50 mg/kg	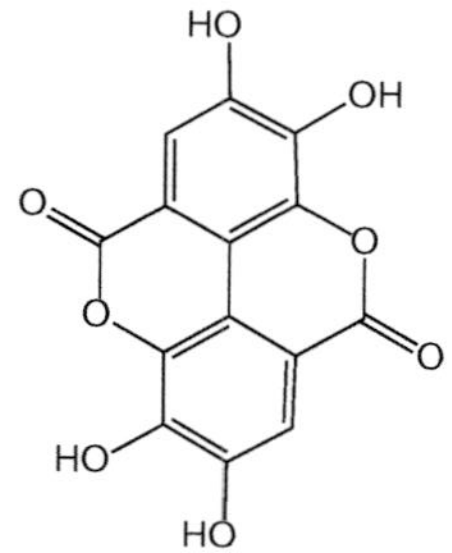	↓ Contralateral rotations ↓TNF-α and IL-1β levels in brain [24]
Esculetin	MPTP-induced mice	0.75 mg/kg/d	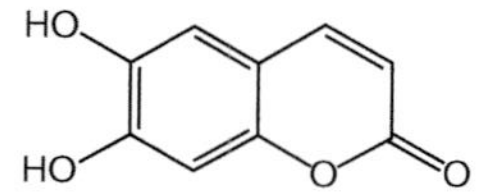	TH-positive neuron protection ↓ Nitrosative stress, and caspase-3 activation [43]
Gallic acid	6-OHDA-induced rat	50, 100, 200 mg/kg	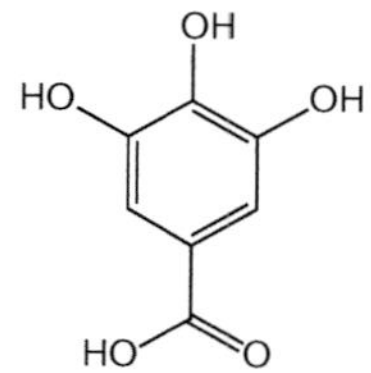	↑ Memory of passive avoidance ↑ Brain total thiol ↑ GPx ↓ MDA level [44]
Gastrodin (*Gastrodia elata Orchidaceae*)	MPTP-exposed SH-SY5Y neuroblastoma	30 and 60 mg/kg	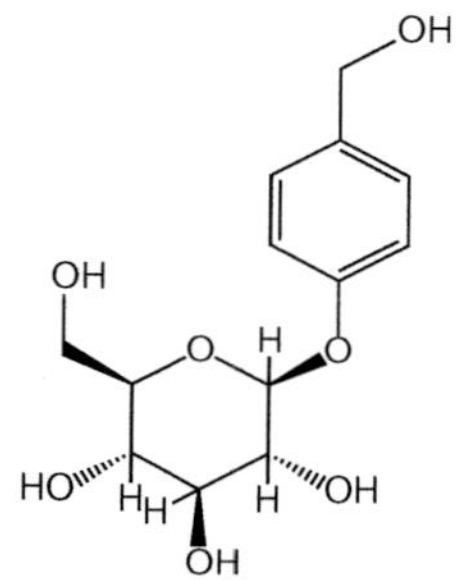	↑ Neuron viability ↓ ROS; ↓ Bax/Bcl-2; ↓ PARP proteolysis [27]

(*Continued*)

Table 2.1 (Continued)

Name of compound	Model	Dose	Structure	Effect
Genistein	6-OHDA neurotoxicity in human neuroblastoma SK-N-SH cells	0.01 μM	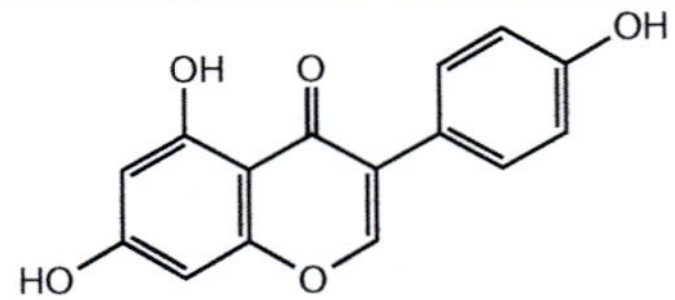	↑ Neuronal cell survival restores Bax/Bcl2 mRNA expression ↓ Apoptosis [24]
Ginkgolide B	6-OHDA-induced PC12 cells in rat	10–100 μM	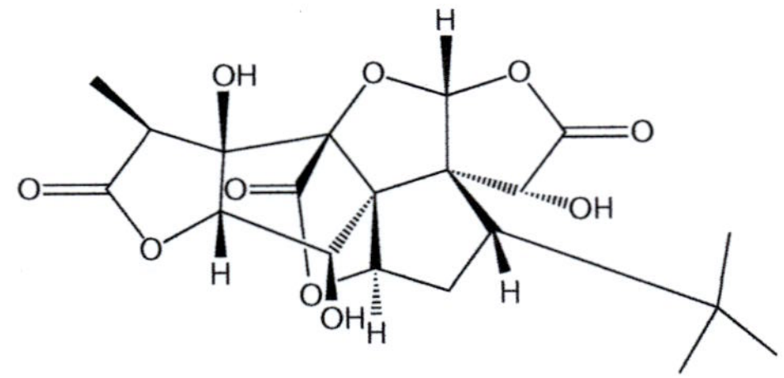	↑ Neuron cell viability ↓ Calcium in the cell ↓ Caspase-3 function ↑ CaB D28K mRNA [28]
Ginsenoside Rb1 (*Panax ginseng*)	Neuroblastoma cells BE(2)-M17 model	10 or 50 μM	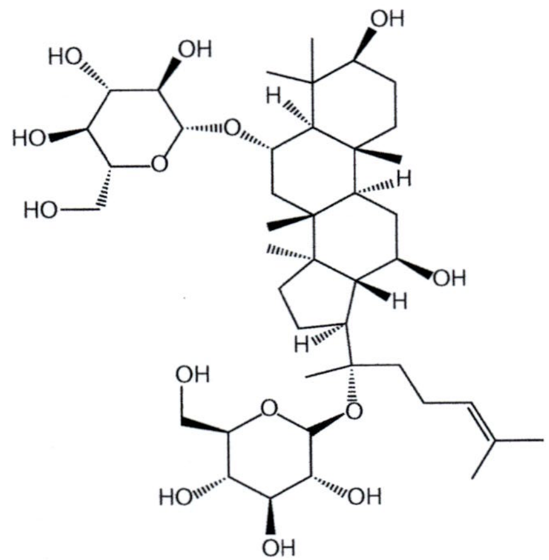	↑ Neuronal cell viability ↓ α-Syn fibrillation and oligomerization ↑ Oligomeric intermediary molecule durability ↑ PK digests α-syn aggregates ↓ Complex IV insufficiency ↑ LRPPRC Hsp90 and Hsp60 [20, 30]

Luteolin (perilla leaf, chamomile tea, and celery)	LPS-induced neuron-glial cell death	228 and 58.3 μl/kg	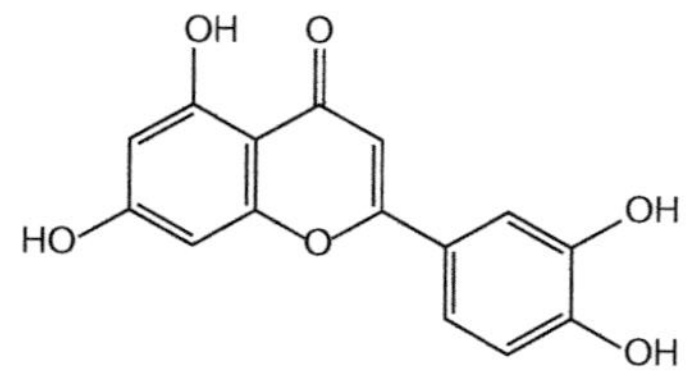	↓ LPS-induced microglial activation and excessive TNF-NO generation Dopaminergic neuron and neurite lengthening [44]
Magnolol	6-OHD- induced mice LP+ IF-γ BV-2 stimulated and HAPI	30 mg/kg	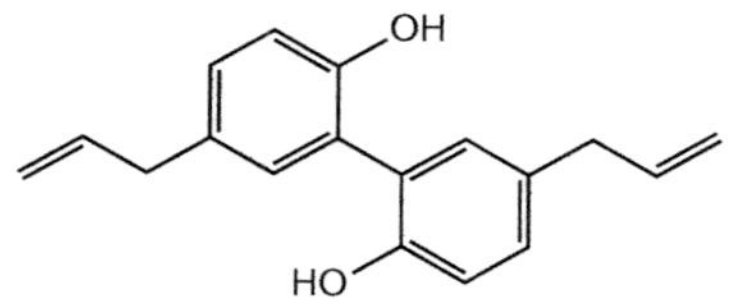	↓ Motion and ↑ TH protein in striatum ↓ NO, cytokine, and ROS expression [24]
Morin	MPP+ toxicity in PC12 cells MPTP-induced mice	5, 20, 40, or 100 mg/kg	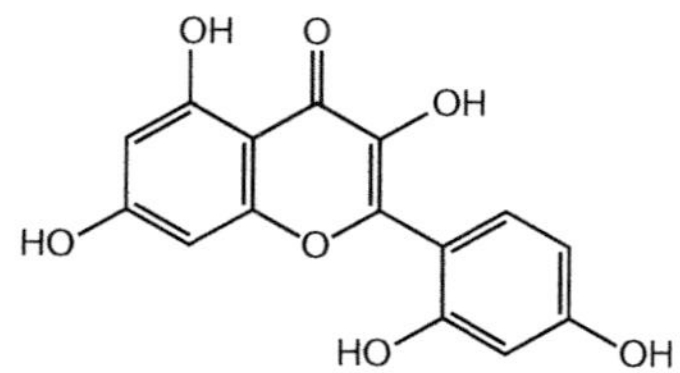	↓ Cell viability, ROS, and mortality in PC12 cells ↓ Behavioral deficits and striatal dopamine loss [43]
Paeoniflorin (*Paeoniae alba, Paeonia lactiflora*, or *Paeonia veitchii*)	PC12 cells exposed to MPTP	50 μM	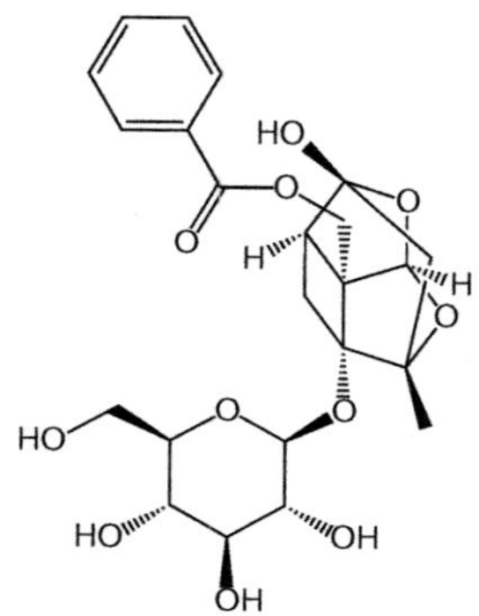	↓ α-Syn accumulation and protein expression [45]

(*Continued*)

Table 2.1 (Continued)

Name of compound	Model	Dose	Structure	Effect
Quercetin	6-OHDA-induced neurotoxicity in rat	50 mg/kg	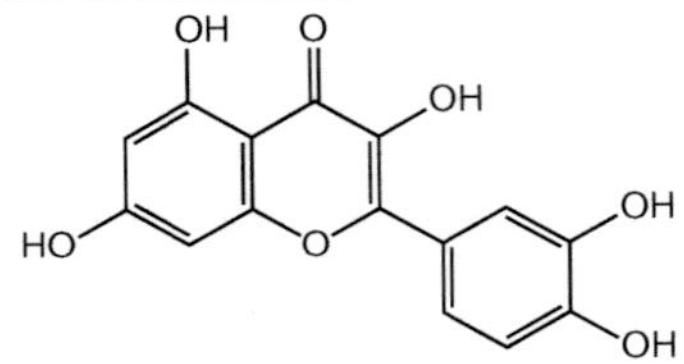	↑ Dopamine level ↓ Protein carbonyl content [9, 10]
Resveratrol	SH-SY5Y neuroblastoma cells 6-OHDA-induced neurotoxicity in rat	50 mg/kg/d	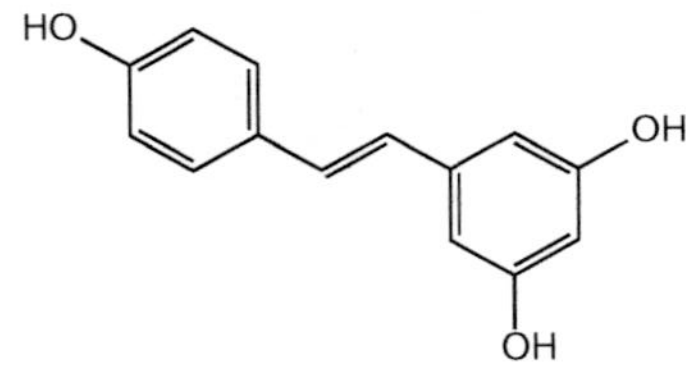	↓ LDH and enzyme of caspase-3 activity [9, 33]
Rosmarinic acid	MPTP-treated MES23.5 cell	16 mg/kg	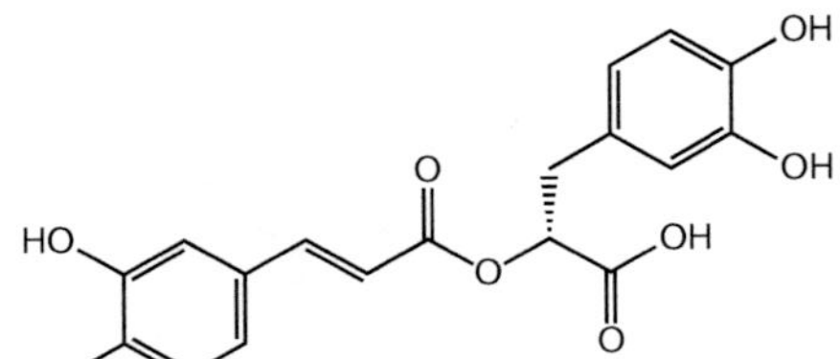	↑ Cell survival ↓ cell turnover Protect MMP ↓ ROS ↑ Dopamine Bcl-2/Bax Boost caspase-3 function [44]
S-allylcysteine	MPTP-induced mice Striatal injury by MPP+ in mice	40 mg/kg	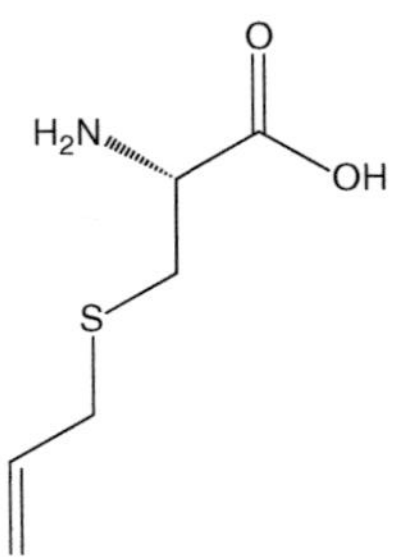	↓ DA depletion, oxidative stress, and behavioral deficits ↓ TNF-α, iNOS, and GFAP expression reduction [24]

Sulforaphane	6-OHDA-induced rat	5 mg/kg	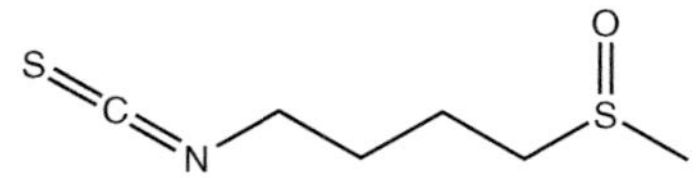	Enhance motor coordination Block DNA fragmentation ↑ GSH [43]
Theaflavin	MPTP-induced mice Chronic MPTP/probenecid-induced mice	5, 10, and 20 mg/kg	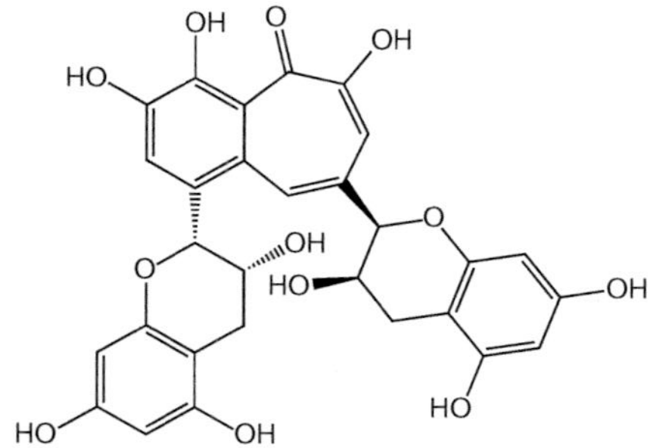	Stimulates α-syn assembly into non-toxic, spherical aggregates Reduced oxidative stress, motor dysfunction, and increased expression of DAT and VMAT2 in striatum and SN [44]
Thymoquinone	6-OHDA-induced neurotoxicity in rats	5 and 10 mg/kg	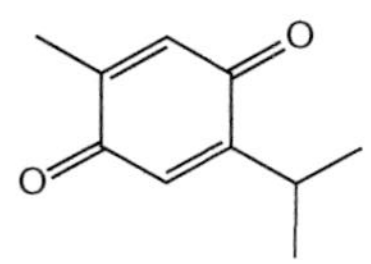	Improve turnover ↓ MDA level ↑ SOD
Tripchlorolide	MPTP-lesioned C57BL/6 mice	1 μg/kg	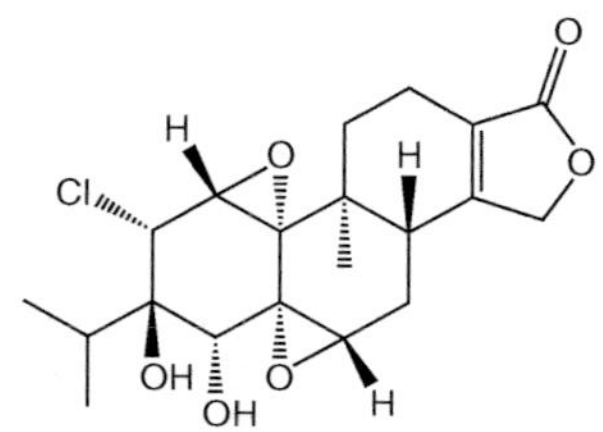	↓ Loss of SNC neuron [43] ↑ Dopamine level in SN and striatum [44]

6-OHDA, 6-hydroxydopamine; COX-2, cyclooxygenase-2; DA, dopamine; DAT, dopamine transporter; DN, diabetic nephropathy; DOPAC, 3,4-dihydroxyphenylacetic acid; GFAP, glial fibrillary acidic protein; GPx, glutathione peroxidase; GR, glutamate reductase; GSH, glutathione; GSK-3β, glycogen synthase kinase-3β; IF-γ, interferon-γ; IL-6, interleukin-6; iNOS, inducible nitric oxide; IV, intravenous; LDH, lactate dehydrogenase; LRPPRC, leucine rich pentatricopeptide repeat containing; LPS, lipopolysaccharide; MDA, malondialdehyde; MPP, 1-methyl-4-phenylpyridinium; MPTP, 1-methyl-4-phenyl-1,2,3,6-tetrahydropyridine; mTOR, mammalian target of rapamycin; NO, nitric oxide; PARP, poly-ADP ribose polymerase; PC12, pheochromocytoma; PD, Parkinson's disease; PK, protein kinase; PGE2, prostaglandin E2; SN, substantia nigra; SNC, substantia nigra pars compacta; SOD, superoxide dismutase; TH, tyrosine hydroxylase; TNF-α, tumor necrosis factor-α; TRKB, tyrosine kinase B; VMAT2, vesicular monoamine transporter 2.

Table 2.2 Medicinal plants used for Parkinson's.

Plants	Model used	Dose mg/kg	Effect	References
Tinospora cordifolia	6-OHDA-induced rat	200, 400	↑ DA levels ↓ Fe asymmetry ratio ↓ MDA ↑ Mitochondrial complex I action Enhanced locomotor function	[35]
Sesame seed oil	6-OHDA-induced rat		↑ GR, ↑ GST, ↑ GPx, CAT, GSH, and TBARS Block NOX-2 and COX-2 activation Activated MnSOD	[46]
Carthamus tinctorius L.	MPTP-induced rat	70, 35	Boost behavioral results α-Synuclein overexpression or aggregation Astrogliosis suppression	[40]
Chaenomeles speciosa	CHO cells; 6-OHDA-induced rat ; MPTP-induced mice	250, 500, 1000	↑ TH-positive neurons Viability of D8 cells ↓ Apomorphine-induced rotational twisting abnormalities	[47]
Portulaca oleracea	6-OHDA-induced rat	200, 400	Open field test: increased crossing and rearing	[48]
Paeonia suffruticosa	MPTP-induced mice	25, 50	Extended movement distance in open field ↑ Striatal dopamine ↓ Dopaminergic neuron loss Upregulation of Akt and mitochondrial OXPHOS subunits	[45]
Mucuna pruriens	6-OHDA-induced mice	40, 80, 120	Drug-induced dyskinesias are less likely Nigerstriatal catecholamine content reduced	[36]
Hyoscyamus niger seeds	Rotenone intrastriatal injection in rats	125, 250, 500	Enhance motor function ↑ GSH, GPx, SOD, and CAT function	[38]
Hibiscus asper leaves	6-OHDA-induced mice	50, 100	↑ SOD, GPx, and CAT activity, ↑ GSH content ↓ MDA level	[49]
Ginkgo biloba	6-OHDA-induced rats	50, 100, 150	↓ Rotation ↑ Locomotor action and neuromuscular coordination ↑ GSH level ↓ TBARS production ↑ SOD and CAT activity ↑ Dopamine D2 receptor in the striatum	[28]
Fructus Alpinia oxyphylla	ZebrafishPC12 cells	20% solution	Restore DA neuron degeneration Reduce lack of locomotion Improve 6-OHDA-treated PC12 cell viability Reduce apoptosis	[22, 50]
Delphinium denudatum	6-OHDA-induced rats	200, 400, 600	↓ MDA levels ↑ GSH content ↑ SOD, CAT function ↑ DA level	[51]

Table 2.2 (Continued)

Plants	Model used	Dose mg/kg	Effect	References
Bacopa monniera L.	6-OHDA-induced rats	20, 40	↓ MDA levels ↑ GSH content ↑ SOD, CAT function	[52]
Althaea officinalis L.	6-OHDA-induced rats	10	Rotational behavior slowed Lower MDA levels	[53]
Albizia adianthifolia	6-OHDA-induced rats	150, 300	Better working and reference memory Reduce asymmetry of contralateral rotation	[54]
Valeriana officinalis	6-OHDA-induced rats	0.049, 0.098, and 0.195	Enhance cell viability	[55]
Black tea	6-OHDA-induced rats	1.5%	Amphetamine circling and voluntary locomotor function recovery Recovery of D2 receptor binding, striatal DA, and DOPAC levels Lower MDA levels Higher GSH level Activation of SOD and CAT Enhance TH protein and mRNA expression in the substantia nigra	[56]
Panax ginseng	Rat received β-sitosterol β-D-glucoside	100	↓ Dopaminergic neuron loss, microgliosis, and accumulation of α-synuclein aggregates	[30]
Safflower	MPTP-induced mouse	35, 70	Restored TH, DA transporter, and DJ-1 protein expression DA levels Low acetylcholine	[15]
Hypericum perforatum	6-OHDA-induced rat	200	↓ Apomorphine rotational behavior ↓ Time to commence and complete the narrow beam task ↓ MDA level ↑ CAT activity	[57]

6-OHDA, 6-hydroxydopamine; CAT, choline acetyltransferase; CHO, carbohydrate; COX-2, cyclooxygenase-2; DA, dopamine; DOPAC, 3,4-dihydroxyphenylacetic acid; GPx, glutathione peroxidase; GR, glutamate reductase; GSH, glutathione; GST, glutathione S-transferase; MDA, malondialdehyde; MnSOD, manganese superoxide dismutase; MPTP, 1-methyl-4-phenyl-1,2,3,6-tetrahydropyridine; NOX-2, NADPH oxidase 2; OXPHOS, oxidative phosphorylation; SOD, superoxide dismutase; TBARS, thiobarbituric acid reactive substance; TH, tyrosine hydroxylase.

References

1 Moser, S.S. and Besler-Panos, W. (2006). Pharmacotherapy for Parkinson's disease: current options, promising future therapies. *JAAPA* 19 (8): 38–44.

2 Jellinger, K.A. (2018). Dementia with Lewy bodies and Parkinson's disease-dementia: current concepts and controversies. *J. Neural Transm. (Vienna)* 125 (4): 615–650.

3 Liu, M., Yu, S., Wang, J. et al. (2020). Ginseng protein protects against mitochondrial dysfunction and neurodegeneration by inducing mitochondrial unfolded protein response in *Drosophila melanogaster* PINK1 model of Parkinson's disease. *J. Ethnopharmacol.* 247: 112213.

4 Perez-Pardo, P., Dodiya, H.B., and Engen, P.A. (2019). Role of TLR4 in the gut-brain axis in Parkinson's disease: a translational study from men to mice. *Gut* 68 (5): 829–843.

5 Angelova, P.R., Choi, M.L., Berezhnov, A.V. et al. (2020). Correction: alpha synuclein aggregation drives ferroptosis: an interplay of iron, calcium and lipid peroxidation. *Cell Death Differ.* 27 (9): 2747.

6 Dolgacheva, L.P., Berezhnov, A.V., Fedotova, E.I. et al. (2019). Role of DJ-1 in the mechanism of pathogenesis of Parkinson's disease. *J. Bioenerg. Biomembr.* 51 (3): 175–188.

7 Demmings, M.D., Tennyson, E.C., Petroff, G.N. et al. (2021). Activating transcription factor-4 promotes neuronal death induced by Parkinson's disease neurotoxins and α-synuclein aggregates. *Cell Death Differ.* 28 (5): 1627–1643.

8 Yin, R., Xue, J., Tan, Y. et al. (2021). The positive role and mechanism of herbal medicine in Parkinson's disease. *Oxid. Med. Cell. Longev.* 2021: 9923331.

9 Essa, M.M., Braidy, N., Bridge, W. et al. (2014). Review of natural products on Parkinson's disease pathology. *J. Aging Res. Lifestyle* 3 (1): 1–10.

10 Koehn, F.E. and Carter, G.T. (2005). The evolving role of natural products in drug discovery. *Nat. Rev. Drug Discov.* 4 (3): 206–220.

11 Khuwaja, G., Khan, M.M., Ishrat, T. et al. (2011). Neuroprotective effects of curcumin on 6-hydroxydopamine-induced parkinsonism in rats: behavioral, neurochemical and immunohistochemical studies. *Brain Res.* 1368: 254–263.

12 Park, G., Kim, H.G., Ju, M.S. et al. (2013). 6-Shogaol, an active compound of ginger, protects dopaminergic neurons in Parkinson's disease models via antineuroinflammation. *Acta Pharmacol. Sin.* 34: 1131–1139.

13 Park, B.C., Lee, Y.S., Park, H.J. et al. (2007). Protective effects of fustin, a flavonoid from *Rhus verniciflua* stokes, on 6-hydroxydopamine-induced neuronal cell death. *Exp. Mol. Med.* 39: 316–326.

14 Chen, H.-Q., Jin, Z.-Y., and Li, G.-H. (2007). Biochanin A protects dopaminergic neurons against lipopolysaccharide-induced damage through inhibition of microglia activation and proinflammatory factors generation. *Neurosci. Lett.* 417 (2): 112–117.

15 Kim, H.G., Ju, M.S., Ha, S.K. et al. (2012). Acacetin protects dopaminergic cells against 1-methyl-4-phenyl-1,2,3,6-tetrahydropyridine-induced neuroinflammation in vitro and in vivo. *Biol. Pharm. Bull.* 35: 1287–1294.

16 Mu, X., He, G., Cheng, Y. et al. (2009). Baicalein exerts neuro-protective effects in 6-hydroxydopamine-induced experimental Parkinsonism in vivo and in vitro. *Pharmacol. Biochem. Behav.* 92: 642–648.

17 Zhang, Y., Shi, K., Wen, J. et al. (2012). Chiral HPLC determination and stereoselective pharmacokinetics of tetrahydroberberine enantiomers in rats. *Chirality* 24: 239–244.

18 Kannaiyan, R., Shanmugam, M.K., and Sethi, G. (2011). Molecular targets of celastrol derived from Thunder of God Vine: potential role in the treatment of inflammatory disorders and cancer. *Cancer Lett.* 303 (1): 9–20.

19 Zhang, Z.G., Wu, L., Wang, J.L. et al. (2012). Astragaloside IV prevents MPP(+)-induced SH-SY5Y cell death via the inhibition of Bax-mediated pathways and ROS production. *Mol. Cell. Biochem.* 364: 209–216.

20 Radad, K., Gille, G., Moldzio, R. et al. (2004). Ginsenosides Rb1 and Rg1 effects on mesencephalic dopaminergic cells stressed with glutamate. *Brain Res.* 1021 (1): 41–53.

21 Fujikawa, T., Miguchi, S., and Kanada, N. (2005). Acanthopanax senticosus Harms as a prophylactic for MPTP-induced Parkinson's disease in rats. *J. Ethnopharmacol.* 97 (2): 375–381.

22 Li, G., Zhang, Z., Quan, Q. et al. (2016). Discovery, synthesis, and functional characterization of a novel neuroprotective natural product from the fruit of *Alpinia oxyphylla* for use in Parkinson's disease through LC/MS-based multivariate data analysis-guided fractionation. *J. Proteome Res.* 15 (8): 2595–2606.

23 Pinto, B., Da Silva, A.N., Neves, B. et al. (2015). Neuroprotective properties of the standardized extract from *Camellia sinensis* (green tea) and its main bioactive components, epicatechin and epigallocatechin gallate, in the 6-OHDA model of Parkinson's disease. *Evid. Based Complement. Alternat. Med.* 2015: 1–12.

24 More, S.V., Kumar, H., Kang, S.M. et al. (2013). Advances in neuroprotective ingredients of medicinal herbs by using cellular and animal models of Parkinson's disease. *Evid. Based Complement. Alternat. Med.* 2013: 957875.

25 Ju, M.S., Kim, H.G., and Choi, J.G. (2010). Cassiae semen, a seed of *Cassia obtusifolia*, has neuroprotective effects in Parkinson's disease models. *Food Chem. Toxicol.* 48 (8–9): 2037–2044.

26 Xu, Q., Fan, W., Ye, S.-F. et al. (2016). *Cistanche tubulosa* protects dopaminergic neurons through regulation of apoptosis and glial cell-derived neurotrophic factor: in vivo and in vitro. *Front. Aging Neurosci.* 8: 295.

27 Huang, Y. and Yuan, J.Q. (2016). 20C, a bibenzyl compound isolated from *Gastrodia elata*, protects PC12 cells against rotenone-induced apoptosis via activation of the Nrf2/ARE/HO-1 signaling pathway. *Acta Pharmacol. Sin.* 37 (6): 731–740.

28 Mohammed, N., Abdou, H., Essawy, A. et al. (2020). Oral supplements of *Ginkgo biloba* extract alleviate neuroinflammation, oxidative impairments and neurotoxicity in rotenone-induced Parkinsonian rats. *Curr. Pharm. Biotechnol.* 21 (12): 1259–1268.

29 Park, H.J., Zhao, T.T., Kim, S.H. et al. (2020). Ethanol extract from *Gynostemma pentaphyllum* ameliorates dopaminergic neuronal cell death in transgenic mice expressing mutant A53T human alpha-synuclein. *Neural Regen. Res.* 15 (2): 361–368.

30 Van Kampen, J.M., Baranowski, D.B., Shaw, C.A., and Kay, D.G. (2014). *Panax ginseng* is neuroprotective in a novel progressive model of Parkinson's disease. *Exp. Gerontol.* 50: 95–105.

31 Choi, J.G., Kim, H.G., and Kim, M.C. (2011). Polygalae radix inhibits toxin-induced neuronal death in the Parkinson's disease models. *J. Ethnopharmacol.* 134 (2): 414–421.

32 Wang, Y., Xu, H., Fu, Q. et al. (2011). Protective effect of resveratrol derived from *Polygonum cuspidatum* and its liposomal form on nigral cells in parkinsonian rats. *J. Neurol. Sci.* 304 (1–2): 29–34.

33 Jing, H., Wang, S., Wang, M. et al. (2017). Isobavachalcone attenuates MPTP-induced Parkinson's disease in mice by inhibition of microglial activation through NF-κB pathway. *PLoS One* 12 (1).

34 Prakash, J., Chouhan, S., Yadav, S.K. et al. (2014). *Withania somnifera* alleviates parkinsonian phenotypes by inhibiting apoptotic pathways in dopaminergic neurons. *Neurochem. Res.* 39 (12): 2527–2536.

35 Kosaraju, J., Chinni, S., Roy, P.D. et al. (2014). Neuroprotective effect of *Tinospora cordifolia* ethanol extract on 6-hydroxy dopamine induced Parkinsonism. *Indian J. Pharm.* 46 (2): 176–180.

36 Nayak, V.S., Kumar, N., D'Souza, A.S. et al. (2017). The effects of *Mucuna pruriens* extract on histopathological and biochemical features in the rat model of ischemia. *Neuroreport* 28 (18): 1195–1201.

37 Wang, Y., Xie, C.-L., Wang, W.-W. et al. (2013). Epidemiology of complementary and alternative medicine use in patients with Parkinson's disease. *J. Clin. Neurosci.* 20 (8): 1062–1067.

38 Koppula, S., Kumar, H., More, S.V. et al. (2012). Recent updates in redox regulation and free radical scavenging effects by herbal products in experimental models of Parkinson's disease. *Molecules* 17 (10): 11391–11420.

39 Gautam, S., Karmakar, S., Batra, R. et al. (2017). Polyphenols in combination with β-cyclodextrin can inhibit and disaggregate α-synuclein amyloids under cell mimicking conditions: a promising therapeutic alternative. *Biochim. Biophys. Acta* 1865 (5): 589–603.

40 Jadiya, P., Khan, A., Sammi, S.R. et al. (2011). Anti-Parkinsonian effects of *Bacopa monnieri*: insights from transgenic and pharmacological *Caenorhabditis elegans* models of Parkinson's disease. *Biochem. Biophys. Res. Commun.* 413 (4): 605–610.

41 Berrocal, R., Vasudevaraju, P., Indi, S.S. et al. (2014). In vitro evidence that an aqueous extract of *Centella asiatica* modulates α-synuclein aggregation dynamics. *J. Alzheimers Dis.* 39 (2): 457–465.

42 Xu, Q., Langley, M., Kanthasamy, A.G., and Reddy, M.B. (2017). Epigallocatechin gallate has a neurorescue effect in a mouse model of Parkinson disease. *J. Nutr.* 147 (10): 1926–1931.

43 Balakrishnan, R., Azam, S., Cho, D.-Y. et al. (2021). Natural phytochemicals as novel therapeutic strategies to prevent and treat Parkinson's disease: current knowledge and future perspectives. *Oxid. Med. Cell. Longev.* 2021: 6680935.

44 Rabiei, Z., Solati, K., and Amini-Khoei, H. (2019). Phytotherapy in treatment of Parkinson's disease: a review. *Pharm. Biol.* 57 (1): 355–362.

45 Xue, J., Liu, T., Liu, Y. et al. (2019). Neuroprotective effect of biosynthesised gold nanoparticles synthesised from root extract of *Paeonia moutan* against Parkinson disease – in vitro & in vivo model. *J. Photochem. Photobiol. B* 200 (111635): 111635.

46 Ahmad, S., Khan, M.B., Hoda, M.N. et al. (2012). Neuroprotective effect of sesame seed oil in 6-hydroxydopamine induced neurotoxicity in mice model: cellular, biochemical and neurochemical evidence. *Neurochem. Res.* 37 (3): 516–526.

47 Huang, W., He, J., Nisar, M.F. et al. (2018). Phytochemical and pharmacological properties of *Chaenomeles speciosa*: an edible medicinal Chinese mugua. *Evid. Based Complement. Alternat. Med.* 2018: 9591845.

48 Truong, H.K.T., Huynh, M.A., Vuu, M.D., and Dang, T.P.T. (2019). Evaluating the potential of *Portulaca oleracea* L. for Parkinson's disease treatment using a Drosophila model with dUCH-knockdown. *Parkinsons Dis.* 2019: 1818259.

49 Hritcu, L., Foyet, H.S., Stefan, M. et al. (2011). Neuroprotective effect of the methanolic extract of Hibiscus asper leaves in 6-hydroxydopamine-lesioned rat model of Parkinson's disease. *J. Ethnopharmacol.* 137 (1): 585–591.

50 Zhang, Z.-J., Cheang, L.C.V., Wang, M.-W. et al. (2012). Ethanolic extract of fructus *Alpinia oxyphylla* protects against 6-hydroxydopamine-induced damage of PC12 cells in vitro and dopaminergic neurons in zebrafish. *Cell. Mol. Neurobiol.* 32 (1): 27–40.

51 Ahmad, M., Yousuf, S., Khan, M.B. et al. (2006). Protective effects of ethanolic extract of *Delphinium denudatum* in a rat model of Parkinson's disease. *Hum. Exp. Toxicol.* 25 (7): 361–368.

52 Shobana, C., Kumar, R.R., and Sumathi, T. (2012). Alcoholic extract of *Bacopa monniera* Linn. protects against 6-hydroxydopamine-induced changes in behavioral and biochemical aspects: a pilot study. *Cell. Mol. Neurobiol.* 32 (7): 1099–1112.

53 Rezaei, M. and Alirezaei, M. (2014). Protective effects of *Althaea officinalis* L. extract in 6-hydroxydopamine-induced hemi-Parkinsonism model: behavioral, biochemical and histochemical evidence. *J. Physiol. Sci.* 64 (3): 171–176.

54 Beppe, G.J., Dongmo, A.B., Foyet, H.S. et al. (2015). The aqueous extract of Albizia adianthifolia leaves attenuates 6-hydroxydopamine-induced anxiety, depression and oxidative stress in rat amygdala. *BMC Complement. Altern. Med.* 15 (1): 374.

55 de Oliveria, D.M., Barreto, G., De Andrade, D.V.G. et al. (2009). Cytoprotective effect of Valeriana officinalis extract on an in vitro experimental model of Parkinson disease. *Neurochem. Res.* 34 (2): 215–220.

56 Tan, L.C., Koh, W.-P., Yuan, J.-M. et al. (2008). Differential effects of black versus green tea on risk of Parkinson's disease in the Singapore Chinese Health Study. *Am. J. Epidemiol.* 167 (5): 553–560.

57 Kiasalari, Z., Baluchnejadmojarad, T., and Roghani, M. (2016). Hypericum perforatum hydroalcoholic extract mitigates motor dysfunction and is neuroprotective in intrastriatal 6-hydroxydopamine rat model of Parkinson's disease. *Cell. Mol. Neurobiol.* 36 (4): 521–530.

3

Drug Discovery from Medicinal Plants against Alzheimer's Disease

Mona M. Marzouk[1], *Mai M. Farid*[1], *Nesrine M. Hegazi*[1], *and Shahira M. Ezzat*[2,3]

[1] *Department of Phytochemistry and Plant Systematics, National Research Centre, Giza, Egypt*
[2] *Department of Pharmacognosy, Faculty of Pharmacy, Cairo University, Cairo, Egypt*
[3] *Department of Pharmacognosy, Faculty of Pharmacy, October University for Modern Sciences and Arts (MSA), Giza, Egypt*

Alzheimer's disease (AD) is a neurodegenerative disorder that progresses gradually and sequentially worsens. AD is associated with memory and cognition impairment, often causing death [1]. According to the World Health Organization, it contributes to 60–70% of dementia cases [2] and is increasing proportionally with population age globally. In addition to dementia, symptoms include disorientation, speech difficulties, mood swings, and behavioral problems [3]. Globally, 24 million people are suffering from dementia, with a twofold increase every 20 years [4].

The exact cause of developing AD is still not fully understood [3]. Several genetic and acquired risk factors are associated with its development. As reported by Ballaed et al., genetic risk factors contribute to 70% of the risk of developing AD [5]. Mutations in genes of amyloid precursor protein (APP), presenilin 1 (PSEN1), and presenilin 2 (PSEN 2) contribute to early AD cases, whereas polymorphism in the apolipoprotein E gene (APOE) is associated with late forms of AD [6].

Besides the genetic factors, numerous acquired risk factors heighten the liability for developing AD (Figure 3.1). One of the most common factors is cerebrovascular changes (CVC), including ischemic cortical infarcts, hemorrhagic infarcts, vasculopathy, and cerebral white matter changes [6]. As previously conveyed, simultaneous development of CVC is associated with dementia in 6–47% of patients [7]. Comparably, some reports correlated hypertension with an increased risk of AD owing to the associated vascular changes, hypoperfusion, ischemia, and cerebral hypoxia [6]. Similarly, a clear linkage between diabetes and AD has been proved through several epidemiologic studies. This could be explained through diverse mechanisms such as insulin deficiency and resistance, compromised insulin receptors, hyperglycemia, and diabetes-linked complications like advanced glycation products, cerebrovascular damage, and vascular inflammation [8]. Alternatively, controversial reports link obesity with AD, as both underweight and obese individuals are vulnerable to developing AD [9, 10]. Similarly, hypercholesterolemia increases the risk of the development of both atherosclerosis and neurodegenerative diseases (i.e. AD) [11]. Other factors include compromised marital status (e.g. widowhood), stress, sleep deficiency, depression, and smoking [6].

Phytochemical Drug Discovery for Central Nervous System Disorders: Biochemistry and Therapeutic Effects, First Edition. Edited by Chukwuebuka Egbuna and Mithun Rudrapal.

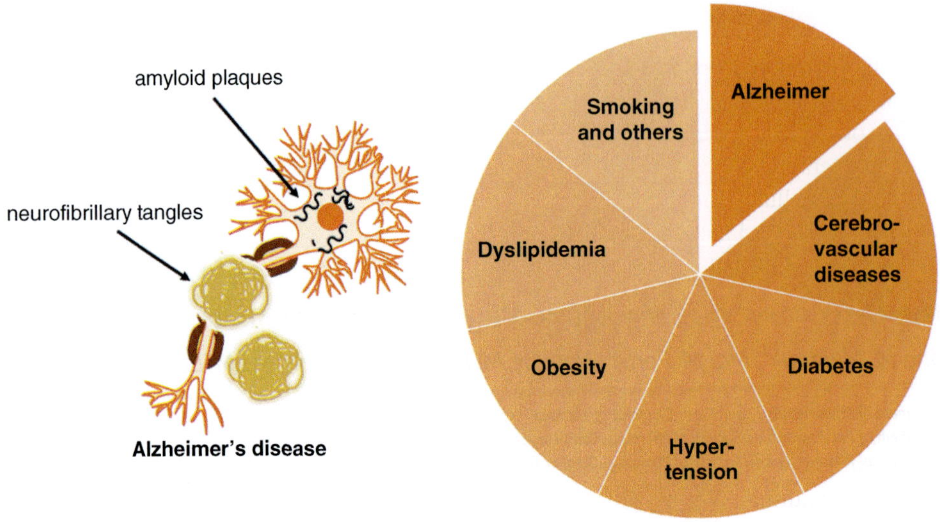

Figure 3.1 Acquired risk factors for developing Alzheimer's disease.

Neurodegenerative disorders are a great burden on public health and represent a socioeconomic problem owing to the increased mortality and expenditure on treatment and personnel assistance. Additionally, the available treatments provide only symptomatic relief and do not help in arresting disease progression [12]. This necessitates ongoing research to uncover the underlying processes in neurodegeneration and to discover new medications to assess neuronal cell loss and restore normal brain functions.

Currently, more than 50 million patients suffer from AD and that figure is presumed to reach more than 150 million by the year 2050, with no real cure for the disease, albeit there are medications for symptom relief. Thus, the purpose of this chapter is to briefly describe AD pathogenicity and treatment strategies, with a special focus on plants and natural products with anti-AD activity.

3.1 Pathogenesis

The pathogenesis of AD involves two forms of neuropathologic changes, which could be described as positive and negative lesions. Positive lesions refer to the extracellular deposition of amyloid plaques, intracellular neurofibrillary tangles, and other deposits. These depositions represent the most significant histopathologic highlights, while the negative lesions are signified by the loss of neurons, neuropils, and synapses [13]. In addition, other factors can cause neurodegeneration, such as injury of cholinergic neurons, oxidative stress, and neuroinflammation [14]. The following is a brief discussion of the common pathologic changes associated with the development of AD.

3.1.1 Amyloid Plaques

Pathologically, AD is associated with the formation of amyloid plaque in the extracellular spaces. This is attributed to proteolytic breakage of APP by β-secretase and γ-secretase enzymes. Deposition of the amyloid-β (Aβ) peptide consequently provokes oxidative stress, inflammation, and mitochondrial dysfunction concomitant with apoptosis [15].

3.1.2 Neurofibrillary Tangles

Intracellularly, particularly in the hippocampus and cerebral cortex regions of the brain, neurofibrillary tangles (NFTs) are formed from hyperphosphorylated tau proteins. The buildup of tau protein in the cytoplasm of neurons negatively affects axonal transmission owing to alterations in the structure of the microtubule [16].

These extracellular and intracellular changes could be correlated to the decreased expression of brain-derived neurotrophic factor (BDNF), which is crucial for neuronal growth and memory functions [17].

3.1.3 Synaptic Damage

Comparable to the positive lesions exemplified by extracellular plaques and intracellular NFTs, synaptic damage is witnessed in the neocortex and limbic system, causing memory deficits earlier in AD. The underlying mechanisms encompass deficiency in axonal transport, mitochondrial injury, oxidative stress, and others. Ultimately, this results in damage to dendritic spines and presynaptic terminals, and axonal atrophy [13, 18].

3.1.4 Cholinergic Functions

The previously mentioned pathologic changes that reduce cholinergic function in different parts of the brain, especially the cortex and hippocampus, are associated with the rigorousness of AD. This decreased activity involves alterations in choline acetyltransferase and acetylcholinesterase functions, lower levels of acetylcholine, and reduced nicotinic and muscarinic receptors in the brain [19]. In mild to moderate AD patients, the binding capacity of the cholinergic receptors is reduced and is correlated with neuropsychiatric manifestations.

3.1.5 Oxidative Stress

Another crucial pathologic change associated with AD is oxidative stress and the linked metabolic reactions. Oxidative stress implies a disparity between prooxidant and antioxidant levels, mainly attributed to the high levels of reactive oxygen species (ROS). Several studies have reported increased oxidation of macromolecules (i.e. lipids, proteins, RNA, and DNA) in the brains of AD patients [20]. Sequentially, the synaptic functions and neurotransmission in the neuronal cells are negatively affected by the increased ROS production [21].

3.1.6 Neuroinflammation

Lastly, neuroinflammation is recognized as one of the primary causes leading to AD. As the inflammation occurs, chronic inflammation and nerve injury follow, triggering a persistent immune response in the brain [22]. Several factors contribute to the occurrence of neuroinflammation, including amyloid plaques and NFTs, or the existence of a proinflammatory profile provoked by genetics, obesity, lifestyle, metabolic syndrome, and aggravated by infections and immune response [16]. Sequentially, neuronal damage and death follow as a result of the release of cytotoxic cytokines by the exposed microglia [23].

3.2 Treatment Strategies for Alzheimer's Disease

AD treatment strategies can be structured according to a number of hypotheses aimed at explaining disease origins, such as the amyloid cascade hypothesis, cholinergic hypothesis, tau protein hypothesis, dendritic hypothesis, 5-hydroxytryptamine receptor (5-HT_6) hypothesis, and metabolic hypothesis.

3.2.1 Amyloid Hypothesis

For over 20 years, the main focus of research into new pharmacologic therapies for AD has been based on the amyloidogenic hypothesis, which postulates that Aβ peptide is primarily responsible for neuronal death. The objective of these therapies is to reduce the production of Aβ by inhibiting secretase enzymes (β and γ) and promoting the degeneration of existing cerebral Aβ plaques [24]. However, this approach has been shown to be only barely effective. Both β and γ secretase inhibitors have been the goals of new drug development [25, 26] (Table 3.1). Also, Aβ aggregation is degraded by a few enzymes and has been considered for new drug development [87, 88]. Another equitable treatment choice is to promote the removal of Aβ deposits and aggregates through three different strategies. They are represented as strategies for activation of enzymes that reduce Aβ plaques, and modulation of Aβ carriage between the brain and the peripheral circulation, as well as an antiamyloid immunotherapy strategy whose objective is to stimulate Aβ clearance with the aim of reducing the amyloid load in AD patients (Table 3.1) [24].

In 2019, the number of antiamyloid trials declined, which could be a tipping point. Thus, the Aβ hypothesis may not be totally practicable. This is due to many international pharmaceutical companies having conducted various clinical trials on amyloid clearance therapy but without success [48, 89].

3.2.2 Tau Proteins Hypothesis

As a replacement for the amyloid cascade hypothesis, tau-focused therapies aim to inhibit phosphorylation and/or accumulation of the tau protein. Furthermore, microtubule-stabilizing drugs could be used as a disease-modifying strategy in AD [90]. Recently, immunomodulation was recommended as a sustainable choice for stimulating active clearance of tau accumulation. Meanwhile, tau phosphorylation is facilitated by different kinases and phosphatases; the disturbance of their normal physiologic purposes would lead to tau hyperphosphorylation. Thus, excessive activation of kinases and downregulation of phosphatases have been linked to AD. Glycogen synthase kinase 3 beta (GSK-3β) and cyclin-dependent kinase 5 (CDK-5) [91, 92] are two proline-directed protein kinases (PDPKs) that facilitate tau hyperphosphorylation at proline sites. GSK-3β inhibitors are debatably in the most advanced stages of clinical trials for AD, among which is tideglusib that has recently completed Phase II trials (Table 3.1). Nonetheless, other non-PDPKs mediate tau hyperphosphorylation at various sites, including calcium/calmodulin-activated protein kinase II (CaMK-II) [93], casein kinase 1 (CK-1) [94], dual-specificity tyrosine-phosphorylation-regulated kinase 1A (DYRK-1A) [95], microtubule affinity-regulated kinases (MARK) [96], and protein kinase A (PKA) [97].

Table 3.1 Treatment strategies and clinical trials for Alzheimer's disease.

Target	Inhibitors/vaccines/antibodies	Clinical trial ID	Activities	References
The amyloid hypothesis				
β-Secretase inhibitors	E2609	NCT01600859	Shown to be effective in reducing production of Aβ up to 80–90% in CSF	[27]
	MK-8931	NCT01739348		[28]
	LY2886721	NCT01561430 and NCT01807026		[29]
γ-Secretase inhibitors	LY450139 (semagacestat)	NCT00762411 and NCT01035138	LY450139 decreased Aβ levels in human blood and CSF. The results of the clinical study (more than 3000 AD patients) are an example of the worst possible outcome. The treatment was associated with deterioration in cognition and the ability to carry out activities of daily living. Further side effects include increased incidence of skin cancer, weight loss, and increased risk of infection	[30]
	Avagacestat	NCT00810147, NCT00890890, and NCT01079819	Development halted due to lack of efficacy	[31, 32]
Selective γ-secretase modulators	NSAIDs, drugs that are devoid of COX-inhibitory activity, such as ibuprofen, indomethacin, R-flurbiprofen and sulindac	NCT00105547 and NCT00322036	Reduced Aβ(1-42) peptide levels in vitro and in vivo. Nevertheless, ibuprofen and R-flurbiprofen did not show efficacy for treatment of AD	[33][34]
	NSAIDs; CHF5074	NCT00954252, NCT01203384, and NCT01303744	Inhibited production of Aβ(1-42) by blocking γ-secretase aggregation in vitro. Has the ability to reduce amyloid burden and prevent microglial stimulation	[35]
		Phase II trial (patients with mild cognitive impairment)	Improves many cognitive actions and reduces levels of inflammatory markers in CSF	[36, 37]
	NIC5-15 (pinitol)		Assumed to modulate γ-secretase and said to be able to reduce Aβ production, while not affecting substrate cleavage of Notch	[24]

(*Continued*)

Table 3.1 (Continued)

Target	Inhibitors/vaccines/antibodies	Clinical trial ID	Activities	References
Inhibitors of Aβ peptide accumulation	3-amino-1-propaneosulfonic acid	Phase III trials	Treatment designed to interfere with or counteract contact of Aβ with endogenous GAG. It has been shown that GAG enhances accumulation and deposition of Aβ fibrils	[38–40]
	Colostrinin (complex of proline-rich polypeptides)	Phase II trial	Showed insignificant enhancements in patients with mild AD (15 months' treatment), effects not preserved after additional 15 months' treatment	[41]
	ELND005 (scyllo-inositol)	Phase II trial	Dose safety and efficiency did not meet primary clinical effectiveness outcomes in patients with mild to moderate AD (18 months' treatment)	[42]
	Metal chelating 8-hydroxiquinoline; PBT1 (clioquinol) and PBT2	Phase II and III trials	Failed clinical development due to lack of efficacy	[43, 44]
	GV-971 (sodium oligomannate)	NCT02293915	Inhibits Aβ aggregation and can redesign gut microbiota and destroy dysbiosis-induced neuroinflammation	[45]
			Provides significant cognitive benefits and has been approved for treatment of mild to moderate AD by CNMPA	[46]
	Nilvadipine	NILVAD trial	No change in outcome measures in mild to moderate AD	[47]
	E2609 (elenbecestat)	NCT02956486 and NCT03036280	Active, not recruiting	[48]
Modulation of Aβ transport from brain to peripheral circulation	PF-0449470052	Phase II	Failed	[24]
	TTP4000	Phase I (NCT01548430)	Completed	[24]

Active immunotherapy	Plasma exchange with albumin 1 immunoglobulin	NCT01561053	Completed	[48]
	AN1792	Phase II trial	6% of patients developed cerebral inflammation that turned out to be sterile meningoencephalitis	[49]
	CAD 106	Phase II trial	Aβ-specific antibody response in 75% of treated patients, without causing adverse inflammatory reactions	[50]
	CNP520 (umibecestat)	Phase II trial	Showed worsening of cognitive function	[48]
	ACC-001	Phase II trials (NCT01284387 and NCT00479557)Phase II trial NCT01227564 (ongoing)	Pharmaceutical company abandoned plan to develop this vaccine	[24]
	ACI-24, AF205, and MER5101		Presently in several stages of preclinical examination	[51–53]
	Bapineuzumab		Showed no treatment effect on either functional or cognitive behaviors	[54]
	Aducanumab (BIIB037), crenezumab, gantenerumab, solanezumab, and combination of gantenerumab and solanezumab	NCT01760005	All treatment trials (26 weeks) showed reduction of amyloid deposition	[55]
	MABT5102A (crenezumab)	Phase II trial (NCT01343966 and NCT01998841)	Assesses safety and efficacy in mild to moderate AD patients	[24, 56]
	Gantenerumab	Phase III trials mild AD (NCT02051608) and prodromal AD (NCT01224106)		[57, 58]
	Bapineuzumab	Phase III trials	Treatment failed to produce substantial enhancements of cognitive function	[48]
	Solanezumab	Phase III trials (NCT01127631, NCT01900665, and NCT02008357)	Trial proposed that solanezumab may increase cognition in mild to moderate AD patients (no statistical significance achieved)	[59, 60]

(*Continued*)

Table 3.1 (Continued)

Target	Inhibitors/vaccines/antibodies	Clinical trial ID	Activities	References
	E2609 (elenbecestat)	Phase II trial	Showed decreased Aβ levels in dose-dependent manner in prodromal or mild AD, but no significant improvements in AD composite score	[61]
	PF-04360365 (ponezumab), NI-101, SAR-228810, and BAN-2401	Phase I trial		[56]
	Gammagard	NCT00818662	May possess immunomodulatory properties that could improve microglialphagocytosis	[62]
Tau proteins hypothesis				
Tau hyperphosphorylation inhibitors	Tideglusib (irre-versible inhibitor of GSK-3β)	Phase II trial (NCT01350362)	Administration (26 weeks) to mild to moderate AD patients did not show clinical efficacy	[63]
	NP031112	Phase II trial (NCT00948259)	Evaluated safety and tolerability of drug administration (20 weeks) compared with placebo in AD patients	
	VEL015 (sodium selenite)	Phase II trial (ACTRN12611001200976)	Reduces tau phosphorylation in vitro and in vivo. Administration to AD mouse models resulted in consequential mental enhancements and significant decrease of neurodegenerative phenotype	[64, 65]
Tau aggregation inhibitors	Rember TM		It appeared to stabilize AD progression in 50-week clinical trial	[66]
	TRx0237	Phase III trial (NCT01626391, NCT01689233, NCT01689246, and NCT01626378)	Primary outcome revealed that TRx0237 failed as add-on treatment for mild AD. Ongoing trial intended to compare efficiency in different doses in patients with early AD	[67]

Microtubule stabilizers	Paclitaxel		Use associated with serious adverse events, limiting its usefulness in AD	[68]
	TPI 287	NCT01966666	Evaluated stabilizer safety and pharmacokinetic properties in mild to moderate AD patients	[69]
	Epothilone D		Drug development for AD unobtainable in 2013 after failed clinical trial	[70]
Antitau immunotherapy	AADvac1	Phase II trials (NCT01850238 and NCT02031198)	Primary outcome measures were tolerability and safety of AADvac1 in mild AD patients (24 months' treatment) based on vital signs, adverse effects, electrocardiogram, laboratory data, physical and neurologic examination, brain MRI, and review of patient diary. Trial strategy still indistinct	[71]
	LY3303560 (zagotenemab)	Phase I and II trials	Phase I trial completed in 2019, but not yet reported. Ongoing phase II trial assessing efficacy and safety of treatment in early AD patients. Primary result measure is modification in integrated AD rating scale. Trial completed in 2021.	[48]
Dendritic hypothesis				
PrPC–mGluR5–Fyn signaling	Masitinib (AB1010)	Phase II trial (NCT00976118)	Oral administration of AB1010 (24 weeks) together with one of the AChE inhibitors (donepezil, galantamine, or rivastigmine) and/or memantine reported significant improvement in response to ADAS-Cog test	[72, 73]
	Saracatinib (AZD0530) and AB1010	Phase II and III clinical trials (NCT01864655, NCT01872598, and NCT02167256)	AZD0530 and AB1010 used as Fyn kinase inhibitors for mild to moderate AD	[74, 75]

(Continued)

Table 3.1 (Continued)

Target	Inhibitors/vaccines/antibodies	Clinical trial ID	Activities	References
Cholinergic hypothesis				
AChE inhibitors	Donepezil, galantamine, and rivastigmine	NCT00976118	Separately administered orally along with AB1010 and/or memantine, each showed significant improvement in response to ADAS-Cog test	[76]
	Ladostigil (TV3326)	Phase II trials (NCT01429623 and NCT01354691)	TV3326 has potent antioxidant and neuroprotective effects. Clinical trials currently underway	[77]
5-HT_6 obstruction				
5-HT6 receptor blockage	Intepirdine (RVT-101)	Phase III trial (MINDSET)	Primary outcome measures included changes in scales of ADAS-Cog 11 and ADCS-ADL 23 in mild to moderate AD patients. Trial failed to achieve primary endpoints. Secondary outcome measures included caregiver interview and improvement in clinician interview-based impression of change	[78]
		Phase III trial; MINDSET (for AD patients who had completed RVT-101-3001 study)	Primary endpoints included occurrence of adverse events and changes in vital signs, electrocardiograms, physical examinations, and routine laboratory assessments. Trial was terminated in March 2018 because did not reach primary endpoints in study RVT-101-3001	[79]
	EVP-6124	Phase III trials	Primary outcomes included changes in ADAS-Cog 13 and CDR-SB in mild to moderate AD patients. In final outcomes, FDA issued clinical hold (still in place) on these trials due to gastrointestinal adverse effect	[80]

Metabolic hypothesis				
Antidiabetic drugs	Thiazolidinediones		Increased insulin sensitivity and decreased cytokine-dependent inflammation	[81]
	Pioglitazone		Enhanced memory and cognition in mild to moderate AD patients	[82]
	Rosiglitazone	Phase II trial (NCT00348140)	Trial proved enhancements in memory with 6 months' treatment in patients who did not possess ApoE4 allele	[83]
		Phase III trial (NCT00550420)	Failed to show efficiency in AD	[84]
	Intranasal insulin (Humulin R U-100)	NCT01767909	Evaluated effectiveness in mild AD patients	[24]
Pancreatic hormones	Amylin and pramlintide	Phase II trials (NCT01429623 and NCT01354691)	Decreased level of Aβ and developed cognition in brain of AD models. Also amylin significantly inhibited β-secretase enzyme, while pramlintide did not	[85, 86]

Aβ, amyloid beta; AChE, acetylcholinesterase; AD, Alzheimer's disease; CNMPA, China National Medical Product Administration; COX, cyclooxygenase; CSF, cerebrospinal fluid; FDA, Food and Drug Administration; GAG, glycosaminoglycan; GSK-3β, glycogen synthase kinase 3 beta; MRI, magnetic resonance imaging; NSAIDs, non-steroidal anti-inflammatory drugs.

Also, the activation of protein phosphatases (PP1, PP2A, and PP5) has been considered as a possible drug target. Among these, PP2A is thought to play a dominant role in the pathogenesis of AD [98]. Inhibition of PP2A could prompt tau hyperphosphorylation, indicating that it might regulate normal tau phosphorylation by preventing overactivation of tau protein kinases [43]. Consequently, tau phosphorylation could be the outcome of a slight equilibrium among tau kinases and phosphatases. Once this balance is disrupted, the resulting imbalance will lead to tau hyperphosphorylation.

Rember TM and TRx0237 (methylene blue dye derivatives) are two tau aggregation inhibitors that decrease the level of accumulated tau proteins to improve tau-related neuronal impairment [99] (Table 3.1).

Microtubule stabilizers may hypothetically accomplish similar outcomes to those seen with the tau hyperphosphorylation and aggregation inhibitors. They decreased axonal dystrophy, enhanced axonal transport, reduced tau neuropathology, and decreased hippocampal neuron loss [70].

In conjunction with the immunotherapies targeting Aβ, both passive and active immunization strategies against tau have been considered. AADvac1 is an active vaccine that induces the immune comeback by targeting numerous key epitopes in pathologic forms of tau, by this means reducing tau aggregation and inhibiting NFT formation [100]. The passive immunotherapy LY3303560 (zagotenemab) is an antitau antibody that aimed to neutralize and imprison tau aggregate [101]. Phase I and II trials to investigate the safety of AADvac1 and LY3303560 in patients with mild AD were considered (Table 3.1).

3.2.3 Dendritic Hypothesis

Dendritic aberrations appear in the relatively early stages of AD. While dystrophic neurons, decreased dendritic complexity, and loss of dendritic spines are all recognized features of AD, the fundamental molecular changes that occur in the dendrite on the postsynaptic side have only recently been understood [74]. Some data propose that soluble Aβ oligomers are the main neurotoxic types responsible for dendritic pathology. Aβ oligomers may cause abnormal N-methyl-D-aspartate receptor (NMDAR) activation postsynaptically by forming complexes with PrPC (the cell-surface prion protein). PrPC is enriched at the neuronal postsynaptic density, where it interacts with Fyn tyrosine-kinase-metabotropic glutamate receptor 5 complex (Fyn-mGluR5). Its activation occurs when the PrPC-Fyn-mGluR5 complex is bound to Aβ, where Fyn can cause tyrosine phosphorylation of the subunit NR2B of NMDAR. This results in an initial increase and then loss of NMDAR cell surface [75]. Fyn overexpression enhanced synapse loss and the onset of cognitive impairment in the J9 (APPswe/Ind) transgenic AD mouse model, while its inhibition produced an opposite effect [102]. Furthermore, Fyn can also contribute to tau hyperphosphorylation. Previous reports have shown raised levels of Fyn in the brains of AD models. Also, Fyn was revealed to phosphorylate tau at Tyr18 residue [103].

Two Fyn kinase inhibitors, saracatinib (AZD0530) and masitinib (AB1010), are being used in clinical trials (Phase II and III) for mild to moderate AD (NCT00976118, NCT01864655, NCT01872598, and NCT02167256) [56, 72]. Both inhibitors are effective for obstructive Fyn in a nanomolar range. In one clinical trial (NCT00976118), oral administration of AB1010 along with one of the acetylcholinesterase (AChE) inhibitors (donepezil, galantamine or rivastigmine) and/or memantine for 24 weeks, a significant improvement in response to the ADAS-Cog test was also reported [24]. The results obtained are hopeful; nevertheless, it is clear that the small patient group (26 individuals) in the memantine Phase II clinical trial is not sufficient to draw conclusions on the potential efficacy of AB1010. Apart from blocking Fyn, AB1010 may prevent neuroinflammation by blocking activated mast cells (MCs)–microglia interactions through inhibiting stem cell factor receptor/c-Kit inhibitor (SCF/c-Kit) signaling on MCs [74]. Additionally, AZD0530 acts through rescindable adenosine triphosphate (ATP) inhibition, and in enzymatic assays inhibits both Fyn isoforms with IC_{50} (8–10 nM) [104]. Consequently, Fyn seems to be a possible target in AD treatments, where the kinase blockers AB1010 and AZD0530 have been shown to be effective in treating AD symptoms in experimental AD mouse models.

3.2.4 Cholinergic Hypothesis

Numerous studies have reported reductions in choline acetyl-transferase (ChAT), acetylcholine (ACh) release, as well as decreases in muscarinic and nicotinic receptors in the hippocampus and cerebral cortex of AD brains under medical examination [105]. AChE inhibitors, one of two classes of drugs currently approved for the treatment of AD, perform via increasing ACh bioavailability. Inopportunely, none of them is capable of reversing the course of AD or of even noticeably slowing down the rate of disease development [76]. Their clinical effect is greatly comforting; nevertheless, their probable use in combination therapy with other compounds should not be excluded. Some AChE inhibitors improve extrapyramidal signs and provide an antidepressant effect [77] (Table 3.1). In a clinical trial (NCT00976118), one of the AChE inhibitors (donepezil, galantamine, or rivastigmine) was orally administrated along with AB1010 and/or memantine (24 weeks of treatment). A significant improvement in response to the ADAS-Cog test was reported.

3.2.5 Strategies for 5-HT_6 Obstruction

5-HT_6 is a subtype of serotonin receptor, expressed in areas of the central nervous system involved in learning and memory. Inhibition of the 5-HT_6 receptor was shown to promote ACh release. In patients with moderate AD, 5-HT_6 antagonism may lead to the restitution of ACh levels [79]. This supposition is supported by evidence that the 5-HT_6 receptor improves spatial learning and memory in normal rats using the Morris water maze test [78]. 5-HT_6 inhibitors may be useful in combination therapy, together with AChE inhibitors [24, 56]. 5HT_6 antagonists are theorized to improve learning, cognition, and memory [106]. Compounds such as idalopirdine (Lu AE58054), cerlapirdine (PF-05212365), intepirdine (SB-742457, RVT-101), and latrepirdine (Dimebon) were evaluated as novel possible treatments for mild to moderate AD [79]. However, their Phase III trials have failed to demonstrate efficacy. Other agents that are in various stages of clinical research are AVN-322, EVP-6124, PRX-07034, and SUVN-502 [24, 56].

3.2.6 Metabolic Hypothesis

Clinical studies indicate that diabetes is a major risk factor contributing to AD. Previous reports indicate a close association between insulin-deficient diabetes and cerebral amyloidosis [107]. It is possible that both diseases have peripheral and central insulin signaling deficiencies. Consequently, the hypothesis of type 3 diabetes in AD is established, which challenges the ability to link the metabolic phenotypes found in diabetes disorder and AD into a comprehensible outline. Intranasal insulin had also been considered as an AD treatment possibility, as a currently ongoing clinical trial (NCT01767909) is assessing long-term (12 months) efficacy (Humulin R U-100) in mild AD [24].

Some antidiabetic drugs such as pioglitazone, rosiglitazone, and thiazolidinediones regulate glucose homeostasis by reducing blood glucose levels, improving insulin sensitivity, and increasing the metabolism of lipids. These compounds have been investigated as potential therapeutic treatment strategies for AD [84, 108]. Rosiglitazone treatment improved memory retention and attention in patients treated for six months (Phase II trial) [83]. Nevertheless, rosiglitazone failed to show efficacy in AD in a Phase III trial (NCT00550420) [84]. Furthermore, no outcomes were reported in a recently completed clinical trial (NCT00348140) through the administration of rosiglitazone combined with AChE inhibitors in patients with mild to moderate AD [24]. Moreover, in a small clinical trial, treatment with pioglitazone improved memory and awareness in mild to moderate AD patients [82].

Amylin and its analogue (pramlintide) are pancreatic hormones and may play a possible role in AD treatments. It was shown that they decreased the level of Aβ and developed cognition in the brain of AD models. Also, amylin significantly inhibited β-secretase enzyme, while pramlintide did not [85, 86].

3.3 Medicinal Plants Having Effects against Alzheimer's Disease

3.3.1 Medicinal Plants Targeting the Cholinergic System

Several species of the family Lamiaceae are significant historical medicinal plants, and some of them could be used as tea supplements or flavorings. The anti-AChE activity of the hydroalcoholic extract of *Salvia triloba* L. has been compared with that of the other two Lamiaceae species, *Teucrium polium* L. and *Melissa officinalis* L. [109]. *S. triloba* and *T. polium* extracts exhibit significant potential with IC_{50} values of 0.71 and 0.55 mg/ml, respectively, while the percentage rate of inhibition of AChE of *M. officinalis* ranged from 35 to 40% (0.25–2 mg/ml). Meanwhile, these plant species are traditionally used in the Anatolian peninsula for memory enhancement. Their antiamnesic function was further investigated in a scopolamine-induced amnesia mouse model. It is noteworthy that *S. triloba* and *T. polium* could relieve amnesia in a dose-dependent manner, which might be related to their anti-AChE activity [109]. An additional in vivo anti-AD evaluation in an $AlCl_3$-induced AD rat model was subjected to *S. triloba* [110]. The results showed inhibition of brain and serum AChE activity (in a dose-dependent manner) along with increased downregulation of brain ACh levels. The essential oil obtained from steam distillation of another member of the Lamiaceae family (*Lavandula angustifolia* Mill.) was also investigated in a scopolamine-induced amnesia mouse model. It could release forgetfulness in a dose-dependent manner, which also could be related to its anti-AChE activity [111] (Table 3.2).

Two additional Lamiaceae species, *Foeniculum vulgare* L. (fennel) and *Ocimum sanctum* L. (basil), exhibited inhibition of brain AChE activity (in a dose-dependent manner) in scopolamine and aging-induced amnesia in Swiss mice [113, 114].

According to the different tea-making procedures, *Camellia sinensis* (L.) Kuntze (tea plant) can be classified as green tea, black tea, white tea, and so on. The water extracts of green tea and white tea have been treated through simulated gastrointestinal digestion to obtain post-gastric or colon-available digested extracts for each. The six extracts obtained exhibited

Table 3.2 Medicinal plants targeting the cholinergic system.

Plant species	Tested extract	Doses	In vitro/ in vivo	Route of administration	Model	Target	Effect	References
Salvia triloba L.	75% ethanol		In vitro		Enzymatic assay	AChE	IC_{50}: 0.71 mg/ml	[109]
	70% methanol	375 mg/kg BW/d for 3 months	In vivo	Orally	$AlCl_3$-induced AD in adult male Sprague-Dawley rats	AChE	% inhibition in brain (10.73%) and serum (7.32%)	[110]
		750 mg/kg BW/d for 3 months					% inhibition in brain (16.3%) and serum (13.1%)	
Melissa officinalis L.	75% ethanol		In vitro		Enzymatic assay	AChE	No effect	[109]
Teucrium polium L.	75% ethanol		In vitro		Enzymatic assay	AChE	IC_{50}: 0.55 mg/ml	[109]
Eremobium aegyptiacum (Spreng.) Schweinf. et Asch. ex Boiss	Water	156 mg/kg BW	In vivo	Orally	$AlCl_3$ female rats	AChE	μMSH/min/mg protein: 58.57 ± 9 compared to $AlCl_3$ rats (97.88 ± 9.87) and donepezil (35.89 ± 4.39)	[112]
Piper nigrum L.	70% methanol	93.75 mg/kg BW/d for 3 months	In vivo	Orally	$AlCl_3$-induced AD in adult male Sprague-Dawley rats	AChE	% inhibition in brain (5.16%) and serum (4.73%)	[110]
		187.5 mg/kg BW/d for 3 months					% inhibition in brain (10.81%) and serum (8.7%)	

(*Continued*)

Table 3.2 (Continued)

Plant species	Tested extract	Doses	In vitro/ in vivo	Route of administration	Model	Target	Effect	References
Foeniculum vulgare L.	90% methanol fruit	(50, 100, and 200 mg/kg PO)	In vivo		Scopolamine (0.4 mg/kg IP) and aging- induced amnesia in Swiss mice	AChE	Inhibition of brain AChE activity (dose-dependent manner)	[113]
Ocimum sanctum L.	95% ethanol	500 mg/kg PO	In vivo		Wistar male rats	AChE	% inhibition in cortex (21%), cerebellum (21%), medulla oblongata (25%), and midbrain (30%)	[114]
Lavandula angustifolia Mill.	Essential oil		In vivo		Scopolamine (1 mg/kg IP) induced AD in C57BL/6J mice	AChE	Significant inhibition of AChE activity	[111]
Zea mays L.	Corn oil (rich in linoleic acid)	500 mg/kg PO	In vivo	Orally	Intracerebroventricular injection of Aβ25–35 into brain of ICR male mice	AChE	Significant inhibition of AChE activity	[115]
Perilla frutescens (L.) Britton	Perilla oil (rich in α-linolenic acid)	500 mg/kg PO	In vivo	Orally	Intracerebroventricular injection of Aβ25–35 into brain of ICR male mice	AChE	Significant inhibition of AChE activity	[115]
Olea europaea L.	Olive oil (rich in oleic acid)	500 mg/kg PO	In vivo	Orally	Intracerebroventricular injection of Aβ25–35 into brain of ICR male mice	AChE	Significant inhibition of AChE activity	[115]

Camellia sinensis (L.) Kuntze (green tea)	Water	In vitro	Enzymatic assay	AChE	IC_{50}: 0.03 ± 0.004 mg/ml	[116]
	Water	In vitro	Enzymatic assay	BuChE	IC_{50}: 0.05 ± 0.005 mg/ml	[116]
	Water	In vitro	Enzymatic assay	AChE	IC_{50}: 7.2 µg/ml	[117]
	Postgastric digested water extract	In vitro	Enzymatic assay	AChE	IC_{50}: 17.84 µg/ml	[117]
	Colon-available digested water extract	In vitro	Enzymatic assay	AChE	IC_{50}: 9.59 µg/ml	[117]
Camellia sinensis (L.) Kuntze White tea	Water	In vitro	Enzymatic assay	AChE	IC_{50}: 8.06 µg/ml	[117]
	Postgastric digested water extract	In vitro	Enzymatic assay	AChE	IC_{50}: 16.1 µg/ml	[117]
	Colon-available digested water extract	In vitro	Enzymatic assay	AChE	IC_{50}: 4.22 µg/ml	[117]
Camellia sinensis (L.) Kuntze (black tea)	Water	In vitro	Enzymatic assay	AChE	IC_{50}: 0.06 ± 0.004 mg/ml	[116]
	Water	In vitro	Enzymatic assay	BuChE	IC_{50}: 0.05 ± 0.005 mg/ml	[116]
	Water	In vivo	$AlCl_3$ (100 mg/kg IP/60 d) induced AD in Wistar male rats	AChE	Promote AChE activities	[118]
Coffee	Water	In vitro	Enzymatic assay	AChE	IC_{50}: 0.41 ± 0.004 mg/ml	[116]

(*Continued*)

Table 3.2 (Continued)

Plant species	Tested extract	Doses	In vitro/ in vivo	Route of administration	Model	Target	Effect	References
Citrus maxima Merr. (pomelo)	Fruit juice		In vitro			AChE	% inhibition rate 60.39–28% at 66.68–16.67 ml/l	[119]
Citrus × *paradisi* Macfad. (grapefruit)	Fruit juice		In vitro			AChE	% inhibition rate 52–29% at 66.68–16.67 ml/l	[119]
Citrus limon (L.) Osbeck (lemon)	Fruit juice		In vitro			AChE	% inhibition rate 48–22% at 66.68-16.67 ml/l	[119]
Citrus × *sinensis* (L.) Osbeck (orange)	Fruit juice		In vitro			AChE	% inhibition rate 50–30.89% at 66.68–16.67 ml/l	[119]
Citrus reticulata Blanco (tangerine)	Fruit juice		In vitro			AChE	% inhibition rate 57–20% at 66.68–16.67 ml/l	[119]

Aβ, amyloid beta; AChE, acetylcholinesterase; AD, Alzheimer's disease; BuChE, butyrylcholinesterase; BW, body weight; GSK-3β, glycogen synthase kinase 3 beta; IP, intraperitoneal; PO, orally.

inhibition of AChE in a dose-dependent manner with an IC_{50} value ranging from 4.22 to 17.84 μg/ml. Nevertheless, it was shown that the inhibition of AChE activity of both green and white tea extracts reduced after gastric digestion, and recovered after pancreatic digestion, which could be related to the changes in phenolic content, especially caffeine [117].

Another early study has shown the anti-AChE activity of black and green tea extracts, with significant IC_{50} (mg/ml) values of 0.06 ± 0.005 and 0.03 ± 0.004, respectively [116]. In more recent in vivo studies, green and black tea water extracts have demonstrated the ability to reduce AChE activity in $AlCl_3$-induced AD in Wistar rats [118, 120].

Moreover, the fruit juices of five common *Citrus* species, namely pomelo, grapefruit, lemon, orange, and tangerine, have shown considerable dose-dependent AChE inhibitory activity in vitro. Orange juice exhibits the strongest activity at the lowest concentration (16.67 ml/l) with 30.89% inhibition of AChE, while pomelo juice displays the highest AChE inhibition (60.39%) at the highest concentration (66.68 ml/l) (Table 3.2) [119].

A more recent study was conducted on the aqueous extract of *Eremobium aegyptiacum* (Spreng.) Schweinf. et Asch. ex Boiss (Brassicaceae) on $AlCl_3$-induced AD in a female rat model [112]. The extract exhibited a significant reduction in AChE level (58.57 ± 9 μMSH/min/mg protein) in comparison with the $AlCl_3$ group (97.88 ± 9.87 μMSH/min/mg protein).

3.3.2 Medicinal Plants Targeting Amyloid Beta

Traditional herbal medicines can be harmless, cheap, and may have fewer side effects than synthetic drugs. Some plants have medicinal properties as neuroprotective agents for treating AD. AD is related to major cognitive deficits [121, 122]. Two important pathologies are the main factors in this disease. First, the formation of amyloid plaques in the brain (extracellular deposits of Aβ, an amino acid peptide, which consists of 39–43 amino acid residues and is considered critical in the early onset of AD) [123]. The aggregation of Aβ peptide with the formation of insoluble fibrils is the main component of the senile plaques. The extent of brain Aβ deposition is connected with the degree of cognitive deficit in AD patients [124].

The commonly used forms of Aβ are Aβ1-42 and Aβ25-35. Aβ1-42 is an antiparallel β-pleated sheet peptide that is a major part of diffused plaques in AD patients' brains and is considered as an important factor in the pathogenesis of AD [125, 126]. Also, the increased production of Aβ1-42 in AD patients and the high tendency of the peptide for amyloid fibrillogenesis accelerate neurodegeneration, leading to the formation of bioactive amyloid plaque species [125, 127–129]. The shorter-length Aβ25-35 peptide is highly toxic and is considered the core fragment of the full-length Aβ1-42 peptide. Numerous studies have revealed that it triggers neurotoxicity similar to that caused by Aβ1-42 peptide [130–132].

Prevention of Aβ aggregation or decreased Aβ levels is the major therapeutic goal for AD prevention. In the case of in vitro experiments, different cell lines have been used as neuronal cell models to determine the Aβ inhibitory activity of plant extracts, as shown in Table 3.3. However, previous literature showed that the most commonly used cell lines are primary cortical neurons, the rat pheochromocytoma (PC12) cell line, and the human neuroblastoma (SH-SY5Y) cell line.

Several animal models for in vivo experiments have been used to determine the effect of medicinal plants on Aβ-induced neurodegeneration, in which animals are injected with Aβ peptide either before or after administration of the tested extract. The preferred model seems to be ICR mice, which belong to the Institute of Cancer Research. Open-field tests, the Y-maze test, the Morris water maze test, and the passive avoidance test are the most commonly used behavior tests [157].

Table 3.3 Medicinal plants targeting amyloid beta (Aβ).

Plant species	Tested extract	Doses	In vitro/ in vivo	Route of administration	Model/assay	Target	Effect	Reference
Achyranthes bidentata Blume	Ethanolic extract	15 and 30 μg/ml	In vitro		Thioflavin-T fluorescence	β-amyloid protein 42	Prevention of Aβ polymerization	[133]
Alpinia galanga (L.) Willd.	n-Hexane, chloroform, and ethyl acetate fractions of rhizomes	200 and 400 mg/kg	In vivo	Orally	Swiss albino mice	Aβ	Cognitive enhancement with increase in Na^+/K^+-ATPase and antioxidant activity	[134]
Caesalpinia crista L.	Leaves aqueous extract	100 μg	In vitro		Thioflavin-T assay	β-amyloid protein 42	Inhibition of Aβ(42) aggregation and disaggregate preformed fibrils	[135]
Centella asiatica (L.) Urban		2.5 or 5.0 g/kg/d	In vivo	Orally	PSAPP mice	Aβ	Alteration in Aβ pathology	[136]
Cinnamomum zeylanicum L.	Bark aqueous extract	0.01–100 μg/ml	In vitro		Thioflavin-T binding fluorescence	40 μg/ml Aβ40 or Aβ42 solution	Inhibition of Aβ fibril formation	[137]
C. zeylanicum	Bark aqueous extract	100 μg/ml	In vivo	Orally	5XFAD mice	Aβ	Marked decrease in Aβ plaques and improvement in cognitive behavior	[137]
Eleutherococcus senticosus (Rupr. & Maxim.) Maxim.	Rhizomes ethyl acetate, n-butanol, and water fractions	0.1, 1, 10, 100 μg/ml	In vitro		Rat cortical neurons	Aβ25-35	Protective effects against Aβ-induced neuritic degeneration	[138]
Magnolia officinalis Rehder & Wilson	Bark ethanol extract	2.5, 5, and 10 mg/kg	In vivo	Orally	ICR mice	Aβ	Inhibition of Aβ1–42-induced neuronal cell death	[139]

Allium sativum L.	20% aqueous ethanol extract	2% aged garlic extract	In vivo	Diet	Mutant mouse model (Tg2576)	Aβ	Significant reduction of Aβ plaques	[140]
Crocus sativus L.	Aqueous methanol extract	100 μg/ml	In vitro		Thioflavine-T-based fluorescence assay	Aβ1-40	Inhibition of aggregation and deposition of Aβ	[141]
Citrus junos Siebold ex Tanaka	Ethanolic extract	3% lyophilized extract	In vivo	Diet	Male Sprague-Dawley rats	Aβ	Prevention of cognitive dysfunction and impaired energy and glucose homeostasis induced by β-amyloid	[142]
Morus atropurpurea L.	Methanol extract containing 0.1% HCl	0.18% and 0.9%	In vivo	Diet	SAMR1 and SAMP8 mice	Aβ	Decrease in Aβ Improvement in learning and memory ability	[143]
Olea europaea L.	Extra-virgin olive oil	0.7 g/d	In vivo	Diet	TgSwDI mice	Aβ	Reduction of total Aβ40 and Aβ42 in brains	[144]
Triticum aestivum L.	Water extract	0.5 and 1 mg/ml200 mg/kg/d	In vitro, in vivo	Orally	SH-SY5Y cells Sprague-Dawley rats	Aβ25-35	Reduction of Aβ and improvement in cognitive behavioral performance	[145]
Paeonia suffruticosa Andrews	Water, methanol, and ethanol extracts	10 μg/ml80 mg/kg/d	In vitro, in vivo	Orally	Thioflavin-T assay Tg2576 transgenic mice	Aβ1–40, Aβ1–42	Inhibition of Aβ fibril formation Improvement in long-term memory impairment Inhibition of accumulated Aβ in brain	[146]
Ptychopetalum olacoides Bentham	Ethanolic extract	800 mg/kg	In vivo	Orally	CF1 albino mice	Aβ	Prevention of Aβ-induced cognitive impairment	[147]

(*Continued*)

Table 3.3 (Continued)

Plant species	Tested extract	Doses	In vitro/ in vivo	Route of administration	Model/assay	Target	Effect	Reference
Schisandra chinensis (Trucz.) Baill.	Fruit hexane ethanol extract	100 mg/kg	In vivo	Orally	ICR mice	Aβ	Significantly attenuated Aβ1-42-induced memory impairment evaluated by behavioral tests	[148]
Uncaria rhynchophylla (Miq.) Jacks.	Aqueous methanol and ethanol extracts	10 μg/ml	In vitro		Thioflavin-T binding assay and atomic force microscopic imaging	Aβ1–40, Aβ1–42	Inhibition of Aβ aggregation	[149]
Dipsacus asper Wall.	Water extract and saponins fraction	0.1–1 mg/ml	In vitro		PC12 cells	Aβ25-35	Antagonize Aβ25-35-induced cytotoxicity in PC12 cells	[150]
Acorus gramineus Aiton	Water extract	×400 dilution	In vitro		PC12 cells	Aβ1-40	Significantly decreased Aβ1-40-induced cell death	[151]
Allium roseum L.	Ethanolic extract	10, 25, 50, and 100 μg/ml	In vitro		Thioflavine-T assay SH-SY5Y	Aβ42	Inhibition of amyloid fibril formation Decreased oxidative stress and intracellular Ca2+	[152]
Salvia officinalis L.	Leaves hydroalcoholic extract	0.01–100 g/ml	In vitro		PC12 cells	Aβ1–42	Prevents neurotoxicity induced by Aβ	[153]
Ginko biloba L.	Extract rich in flavonoids and terpenoids (EGb 761)	10–100 μg/ml	In vitro		Hippocampal primary cultured cells	Aβ25–35, Aβ1–40, and Aβ1–42	Protection of hippocampal neurons against Aβ toxicity	[154]
Curcuma longa L.	Turmeric extract (HSS-888)	0.1% w/w	In vivo	Orally	Tg2576 mice	Aβ	Reduction of Aβ deposition	[155]
Smallanthus sonchifolius (Poepp.) H. Rob.	Hydroalcoholic leaves and root extracts	40 mg/kg	In vivo	Orally	Wistar male rats	Aβ	Prevention of β-amyloid-induced memory deficits	[156]

Previous reports of preclinical studies showed that *Ginkgo biloba* L. (family Ginkgoaceae) could be used as a therapeutic drug for AD and other neurologic disorders, as it can prevent Aβ aggregation and decrease Aβ fibrillogenesis in vitro, while in vivo experiments indicate that *G. biloba* can improve impairment in cognition and memory in an AD rat model [158]. Also, *Rhinacanthus nasutus* root and leaf ethanolic extracts showed neuroprotective effects against cell death induced by Aβ in neuron cell culture [159]. Furthermore, a previous study by Sehgal et al. [160] showed that mice administrated Ashwagandha extract (*Withania somnifera* L.; Solanaceae) could eliminate amyloid plaques in mice cortex and hippocampus and could improve the behavior and cognitive functioning of the mice significantly, while a clinical study reported with the same plant on healthy humans showed enhancement in psychomotor performance reaction time [161]. Moreover, other species of the Solanaceae family (wolfberry; *Lycium barbarum* L.) blocked Aβ activity in rat cortical neurons in an in vitro experiment [162].

Over the past decades, different studies have been reported on plant extracts that showed inhibitory behavior against Aβ leading to improvement of cognitive function, such as *Achyranthes bidentata* Blume., *Alpinia galanga* (L.) Willd., *Caesalpinia crista* L., *Centella asiatica* (L.), *Cinnamomum zeylanicum* L. (Ceylon cinnamon), *Eleutherococcus senticosus* (Rupr. & Maxim.) Maxim. (Siberian ginseng), *Magnolia officinalis* Rehder & E. Wilson, *Allium sativum* L. (aged garlic), *Crocus sativus* L., *Citrus junos* Siebold ex *Tanaka*, *Morus atropurpurea* L. (mulberry), *Olea europaea* L. (extra-virgin olive oil), *Triticum aestivum* L., *Paeonia suffruticosa* Andrews, *Ptychopetalum olacoides* Bentham (Marapuama), *Schisandra chinensis* Trucz., *Uncaria rhynchophylla* (Miq.) Jacks, *Dipsacus asper* Wall, *Acorus gramineus* Aiton Rhizoma, *Allium roseum* L., *Salvia officinalis* L. (sage), *Curcuma longa* L., and *Smallanthus sonchifolius* (Poepp.) H. Rob. (Yacon), as shown in Table 3.3.

Furthermore, other plant decoctions that are made up of several traditional herbs, such as the Korean Seonghyangjeongkisan, were used for protection against Aβ-induced cytotoxicity: *Areca catechu* L., *Perilla frutescens* (L.) Britton, *Agastache rugosa* (Fisch. & C.A. Mey.) Kuntze, *Atractylodes macrocephala* Koidz., *Pinellia ternata* (Thunb.) Makino, *Atractylodes chinensis* (Bunge) Koidz., *Citrus reticulata* Blanco, *Cinnamomum cassia* (L.) J. Presl., *Zingiber officinale* Roscoe, *Alpinia oxyphylla* Miq., *Glycyrrhiza uralensis* Fisch., *Zizyphus jujuba* Mill., *Aucklandia lappa* DC., and *Arisaema erubescens* (Wall.) Schott. A decoction known as LMK02-Jangwonhwan administered to transgenic-APPswe/PS1 dE9 mice decreased Aβ1-42 and Aβ1-40 levels and plaque deposition in mice brains [163].

Also, in Japanese kampo medicine, different decoctions made up of different medicinal herbs have been used for inhibition of Aβ deposition, such as Zokumeito and Yokukansan [157].

3.3.3 Medicinal Plants Targeting Tau-Related Pathways

The intracellular hyperphosphorylated NFT aggregates are an important factor in AD drug discovery research, which has increasingly shifted from Aβ to tau protein [164]. Tau is a set of microtubule-associated proteins mainly expressed in axons [165, 166]. Hyperphosphorylated tau is recognized in the AD brain and in the developing fetal brain [167, 168], leading to the degeneration of some neurons, indicating that the phosphorylation of tau may have a direct or indirect role in neuronal cell death through the destabilization of microtubules.

The somatodendritic buildup of hyperphosphorylated tau stimulates tau-related neuropathologic alterations, which prevent tau assembly onto microtubules and result in intracellular insoluble NFTs [169]. Also, hyperphosphorylation and aggregation of tau protein on chromosome 17 may be the sign of a number of neurodegenerative diseases pathologies like AD.

Tau protein is an important biomarker of AD. The use of plant extracts as a treatment strategy for prevention of the hyperphosphorylation and aggregation of those proteins has been investigated in various studies [140, 144, 153, 170–172].

Preclinical studies showed that red ginseng was reported for inhibition of tau aggregation in vitro [173], since ginseng contains flavonoids and saponin [174]. In addition, the effects of different plant extracts in tau-related pathways include *A. sativum* L. (aged garlic), *Dendrobium nobile* Lindl., *O. europaea* L. (extra-virgin olive oil), *C. zeylanicum* L. (Ceylon cinnamon), *S. officinalis* L. (sage), *C. longa* L. (turmeric), and *Morus alba* L. fruit (mulberry), as shown in (Table 3.4).

Previous reports showed that after one month of drinking the juice of young coconut (*Cocos nucifera*) by ovariectomized rats, tau levels were decreased [176]. The oral administration of tannoid fractions of *Emblica officinalis* to rats for two months could prevent tau phosphorylation by targeting the GSK-3β/Akt pathway [177].

Most of these plants are edible, which spotlights the effect of a balanced diet on a healthy life and delaying the onset of AD. Certainly, bad dietary habits play an important role in side effects in AD patients, as reported previously. Mice fed with a high-fat diet containing 21% fat and 0.2% cholesterol (APdE9 transgenic AD mouse model) showed an increase in tau expression and tau exon 10 inclusion. The mRNA levels of both 3R and 4R tau in the temporal cortex and their mRNA ratio are also increased by a high-fat diet, which may be connected to the impaired memory of AD mice [178].

Some of these natural plant extracts that showed inhibitory action against Aβ and tau proteins have been subjected to clinical trials for AD therapy, such as *G. biloba*, *C. longa*, and *S. officinalis*, and showed improvement in cognitive functions [179].

3.4 Natural Products with Proven Anti-Alzheimer's Activity

Many classes of natural products such as flavonoids, phenolic acids, alkaloids, and naphthoquinones have been examined and tested for different models of anti-Alzheimer's activity. Table 3.5 shows different examples of these natural products and the mechanisms of their activities.

3.5 Conclusion

This chapter attempts to provide a comprehensive account of various medicinal plants and natural compounds that can be used for the treatment of AD through different mechanisms, including ACh inhibitory activity, targeting tau-related pathways, or Aβ. The therapeutic effects of medicinal plants with few side effects can be attributed to the synergism present between the phytochemicals of such plants; this synergism may be of great benefit as it can be like an army against the complex and multifactorial pathogenesis of AD. On the other hand, conventional anti-AD drugs only offer symptomatic and/or partial relief. For this reason, complementary and alternative therapies are being vigorously investigated in a number of preclinical and clinical trials. We provide here a number of plants and phytoconstituents that could be promising candidates against AD and thus extensive preclinical and clinical trials of these drugs should be conducted.

Table 3.4 Medicinal plants targeting tau-related pathways.

Plant species	Tested extract	Doses	In vitro/ in vivo	Route of administration	Model/assay	Target	Effect	References
Allium sativum L.	20% aqueous ethanol extract	2%	In vivo	Diet	Mutant mouse model (Tg2576)	Tau proteins	Reduced phosphorylation-induced conformational change in tau protein	[140]
Dendrobium nobile Lindl.	Methanol acid alkaline total extract	20, 40 mg/kg	In vivo	Intragastrically	Sprague-Dawley rats	Tau proteins	Attenuates LPS-induced hyperphosphorylation of tau protein in rat's hippocampus Protects against LPS-induced apoptosis in rat brain	[170]
Olea europaea L.	Extra-virgin olive oil	0.7 g/d	In vivo	Diet	TgSwDI mice	Tau proteins	Reduction of total tau levels and reduction of its hyperphosphorylation	[144]
C. zeylanicum L.	Aqueous extract	0.22 mg/ml	In vitro		Thioflavin-S fluorescence tau aggregation assay	Tau proteins	*Centella asiatica* treatment alters Aβ pathology in PSAPP mice Inhibition of tau aggregation	[171]
Salvia officinalis L.	Leaves hydroalcoholic extract	0.01–100 g/ml	In vitro		PC12 cells	Tau proteins	Rosamirinic acid in extract reduced hyper phosphorylation of tau proteins	[153]
Panax ginseng Mey.	Total ginsenoside extracts from stems and leaves	0.100 and 0.250 mg/ml	In vitro		Human neuroblastoma cells SY5Y	Purified calcineurin/tau	Regulating phosphatase activity of purified calcineurin and tau phosphorylation	[172]
Curcuma longa L.	Turmeric extract (HSS-888)	0.1% w/w	In vivo	Orally	Tg2576 mice	Tau proteins	Treatment with HSS-88 showed 80% decrease in pTau to actin	[155]
Morus alba L.	Ethanol extract	0.1, 1, 10, and 100 μg/ml 20, 100, and 500 mg/kg	In vitro, in vivo	Orally	Rat hippocampal neuronal cells ICR mice	Aβ25–35/tau	Protection against Aβ by increasing cell viability Reduction of tau phosphorylation and apoptosis	[175]

Aβ, amyloid beta; LPS, lipopolysaccharide.

Table 3.5 Examples of natural products with anti-Alzheimer's activity.

Compound	Class	Mechanism of action	Model	References
Apigenin	Flavone	Enhancing memory and decreasing accumulation of fibrillar amyloid	APP/PS1 double transgenic AD mouse	[180]
Baicalein	Flavone	Inhibits tau phosphorylation Decreases Aβ	APP/PS1 mice Tg2576 transgenic mice	[181, 182]
Luteolin	Flavone	Antioxidant, AChE inhibitor, and decreases accumulation of Aβ	AD flies	[183]
Quercetin	Flavonol	Antioxidant Reducing extracellular Aβ	Zebrafish AD model 3xTg-AD mice and non-transgenic (non-Tg) mice	[184]
Huperzine A	Alkaloid	Decreases accumulation of Aβ plaque Enhances function of mitochondria AChE inhibitor	Transgenic-APPswe/PS1dE9 AD mice	[185]
Galantamine	Alkaloid	AChE inhibitor Reduces oxidative stress	Male C57BL/6Hsd mice APPswe/PS1d E9 mouse	[186]
Berberine	Alkaloid	Inhibits amyloid plaque accumulation and reduces AD phenotypic pathology	TgCRND8 mouse	[187]
Caffeine	Alkaloid	Inhibits tau phosphorylation and oxidative stress	THY-Tau22 mice	[188]

Infractopicrin and 10-hydroxy-infractopicrin	Alkaloids	AChE and BuChE inhibitor	In vitro model	[189]
Rotundine	Alkaloid	Neuroprotective agent against AD	Rats subcutaneously injected by D-galactose	[190]
Rhinacanthin S	Naphthoquinone	AChE inhibitor	In vitro model	[191]
Curcumin	Diferuloylmethane	Decreases accumulation of Aβ plaque	APP/PS1 transgenic mice	[192]
Resveratrol	Stilbene	Decreases Aβ plaque accumulation and activates microglia in hippocampus and cortex	Aβ PP/PS-1 transgenic mouse	[193]
Pteryxin	Dihydropyranocoumarin	AChE and BuChE inhibitor	In vitro model	[194]
Caffeoylquinic acid	Phenolic acids	AChE inhibitor and antioxidant	In vitro model	[195]
Safranal, crocetin, and dimethylcrocetin		AChE inhibitor	In vitro model	[196]
Aloe-emodin	Dihydroxyanthraquinone derivative	Ameliorates learning and memory performance	Scopolamine induced AD in mice	[197]
Sinapic acid	Phenolic acid	Attenuates intracerebroventricular–STZ-caused cognitive impairment	STZ Alzheimer rat model	[198]

Aβ, amyloid beta; AChE, acetylcholinesterase; AD, Alzheimer's disease; BuChE, butyrylcholinesterase.

References

1 Dey, A., Bhattacharya, R., Mukherjee, A., and Pandey, D.K. (2017). Natural products against Alzheimer's disease: pharmaco-therapeutics and biotechnological interventions. *Biotechnol. Adv.* 35 (2): 178–216.

2 World Health Organization (2020). Dementia. https://www.who.int/news-room/fact-sheets/detail/dementia (accessed 8 March 2023).

3 Burns, A. and Iliffe, S. (2009). Alzheimer's disease. *BMJ* 338: b158.

4 Mayeux, R. and Stern, Y. (2012). Epidemiology of Alzheimer disease. *Cold Spring Harb. Perspect. Med.* 2 (8): a006239.

5 Ballaed, C., Gauthier, S., Corbett, A. et al. (2011). Alzheimer's disease. *Lancet* 377: 1019–1031.

6 Silva, M.V.F., Loures, C.M.G., Alves, L.C.V. et al. (2019). Alzheimer's disease: risk factors and potentially protective measures. *J. Biomed. Sci.* 26 (1): 1–11.

7 Liu, W., Wong, A., Law, A.C., and Mok, V.C. (2015). Cerebrovascular disease, amyloid plaques, and dementia. *Stroke* 46 (5): 1402–1407.

8 Li, X., Song, D., and Leng, S.X. (2015). Link between type 2 diabetes and Alzheimer's disease: from epidemiology to mechanism and treatment. *Clin. Interv. Aging* 10: 549.

9 Fitzpatrick, A.L., Kuller, L.H., Lopez, O.L. et al. (2009). Midlife and late-life obesity and the risk of dementia: cardiovascular health study. *Arch. Neurol.* 66 (3): 336–342.

10 Anstey, K., Cherbuin, N., Budge, M., and Young, J. (2011). Body mass index in midlife and late-life as a risk factor for dementia: a meta-analysis of prospective studies. *Obes. Rev.* 12 (5): e426–e437.

11 Xue-Shan, Z., Qi, W., Zhong, R. et al. (2016). Imbalanced cholesterol metabolism in Alzheimer's disease. *Clin. Chim. Acta* 456: 107–114.

12 Sosa-Ortiz, A.L., Acosta-Castillo, I., and Prince, M.J. (2012). Epidemiology of dementias and Alzheimer's disease. *Arch. Med. Res.* 43 (8): 600–608.

13 Breijyeh, Z. and Karaman, R. (2020). Comprehensive review on Alzheimer's disease: causes and treatment. *Molecules* 25 (24): 5789.

14 Singh, S.K., Srivastav, S., Yadav, A.K. et al. (2016). Overview of Alzheimer's disease and some therapeutic approaches targeting Aβ by using several synthetic and herbal compounds. *Oxidative Med. Cell. Longev.* 2016: 7361613.

15 Armstrong, R.A. (2009). The molecular biology of senile plaques and neurofibrillary tangles in Alzheimer's disease. *Folia Neuropathol.* 47 (4): 289–299.

16 Noori, T., Dehpour, A.R., Sureda, A. et al. (2021). Role of natural products for the treatment of Alzheimer's disease. *Eur. J. Pharmacol.* 898: 173974.

17 Amato, A., Terzo, S., and Mulè, F. (2019). Natural compounds as beneficial antioxidant agents in neurodegenerative disorders: a focus on Alzheimer's disease. *Antioxidants.* 8 (12): 608.

18 Overk, C.R. and Masliah, E. (2014). Pathogenesis of synaptic degeneration in Alzheimer's disease and Lewy body disease. *Biochem. Pharmacol.* 88 (4): 508–516.

19 Mulugeta, E., Karlsson, E., Islam, A. et al. (2003). Loss of muscarinic M4 receptors in hippocampus of Alzheimer patients. *Brain Res.* 960 (1-2): 259–262.

20 Tramutola, A., Lanzillotta, C., Perluigi, M., and Butterfield, D.A. (2017). Oxidative stress, protein modification and Alzheimer disease. *Brain Res. Bull.* 133: 88–96.

21 Vogel, J.W., Iturria-Medina, Y., Strandberg, O.T. et al. (2020). Spread of pathological tau proteins through communicating neurons in human Alzheimer's disease. *Nat. Commun.* 11 (1): 1–15.

22 Ardura-Fabregat, A., Boddeke, E., Boza-Serrano, A. et al. (2017). Targeting neuroinflammation to treat Alzheimer's disease. *CNS Drugs* 31 (12): 1057–1082.
23 Walters, A., Phillips, E., Zheng, R. et al. (2016). Evidence for neuroinflammation in Alzheimer's disease. *Prog. Neurol. Psychiatry* 20 (5): 25–31.
24 Folch, J., Petrov, D., Ettcheto, M. et al. (2016). Current research therapeutic strategies for Alzheimer's disease treatment. *Neural Plast.* 2016: 8501693.
25 May, P.C., Dean, R.A., Lowe, S.L. et al. (2011). Robust central reduction of amyloid-β in humans with an orally available, non-peptidic β-secretase inhibitor. *J. Neurosci.* 31 (46): 16507–16516.
26 Doody, R.S., Raman, R., Farlow, M. et al. (2013). A phase 3 trial of semagacestat for treatment of Alzheimer's disease. *N. Engl. J. Med.* 369 (4): 341–350.
27 Chiang, K. and Koo, E.H. (2014). Emerging therapeutics for Alzheimer's disease. *Annu. Rev. Pharmacol. Toxicol.* 54: 381–405.
28 Menting, K.W. and Claassen, J.A. (2014). β-secretase inhibitor; a promising novel therapeutic drug in Alzheimer's disease. *Front. Aging Neurosci.* 6: 165.
29 Yan, R. and Vassar, R. (2014). Targeting the β secretase BACE1 for Alzheimer's disease therapy. *Lancet Neurol.* 13 (3): 319–329.
30 Coric, V., van Dyck, C.H., Salloway, S. et al. (2012). Safety and tolerability of the γ-secretase inhibitor avagacestat in a phase 2 study of mild to moderate Alzheimer disease. *Arch. Neurol.* 69 (11): 1430–1440.
31 Dockens, R., Wang, J.-S., Castaneda, L. et al. (2012). A placebo-controlled, multiple ascending dose study to evaluate the safety, pharmacokinetics and pharmacodynamics of avagacestat (BMS-708163) in healthy young and elderly subjects. *Clin. Pharmacokinet.* 51 (10): 681–693.
32 Tong, G., Castaneda, L., Wang, J.-S. et al. (2012). Effects of single doses of avagacestat (BMS-708163) on cerebrospinal fluid Aβ levels in healthy young men. *Clin. Drug Investig.* 32 (11): 761–769.
33 Pasqualetti, P., Bonomini, C., Dal Forno, G. et al. (2009). A randomized controlled study on effects of ibuprofen on cognitive progression of Alzheimer's disease. *Aging Clin. Exp. Res.* 21 (2): 102–110.
34 Miguel-Álvarez, M., Santos-Lozano, A., Sanchis-Gomar, F. et al. (2015). Non-steroidal anti-inflammatory drugs as a treatment for Alzheimer's disease: a systematic review and meta-analysis of treatment effect. *Drugs Aging* 32 (2): 139–147.
35 Ross, J., Sharma, S., Winston, J. et al. (2013). CHF5074 reduces biomarkers of neuroinflammation in patients with mild cognitive impairment: a 12-week, double-blind, placebo-controlled study. *Curr. Alzheimer Res.* 10 (7): 742–753.
36 Imbimbo, B.P., Frigerio, E., Breda, M. et al. (2013). Pharmacokinetics and pharmacodynamics of CHF5074 after short-term administration in healthy subjects. *Alzheimer Dis. Assoc. Disord.* 27 (3): 278–286.
37 Ronsisvalle, N., Di Benedetto, G., Parenti, C. et al. (2014). CHF5074 protects SH-SY5Y human neuronal-like cells from amyloidbeta 25-35 and tumor necrosis factor related apoptosis inducing ligand toxicity in vitro. *Curr. Alzheimer Res.* 11 (7): 714–724.
38 Gauthier, S., Aisen, P., Ferris, S. et al. (2009). Effect of tramiprosate in patients with mild-to-moderate Alzheimer's disease: exploratory analyses of the MRI sub-group of the Alphase study. *J. Nutr. Health Aging* 13 (6): 550–557.
39 Aisen, P.S., Gauthier, S., Ferris, S.H. et al. (2011). Tramiprosate in mild-to-moderate Alzheimer's disease–a randomized, double-blind, placebo-controlled, multi-centre study (the Alphase Study). *Arch. Med. Sci.* 7 (1): 102.

40 Pitt, J., Thorner, M., Brautigan, D. et al. (2013). Protection against the synaptic targeting and toxicity of Alzheimer's-associated Aβ oligomers by insulin mimetic chiro-inositols. *FASEB J.* 27 (1): 199–207.

41 Bilikiewicz, A. and Gaus, W. (2004). Colostrinin (a naturally occurring, proline-rich, polypeptide mixture) in the treatment of Alzheimer's disease. *J. Alzheimers Dis.* 6 (1): 17–26.

42 Salloway, S., Sperling, R., Keren, R. et al. (2011). A phase 2 randomized trial of ELND005, scyllo-inositol, in mild to moderate Alzheimer disease. *Neurology* 77 (13): 1253–1262.

43 Wang, Z., Wang, Y., Li, W. et al. (2014). Design, synthesis, and evaluation of multitarget-directed selenium-containing clioquinol derivatives for the treatment of Alzheimer's disease. *ACS Chem. Neurosci.* 5 (10): 952–962.

44 Ryan, T.M., Roberts, B.R., McColl, G. et al. (2015). Stabilization of nontoxic Aβ-oligomers: insights into the mechanism of action of hydroxyquinolines in Alzheimer's disease. *J. Neurosci.* 35 (7): 2871–2884.

45 Wang, X., Sun, G., Feng, T. et al. (2019). Sodium oligomannate therapeutically remodels gut microbiota and suppresses gut bacterial amino acids-shaped neuroinflammation to inhibit Alzheimer's disease progression. *Cell Res.* 29 (10): 787–803.

46 Green Valley (2019). Green Valley announces NMPA approval of oligomannate for mild to moderate Alzheimer's disease. https://www.prnewswire.com/news-releases/green-valley-announces-nmpa-approval-of-oligomannate-for-mild-to-moderate-alzheimers-disease-300950349.html (accessed 8 March 2023).

47 Lawlor, B., Segurado, R., Kennelly, S. et al. (2018). Nilvadipine in mild to moderate Alzheimer disease: a randomised controlled trial. *PLoS Med.* 15 (9): e1002660.

48 Huang, L.-K., Chao, S.-P., and Hu, C.-J. (2020). Clinical trials of new drugs for Alzheimer disease. *J. Biomed. Sci.* 27 (1): 1–13.

49 Gilman, S., Koller, M., Black, R. et al. (2005). Clinical effects of Aβ immunization (AN1792) in patients with AD in an interrupted trial. *Neurology* 64 (9): 1553–1562.

50 Wiessner, C., Wiederhold, K.-H., Tissot, A.C. et al. (2011). The second-generation active Aβ immunotherapy CAD106 reduces amyloid accumulation in APP transgenic mice while minimizing potential side effects. *J. Neurosci.* 31 (25): 9323–9331.

51 Liu, B., Frost, J.L., Sun, J. et al. (2013). MER5101, a novel Aβ1-15: DT conjugate vaccine, generates a robust anti-Aβ antibody response and attenuates Aβ pathology and cognitive deficits in APPswe/PS1ΔE9 transgenic mice. *J. Neurosci.* 33 (16): 7027–7037.

52 Panza, F., Solfrizzi, V., Imbimbo, B.P., and Logroscino, G. (2014). Amyloid-directed monoclonal antibodies for the treatment of Alzheimer's disease: the point of no return? *Expert. Opin. Biol. Ther.* 14 (10): 1465–1476.

53 Panza, F., Solfrizzi, V., Imbimbo, B.P. et al. (2014). Amyloid-based immunotherapy for Alzheimer's disease in the time of prevention trials: the way forward. *Expert. Rev. Clin. Immunol.* 10 (3): 405–419.

54 Salloway, S., Sperling, R., Fox, N.C. et al. (2014). Two phase 3 trials of bapineuzumab in mild-to-moderate Alzheimer's disease. *N. Engl. J. Med.* 370 (4): 322–333.

55 Jacobsen, H., Ozmen, L., Caruso, A. et al. (2014). Combined treatment with a BACE inhibitor and anti-Aβ antibody gantenerumab enhances amyloid reduction in APPLondon mice. *J. Neurosci.* 34 (35): 11621–11630.

56 Folch, J., Ettcheto, M., Petrov, D. et al. (2018). Review of the advances in treatment for Alzheimer disease: strategies for combating β-amyloid protein. *Neurología.* 33 (1): 47–58.

57 Bohrmann, B., Baumann, K., Benz, J. et al. (2012). Gantenerumab: a novel human anti-Aβ antibody demonstrates sustained cerebral amyloid-β binding and elicits cell-mediated removal of human amyloid-β. *J. Alzheimers Dis.* 28 (1): 49–69.

58 Novakovic, D., Feligioni, M., Scaccianoce, S. et al. (2013). Profile of gantenerumab and its potential in the treatment of Alzheimer's disease. *Drug Des. Devel. Ther.* 7: 1359.

59 Tayeb, H.O., Murray, E.D., Price, B.H., and Tarazi, F.I. (2013). Bapineuzumab and solanezumab for Alzheimer's disease: is the "amyloid cascade hypothesis" still alive? *Expert. Opin. Biol. Ther.* 13 (7): 1075–1084.

60 Doody, R.S., Thomas, R.G., Farlow, M. et al. (2014). Phase 3 trials of solanezumab for mild-to-moderate Alzheimer's disease. *N. Engl. J. Med.* 370 (4): 311–321.

61 Murakami, K. and Irie, K. (2019). Three structural features of functional food components and herbal medicine with amyloid β42 anti-aggregation properties. *Molecules* 24 (11): 2125.

62 Dodel, R., Rominger, A., Bartenstein, P. et al. (2013). Intravenous immunoglobulin for treatment of mild-to-moderate Alzheimer's disease: a phase 2, randomised, double-blind, placebo-controlled, dose-finding trial. *Lancet Neurol.* 12 (3): 233–243.

63 Lovestone, S., Boada, M., Dubois, B. et al. (2015). A phase II trial of tideglusib in Alzheimer's disease. *J. Alzheimers Dis.* 45 (1): 75–88.

64 Corcoran, N.M., Martin, D., Hutter-Paier, B. et al. (2010). Sodium selenate specifically activates PP2A phosphatase, dephosphorylates tau and reverses memory deficits in an Alzheimer's disease model. *J. Clin. Neurosci.* 17 (8): 1025–1033.

65 Van Eersel, J., Ke, Y.D., Liu, X. et al. (2010). Sodium selenate mitigates tau pathology, neurodegeneration, and functional deficits in Alzheimer's disease models. *Proc. Natl. Acad. Sci.* 107 (31): 13888–13893.

66 Wischik, C.M., Staff, R.T., Wischik, D.J. et al. (2015). Tau aggregation inhibitor therapy: an exploratory phase 2 study in mild or moderate Alzheimer's disease. *J. Alzheimers Dis.* 44 (2): 705–720.

67 Novak, P., Kontsekova, E., Zilka, N., and Novak, M. (2018). Ten years of tau-targeted immunotherapy: the path walked and the roads ahead. *Front. Neurosci.* 12: 798.

68 Shemesh, O.A. and Spira, M.E. (2011). Rescue of neurons from undergoing hallmark tau-induced Alzheimer's disease cell pathologies by the antimitotic drug paclitaxel. *Neurobiol. Dis.* 43 (1): 163–175.

69 Wisniewski, T. and Goñi, F. (2015). Immunotherapeutic approaches for Alzheimer's disease. *Neuron* 85 (6): 1162–1176.

70 Zhang, B., Carroll, J., Trojanowski, J.Q. et al. (2012). The microtubule-stabilizing agent, epothilone D, reduces axonal dysfunction, neurotoxicity, cognitive deficits, and Alzheimer-like pathology in an interventional study with aged tau transgenic mice. *J. Neurosci.* 32 (11): 3601–3611.

71 Kontsekova, E., Zilka, N., Kovacech, B. et al. (2014). First-in-man tau vaccine targeting structural determinants essential for pathological tau–tau interaction reduces tau oligomerisation and neurofibrillary degeneration in an Alzheimer's disease model. *Alzheimers Res. Ther.* 6 (4): 1–12.

72 Folch, J., Petrov, D., Ettcheto, M. et al. (2015). Masitinib for the treatment of mild to moderate Alzheimer's disease. *Expert. Rev. Neurother.* 15 (6): 587–596.

73 Piette, F., Belmin, J., Vincent, H. et al. (2011). Masitinib as an adjunct therapy for mild-to-moderate Alzheimer's disease: a randomised, placebo-controlled phase 2 trial. *Alzheimers Res. Ther.* 3 (2): 1–11.

74 Nygaard, H.B., van Dyck, C.H., and Strittmatter, S.M. (2014). Fyn kinase inhibition as a novel therapy for Alzheimer's disease. *Alzheimers Res. Ther.* 6 (1): 1–8.

75 Yang, K., Belrose, J., Trepanier, C.H. et al. (2011). Fyn, a potential target for Alzheimer's disease. *J. Alzheimers Dis.* 27 (2): 243–252.

76 Wallace, T. and Bertrand, D. (2013). Importance of the nicotinic acetylcholine receptor system in the prefrontal cortex. *Biochem. Pharmacol.* 85 (12): 1713–1720.

77 Weinreb, O., Amit, T., Bar-Am, O. et al. (2012). Ladostigil: a novel multimodal neuroprotective drug with cholinesterase and brain-selective monoamine oxidase inhibitory activities for Alzheimer's disease treatment. *Curr. Drug Targets* 13 (4): 483–494.

78 Woolley, M., Bentley, J., Sleight, A. et al. (2001). A role for 5-ht6 receptors in retention of spatial learning in the Morris water maze. *Neuropharmacology* 41 (2): 210–219.

79 Wilkinson, D., Windfeld, K., and Colding-Jørgensen, E. (2014). Safety and efficacy of idalopirdine, a 5-HT6 receptor antagonist, in patients with moderate Alzheimer's disease (LADDER): a randomised, double-blind, placebo-controlled phase 2 trial. *Lancet Neurol.* 13 (11): 1092–1099.

80 Ahmed, M., Malik, M., Teselink, J. et al. (2019). Current agents in development for treating behavioral and psychological symptoms associated with dementia. *Drugs Aging* 36 (7): 589–605.

81 Lourenco, M.V., Ferreira, S.T., and De Felice, F.G. (2015). Neuronal stress signaling and eIF2α phosphorylation as molecular links between Alzheimer's disease and diabetes. *Prog. Neurobiol.* 129: 37–57.

82 Sato, T., Hanyu, H., Hirao, K. et al. (2011). Efficacy of PPAR-γ agonist pioglitazone in mild Alzheimer disease. *Neurobiol. Aging* 32 (9): 1626–1633.

83 Risner, M., Saunders, A., Altman, J. et al. (2006). Efficacy of rosiglitazone in a genetically defined population with mild-to-moderate Alzheimer's disease. *Pharmacogenomics J.* 6 (4): 246–254.

84 Gold, M., Alderton, C., Zvartau-Hind, M. et al. (2010). Rosiglitazone monotherapy in mild-to-moderate Alzheimer's disease: results from a randomized, double-blind, placebo-controlled phase III study. *Dement. Geriatr. Cogn. Disord.* 30 (2): 131–146.

85 Qiu, W.Q. and Zhu, H. (2014). Amylin and its analogs: a friend or foe for the treatment of Alzheimer's disease? *Front. Aging Neurosci.* 6: 186.

86 Adler, B.L., Yarchoan, M., Hwang, H.M. et al. (2014). Neuroprotective effects of the amylin analogue pramlintide on Alzheimer's disease pathogenesis and cognition. *Neurobiol. Aging* 35 (4): 793–801.

87 Mohajeri, M.H., Wollmer, M.A., and Nitsch, R.M. (2002). Aβ42-induced increase in neprilysin is associated with prevention of amyloid plaque formation in vivo. *J. Biol. Chem.* 277 (38): 35460–35465.

88 El-Amouri, S.S., Zhu, H., Yu, J. et al. (2008). Neprilysin: an enzyme candidate to slow the progression of Alzheimer's disease. *Am. J. Pathol.* 172 (5): 1342–1354.

89 Peng, Y., Tao, H., Wang, S. et al. (2021). Dietary intervention with edible medicinal plants and derived products for prevention of Alzheimer's disease: a compendium of time-tested strategy. *J. Funct. Foods* 81: 104463.

90 Shefet-Carasso, L. and Benhar, I. (2015). Antibody-targeted drugs and drug resistance – challenges and solutions. *Drug Resist. Updat.* 18: 36–46.

91 Takahashi, S., Saito, T., Hisanaga, S.-i. et al. (2003). Tau phosphorylation by cyclin-dependent kinase 5/p39 during brain development reduces its affinity for microtubules. *J. Biol. Chem.* 278 (12): 10506–10515.

92 Li, T. and Paudel, H.K. (2006). Glycogen synthase kinase 3β phosphorylates Alzheimer's disease-specific Ser396 of microtubule-associated protein tau by a sequential mechanism. *Biochemistry* 45 (10): 3125–3133.

93 Yoshimura, Y., Ichinose, T., and Yamauchi, T. (2003). Phosphorylation of tau protein to sites found in Alzheimer's disease brain is catalyzed by Ca2+/calmodulin-dependent protein kinase II as demonstrated tandem mass spectrometry. *Neurosci. Lett.* 353 (3): 185–188.

94 Hanger, D.P., Byers, H.L., Wray, S. et al. (2007). Novel phosphorylation sites in tau from Alzheimer brain support a role for casein kinase 1 in disease pathogenesis. *J. Biol. Chem.* 282 (32): 23645–23654.

95 Yin, X., Jin, N., Gu, J. et al. (2012). Dual-specificity tyrosine phosphorylation-regulated kinase 1A (Dyrk1A) modulates serine/arginine-rich protein 55 (SRp55)-promoted Tau exon 10 inclusion. *J. Biol. Chem.* 287 (36): 30497–30506.

96 Timm, T., Marx, A., Panneerselvam, S. et al. (2008). Structure and regulation of MARK, a kinase involved in abnormal phosphorylation of Tau protein. *BMC Neurosci.* 9 (2): 1–6.

97 Andorfer, C.A. and Davies, P. (2000). PKA phosphorylations on tau: developmental studies in the mouse. *Dev. Neurosci.* 22 (4): 303–309.

98 Liu, F., Grundke-Iqbal, I., Iqbal, K., and Gong, C.X. (2005). Contributions of protein phosphatases PP1, PP2A, PP2B and PP5 to the regulation of tau phosphorylation. *Eur. J. Neurosci.* 22 (8): 1942–1950.

99 Relkin, N.R., Szabo, P., Adamiak, B. et al. (2009). 18-Month study of intravenous immunoglobulin for treatment of mild Alzheimer disease. *Neurobiol. Aging* 30 (11): 1728–1736.

100 Congdon, E.E. and Sigurdsson, E.M. (2018). Tau-targeting therapies for Alzheimer disease. *Nat. Rev. Neurol.* 14 (7): 399–415.

101 Wilkinson, S.T. and Sanacora, G. (2019). A new generation of antidepressants: an update on the pharmaceutical pipeline for novel and rapid-acting therapeutics in mood disorders based on glutamate/GABA neurotransmitter systems. *Drug Discov. Today* 24 (2): 606–615.

102 Shirazi, S.K. and Wood, J.G. (1993). The protein tyrosine kinase, fyn, in Alzheimer's disease pathology. *Neuroreport* 4 (4): 435–437.

103 Cochran, J.N., Hall, A.M., and Roberson, E.D. (2014). The dendritic hypothesis for Alzheimer's disease pathophysiology. *Brain Res. Bull.* 103: 18–28.

104 Kaufman, A.C., Salazar, S.V., Haas, L.T. et al. (2015). Fyn inhibition rescues established memory and synapse loss in A lzheimer mice. *Ann. Neurol.* 77 (6): 953–971.

105 Tata, A.M., Velluto, L., D'Angelo, C., and Reale, M. (2014). Cholinergic system dysfunction and neurodegenerative diseases: cause or effect? *CNS Neurol Disord Drug Targets.* 13 (7): 1294–1303.

106 Khoury, R., Grysman, N., Gold, J. et al. (2018). The role of 5 HT6-receptor antagonists in Alzheimer's disease: an update. *Expert Opin. Investig. Drugs* 27 (6): 523–533.

107 De Felice, F.G. (2013). Alzheimer's disease and insulin resistance: translating basic science into clinical applications. *J. Clin. Invest.* 123 (2): 531–539.

108 Blalock, E.M., Phelps, J.T., Pancani, T. et al. (2010). Effects of long-term pioglitazone treatment on peripheral and central markers of aging. *PLoS One* 5 (4): e10405.

109 Orhan, I. and Aslan, M. (2009). Appraisal of scopolamine-induced antiamnesic effect in mice and in vitro antiacetylcholinesterase and antioxidant activities of some traditionally used Lamiaceae plants. *J. Ethnopharmacol.* 122 (2): 327–332.

110 Ahmed, H.H., Salem, A.M., Sabry, G.M. et al. (2013). Possible therapeutic uses of Salvia triloba and Piper nigrum in Alzheimer's disease–induced rats. *J. Med. Food* 16 (5): 437–446.

111 Xu, P., Wang, K., Lu, C. et al. (2016). Protective effect of lavender oil on scopolamine induced cognitive deficits in mice and H_2O_2 induced cytotoxicity in PC12 cells. *J. Ethnopharmacol.* 193: 408–415.

112 Marzouk, M.M., Ibrahim, L.F., El-Hagrassi, A.M. et al. (2020). Phenolic profiling and anti-Alzheimer's evaluation of Eremobium aegyptiacum. *Adv. Trad. Med.* 20: 233–241.

113 Joshi, H. and Parle, M. (2006). Cholinergic basis of memory-strengthening effect of Foeniculum vulgare Linn. *J. Med. Food* 9 (3): 413–417.

114 Giridharan, V.V., Thandavarayan, R.A., Mani, V. et al. (2011). Ocimum sanctum Linn. leaf extracts inhibit acetylcholinesterase and improve cognition in rats with experimentally induced dementia. *J. Med. Food* 14 (9): 912–919.

115 Lee, A.Y., Choi, J.M., Lee, J. et al. (2016). Effects of vegetable oils with different fatty acid compositions on cognition and memory ability in A β 25–35-induced Alzheimer's disease mouse model. *J. Med. Food* 19 (10): 912–921.

116 Okello, E.J., Savelev, S.U., and Perry, E.K. (2004). In vitro anti-β-secretase and dual anti-cholinesterase activities of Camellia sinensis L.(tea) relevant to treatment of dementia. *Phytother. Res.* 18 (8): 624–627.

117 Okello, E.J., Leylabi, R., and McDougall, G.J. (2012). Inhibition of acetylcholinesterase by green and white tea and their simulated intestinal metabolites. *Food Funct.* 3 (6): 651–661.

118 Mathiyazahan, D.B., Thenmozhi, A.J., and Manivasagam, T. (2015). Protective effect of black tea extract against aluminium chloride-induced Alzheimer's disease in rats: A behavioural, biochemical and molecular approach. *J. Funct. Foods* 16: 423–435.

119 Ademosun, A.O. and Oboh, G. (2012). Inhibition of acetylcholinesterase activity and Fe^{2+}-induced lipid peroxidation in rat brain in vitro by some citrus fruit juices. *J. Med. Food* 15 (5): 428–434.

120 Jelenković, A., Jovanović, M.D., Stevanović, I. et al. (2014). Influence of the green tea leaf extract on neurotoxicity of aluminium chloride in rats. *Phytother. Res.* 28 (1): 82–87.

121 Karthivashan, G., Ganesan, P., Park, S.-Y. et al. (2018). Therapeutic strategies and nano-drug delivery applications in management of ageing Alzheimer's disease. *Drug Deliv.* 25 (1): 307–320.

122 Van Dam, D. and Deyn, D. (2011). Animal models in the drug discovery pipeline for Alzheimer's disease. *Br. J. Pharmacol.* 164 (4): 1285–1300.

123 Siddiqi, K.S., Husen, A., Sohrab, S.S., and Yassin, M.O. (2018). Recent status of nanomaterial fabrication and their potential applications in neurological disease management. *Nanoscale Res. Lett.* 13 (1): 1–17.

124 Cummings, B.J. and Cotman, C.W. (1995). Image analysis of β-amyloid load in Alzheimer's disease and relation to dementia severity. *Lancet* 346 (8989): 1524–1528.

125 Moongkarndi, P., Srisawat, C., Saetun, P. et al. (2010). Protective effect of mangosteen extract against β-amyloid-induced cytotoxicity, oxidative stress and altered proteome in SK-N-SH cells. *J. Proteome Res.* 9 (5): 2076–2086.

126 Hardy, J. and Selkoe, D.J. (2002). The amyloid hypothesis of Alzheimer's disease: progress and problems on the road to therapeutics. *Science* 297 (5580): 353–356.

127 Wogulis, M., Wright, S., Cunningham, D. et al. (2005). Nucleation-dependent polymerization is an essential component of amyloid-mediated neuronal cell death. *J. Neurosci.* 25 (5): 1071–1080.

128 Deshpande, A., Mina, E., Glabe, C., and Busciglio, J. (2006). Different conformations of amyloid β induce neurotoxicity by distinct mechanisms in human cortical neurons. *J. Neurosci.* 26 (22): 6011–6018.

129 Lesné, S., Koh, M.T., Kotilinek, L. et al. (2006). A specific amyloid-β protein assembly in the brain impairs memory. *Nature* 440 (7082): 352–357.

130 Pike, C.J., Walencewicz-Wasserman, A.J., Kosmoski, J. et al. (1995). Structure-activity analyses of β-amyloid peptides: contributions of the β25–35 region to aggregation and neurotoxicity. *J. Neurochem.* 64 (1): 253–265.

131 Harada, J. and Sugimoto, M. (1999). Activation of caspase-3 in β-amyloid-induced apoptosis of cultured rat cortical neurons. *Brain Res.* 842 (2): 311–323.

132 Ho, Y.-S., Yu, M.-S., Lai, C.S.-W. et al. (2007). Characterizing the neuroprotective effects of alkaline extract of Lycium barbarum on β-amyloid peptide neurotoxicity. *Brain Res.* 1158: 123–134.

133 Luo, H., Gu, F., and Li, X. (2003). Inhibiting effect of ethanol extract from Achyranthes bidentata on A beta 42 aggregation. *Zhong Yao Cai.* 26 (6): 412–415.

134 Singh, J.H., Alagarsamy, V., Diwan, P.V. et al. (2011). Neuroprotective effect of Alpinia galanga (L.) fractions on Aβ (25–35) induced amnesia in mice. *J. Ethnopharmacol.* 138 (1): 85–91.

135 Ramesh, B., Indi, S., and Rao, K. (2010). Anti-amyloidogenic property of leaf aqueous extract of Caesalpinia crista. *Neurosci. Lett.* 475 (2): 110–114.

136 Dhanasekaran, M., Holcomb, L.A., Hitt, A.R. et al. (2009). Centella asiatica extract selectively decreases amyloid β levels in hippocampus of Alzheimer's disease animal model. *Phytother. Res.* 23 (1): 14–19.

137 Frydman-Marom, A., Levin, A., Farfara, D. et al. (2011). Orally administrated cinnamon extract reduces β-amyloid oligomerization and corrects cognitive impairment in Alzheimer's disease animal models. *PLoS One* 6 (1): e16564.

138 Bai, Y., Tohda, C., Zhu, S. et al. (2011). Active components from Siberian ginseng (Eleutherococcus senticosus) for protection of amyloid β (25–35)-induced neuritic atrophy in cultured rat cortical neurons. *J. Nat. Med.* 65 (3): 417–423.

139 Lee, J.W., Lee, Y.K., Lee, B.J. et al. (2010). Inhibitory effect of ethanol extract of Magnolia officinalis and 4-O-methylhonokiol on memory impairment and neuronal toxicity induced by beta-amyloid. *Pharmacol. Biochem. Behav.* 95 (1): 31–40.

140 Chauhan, N.B. (2006). Effect of aged garlic extract on APP processing and tau phosphorylation in Alzheimer's transgenic model Tg2576. *J. Ethnopharmacol.* 108 (3): 385–394.

141 Papandreou, M.A., Kanakis, C.D., Polissiou, M.G. et al. (2006). Inhibitory activity on amyloid-β aggregation and antioxidant properties of Crocus sativus stigmas extract and its crocin constituents. *J. Agric. Food Chem.* 54 (23): 8762–8768.

142 Yang, H.J., Hwang, J.T., Kwon, D.Y. et al. (2013). Yuzu extract prevents cognitive decline and impaired glucose homeostasis in β-amyloid–infused rats. *J. Nutr.* 143 (7): 1093–1099.

143 Shih, P.-H., Chan, Y.-C., Liao, J.-W. et al. (2010). Antioxidant and cognitive promotion effects of anthocyanin-rich mulberry (Morus atropurpurea L.) on senescence-accelerated mice and prevention of Alzheimer's disease. *J. Nutr. Biochem.* 21 (7): 598–605.

144 Qosa, H., Mohamed, L.A., Batarseh, Y.S. et al. (2015). Extra-virgin olive oil attenuates amyloid-β and tau pathologies in the brains of TgSwDI mice. *J. Nutr. Biochem.* 26 (12): 1479–1490.

145 Jang, J.H., Kim, C.Y., Lim, S.H. et al. (2010). Neuroprotective effects of Triticum aestivum L. against β-Amyloid-induced cell death and memory impairments. *Phytother. Res.* 24 (1): 76–84.

146 Fujiwara, H., Tabuchi, M., Yamaguchi, T. et al. (2009). A traditional medicinal herb Paeonia suffruticosa and its active constituent 1, 2, 3, 4, 6-penta-O-galloyl-β-d-glucopyranose have potent anti-aggregation effects on Alzheimer's amyloid β proteins in vitro and in vivo. *J. Neurochem.* 109 (6): 1648–1657.

147 Figueiró, M., Ilha, J., Linck, V. et al. (2011). The Amazonian herbal Marapuama attenuates cognitive impairment and neuroglial degeneration in a mouse Alzheimer model. *Phytomedicine* 18 (4): 327–333.

148 Jeong, E.J., Lee, H.K., Lee, K.Y. et al. (2013). The effects of lignan-riched extract of Shisandra chinensis on amyloid-β-induced cognitive impairment and neurotoxicity in the cortex and hippocampus of mouse. *J. Ethnopharmacol.* 146 (1): 347–354.

149 Fujiwara, H., Iwasaki, K., Furukawa, K. et al. (2006). Uncaria rhynchophylla, a Chinese medicinal herb, has potent antiaggregation effects on Alzheimer's β-amyloid proteins. *J. Neurosci. Res.* 84 (2): 427–433.

150 Zhou, Y.-Q., Yang, Z.-L., Xu, L. et al. (2009). Akebia saponin D, a saponin component from Dipsacus asper Wall, protects PC 12 cells against amyloid-β induced cytotoxicity. *Cell Biol. Int.* 33 (10): 1102–1110.

151 Irie, Y. and Keung, W.M. (2003). Rhizoma acori graminei and its active principles protect PC-12 cells from the toxic effect of amyloid-β peptide. *Brain Res.* 963 (1-2): 282–289.

152 Boubakri, A., Leri, M., Bucciantini, M. et al. (2020). Allium roseum L. extract inhibits amyloid beta aggregation and toxicity involved in Alzheimer's disease. *PLoS One* 15 (9): e0223815.

153 Iuvone, T., De Filippis, D., Esposito, G. et al. (2006). The spice sage and its active ingredient rosmarinic acid protect PC12 cells from amyloid-beta peptide-induced neurotoxicity. *J. Pharmacol. Exp. Ther.* 317 (3): 1143–1149.

154 Bastianetto, S., Ramassamy, C., Doré, S. et al. (2000). The Ginkgo biloba extract (EGb 761) protects hippocampal neurons against cell death induced by beta-amyloid. *Eur. J. Neurosci.* 12 (6): 1882–1890.

155 Shytle, R.D., Tan, J., Bickford, P.C. et al. (2012). Optimized turmeric extract reduces β-Amyloid and phosphorylated Tau protein burden in Alzheimer's transgenic mice. *Curr. Alzheimer Res.* 9 (4): 500–506.

156 Martinez-Oliveira, P., de Oliveira, M.F., Alves, N. et al. (2018). Yacon leaf extract supplementation demonstrates neuroprotective effect against memory deficit related to β-amyloid-induced neurotoxicity. *J. Funct. Foods* 48: 665–675.

157 Adewusi, E.A. and Steenkamp, V. (2015). Medicinal plants and their derivatives with amyloid beta inhibitory activity as potential targets for drug discovery. *Asian Pac. J. Trop. Dis.* 5 (6): 430–440.

158 Zhang, L.-D., Ma, L., Zhang, L. et al. (2015). Hyperbaric oxygen and ginkgo biloba extract ameliorate cognitive and memory impairment via nuclear factor kappa-b pathway in rat model of alzheimer's disease. *Chin. Med. J.* 128 (22): 3088.

159 Brimson, J.M., Brimson, S.J., Brimson, C.A. et al. (2012). Rhinacanthus nasutus extracts prevent glutamate and amyloid-β neurotoxicity in HT-22 mouse hippocampal cells: possible active compounds include lupeol, stigmasterol and β-sitosterol. *Int. J. Mol. Sci.* 13 (4): 5074–5097.

160 Sehgal, N., Gupta, A., Valli, R.K. et al. (2012). Withania somnifera reverses Alzheimer's disease pathology by enhancing low-density lipoprotein receptor-related protein in liver. *Proc. Natl. Acad. Sci. U. S. A.* 109 (9): 3510–3515.

161 Pingali, U., Pilli, R., and Fatima, N. (2014). Effect of standardized aqueous extract of Withania somnifera on tests of cognitive and psychomotor performance in healthy human participants. *Pharm. Res.* 6 (1): 12–18.

162 Ratheesh, G., Tian, L., Venugopal, J.R. et al. (2017). Role of medicinal plants in neurodegenerative diseases. *Biomanuf. Rev.* 2 (1): 2.

163 Seo, J.S., Yun, J.H., Baek, I.S. et al. (2010). Oriental medicine Jangwonhwan reduces Abeta(1-42) level and beta-amyloid deposition in the brain of Tg-APPswe/PS1dE9 mouse model of Alzheimer disease. *J. Ethnopharmacol.* 128 (1): 206–212.

164 Giacobini, E. and Gold, G. (2013). Alzheimer disease therapy – moving from amyloid-β to tau. *Nat. Rev. Neurol.* 9 (12): 677–686.

165 Binder, L.I., Frankfurter, A., and Rebhun, L.I. (1985). The distribution of tau in the mammalian central nervous system. *J. Cell Biol.* 101 (4): 1371–1378.

166 Kosik, K.S. and Finch, E.A. (1987). MAP2 and tau segregate into dendritic and axonal domains after the elaboration of morphologically distinct neurites: an immunocytochemical study of cultured rat cerebrum. *J. Neurosci.* 7 (10): 3142–3153.

167 Brion, J.P., Smith, C., Couck, A.M. et al. (1993). Developmental changes in tau phosphorylation: fetal tau is transiently phosphorylated in a manner similar to paired helical filament-tau characteristic of Alzheimer's disease. *J. Neurochem.* 61 (6): 2071–2080.

168 Adelekan, M.L., Abiodun, O.A., Imouokhome-Obayan, A.O. et al. (1993). Psychosocial correlates of alcohol, tobacco and cannabis use: findings from a Nigerian university. *Drug Alcohol Depend.* 33 (3): 247–256.

169 Iqbal, K., Liu, F., Gong, C.-X. et al. (2009). Mechanisms of tau-induced neurodegeneration. *Acta Neuropathol.* 118 (1): 53–69.

170 Yang, S., Gong, Q., Wu, Q. et al. (2014). Alkaloids enriched extract from Dendrobium nobile Lindl. attenuates tau protein hyperphosphorylation and apoptosis induced by lipopolysaccharide in rat brain. *Phytomedicine* 21 (5): 712–716.

171 Peterson, D.W., George, R., Scaramozzino, F. et al. (2009). Cinnamon extract inhibits tau aggregation associated with Alzheimer's disease in vitro. *J. Alzheimer's Dis.* 17 (3): 585–597.

172 Tu, L.H., Ma, J., Liu, H.P. et al. (2009). The neuroprotective effects of ginsenosides on calcineurin activity and tau phosphorylation in SY5Y cells. *Cell. Mol. Neurobiol.* 29 (8): 1257–1264.

173 Shin, S.J., Park, Y.H., Jeon, S.G. et al. (2020). Red ginseng inhibits tau aggregation and promotes tau dissociation in vitro. *Oxidative Med. Cell. Longev.* 2020: 7829842.

174 Kt, C.H.O.I. (2008). Botanical characteristics, pharmacological effects and medicinal components of Korean Panax ginseng CA Meyer. *Acta Pharmacol. Sin.* 29 (9): 1109–1118.

175 Kim, H.G., Park, G., Lim, S. et al. (2015). Mori Fructus improves cognitive and neuronal dysfunction induced by beta-amyloid toxicity through the GSK-3β pathway in vitro and in vivo. *J. Ethnopharmacol.* 171: 196–204.

176 Radenahmad, N., Saleh, F., Sawangjaroen, K. et al. (2011). Young coconut juice, a potential therapeutic agent that could significantly reduce some pathologies associated with Alzheimer's disease: novel findings. *Br. J. Nutr.* 105 (5): 738–746.

177 Thenmozhi, A.J., Dhivyabharathi, M., Manivasagam, T., and Essa, M.M. (2016). Tannoid principles of Emblica officinalis attenuated aluminum chloride induced apoptosis by suppressing oxidative stress and tau pathology via Akt/GSK-3βsignaling pathway. *J. Ethnopharmacol.* 194: 20–29.

178 Takalo, M., Haapasalo, A., Martiskainen, H. et al. (2014). High-fat diet increases tau expression in the brain of T2DM and AD mice independently of peripheral metabolic status. *J. Nutr. Biochem.* 25 (6): 634–641.

179 Akram, M. and Nawaz, A. (2017). Effects of medicinal plants on Alzheimer's disease and memory deficits. *Neural Regen. Res.* 12 (4): 660–670.

180 Balez, R., Steiner, N., Engel, M. et al. (2016). Neuroprotective effects of apigenin against inflammation, neuronal excitability and apoptosis in an induced pluripotent stem cell model of Alzheimer's disease. *Sci. Rep.* 6 (1): 1–16.

181 Gauthier, S., Feldman, H.H., Schneider, L.S. et al. (2016). Efficacy and safety of tau-aggregation inhibitor therapy in patients with mild or moderate Alzheimer's disease: a randomised, controlled, double-blind, parallel-arm, phase 3 trial. *Lancet* 388 (10062): 2873–2884.

182 Zhang, S.Q., Obregon, D., Ehrhart, J. et al. (2013). Baicalein reduces β-amyloid and promotes nonamyloidogenic amyloid precursor protein processing in an Alzheimer's disease transgenic mouse model. *J. Neurosci. Res.* 91 (9): 1239–1246.

183 Ali, F., Jyoti, S., Naz, F. et al. (2019). Therapeutic potential of luteolin in transgenic Drosophila model of Alzheimer's disease. *Neurosci. Lett.* 692: 90–99.

184 Mani, R.J., Mittal, K., and Katare, D.P. (2018). Protective effects of quercetin in zebrafish model of Alzheimer's disease. *Asian J. Pharm.* 12 (2): S660.

185 Chetia, P., Mazumder, M.K., Mahanta, S. et al. (2020). A novel phytochemical from Dipteris wallichii inhibits human β-secretase 1: Implications for the treatment of Alzheimer's disease. *Med. Hypotheses* 143: 109839.

186 Li, D.-D., Zhang, Y.-H., Zhang, W., and Zhao, P. (2019). Meta-analysis of randomized controlled trials on the efficacy and safety of donepezil, galantamine, rivastigmine, and memantine for the treatment of Alzheimer's disease. *Front. Neurosci.* 13: 472.

187 Durairajan, S.S.K., Liu, L.-F., Lu, J.-H. et al. (2012). Berberine ameliorates β-amyloid pathology, gliosis, and cognitive impairment in an Alzheimer's disease transgenic mouse model. *Neurobiol. Aging* 33 (12): 2903–2919.

188 Laurent, C., Eddarkaoui, S., Derisbourg, M. et al. (2014). Beneficial effects of caffeine in a transgenic model of Alzheimer's disease-like tau pathology. *Neurobiol. Aging* 35 (9): 2079–2090.

189 Geissler, T., Brandt, W., Porzel, A. et al. (2010). Acetylcholinesterase inhibitors from the toadstool Cortinarius infractus. *Bioorg. Med. Chem.* 18 (6): 2173–2177.

190 Woo, J.-Y., Gu, W., Kim, K.-A. et al. (2014). Lactobacillus pentosus var. plantarum C29 ameliorates memory impairment and inflammaging in a d-galactose-induced accelerated aging mouse model. *Anaerobe* 27: 22–26.

191 Boonyaketgoson, S., Rukachaisirikul, V., Phongpaichit, S., and Trisuwan, K. (2018). Naphthoquinones from the leaves of Rhinacanthus nasutus having acetylcholinesterase inhibitory and cytotoxic activities. *Fitoterapia* 124: 206–210.

192 Sadegh Malvajerd, S., Izadi, Z., Azadi, A. et al. (2019). Neuroprotective potential of curcumin-loaded nanostructured lipid carrier in an animal model of Alzheimer's disease: behavioral and biochemical evidence. *J. Alzheimers Dis.* 69 (3): 671–686.

193 Ahmed, T., Javed, S., Javed, S. et al. (2017). Resveratrol and Alzheimer's disease: mechanistic insights. *Mol. Neurobiol.* 54 (4): 2622–2635.

194 Orhan, I.E., Senol, F.S., Shekfeh, S. et al. (2017). Pteryxin – a promising butyrylcholinesterase-inhibiting coumarin derivative from Mutellina purpurea. *Food Chem. Toxicol.* 109: 970–974.

195 Zhang, X.D., Liu, X.Q., Kim, Y.H., and Whang, W.K. (2014). Chemical constituents and their acetyl cholinesterase inhibitory and antioxidant activities from leaves of Acanthopanax henryi: potential complementary source against Alzheimer's disease. *Arch. Pharm. Res.* 37 (5): 606–616.

196 Geromichalos, G.D., Lamari, F.N., Papandreou, M.A. et al. (2012). Saffron as a source of novel acetylcholinesterase inhibitors: molecular docking and in vitro enzymatic studies. *J. Agric. Food Chem.* 60 (24): 6131–6138.

197 Drevets, W.C., Zarate, C.A., and Furey, M.L. (2013). Antidepressant effects of the muscarinic cholinergic receptor antagonist scopolamine: a review. *Biol. Psychiatry* 73 (12): 1156–1163.

198 Singh, D., Verma, V., and Kh, R. (2019). Sinapic acid mitigates intracerebroventricular streptozotocin induced oxidative stress and neuro-inflammatory changes in rats. *J. Neurol. Sci.* 405: 109.

4

Effects of Medicinal Plants and Phytochemicals on Schizophrenia

Nithya Rani Raju[1], *S.V. Rashmitha*[1], *S. Pavithra*[1], *Erika Amparo Torres*[2], *M. Kishor*[3], *and Raghu Ram Achar*[1]

[1] *Division of Biochemistry, School of Life Sciences, JSS Academy of Higher Education and Research, Mysuru, Karnataka, India*
[2] *Departament d' Enginyeria Química, Universitat Rovira i Virgili, Av Països Catalans 26, 43007, Tarragona, Spain*
[3] *Department of Psychiatry, JSS Medical College & Hospital, JSS Academy of Higher Education & Research, Mysuru, Karnataka, India*

Schizophrenia is a chronic mental condition characterized by a diverse hereditary and neurobiologic foundation that affects early brain development and manifests as a combination of positive, negative, and cognitive psychotic symptoms [1]. Schizophrenia is referred to as Unmada in Ayurveda. Unmada is the Manasvyadhi, according to Acharya Charak, which is defined as an unsettled state of mind, understanding, consciousness, perception, memory, inclination, character, and behavior [2]. The signs and symptoms of schizophrenia are diverse; they include disorders of perception (hallucinations), inferential thinking (delusions), goal-directed behavior (avolition), and emotional expression (affective blunting), to mention only a few (Figure 4.1) [3], [4].

Schizophrenia is the product of thousands of genes interacting with various environmental risk factors [5]. It is becoming more well recognized as a modest neurodevelopmental issue of brain connection or the wiring of our brain's functioning circuits [6]. Depressive disorders, anxiety disorders, schizophrenia, and bipolar disorder are all on the rise in India. The majority of those suffering from schizophrenia are in the 35–44-yearage group, and the majority of suicidal deaths are linked to schizophrenic and depressive individuals [7].

4.1 Mechanisms of Action Related to Schizophrenia

Schizophrenia is a complex mental disorder with multiple symptoms involved in its diagnosis. Some studies have shown that the development of schizophrenia starts during cortical neurogenesis [8, 9], as a result of maternal immune activation and stressful life events during mid-pregnancy [10, 11], which produce cortical layer disorganization, involving glutamatergic and GABAergic neurons, and reduced γ-aminobutyric acid (GABA) release [12]; it is also related to an excitatory–inhibitory synaptic imbalance, usually with an increased number of inhibitory neurons [13–15]. The disturbances in GABA neurons in the prefrontal cortex have been linked to cognitive deficits in schizophrenia [16].

Phytochemical Drug Discovery for Central Nervous System Disorders: Biochemistry and Therapeutic Effects, First Edition. Edited by Chukwuebuka Egbuna and Mithun Rudrapal.

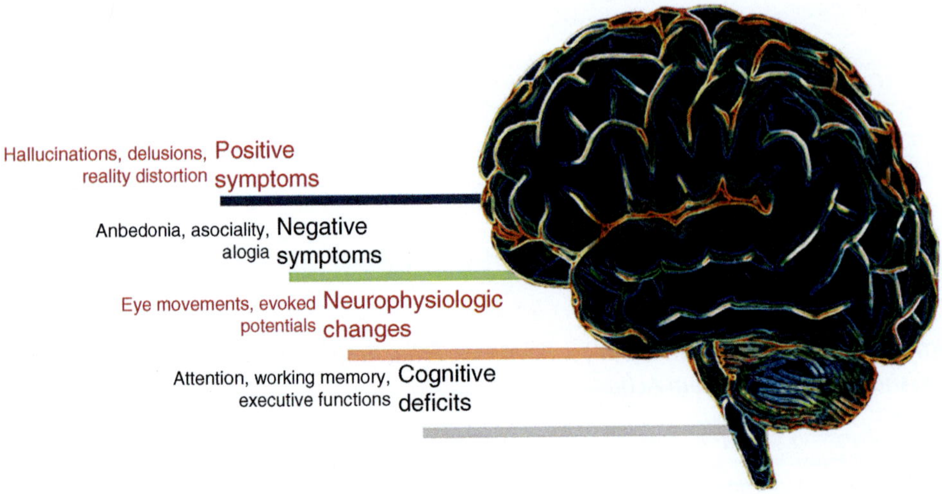

Figure 4.1 Symptoms of schizophrenia.

Some of the mechanisms of action related to schizophrenia are vesicular monoamine transporters (VMATs), dopamine receptors and transporters, serotonin receptors and transporters, vesicular glutamate transporter (VGTs), GABA level and receptors. and gene expression-nuclear factor.

4.1.1 Vesicular Monoamine Transporters

VMATs are responsible for the uptake of cytosolic monoamines into synaptic vesicles in monoaminergic neurons, and there are two identified isoforms, VMAT1 and VMAT2. The presynaptic VMAT1 modulates emotion processing [17, 18] and it can play a key role in the survival of hippocampal neurons, thus might contribute to the neurocognitive deficits observed in neuropsychiatric disorders [19].

On the other hand, VMAT2 plays a major role in the synaptic accumulation and release of monoamines; it is also present in the membrane of secretory vesicles and transports dopamine, norepinephrine (NE), serotonin, histamine, glutamate (Glu), and GABA into vesicles for presynaptic release [20], and is expressed primarily in the central nervous system. In schizophrenic patients, it was found that platelet VMAT2 density was increased [21]. Also, an increase in VMAT2 expression was found in the ventral brainstem of schizophrenic patients [22]. The increase of VMAT2 could lead to a potential functional overactivity of brain dopaminergic, serotonergic, and noradrenergic pathways, and can be related to the hyperactivity symptoms of schizophrenia. New drugs have been developed to inhibit VMAT2, and suppress antipsychotic symptoms and hyperlocomotion/tardive dyskinesia [20, 23]. Also, there is a study that suggests that the synaptic vesicle monoamine transporter gene (SVMT) gene, also known as VMAT2 or SLC18A2, is susceptible to schizophrenia [24]. Both VMAT1 and VMAT2 affect/regulate dopamine release [18, 25].

4.1.2 Dopamine Receptors/Transporters

Schizophrenia is thought to include dopaminergic imbalance, with the dysfunction of dopamine D2 receptors and dopamine transporter (DAT) (Figure 4.2) [26, 27]. The genes *DRD2* and *GRIN2A* are involved in schizophrenia development by the dopamine D2 receptor (DRD2) [28].

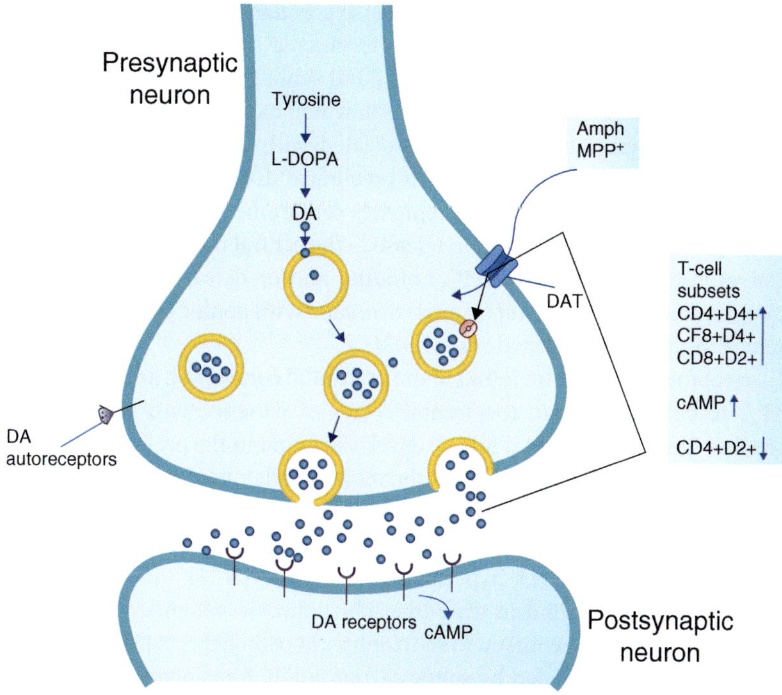

Figure 4.2 Schematic representation of dopamine synaptic terminals. Dopamine transporters are localized to perisynaptic sites, where they are crucial for the termination of dopamine transmission and the maintenance of presynaptic dopamine storage. Amph, amphetamine; cAMP, cyclic adenosine monophosphate; DA, dopamine; DAT, dopamine transporter; L-DOPA, L-3,4-dihydroxyphenylalanine; MPP+, 1-methyl-4-phenylpyridinium.

The modulation of the immune response by neurotransmitters or neuroimmunologic interaction is correlated with the presence of specific receptors on the surface of leukocytes. The expression of dopamine (D2 and D4) was related to T-cell subsets (CD4 and CD8) in patients with schizophrenia, in whom percentages of CD4+D4+, CD8+D4+, and CD8+D2+ cells were higher, while the percentage of CD4+D2 cells was lower. These results were linked with clinical symptoms; the Brief Psychiatric Rating Scale (BPRS), the Positive and Negative Syndrome Scale (PANSS, the PANSS-negative only), and the Abnormal Involuntary Movement Scale (AIMS) were positively related to CD8+D2+ cells, while AIMS was inversely related to CD4+D4+ cells [29]. Another report showed modulatory activity of dopamine over CD4+ and CD8+ cells through the increase of intracellular cyclic adenosine monophosphate (cAMP) by activation of D1, D2, and D3 receptors [30, 31]. It has been found that the DRD2 rs4436578*C genetic variant might have a protective role against schizophrenia in Armenians [32]. The A2B1D2 haplotype of *DRD2* genes (Taq 1 A, B, and D) may be protective against schizophrenia in the North Indian population, while the B2 allele of rs1079597 of *DRD2* genes (Taq 1B) may increase the risk of schizophrenia in the same population [33].

4.1.3 Serotonin Receptors/Transporters

Serotonin or 5-hydroxytryptamine (5-HT) is a monoamine neurotransmitter that regulates the function of the central nervous system, and is essential for learning, cognition, attention, memory, and mood. It was found that schizophrenic patients with severe depression have lower platelet serotonin concentrations, showing an inverse correlation between these two factors [33]. There are 5-HT transporters located on presynaptic serotonergic terminals; 5-HT transporters, as well as the receptors 5-HT_1–5-HT_7 and some genes like *5-HTT*, have been associated with the serotonergic system, alterations in which have been linked to schizophrenia [34–38]. Low brain serotonin levels are associated with cognitive dysfunction, memory deficit, and depression [39]. According to a preclinical assay using a rat model, stimulation of serotonin-1A receptors (5-HT_{1A}) suppresses serotonin, while stimulation of serotonin-1B receptors (5-HT_{1B}) reduces serotonin release in the central nervous system [40]. Decreased frontal serotonin-2A receptor (5-HT_{2A}) binding and/or density was found in patients with schizophrenia, which in one report was correlated with poorer performance in spatial working memory and executive functions [41–43].

A reduced 5-HT_{2A} receptor level was also found in the prefrontal cortex [44], and a reduced expression of 5-HT_{2A} receptor mRNA in the frontal cortex of patients with schizophrenia [45–47]. On the other hand, an increased 5-HT_{1A} level was found in the prefrontal cortex in schizophrenia [44, 45]. The *5-HTT* gene is one of the major regulators of serotonin in the synaptic cleft; an association was found between two STin2-containing haplotype combinations of the *5-HTT* gene and schizophrenia [48], as well as this gene being linked to schizophrenia in previous reports [49, 50]. DNA hypermethylation of the 5-HTT promoter and reduced expression of 5-HTT were found in drug-free schizophrenic patients, which suggests that hypoactivity of 5-HTT may be linked to schizophrenia pathogenesis [51].

The serotonin transporter SERT, encoded by solute carrier family 6 member 4 (SLC6A4), regulates synaptic concentrations of serotonin and is considered an index of serotonergic innervation. A negative correlation was found of SERT availability in the anterior nucleus of the left thalamus with negative symptoms of schizophrenia, which may suggest a role for the anterior thalamic nucleus in the pathophysiology of symptoms of this disease [52]. The rs140700G/A allele frequency of SLC6A4 was lower in patients with schizophrenia, and it was strongly associated with schizophrenia susceptibility and symptoms in the Chinese Han people [53]. Reduced expression of SERT protein mRNA was found in the temporal and frontal cortex, which could produce changes in serotonin neurotransmission (Figure 4.3) [31, 47]. The mitogen-activated protein kinase-3 (*MAPK3*) gene participates significantly in schizophrenia pathogenesis, is directly associated with schizophrenia pathogenesis and the serotonergic system, and is involved in 11 out of 16 schizophrenia-associated pathways [54].

4.1.4 Glutamate Transporters

The serotonin receptors $5HT_{1A}$ and $5HT_4$ were correlated with glutamate receptor genes (*NMDA* and *AMPA*) expression, and hippocampal volume was reduced, especially in the CA4 and dentate gyrus regions, in patients in first-episode psychosis, which reveals the mechanistic relationship between serotonin and glutamate and the hippocampal morphology in patients with schizophrenia (Figure 4.4) [55, 56]. Increased glutamate levels in the left thalamus can be implicated in schizophrenia pathophysiology, and glutamatergic neurotransmission or metabolism may be involved in the cognitive domains of attention and spatial working memory [57].

4.1.5 Gamma-Aminobutyric Acid Level/Receptors

GABA is a primary inhibitory neurotransmitter that mostly coordinates neural activity to facilitate cognitive processes and memory performance. Prefrontal cortical alterations of glutamate and GABA neurotransmission have been associated with schizophrenia [58].

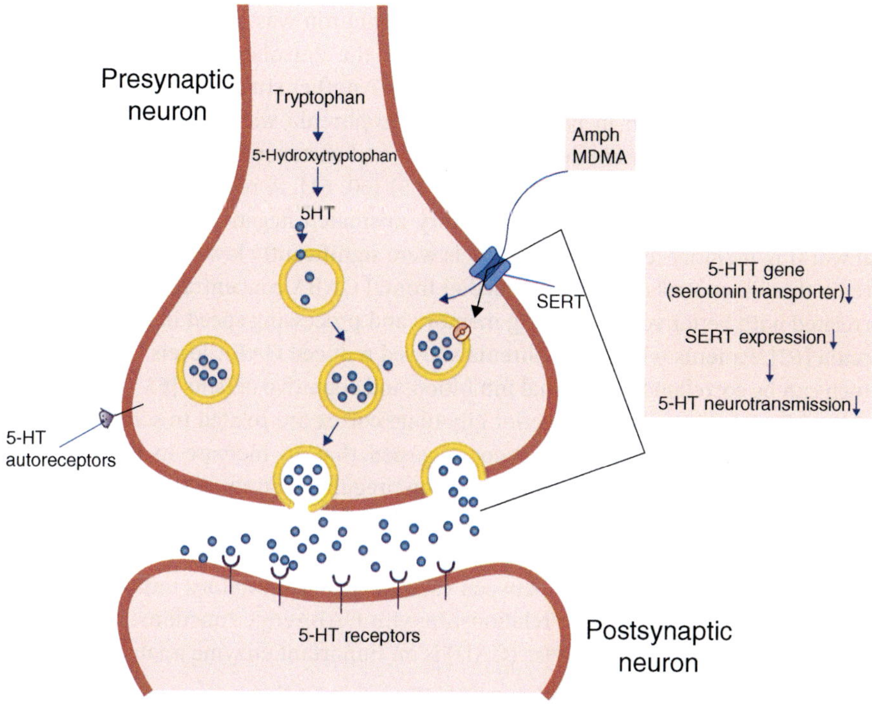

Figure 4.3 Schematic representation of 5-HT synaptic terminals. 5-HT transporters are localized to perisynaptic sites, where they are crucial for the termination of serotonin transmission and the maintenance of presynaptic dopamine storage. 5-HT, 5-hydroxytryptamine; Amph, amphetamine; MDMA, (+)-3,4-methylenedioxymethamphetamine; SERT, 5-HT transporter.

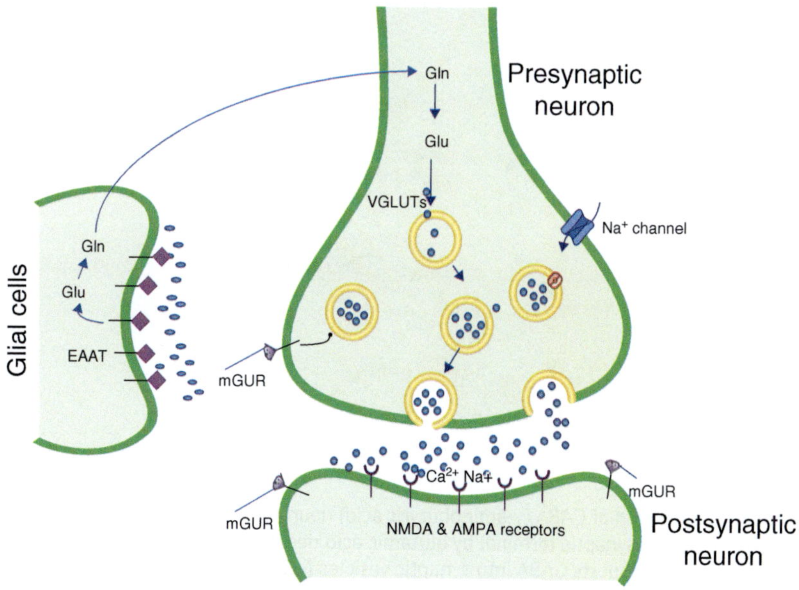

Figure 4.4 Schematic representation of 5-HT synaptic terminals. Glutaminase converts glutamine (Gln) to glutamate (Glu). The vesicular Glu transporters (VGLUTs) package Glu into presynaptic vesicles and release it from the cell in an activity-dependent manner. Excitatory amino acid transporters (EAATs), which are mostly found on glial cells, remove Glu from the extracellular space. Gln synthetase converts Glu to Gln in glial cells. Glu receptors are AMPA (amino-3-hydroxy-5-methyl-4-isoxazole propionic acid), NMDA (N-methyl-d-aspartate), and mGluRs (metabotropic Glu receptors).

The distribution of different classes of GABAergic interneuron was altered in people with schizophrenia, and higher concentrations of GABA in the dorsolateral prefrontal cortex were related to bad working memory performance [59]. Another study reported that working memory dysfunction shown in patients with schizophrenia was mediated by altered GABA neurotransmission in certain dorsolateral prefrontal cortex microcircuits, and it was related to the $GABA_A$ receptor expression (Figure 4.5) [60, 61]. A relationship was found between the glutamatergic system and the auditory mismatch negativity wave amplitude and verbal working memory test; glutamate levels were significantly lower in patients with schizophrenia than in controls, and higher medial frontal GABA concentration was significantly correlated with better verbal working memory and processing speed in patients with schizophrenia [62]. Patients with schizophrenia showed reduced GABA levels in the visual cortex, which can be correlated with visual inhibition and cognitive deficits [63, 64].

Low levels of GABA in the dorsal anterior cingulate cortex are related to schizophrenia and psychotic disorder [57]. It has also been suggested that an increase in the GABAA receptor density and a decreased cortical GABA (downregulated stimulation) in the posterior cingulate cortex of the brain are related to schizophrenia [65]. The high GABA levels of perisylvian cortices (right hemisphere) in patients with schizophrenia could be a compensatory mechanism to obtain the balance between excitatory and inhibitory impulses in the cerebral cortex, and may have a direct relationship with GABAergic functions in auditory cortices [66]. Glutamic acid decarboxylase (GAD) is an important enzyme related to GABA

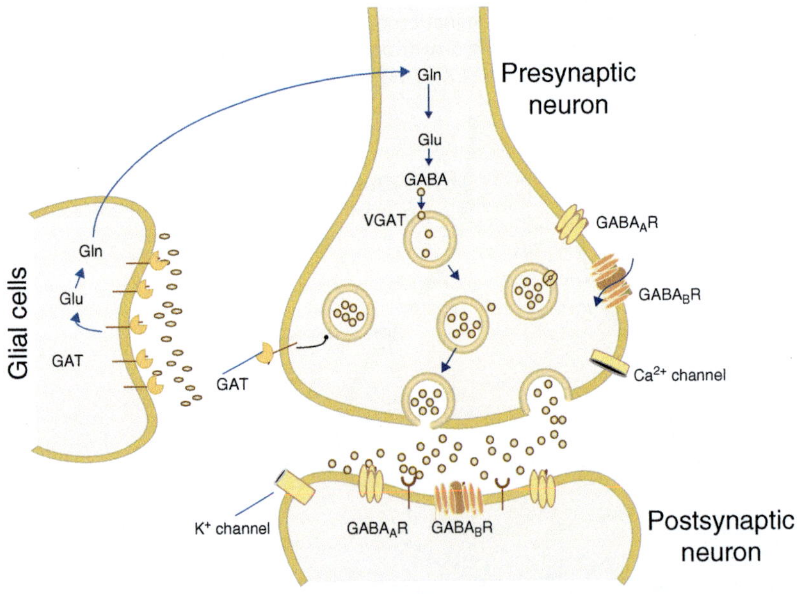

Figure 4.5 Schematic representation of GABA (γ-aminobutyric acid) neurotransmission. GABA is produced from glutamate in the presynaptic terminal by glutamic acid decarboxylase (GAD). The vesicular GABA transporter then transports GABA into synaptic vesicles (vGAT). GABA is released into the synapse during membrane depolarization and can bind to either ionotropic GABAA receptors ($GABA_AR$) or metabotropic GABAB receptors ($GABA_BR$) on the postsynaptic membrane, inhibiting the postsynaptic neuron. Membrane-bound GABA transporters (GATs), found in neurons and astrocytes, clear released GABA from the synapse. GABA is recycled into synaptic vesicles in astrocytes or taken up by mitochondria, where it is converted to glutamine by GABA transaminase (GABA-T) for neuronal absorption.

synthesis; patients with schizophrenia showed a low GAD level in GAD-immunoreactivity assays (GAD-ir) in the hippocampal CA1 field, superior temporal gyrus (STG) layer V, and mediodorsal (MD) region of the brain, and patients with schizophrenia and paranoid symptoms showed even lower GAD-ir levels in the hippocampal CA1 field, STG layers III and V, MD, and dentate gyrus, suggesting a relationship between GABAergic dysfunction and schizophrenia [67, 68].

A reduced expression or absence of the fragile X mental retardation protein (FMRP) in the cerebellum of patients with schizophrenia is probably related to unregulated mGluR5 signaling, resulting in the behavioral and intellectual impairments associated with schizophrenia [69]. The downregulation of three GABA receptor subunits of type A (GABRA1, GABRA2, and GABRB3) may contribute to the pathophysiology and clinical manifestations of schizophrenia by altering GABAergic neurotransmission and oscillation synchronization in the STG [70]. The calcium voltage-gated channel subunit alpha1 C (*CACNA1C*) gene is highly and directly associated with schizophrenia pathogenesis and the GABAergic system [54].

4.1.6 Genes Related to Schizophrenia

Multiple genes have been related to schizophrenia. Some of them are as follows:

- The adenosine triphosphate (ATP) synthase membrane subunit c locus 1 (*ATP5G1* or *ATP5MC1*) gene encodes the c unit of the Fo component of the ATP synthase in oxidative phosphorylation that catalyzes ATP production. A reduced ATP5G1 expression level was found in schizophrenic patients [71].
- There are 98 genes that have shown significant correlations with gray matter volume changes in schizophrenia, including *HAPLN4*, *RILPL2*, and *GRIA1* [72].
- Within the tachykinin-3 (TAC3) pathway, related to the neuroendocrine system, 19 genes may contribute to schizophrenia and may be part of a larger pathway [28].
- Mineralocorticoid receptor (MR, encoded by gene *NR3C2*), a deficient prefrontal MR mRNA expression, and epigenetic aberrations of *NR3C2* that generate a reduced DNA methylation reaction can be related to schizophrenia [73, 74].
- The microRNA (miRNA) gene (*MIR137*) is associated with the regulation of synaptogenesis, synapse maturation, and synaptic transmission. *MIR137* overexpression generated reduced synapse formation and reduced synaptic transmission, which are linked to schizophrenia [75–77].
- The guanine nucleotide-binding protein subunit beta-like protein 1 gene (*GNB1L*) maps to the deleted region on chromosome 22. Region 22q11.2 deletion syndrome [78], reduced expression of *GNB1L* in the prefrontal cortex, or mutations in that gene can increase the risk of developing schizophrenia [79–82].
- Nogo-A protein, encoded by the reticulon 4 gene (*RTN4*), is important for myelination of the neurons and neuron migration during nervous system development [83, 84]. Reduced levels of Nogo-A in serum were found in patients with schizophrenia [85], and different haplotypes in the *RTN4* gene could affect gene transcription and be related to schizophrenia [55, 86].
- The combination of guanine nucleotide-binding protein G(i), the α-1 subunit gene (*GNAI1*), tyrosine-protein kinase genes (*FYN* and *LYN*), protein kinase C α gene (*PRKCA*), protein kinase C β gene (*PRKCB*), and tyrosine 3-monooxygenase/tryptophan 5-monooxygenase activation protein zeta gene (*YWHAZ*), expressed in the peripheral blood, can be used to diagnose schizophrenia [87].

- Neurogenic locus notch homolog protein 4, coded by the *NOTCH4* gene, is located on chromosome 6p, and is relevant in cell signaling pathways and cell differentiation. The *NOTH4* gene has been associated with schizophrenia [88–91].

Neuroactive ligand–receptor interaction is the most central pathway related to schizophrenia [54].

4.2 Ayurvedic Plants Used as Treatment for Schizophrenia and Related Psychoses

4.2.1 *Allium cepa*

The onion (*Allium cepa*), a member of the Amaryllidaceae family and the genus Allium, is a popular spice all over the world. It is thought to have originated in Asia's southwest [92]. Onion contains numerous phyto molecules such as polyphenolic substances, phenolic acids, flavonoids, ascorbic acid, and sulfur compounds that have antitoxic, anticarcinogenic, antiasthmatic, antithrombotic, antiplatelet, antibiotic, antidiabetic, hypotensive, hypolipidemic, and uric acid-lowering properties [92, 93]. In a study, it was discovered that its extract suppresses dopaminergic neurotransmission in Wister rats via acting on D2 receptors [93]. N-acetylcysteine (NAC), an antioxidant derived from Allium sp., has been shown to serve as a glutamatergic neurotransmission antagonist [94]. NAC has been proven to improve clozapine reactions in non-responsive schizophrenic individuals [95].

4.2.2 *Acorus calamus*. Linn

Acorus calamus Linn. (Araceae) is a perennial aromatic herb with creeping rhizomes that grows semi-aquatically. Sweet flag and calamus are common names for this plant. *A. calamus* thrives in temperate and subtropical regions in Asia, North America, and Europe [96]. Pharmacologic studies have revealed that the *Acorus* rhizome and its constituents have a wide range of pharmacologic activities, including sedative, central nervous system (CNS) depressant, behavior-modifying, anticonvulsant, acetylcholinesterase-inhibitory, memory-enhancing, anti-inflammatory, antioxidant, immunosuppressive, cytoprotective, antioxidant, genotoxic, and mutagenic activities [96]. Glycosides, flavonoids, saponins, tannins, polyphenolic compounds, mucilage, volatile oil, and a bitter principle have all been discovered in *A. calamus* phytochemical investigations. There have also been reports of glucosides, alkaloids, and essential oil. The bioactive components of the plant include acorenone, isoacorone, (Z)-sesquilavandulol, dehydroxy-isocalamendiol, b-asarone, isoacorone, crypto-acorone, (Z)-asarone, (Z)-methyl isoeugenol, (E)-caryophyllene, a-humulene, germacrene, linalool, camphor, b-gurjunene, preisocalamenediol, acorenone, shyobunone, cryptoacorone, and acorafuran-a new cadalin-type sesquiterpenoid [97]. The adrenergic, dopaminergic, and GABAergic systems are thought to be involved in the antidepressant effect of methanolic or hydroalcoholic extracts [98, 99]. The antidepressant activity of *A. calamus* could be linked to the presence of flavonoids, which modulate the central neurochemical and hypothalamic–pituitary–adrenal axis [100–102]. The antipsychotic properties of asarone and alstonine have been investigated. Different fractions of *A. calamus* L. are dose-dependently efficient in avoiding memory impairment and stress development through reducing oxidative stress and inflammation [103].

4.2.3 *Bacopa monnieri*

Bacopa monnieri (also known as water hyssop, Brahmi, herb of grace, and Indian pennywort) is a perennial creeping herb endemic to India's moist plains, mainly in the northeast and south [104]. Brahmi appears to have numerous roles in the CNS; this plant shows promise in improving memory [105, 106] and learning skills [107, 108] and to be neuroprotective [109]; also, it is relevant as a possible treatment for schizophrenia [110], depression [111, 112], dementia [113], and neuroinflammation [114] *B. monnieri* has antidepressant properties in addition to its wide role in healing diseased brains, and its plant extracts are thought to have these properties [115]. Some constituents of Brahmi extracts include bacoside A and B, bacopasaponin C, bacopasides I and II, and other phytochemicals such as alkaloids, glycosides, flavonoids, and saponins [116].

A standardized extract of *B. monnieri* containing 20% of bacosides A + B showed neuroprotective effects in a rat model of schizophrenia, acting over the cerebral vesicular glutamate transporter type 3 [110]. An ethanolic extract of *B. monnieri* containing 5% (w/w) of total saponins, a mixture of bacoside A_3, bacopaside II, bacopasaponin X, bacopasaponin C, and bacopaside I, had potential as a cognitive enhancer and neuroprotectant against Alzheimer's disease, showing the ability to mitigate the reduction of neurons and cholinergic neuron densities [117], and affects the nuclear factor erythroid 2 related factor 2 (Nrf2) pathway attenuating memory dysfunction in rats [105]. A methanolic extract of *B. monnieri* whose main components correspond to bacopasides, including the major component bacoside A_3 followed by bacopasaponin C and bacopaside ll, was reported to have a strong antidepressant effect in a subacute dose in mice [118]. This suggests the potential role of Brahmi in clinical treatment for depression, which can be attributed to the bacoside A_3 [2]. This effect is mediated by nitric oxide synthase inhibition and the serotonergic, dopaminergic, and adenosinergic pathways [111, 112]. An extract of *B. monniera* given chronically to humans for 12 weeks improved the speed of early information processing, mental control, memory acquisition, memory consolidation, logical memory, verbal learning rate, paired associate learning, and retention [107, 119, 120]. This effect can be related to *B. monniera*'s capacity to increase cortical, striatal, and hippocampal superoxide dismutase (SOD), catalase (CAT), and glutathione peroxidase (GPx) activities, which are responsible for an increase in antioxidant activity [112, 121]. An ethanolic extract of *B. monnieri* containing a mixture of bacoside A_3, bacopaside II, bacopasaponin X, bacopasaponin C, and bacopaside I can improve attention, cognitive processing, and working memory, partly via the suppression of acetylcholinesterase (AChE) activity in humans. Aqueous extracts of *B. monnieri* have shown anti-inflammatory effects on the CNS by in vitro assays, targeting neuroinflammatory mediators like interleukin (IL)-6 and tumor necrosis factor (TNF)-α [108, 114].

4.2.4 *Carum carvi*

Carum carvi Linn. (Zeera Siyah) is one of the world's oldest cultivated herbs, with origins in Asia, Africa, and Europe. It is a glabrous annual or biennial herb with narrow-tipped cylindrical fruits that grows 30–100 cm tall. *C. carvi* grows wild in India to a limited extent in the temperate western Himalayan districts [122, 123]. According to the Unani school of medicine, the seeds of this plant are cooling to the brain, astringent to the bowel, expectorant, pectoral, tonic, strengthen the sight, and heal eructation. Its oil can be synthesized into thymol, which is used as an anthelmintic, antibiotic, and an ingredient in a variety of patent medicines [124]. The chemical composition of the essential oil extracted from the entire *C. carvi* plant is as follows: α-pinene, β-pinene, camphene, myrcene, δ-3-carene, limonene, γ-terpinene, ρ-cymene, cadinene, myristicin, carvyl and dihydrocarvyl acetate, dihydrocarvone, carvone, terpinen-4-ol, dihydrocaveol, and perillyl alcohol [123]. Caraway fruits include essential oil, fatty acids, protein, glucose, phenolic acids (caffeic acids), and flavonoids (quercetin, kaempferol) [125]. The extract regulates central neurotransmitters and neuromodulators such as dopamine and activates the cholinergic system by lowering AChE levels [123].

4.2.5 *Celastrus panicutalus*

Celastrus panicutalus, often known as jyotishmati or "tree of life," is a very important Ayurvedic herb. This plant can be found up to 2000 m above sea level in Central India, in the western and eastern ghats, and up to 1500 m above sea level in Bihar and Orissa's Rajmahal hills [126, 127]. *C. pinaculatus* has numerous health benefits. The oil derived from the seeds has relaxing characteristics and is utilized in Unani and Ayurvedic medicine as a traditional treatment [128]. Fatty acids such as oleic acid, linoleic acid, palmitic acid, stearic crude lignoceric acid, benzoic acid, and acetic acid are found in oils from *C. paniculatus* seeds [129]. Jyotishmati seed extract's anti-inflammatory and antinociceptive effects are mediated by brain pathways [130]. The plant's seed oil abolished scopolamine's effects on navigational memory [131]. Celastrus oil improved cognitive function and memory in rats. It promotes passive avoidance of learning and memory while lowering biogenic amines, noradrenaline, dopamine, and serotonin levels in the brain [132]. Scopolamine causes a considerable rise in oxidative stress in the brain (increase in thiobarbituric acid-reacting substances [TBARS] and decrease in glutathione [GSH]) [132]. In rats, an aqueous extract of *C. panicutalus* (concentration 200–300 mg/kg body mass) revealed cognitive-enhancing benefits, presumably due to its antioxidant activity [133]. The aqueous extract of *C. panicutalus* was also found to cause a large drop in malondialdehyde (MDA) levels in the brain, as well as significant increases in GSH and CAT levels [134].

4.2.6 *Centella asiatica*

Centella asiatica, a clonal perennial herb in the umbelliferous family (Apiaceae), is widespread across India, growing in humid areas up to 1800 m altitude, from the Himalayan foothills to Sri Lanka (600 m altitude) [135]. It can be found growing in marshy places in most tropical and subtropical nations, including sections of India, Pakistan, Sri Lanka, Madagascar, South Africa, the South Pacific, and Eastern Europe [136]. Mandukaparni is one such valuable medicinal herb. The literal meaning of mandukaparni is a plant with leaves resembling the shape of frogs [137]. It also indicates that the stolon develops in the same way that a jumping frog does. It is a stoloniferous creeping herb with nodes that root. The whole plant is reported to be nervine, cardiotonic, astringent, and diuretic [136, 138]. The herb is utilized in its entirety for medical purposes, as a blood purifier, as well as a treatment for hypertension, memory enhancement, and lifespan. This is a popular herb in Ayurveda for reviving nerves and brain cells [136]. So far, isoprenoids (sesquiterpenes, sterols, pentacyclic triterpenoids, and saponins) and phenylpropanoid derivatives have been found in *C. asiatica* (eugenol derivatives, caffeoylquinic acids, and flavonoids) [139]. The relevance of *C. asiatica* in the Ayurvedic tradition is as a "medhya rasayana" plant, which promotes memory, prevents cognitive deficiencies, and improves brain function. The entire plant, particularly the leaves, is utilized in Ayurvedic medicine [140, 141]. Triterpenes from *C. asiatica* exhibit antidepressant effects [142]. Corticosterone's antidepressant effects were measured in the brain, and corticosterone levels were significantly reduced while monoamine neurotransmitters increased. GABA levels in the brain are also increased by this plant [143].

4.2.7 *Convolvulus pluricaulis*

In Ayurvedic scriptures, *Convolvulus pluricaulis* (Shankhpushpi) is referred to as a medhya rasayana. It is a perennial herb endemic to the southern provinces of India [144]. It occurs in wastelands in the Punjab, Bihar, and Chotanagpur plains. *Convolvulus pluricaulis* is a well-known Ayurvedic medicine that has been used as a brain tonic and nervine tonic [145]. Due to its clinical antianxiety impact and increased mental function, it is also beneficial in anxiety and neurosis. It was highly regarded by ancient physicians as a fantastic nervine tonic and memory invigorator and was utilized in cerebral irregularities, epilepsy, and insomnia [146]. The medicinal effects of the herb are known to extend to all parts of the plant [145]. Shankhpushpi contains carbohydrate-D-glucose, rhamnose, maltose, sucrose, and starch as chemical components [145]. In animal studies, alcoholic and aqueous extracts of *Convolvulus pluricaulis* revealed GABAA agonist characteristics, as well as being neuroprotective, antidepressant, increasing cortical hippocampus GABA levels, and improving cholinergic function. Kaempferol, phytosterol, and β-sitosterol are some of the molecules linked to those functions [144]. This plant regulates stress hormones, adrenaline, and cortisol. It relieves stress by correcting hyperglycemia, plasma corticosterone, creatine kinase, and adrenal hypertrophy, all of which are damaged by stress [140].

4.2.8 *Coriandrum sativum*

Coriander (*Coriandrum sativum* L.) is a member of the Apiaceae family and the genus Coriandrum L. [147]. Historically from Italy, *C. sativum* is now produced across Central and Eastern Europe [148]. Coriander is one of the most often used spices in the world of cooking. It is also used to treat anorexia, dyspepsia, flatulence, diarrhea, discomfort, and vomiting, among other gastrointestinal issues [148, 149]. Antimicrobial, antioxidant, antidiabetic, anxiolytic, antiepileptic, antidepressant, antimutagenic, anticonvulsant, anticancer, analgesic, migraine-relieving, anti-inflammatory, antidyslipidemic, antihypertensive, neuroprotective, and diuretic properties have been attributed to different parts of this herb due to the presence of a multitude of bioactive compounds [149]. The chemical makeup of Argentinean and European coriander found that linalool was 72.3% and 77.7%, respectively, whereas pinene was 5.9% and 4.4%, terpinene 4.7% and 5.6%, camphor 4.6% and 2.4%, and limonene 2.0% and 0.9% [150]. Coriander's leaves and fruits (seeds) are both edible and medicinal. The gene synthesis of the GABAA receptor 1 subunit, which is implicated in the endogenous sleep pathway, was increased by coriander leaf. Gene expression of the GABA transporter and the N-methyl-D-aspartate (NMDA) receptor, on the other hand, was reduced. These data imply that coriander leaf extract therapy raises GABA levels in the synaptic cleft while lowering excitatory activity [151].

4.2.9 *Cuminum cyminum* L.

Cumin (*Cuminum cyminum L.*) is a plant that originated in the Mediterranean region, Turkistan, and Egypt, but has since expanded to dry and semi-arid locations around the world, including the Middle East, India, and Turkey [152]. Cumin is the most extensively utilized Apiaceae plant because of its multifunctional medicinal properties, including antibacterial, antioxidant, and antidiabetic properties, as well as being a good source of vitamin B complex and minerals like iron, calcium, and zinc. Cumin aldehyde, the active ingredient in cumin essential oil, has anti-inflammatory, antiaflatoxigenic, and antifibrillation properties [153]. *C. cyminum* includes alkaloid, anthraquinone, coumarin, flavonoid, glycoside, protein, resin, saponin, tannin, and steroid, according to phytochemical investigation. In addition, the extract has strong memory-enhancing and antistress properties [154].

4.2.10 *Cyperus Rotundus* Linn.

The nutgrass, *Cyperus Rotundus* L. (family Cyperaceae), is a colonial perennial plant considered native to India 2000years ago and widely used in Ayurveda to treat several illnesses [155]. Scientists have reported that rhizomes and tubers of *Cyperus Rotundus* L. possess antidiarrheal, antioxidant, anti-inflammatory, antimutagenic, antiperiodic, anticonvulsant, antisaturative, antipyretic, antifungal, antidiabetic, antimalarial, antilipidemic, antibacterial, antiviral, antitumoral, cardioprotective, and wound-healing properties [156]. Phytochemical studies of *Cyperus Rotundus* L. led to the isolation of sesquiterpenes, flavonoids, phenylpropanoids, phenolic acids, alkaloids, and saponins. Recently, rotunduside, a phenolic compound from methanol extract of *Cyperus Rotundus* L., was isolated and its potential antidepressant activity with murine models was demonstrated [155].

4.2.11 *Eclipta alba*

Eclipta alba (L.), often known as false daisy or Bhringraj, is a medicinal plant that belongs to the Asteraceae family. It is commonly available in India, Brazil, China, and Thailand [157]. *E. alba* is commonly used for the treatment of gastrointestinal disorders, respiratory tract disorders (including asthma), fever, hair loss and graying of hair, liver disorders (including jaundice), skin disorders, spleen enlargement, cuts, and wounds in the traditional medicinal systems of subcontinental Indian countries as well as by tribal practitioners [158]. Wedelolactone, eclabasaponins, ursolic acid, oleanolic acid, luteolin, apigenin, ecliptalbine, luteolin-7-glucoside, 4-hydroxybenzoic acid, protocatechuic acid, and heptacosanol are among the phytocomponents found in the plant [159]. The methanol extract of *E. alba* exhibits high anticonvulsant efficacy due to its positive modulatory effect on GABAA receptors [160].

4.2.12 *Ficus religiosa*

Ficus religiosa L. (Moraceae) is often known as the peepal tree. The fig is a sacred tree, a large perennial, glabrous when young, distributed throughout India's plains up to 170 m above sea level in the Himalayas, primarily planted as an avenue and wayside tree, particularly near temples [161]. It is a well-known bodhi tree that has long held mythical, religious, and medical significance in Indian culture [162]. The abundant bioactive chemicals in this tree, such as flavonoids, alkaloids, tannins, saponins, phenols, and others, are responsible for its medicinal qualities in the treatment of a wide range of ailments [163]. It has various therapeutic purposes in folk medicine, including the treatment of asthma, cough, sexual issues, diarrhea, hematuria, toothache, migraine, eye problems, stomach problems, and scabies with leaf juice [161, 163]. Herbal medicines are made from almost every part of the tree, including the leaves, bark, seeds, and fruits. The impact of this plant on the brain raises thiobarbituric acid levels while lowering SOD and GSH levels [164].

4.2.13 *Glycyrrhiza glabra*

Glycyrrhiza glabra is a member of the Fabaceae family. Black sugar, common liquorice, licorice, and licorice root are its common names. *G. glabra* is found in Eurasia, particularly in central and southwestern Asia (Turkey, Iran, Iraq, northwestern China), as well as the Mediterranean (Spain, Italy) [165]. The plant is used as a frequent cure in traditional Chinese medicine for gastrointestinal issues, cough, bronchitis, and arthritis, for example. It is still commonly used in folk medicine to treat gastritis, peptic ulcers, respiratory infections, and tremors [166]. *G. glabra* has a water-soluble, physiologically active compound that makes up 40–50% of the total dry weight. Triterpene, saponin, flavonoids, polysaccharides, pectins, simple sugars, amino acids, mineral salts, asparagines, bitters, essential oil, fat, estrogen, gums, mucilage (rhizome), protein, resins, starches, sterols, volatile oils, tannins, glycosides, and various other substances are constituents of *G. glabra* [167]. The chalcone structure of isoliquiritigenin isolated from licorice root has a powerful anticancer impact. Other important compounds in this plant are glycyrrhizin, glycyrrhizinic acid, isoliquiritin, and glycyrrhizic acid, which have antiatherogenic, anticancer, antidiabetic, antimicrobial, antispasmodic, anti-inflammatory, and antiasthmatic qualities [168]. Glabridin isolated from *G. glabra* roots has effects on cognitive performance and piracetam on brain cholinesterase activity. As a result, glabridin has been validated as a promising memory-enhancer option. This plant increases dopamine and norepinephrine levels in the brain [169]. The impact was linked to glabridin's antioxidant, neuroprotective, and anticholinesterase capabilities, implying that it could be used to treat dementia in diabetic individuals [170].

4.2.14 *Moringa oleifera*

Moringa oleifera Lam., sometimes known as Munga, is one of the most frequently cultivated plants in India. It is a member of the Moringaceae family [171]. It is found in India, Pakistan, Bangladesh, and Afghanistan in the sub-Himalayan region. This plant is commonly used as a nutritious herb and has antiasthmatic, antidiabetic, hepatoprotective, antioxidant, cardiovascular, antiulcer, antiallergic, wound-healing, analgesic, and antipyretic properties, as well as an impact on CNS activity [172]. Each part of the herb has a valuable therapeutic component. It is high in vitamins A and C, as well as milk protein, phytoactive components such as morphine, rhamnetin, kaempferol, apigenin, myricetin, quercetin, pterygospermin, benzoic acid, gallic acid, benzyl isothiocyanate, naizirin, and niaziminin [173]. The molecules responsible for this activity are quercetin, apigenin, and lupeol, all of which have previously been shown to have antihypertensive properties [174]. This plant's leaf extract raises monoamine levels in the brain, which may aid in the treatment of Alzheimer's disease. It also serves as an antidepressant by bringing serotonin and dopamine levels back to normal [175].

4.2.15 *Nigella sativa*

In many regions of the world, *Nigella sativa* (family Ranunculaceae) is a popular therapeutic herb. It is commonly utilized in traditional medical systems including Unani, Tibb, Ayurveda, and Siddha, and various studies have shown that it has a wide range of therapeutic potential [176]. The common term for *N. sativa* is black seed. This plant is native to Southern Europe, North Africa, and Southwest Asia, and it is grown in countries all over the world, including the Middle East, South Europe, India, Pakistan, Syria, Turkey, and Saudi Arabia [177]. For millennia, *N. sativa* seeds and oil have been used to treat a range of diseases throughout the world. Black cumin seeds have been used for a variety of diseases in traditional herbal systems, including chronic headaches, back pain, paralysis, inflammation, hypertension, blisters, nasal abscesses, orchitis, dermatitis, and swollen joints [176, 178]. Many active compounds have been isolated, identified, and reported from many varieties of black seeds. The most prominent active compounds include thymoquinone (30–40%), thymohydroquinone, dithymoquinone, trans-anethole, p-cymene, carvacrol, G-terpineol, t-anethol, sesquiterpene longifolene, pinene, thymol, β-sitosterol, campesterol, stigmasterol, and 5-avenasterol [178]. Aqueous and methanol extracts of the defatted *N. sativa* seeds were found to exhibit powerful CNS and analgesic actions, with the methanolic extract having a notably depressive impact. Long-term administration of *N. sativa* to rats boosts 5-HT levels in the brain, which improves learning and memory [179, 180].

4.2.16 *Piper longum*

Piper longum is a flowering plant that belongs to the Piperaceae family. *P. longum*, often known as Indian long pepper or pippali, is a species of pepper that grows in India [181]. There are 4000 species in the Piperaceae family, which is divided into five genera. The two most important genera in this family are Piper (2000 species) and Peperomia (1600 species). Linnaeus classified 17 species of this family in his *Species plantarum* in 1753, all of which were placed in the genus Piper. Piper species are important commercially, economically, and medicinally. *Piper nigrum* and *P. longum* are the most economically important species in this family [182]. The fruits of *P. longum* are used as a stomachic, liver tonic, abortifacient, pungent, aphrodisiac, and laxative, has antidiarrhea, antidysentery, antiasthmatic, and antibronchitis properties, and is used for abdominal complaints, urinary discharges, tumors, spleen diseases, pains, inflammation, leprosy, insomnia, jaundice, and hiccups [182]. Irritants, analgesics, asthma, and bronchitis are all treated using the fruit and root of the *P. longum* species [183]. With a concentration of 3–5%, piperine is a major active component in *P. longum*. The fruits of *P. longum* include starch, protein, and alkaloids, as well as volatile oils, saponins, polysaccharides, and amygdalin [184]. Alkaloids and amides are two different types of chemicals found in the human body. In the fruit of *P. longum*, piperine is the most abundant alkaloid and related substance, followed by methyl piperine, pipernonaline, piperettine, asinine, and others [184]. In every section of the brain, the chemical molecule piperine raises dopamine and serotonin levels while decreasing norepinephrine levels [185].

4.2.17 *Rauwolfia serpentina*

Rauwolfia serpentina, also known as Indian snakeroot, sarpagandha, or saranghanda, is a 50-species plant found in India's Himalayan foothills, up to 1300–1400 m in height [186]. Saranghanda is a glabrous perennial plant that is evergreen. The plant can grow up to 60 cm in height. It has tuberous roots that are covered with pale brown cork [187]. *R. serpentina* is classified as endangered or threatened by the rare, endangered, and threatened (RET) category for valuable medicinal plants in the Himalayas region [188]. Antioxidant, antihypertensive, anticancerous, antimalarial, anti-inflammatory, antifibrillar, anthelmintic, antiarrhythmic, anticholinergic, antidysentry, antidiarrhoeal, antihypotensive, anticontractile, antipyretic, antidiuretic, sympathomimetic, and antipsychotic properties have also been reported for its extracts [189–191]. Rauwolfia extract appears to block GABA and α-2-noradrenergic receptors. Alcohols, sugars, and glycosides, fatty acids, flavonoids, phytosterols, oleoresins, steroids, tannins, and alkaloids are among the phytochemicals found in Rauwolfia.

Reserpine is the alkaloid discovered in *R. serpentina* that has been investigated the most [192, 193]. Reserpine blocks the vesicular monoamine channel in an irreversible manner, depleting monoamines in the brain and other tissues, and there is evidence that it also suppresses the dopamine channel. Reserpine has been proven to be a human ether-a-go-go related gene (*hERG*) potassium channel inhibitor in the thallium inflow assay and patch-clamp experiment [27, 194–196]. In the amphetamine-induced hyperactivity mouse paradigm, reserpine showed antipsychotic effect in a dose-dependent manner [197]. Alstonine is an antipsychotic that may be found in a variety of plants and has been used to treat a variety of mental diseases in the past. Memory loss, antipsychotic effects, hyperlocomotion, and social interaction deficits are all addressed by this chemical, which also has anxiolytic properties. Alstonine has been proven to enhance the antioxidant GSH, which is often low in the brains of individuals with schizophrenia. Alstonine also has an indirect influence on dopamine receptors [198–201].

4.2.18 *Sesbania grandiflora*

Sesbania grandiflora is a small tree of the Sesbania genus that is also known as the vegetable hummingbird or hummingbird tree. It has been used as a major source of nutrients in Southeast Asian countries under the names "sesbania" and "agathi." The leaves of *S. grandiflora* are high in amino acids, vital minerals, and vitamins (vitamin A, vitamin C, thiamine, riboflavin, and nicotinic acid) [202, 203]. *S. grandiflora* is a perennial legume tree that grows quickly and can reach a height of 10–15 m. It has a lifespan of 20 years [204]. The Siddha system of Indian traditional medicine uses various portions of this plant to cure a variety of maladies, including inflammation, leprosy, gout, and rheumatism. Anxiolytic, anticonvulsive, and hepatoprotective effects are also present [203]. In folk medicine, all parts of *S. grandiflora* have been used empirically to treat numerous ailments such as catarrh, diarrhea, fevers, headaches, smallpox, sore throat, and stomatitis [205]. The main chemical components are alkaloids, flavonoids, glycosides, tannins, anthraquinones, steroids, pholobatannins, and terpenoids. Isovestitol, medicarpin, sativan (isoflavonoids), and betulinic acid are the major chemicals responsible for antibacterial, antifungal, antioxidant, antiurolithiatic, anticonvulsant, anxiolytic, and hepatoprotective actions (tannin compound). Plant extracts also contain alkaloids, phenolics, tannins, triterpenoids, and sterols [206]. Administration of the extract reduced the activity of AChE and MDA in experimental rats while restoring the antioxidant enzymes SOD, GSH, and CAT, boosting GABA, 5-HT, and improving memory [207, 208].

4.2.19 *Sphaeranthus indicus*

Sphaeranthus indicus Linn. is a member of the Asteraceae family. It is also known by other synonyms like Mundi, Munditika, Shravana, Shravanava, and Shravanashirshaka. It is found throughout India, Sri Lanka, Africa, and Australia, from sea level to 1200 m altitude [209].

It is abundant in plains wetlands, and as a weed in rice fields. Medicinal *S. indicus* has been proven to exhibit ovicidal and hepatoprotective properties, and antitussive, wound-healing, anxiolytic, neuroleptic, immunomodulatory, anthelmintic, antidiabetic, antiviral, antihyperlipidemic, and antioxidant activity [209, 210]. A wide range of phytochemicals have been isolated from this plant, including sesquiterpenes, lactones, eudesmenlides, apigenin 7-O-glycosides, β-sitosterol, n-triacontanol, phenylurethane, n-pentacosane, echinacin, echinopsine, echinozolinone, dihydroquercetin-4-methyl ether, lupeol, echitin, 7-hydroxyechinozolinone, methyl chavicol, δ-cadinene, α-ionone, and para-methoxycinnamaldehyde [210, 211]. In the Indian system of medicine, the whole plant, or its different anatomic parts such as leaf, stem, bark, root, flowers, and seed, are extensively used for curing many diseases [209]. A dose of 300 mg/kg methanolic extract of *S. indicus* increased the GABA level in the brain in a mice model.

4.2.20 *Tinospora cordifolia*

Tinospora cordifolia is a member of the Menispermaceae family (also known as "Guduchi" or "Giloy" in Sanskrit). It is a deciduous climbing shrub with distinctive greenish-yellow flowers that grows at higher elevations in places like India, Myanmar, and Sri Lanka. *Tinospora cordifolia* has been demonstrated to have various differentiating characteristics that set it apart from all other closely connected *Tinospora* species, such as *Tinospora malabarica. Tinospora cordifolia has* a cork that is ash-colored, has more lenticels, nodes, and internodes, and has less mucilage than *Tinospora malabarica* [212]. It has been proven to exhibit antineoplastic, antidiabetic, antiperiodic, antispasmodic, anti-inflammatory, antiarthritic, antioxidant, antiallergic, antistress, antileprotic, antimalarial, hepatoprotective, and immunomodulatory properties [213–215], The reported chemical constituents of this shrub belong to various classes, such as alkaloids, like berberine, palmatine, tembetarine, isocolumbin, and magnoflorine; glycosides, like syringin, tinosporide, furanolactone diterpene, giloin, and cordifolioside A; steroids, such as β-sitosterol and ecdysterone; and sesquiterpenoids, phenols, aliphatic compounds, and polysaccharides [214].

4.2.21 *Valeriana wallichii*

Valeriana wallichii, popularly known as "tagara" in Ayurveda, is a hairy perennial herb belonging to the Valerianaceae family that grows in the Himalayas region. The plants are employed as a diuretic, carminative antispasmodic, anticonstipation, antiscorpion toxin, and blood circulation booster. The roots and rhizomes of the *Valeriana* species contain valerenic acids, valepotriates, baldrinal, homobaldrinal, valeric acid, valechlorine, valerianine, resins, alkaloids, and metabolites such as phenolic acids and flavonoids [216, 217]. *V. wallichii* is one of several medicinal herbs from India that have been demonstrated to have activity using classic psychopharmacologic approaches. It is a natural source of the sedative and tranquilizing valepotriates and is one of the most common medicines for insomnia, particularly insomnia caused by nerve weariness and mental overwork [218]. In rats, *V. wallichii* and its component valeric acid have strong neuroprotective effects on memory and retentive ability via GABA receptor distinctions. The essential oil of the root has antidepressant activity [217]. Valerian extracts have shown anxiolytic and antidepressant effects, they also improve sleep, and it has been reported that valerian extracts have affinity for GABAA receptors. It increases norepinephrine and dopamine levels in the forebrain [219].

4.2.22 *Withania somnifera*

Withania somnifera (Ashwagandha) is commonly known as the Indian winter cherry or Indian ginseng [220]. It is one of the most important herbs in Ayurveda and has been used for millennia as a medhya rasayana for its wide-ranging health benefits, especially as a nervine tonic. In India it is cultivated on a commercial scale [221]. In Ayurvedic medicine, its root is used as an anti-inflammatory drug for swellings, tumors, scrofula, rheumatism, constipation, insomnia, nervous breakdown, goiter, carbuncles, ulcers, painful swellings, leukorrhea, boils, pimples, flatulent colic, worms, and piles, and as a sedative and hypnotic in anxiety neurosis [221–224]. Much of the pharmacologic activity is attributed to the presence of steroidal lactones called withanolides, such as withaferin A, in the Indian chemotype leaves. The roots provide 18 fatty acids, β-sitosterol, polyphenols, and phytosterols [225], and include numerous alkaloids, such as withaninine, whithanine, pseudo-withaninine, somnine, and somniferine. The roots also include a substance that mimics GABA function and activates GABAρ1 in the brain [226].

4.2.23 *Ziziphus mauritiana*

Indian plum, Indian jujube, and dessert apple are all names for *Ziziphus mauritiana*, from the Rhamnaceae family. It is one of India's oldest fruits and is grown in various parts of the country, with India ranking second among jujube-growing countries. The stem, bark, leaves, flowers, fruits, and roots of the plant have all been demonstrated to have medicinal properties [227]. Different components of the plant are used to cure coughs, headaches, boils, dysentery, and tuberculosis, and help with digestion, eye difficulties, and leukorrhea. According to the available information on Indian jujube, it is well known for being a good source of total phenolics and a rich source of ascorbic acid, as well as essential minerals like calcium, phosphorus, and iron [228]. Chemical constituents are kaempferol 3-O-rutinoside, quercetin, quercetin 3-O-galactoside, quercetin 3-O-rutinoside, puerarin, galactose, arabinose, rhamnose, glucose, and xylose [72, 227]. This plant increases the level of AChE in the brain and stimulates the transcriptional expression of GABA receptor subunits in rat hippocampal neurons [227].

4.3 Conclusion

This review emphasizes the involvement of several neurotransporters and phytochemicals in the regulation of schizophrenia, summarized in Table 4.1. In a variety of therapies, phytochemicals may be an alternative to more standard therapy approaches. Phytoconstituents and compounds from a variety of plants, in particular, have been demonstrated to protect against neurologic illnesses. Flavonoids, flavones, sterols, alkaloids, terpenoids, catechins, and phenols are the most frequent bioactive compounds with therapeutic significance. Phytochemicals have the potential to modulate a number of pathogenic pathways. Most prior research found phytochemicals to have similar effects, such as lowering oxidative stress and free radicals via an antioxidant impact, improving immune cells' phagocytic characteristics, increasing neurotransmitter concentrations, and inhibiting AChE activity. Phytochemicals that influence schizophrenia using the ways described are still underrepresented in preclinical in vivo trials. Although there is an increasing demand for phytotherapeutic drugs, scientific confirmation is

Table 4.1 Phytochemicals and their neurologic mechanisms of action.

Plant	Phytochemical	Structure	Target	References
Allium cepa	N-acetylcysteine	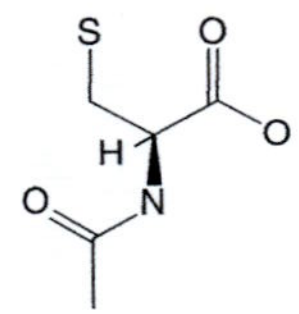	Inhibits glutaminergic and dopaminergic neurotransmission	[93, 95]
Acorus calamus	Acolamone	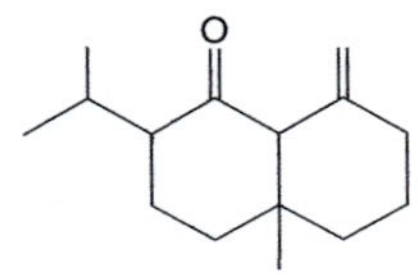	Not identified	
	Acorenone	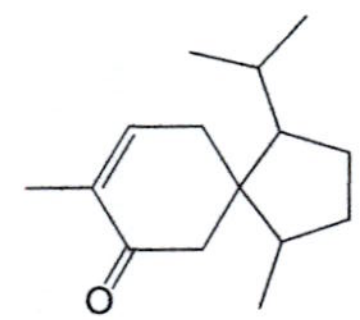	Acetylcholinesterase and butyrylcholinesterase are inhibited	[229]
	Acoragermacrone	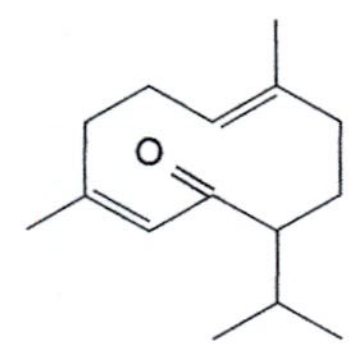	Not identified	
	Cadalene	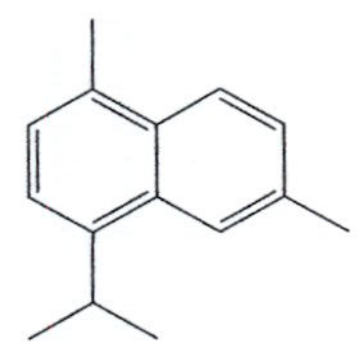	Not identified	

Eugenol	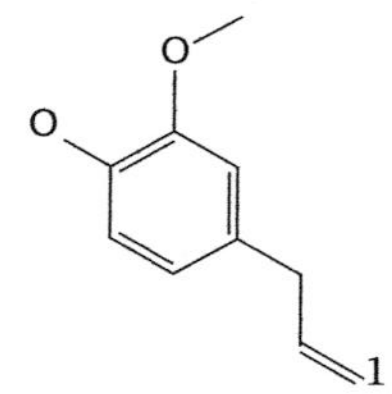	Inhibits monoamine oxidase A and also restores monoamine levels in depressed patients' brains	[230]
Acoric acid	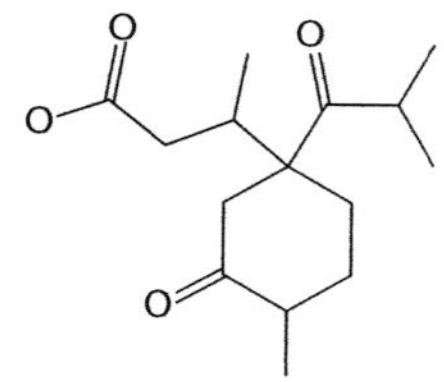	Not identified	
Carveol	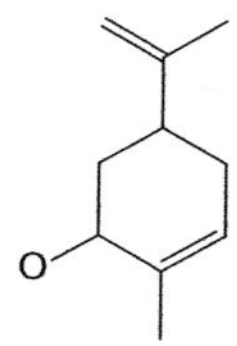	By lowering acetylcholinesterase levels, activates cholinergic system	[229]
Carvone	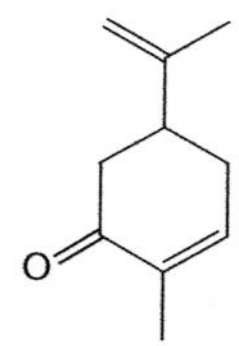	At its non-competitive blocker site, acts with the GABA(A) receptor	[160]
Carvacrol	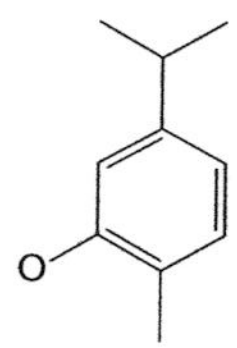	Regulates dopamine and other central neurotransmitters and neuromodulators	[231]

(*Continued*)

Table 4.1 (Continued)

Plant	Phytochemical	Structure	Target	References
Bacopa monneiriv	Bacoside A	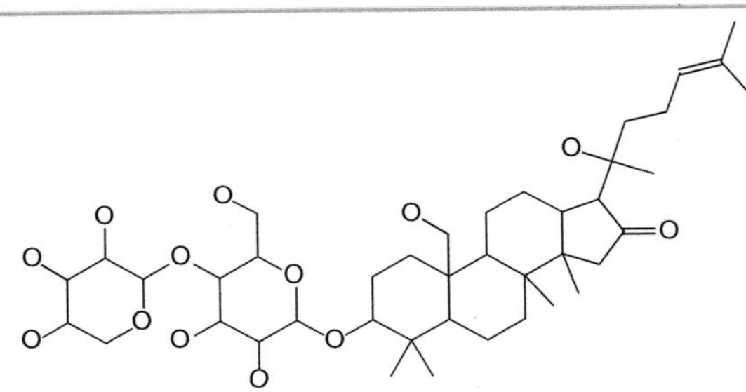	Reduces acetylcholinesterase activity, which improves cholinergic function and attention and memory	[232]
	Bacoside b	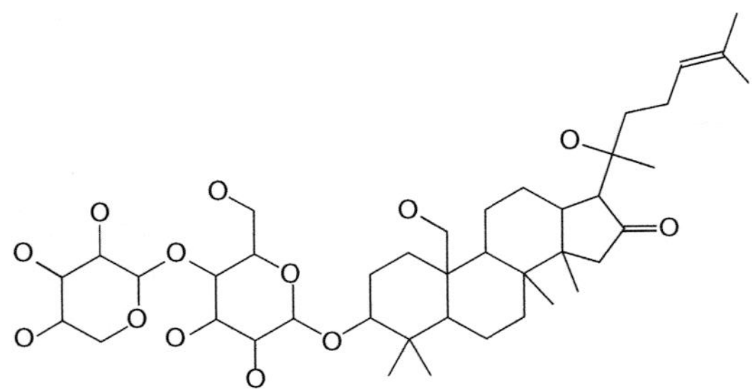	Serotonin, a brain neurotransmitter associated with relaxation, is increased	[232]
	Bacopsaponin c	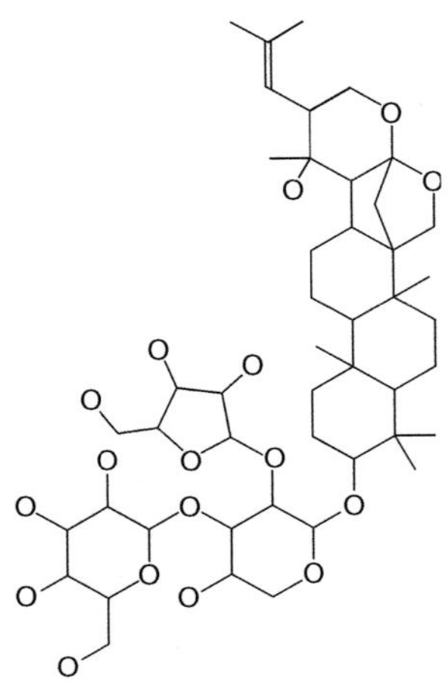	In many areas of brain, enhances neuroreceptor expression	[232]

Bacopaside I	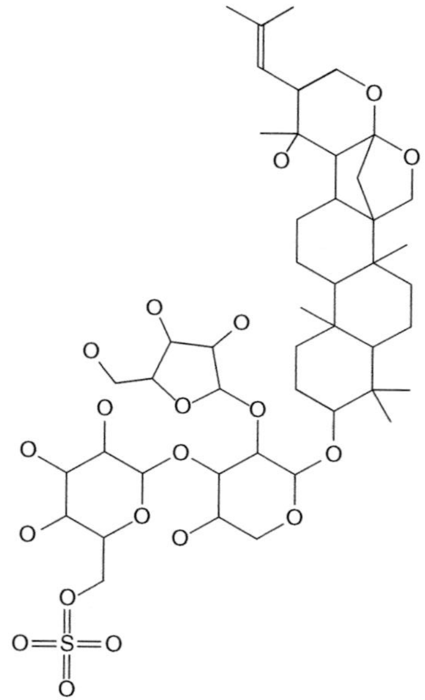	Antioxidant neuroprotection, acetylcholinesterase inhibition, and/or choline acetyltransferase activation, amyloid reduction, increased cerebral blood flow, and neurotransmitter modulation	[232]
Bacopaside II	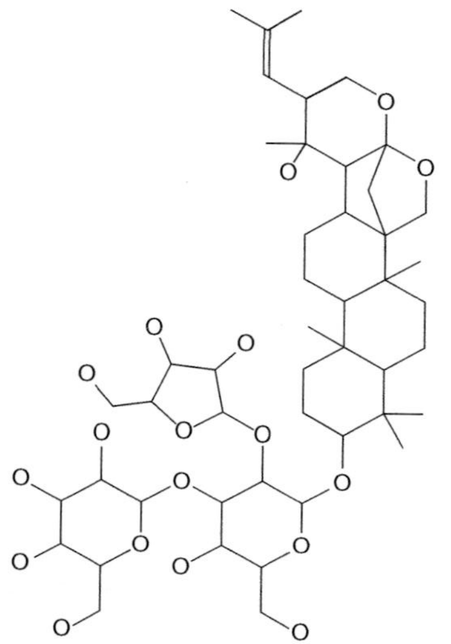	Not identified	

(Continued)

Table 4.1 (Continued)

Plant	Phytochemical	Structure	Target	References
Carum carvi	Myrcene	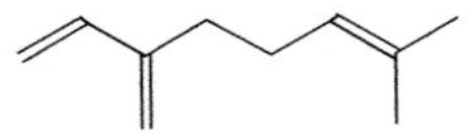	Myrcene and other terpenes affect GABA receptors, which act synergistically with THC	[233]
	Dihydrocarvone	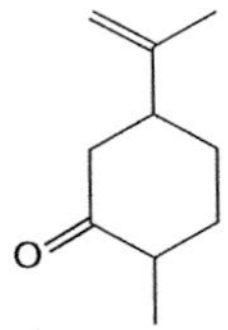	Not identified	
	Isorhamnetin 3-glucoside	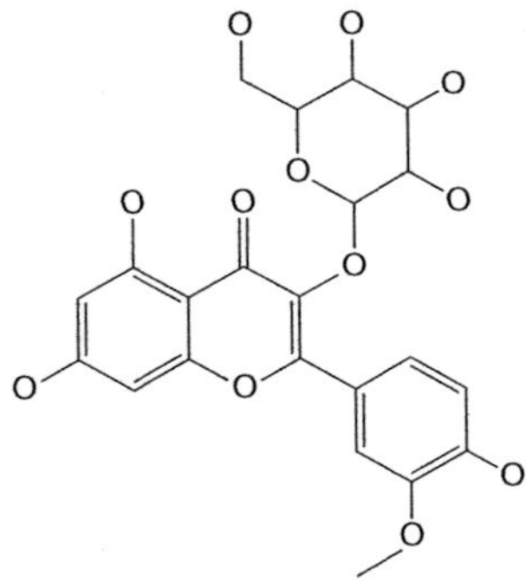	Not identified	
Centella asiatica	Triterpenoid A	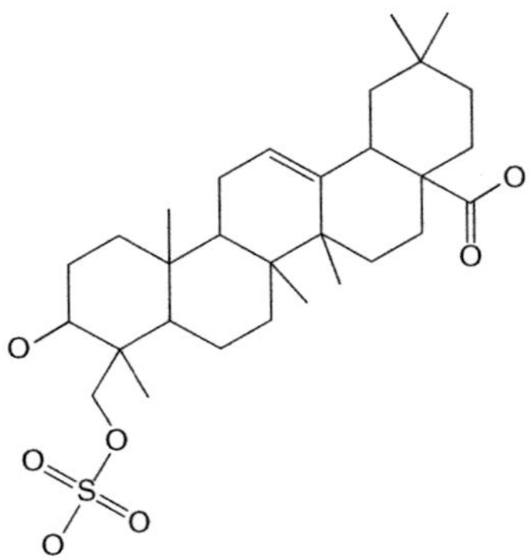	Mitogen-activated protein kinase signaling pathway, phosphotidylinositol 3 kinase/protein kinase B/mammalian target of rapamycin signaling pathway, and nuclear factor kappa–light chain enhancer of activated B cells signaling pathway could all be affected by triterpenoids	[234]

	Asiaticoside B	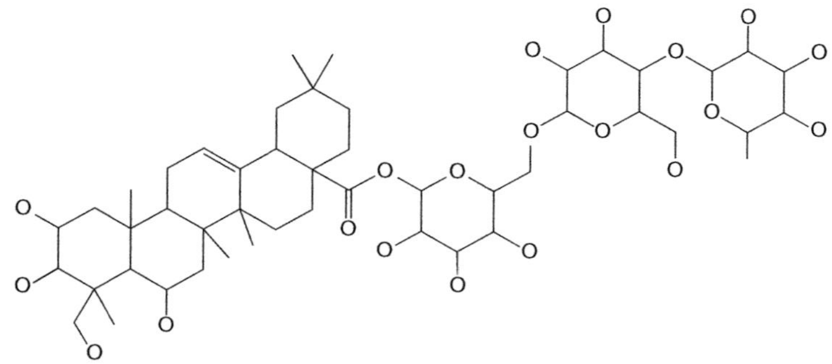	Works by stimulating BDNF signaling in hippocampus to produce antidepressant-like effects	[142]
	Brahmic acid	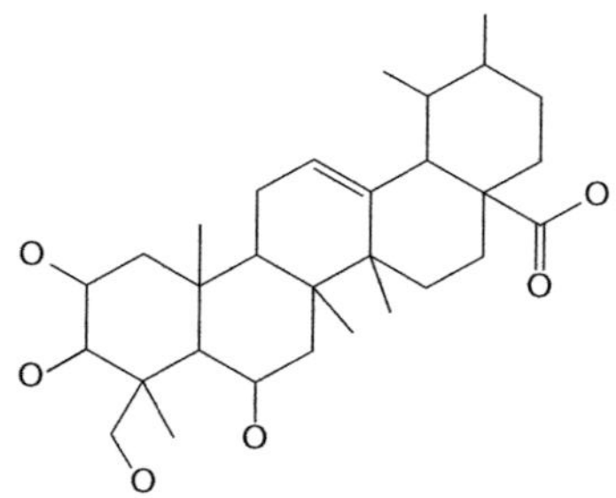	Not identified	
Convolvulus pluricaulis	Kaempferol	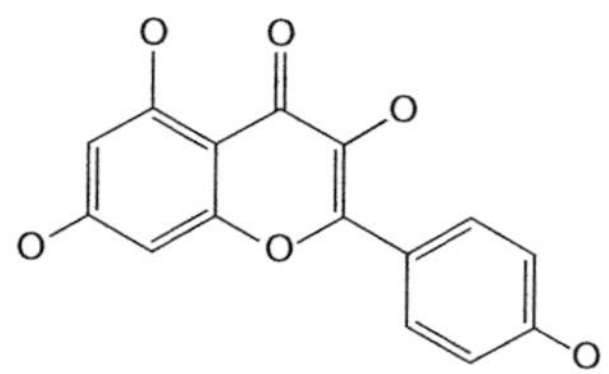	Improves depression-like behaviors by decreasing oxidative stress and neuroinflammation via Akt/β-catenin pathway	[235]
	Hydroxycinnamic acid	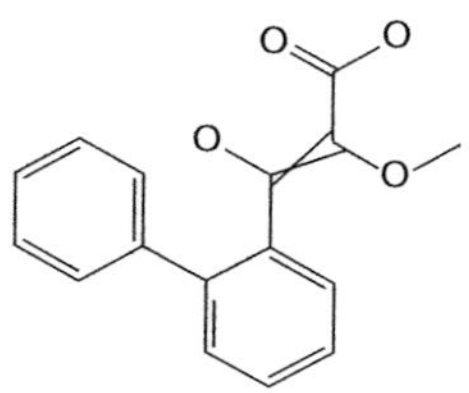	Not identified	
	N-hexacosanol	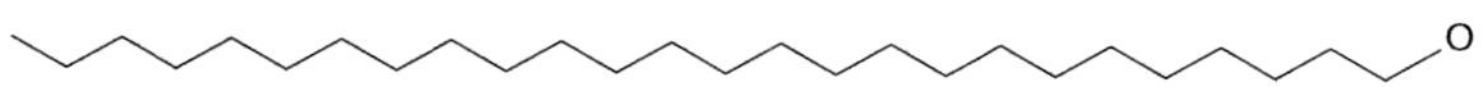	Not identified	

(*Continued*)

Table 4.1 (Continued)

Plant	Phytochemical	Structure	Target	References
Coriander sativum	Lialool	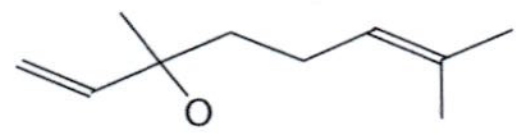	In brain, functions as glutamate-competitive antagonist and non-competitive antagonist of NMDA receptors	[236]
	Linalyl acetate	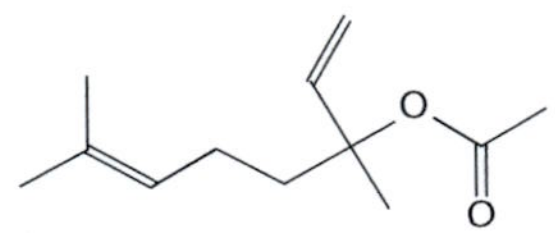	Has inhibitory impact on acetylcholinesterase and aids memory	[237]
	Gallic acid	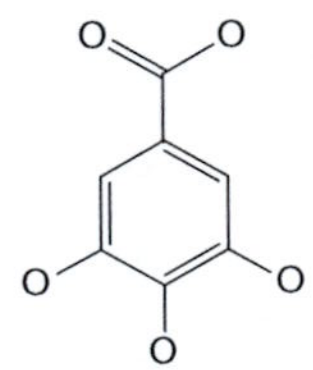	Inhibits acetylcholinesterase activity	[238]
	Caffeic acid	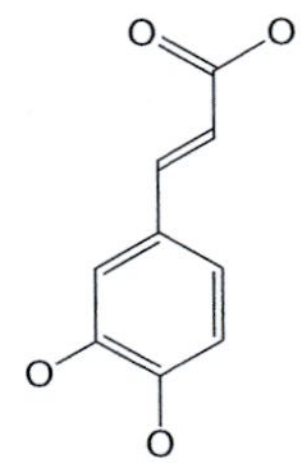	Affects α1A-adenoreceptor system, involved in caffeic acid's antidepressant or anxiolytic-like actions	[239]
	Ferulic acid	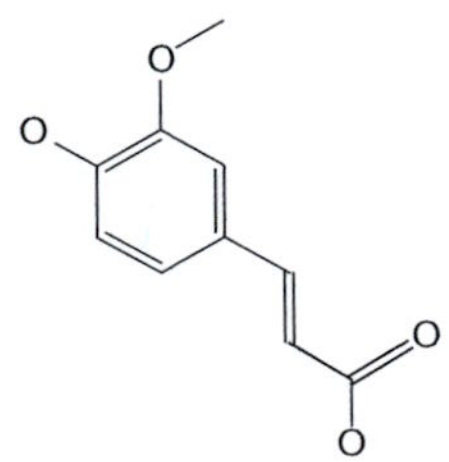	By reducing injury-induced decreases in -enolase production in brain cells	[236]

Chlorogenic acid	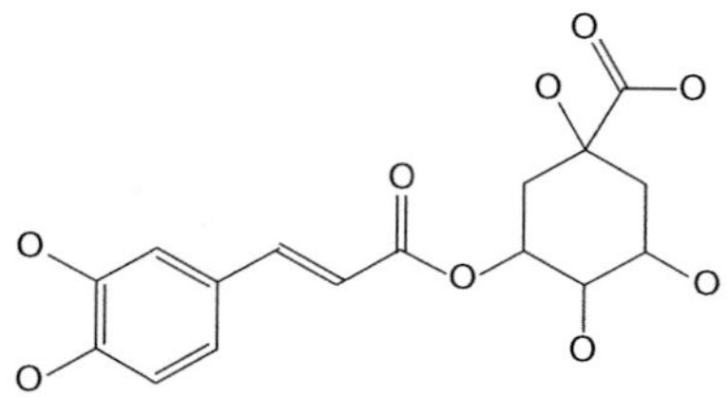	Not identified	
β-sitosterol	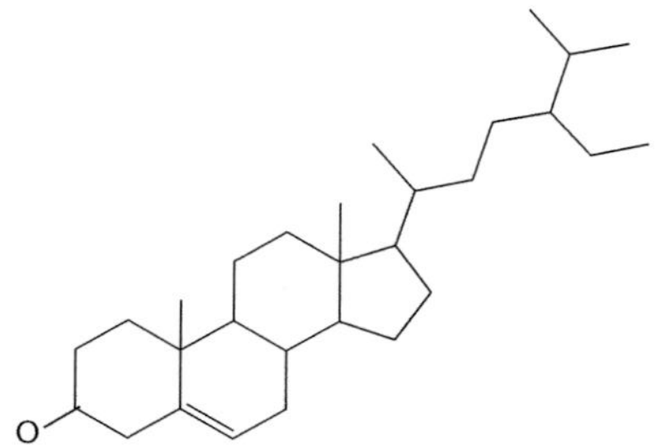	Antidepressant-like effect of Sit-S could be mediated by modification of 5-HT, DA, and GABA-ergic systems	[240]
Stigmasterol	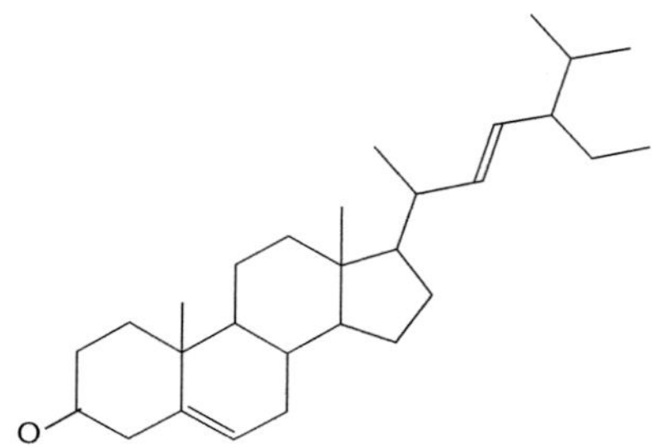	Because of positive regulation of GABA receptors, regarded as promising steroidal medication	[241]

(*Continued*)

Table 4.1 (Continued)

Plant	Phytochemical	Structure	Target	References
Cyperus rotundus	α-Cyperone	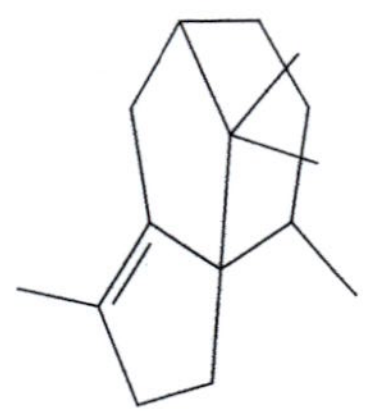	Binds to tubulin and interacts with it, causing microtubule polymerization to be disrupted and inflammation to be reduced	[242]
	Humulene	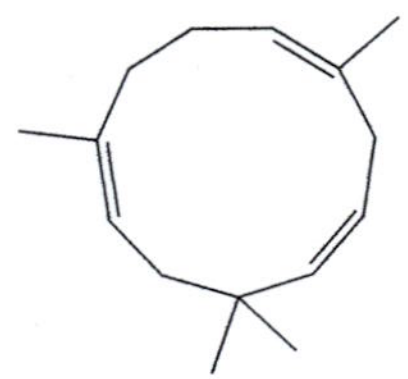	Not identified	
	Selinene	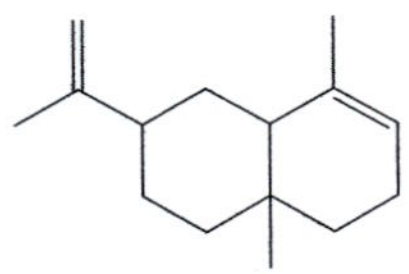	Not identified	
	β-Asarone	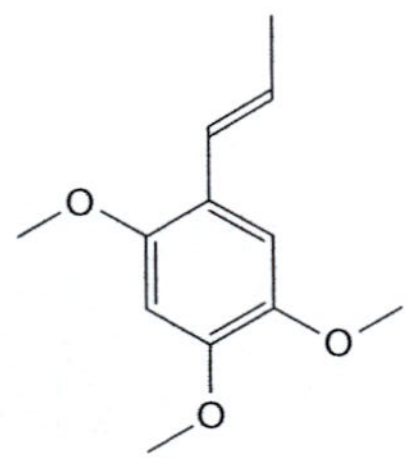	Reduces oxidative stress, amyloid, and neuronal cell apoptosis while increasing BDNF levels, hence boosting learning and memory	[243]

Eclipta alba	Wedelolactone	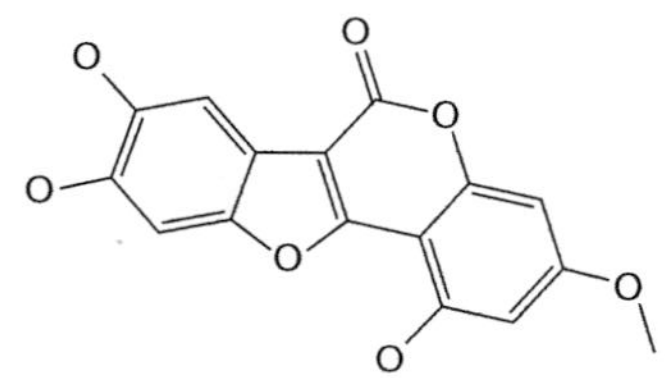	Because of positive modulatory effects on GABAA receptors, has powerful anticonvulsant properties	[160]
	Ursolic acid	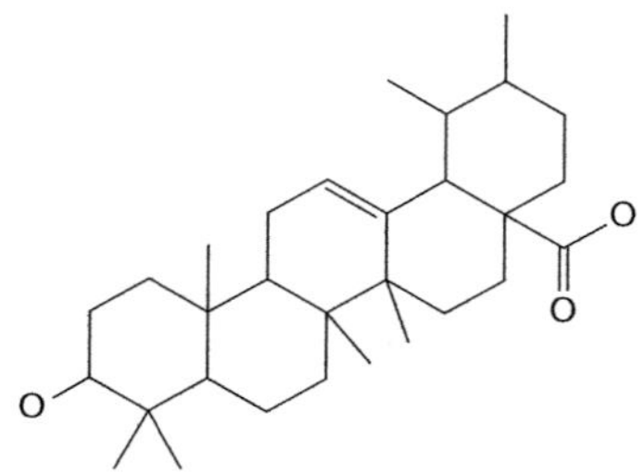	Decreases Aβ25 deposition by significant amount	[244]
	Oleanolic acid	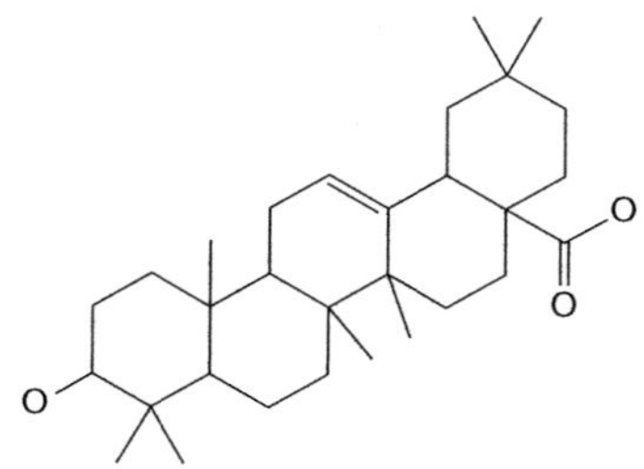	Not identified	
	Luteolin	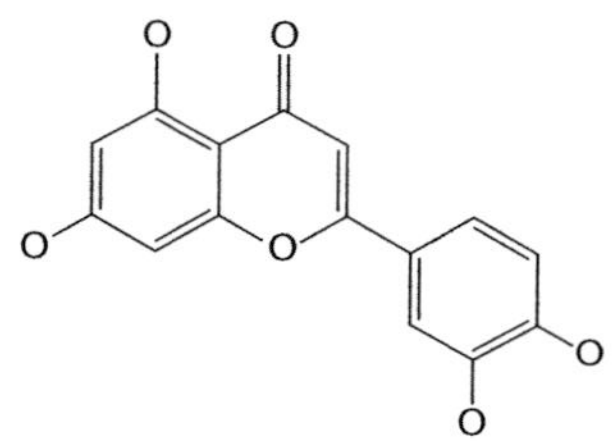	To reduce zinc-induced tau hyperphosphorylation, reduced amyloid precursor protein and Aβ production and modulated tau phosphatase/kinase pathway	[245]

(Continued)

Table 4.1 (Continued)

Plant	Phytochemical	Structure	Target	References
	Apigenin	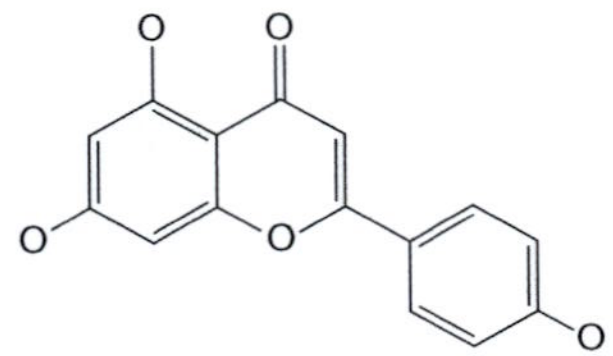	GABAA receptor activity in the brain inhibited	[246]
	Ecliptalbine	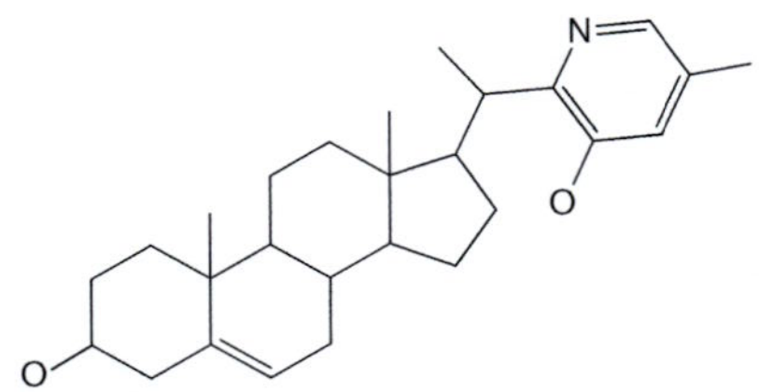	Not identified	
	Luteolin 7 glucoside	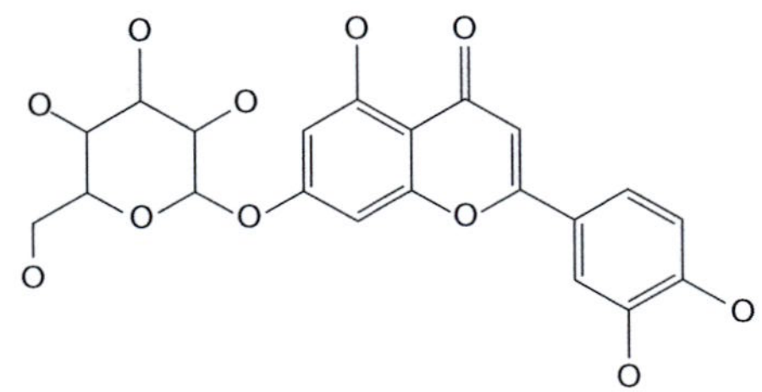	Not identified	
	Protocatechuinc acid	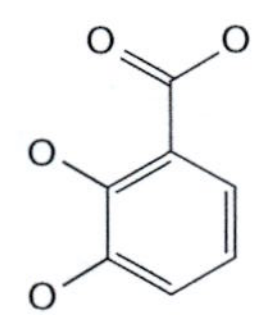	Acts as glutamate receptor inhibitor	[247]

Ficus religiosa	Lanosterol	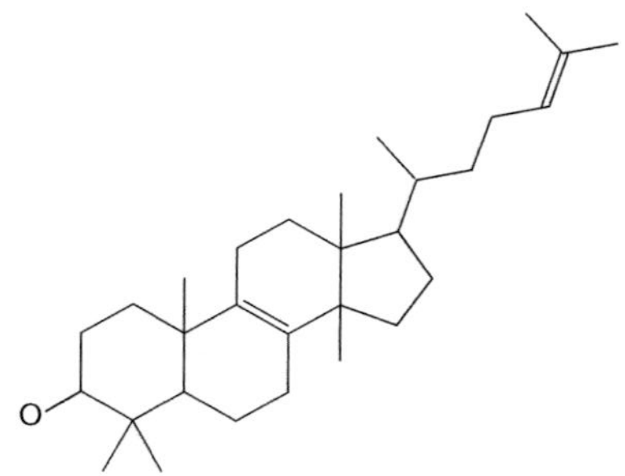	Not identified	
	Sitosterol	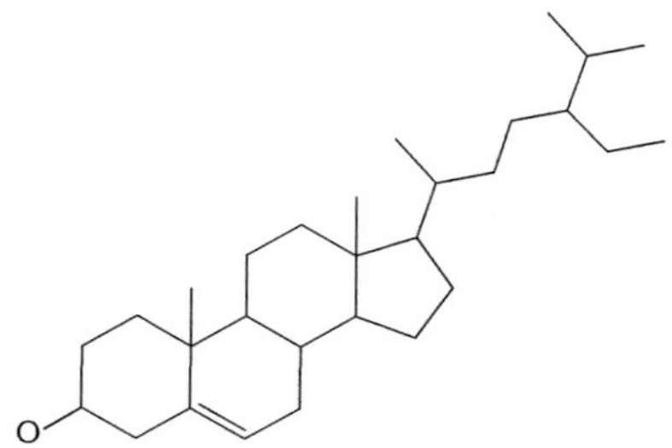	Antidepressant-like action of Sit-S may be mediated through changes in 5-HT, DA, and GABAergic systems	[240]
	Stigmasterol	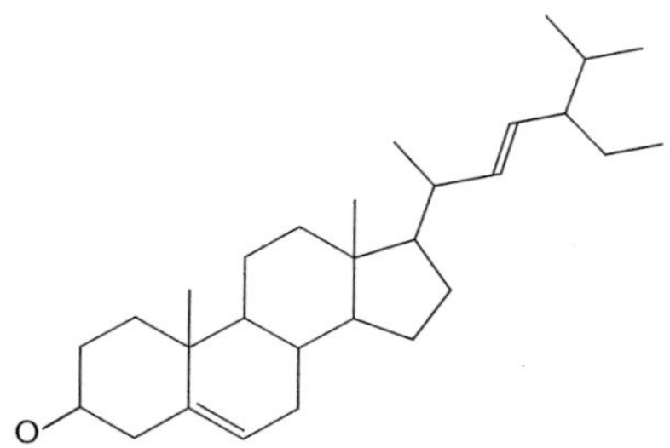	GABA levels in brain increased by stigmasterol	[241]

(*Continued*)

Table 4.1 (Continued)

Plant	Phytochemical	Structure	Target	References
	Bergapten	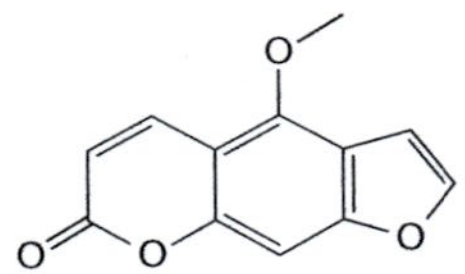	Not identified	
	Lupin 3-One	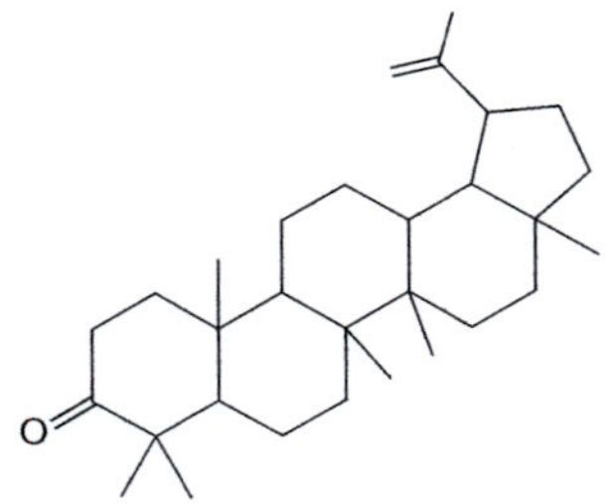	Not identified	
	1-Octacosanol		Engaged in inhibiting MAPK/NF-B/AP-1 signaling pathway and downregulating production of inflammatory factors	[248]
	Inositol	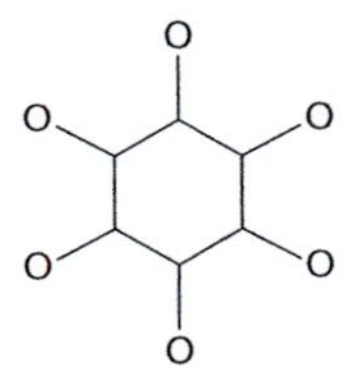	IP3 binds to IP3 receptors, changing calcium concentrations	[249]
	Methyl oleate	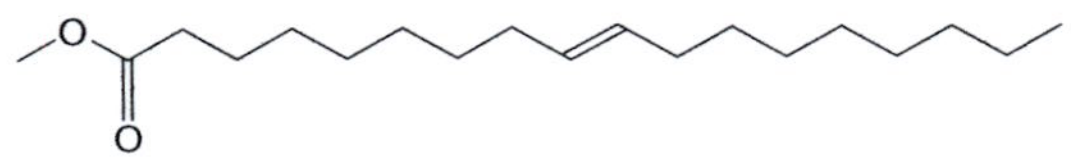	Not identified	

Glycyrrhiza glabra	Glycyrrhizic acid	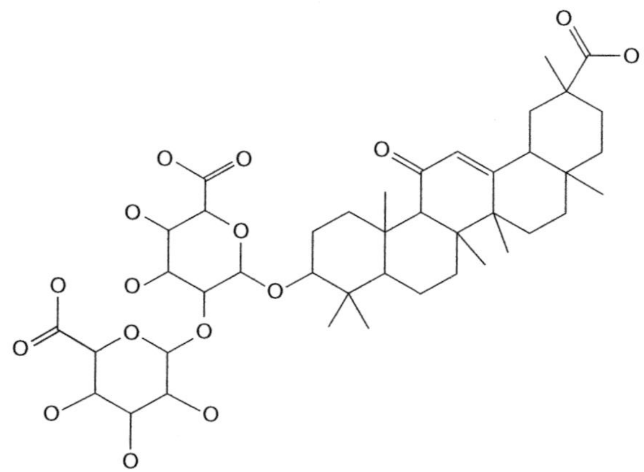	NFκB translocation into nucleus inhibited, and free radicals conjugated. Glycyrrhizic-driven suppression of CD4+ T-cell proliferation via JNK, ERK, and PI3K/Akt seen in several investigations	[250]
Moringa olifera	Morphine	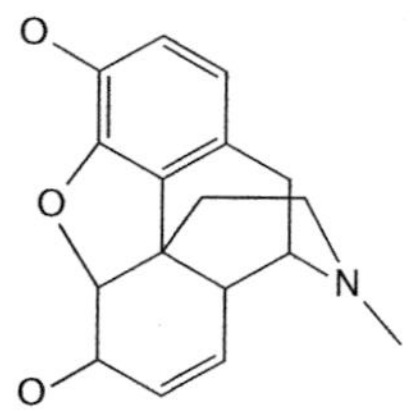	Binds to mu-opioid receptor in CNS and PNS to produce majority of analgesic effects	[251]
	Kaempferol	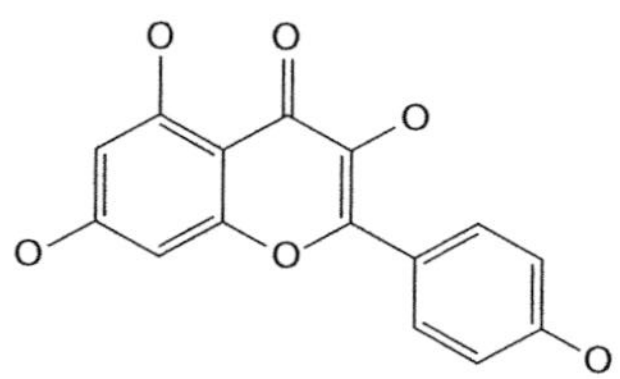	Reduces oxidative stress and neuroinflammation via Akt/catenin pathway, improves depression-like behaviors	[235]

(*Continued*)

Table 4.1 (Continued)

Plant	Phytochemical	Structure	Target	References
	Quercetin	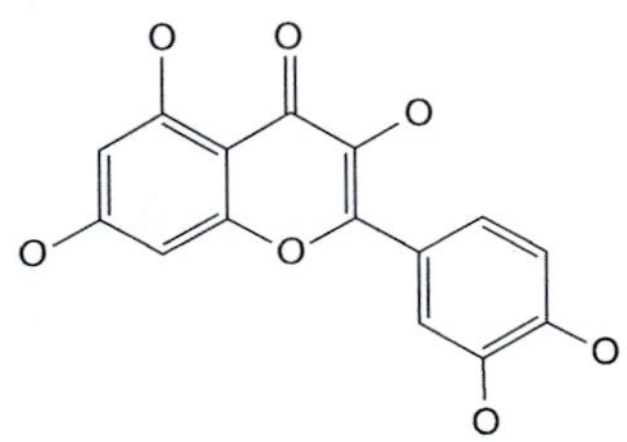	At dopaminergic level, effect on neuroprotective signaling events. In MN9D dopaminergic neuronal cells, activates cells: survival kinases D1 and Akt	[252]
	Pterygospremin	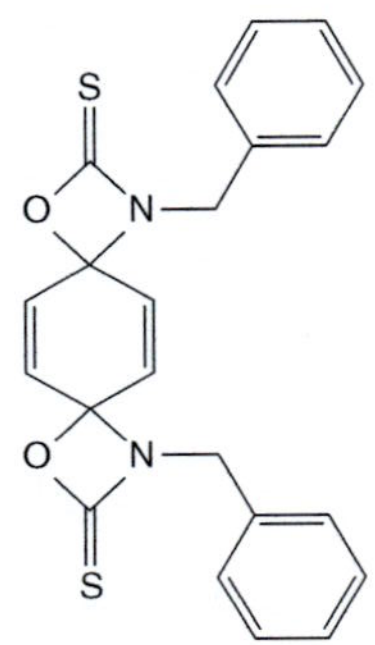	Not identified	
	Benzoic acid	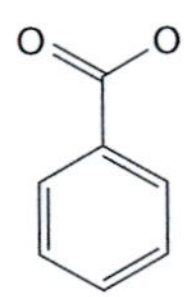	Inhibition of human D-amino acid oxidase. Has high activity toward D- DOPA and contributes to dopamine synthesis	[253]

	Niaziminin	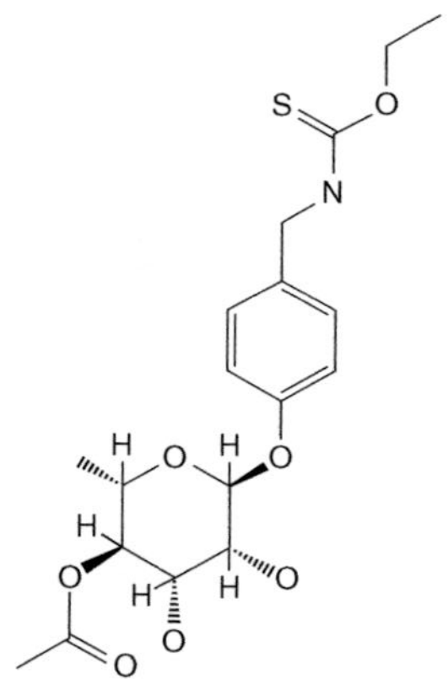	Not identified	
Pipper longum	Methyl piperene	Not available	GABA and serotonin levels in hippocampus nucleus increase	[254]
	Pipernonaline	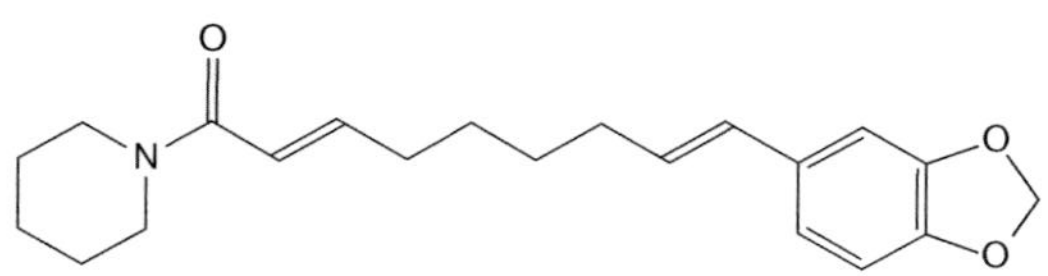	Not identified	
	Linoleic acid	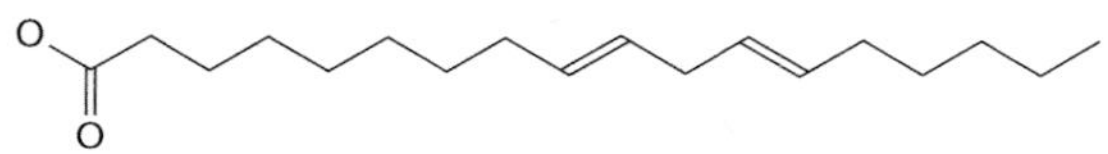	Not identified	
	Palmitic acid	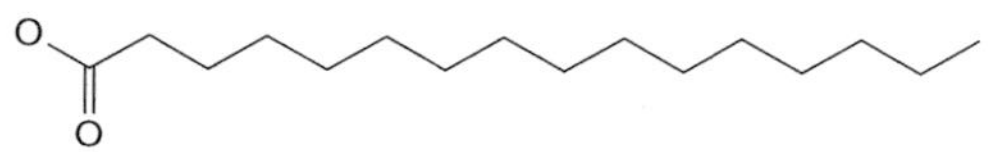	Modifies GAP-43 NMDA and AMPA receptors, both important for learning	[255]
	Lignoceric acid	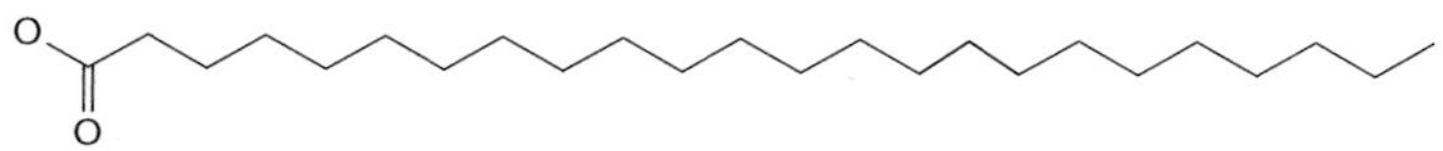	Not identified	

(*Continued*)

Table 4.1 (Continued)

Plant	Phytochemical	Structure	Target	References
Rauwolfia serpentina	Reserpine	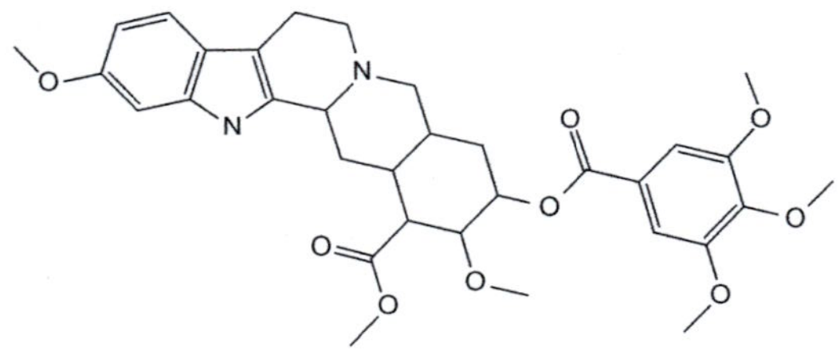	Catecholamines like dopamine and norepinephrine have storage vesicles that this protein binds to. Irreversibly blocks vesicular monoamine transporter-2	[256]
	Rescinnamine	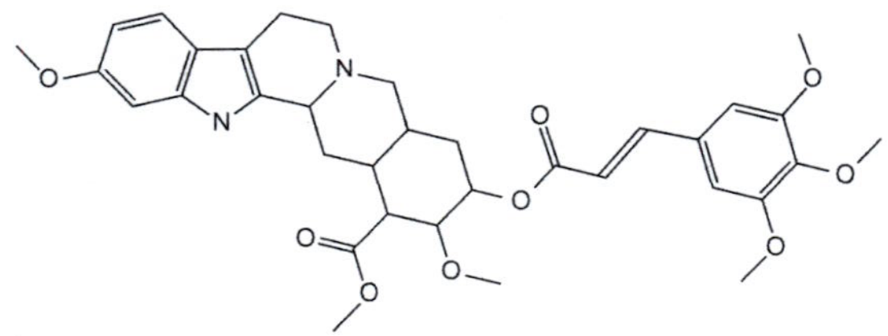	Most potent inhibitor of AChE and BuChE	[199]
	Deserpidine	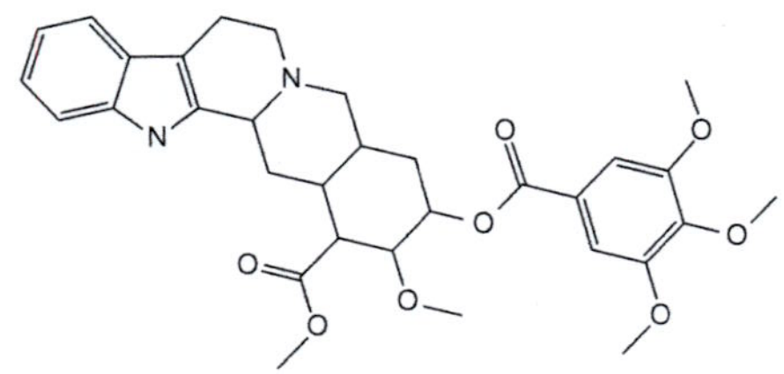	ATP/Mg2+ pump, responsible for sequestering in storage vesicle, inhibited, permitting catecholamines to be rapidly metabolized by MOA, resulting in drop in catecholamines	
	Ajmaline	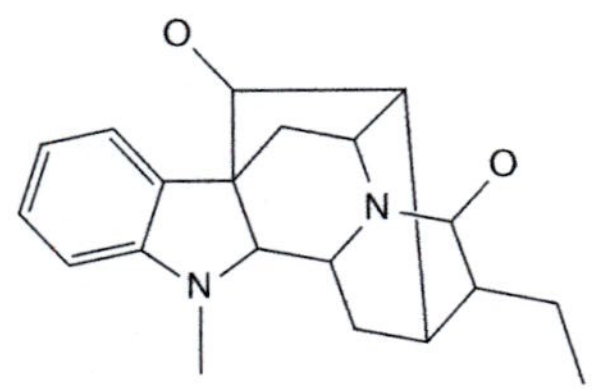	Not identified	

	Serpentine	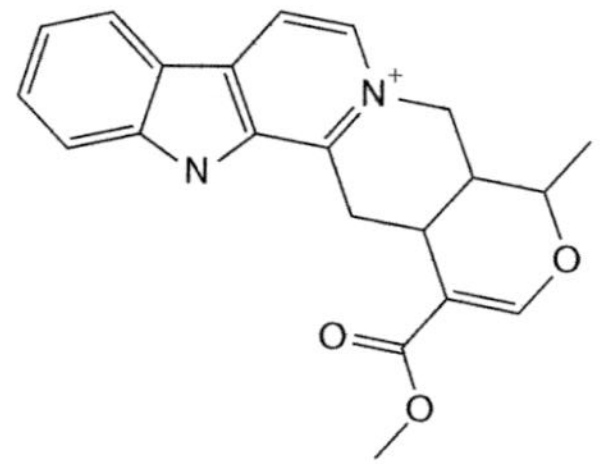	Not identified	
	Alstonine	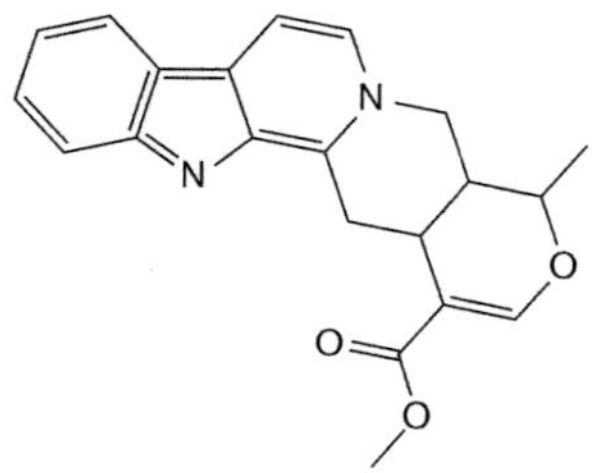	Shows clear anxiolytic effects, mediated by 5-HT2A/2C serotonin receptors, implying efficacy against schizophrenia's negative symptoms	[199]
Sesbania grandiflora	Isovestitol	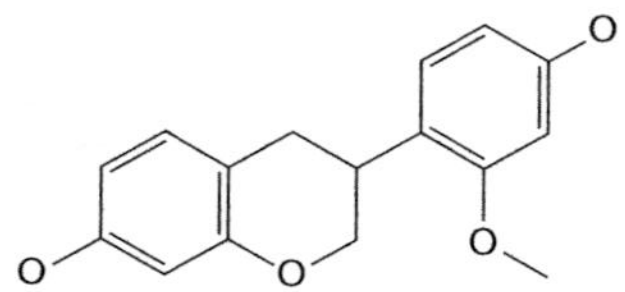	Not identified	
	Medicarpin	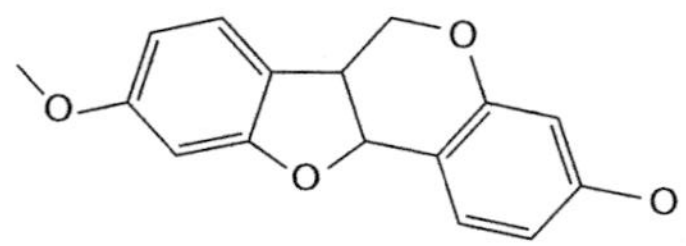	Not identified	
	Sativan	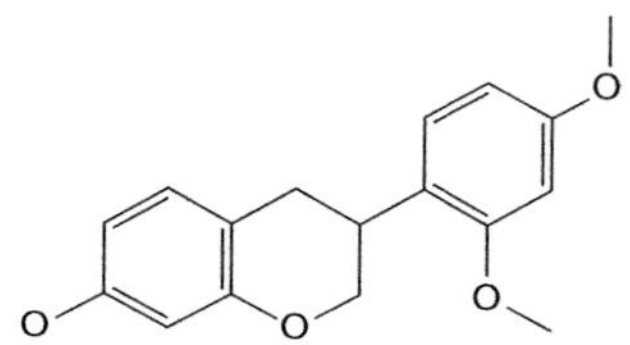	GABA's affinity for its receptor site on same receptor complex increased when attached to specific binding site	[257]

(*Continued*)

Table 4.1 (Continued)

Plant	Phytochemical	Structure	Target	References
	Betulinic acid	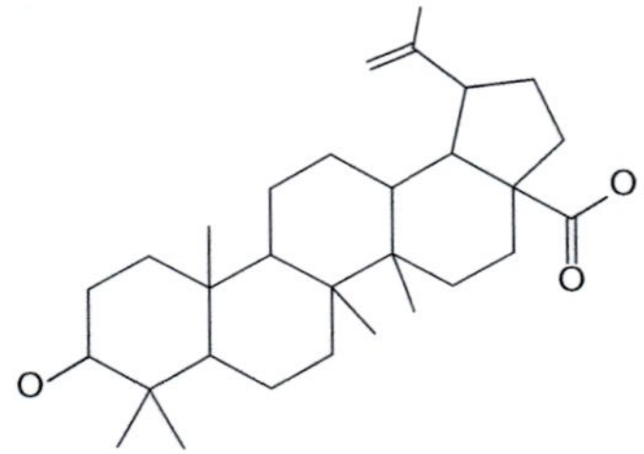	GABAA-BZD receptor thought to be involved in betulinic acid activity	[258]
Sphaeranthus indicus	Sesquiterpenes	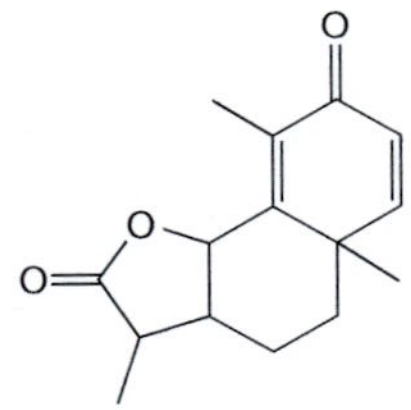	Improves cholinergic transmission by inhibiting AChE	[258]
	Echinacin	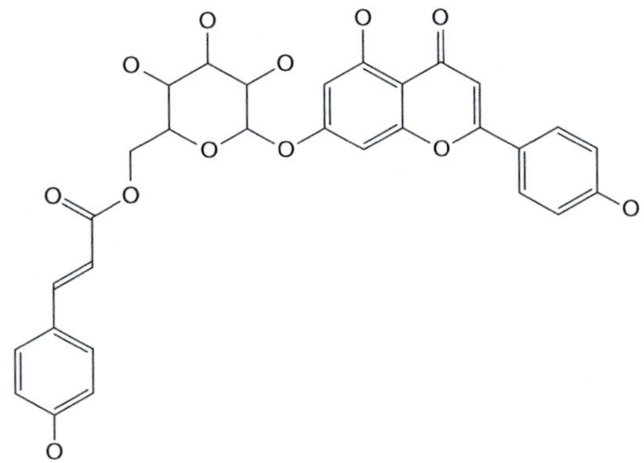	L-DOPA stimulation increased, and antidepressant action observed	[259]
	Echinozolinone	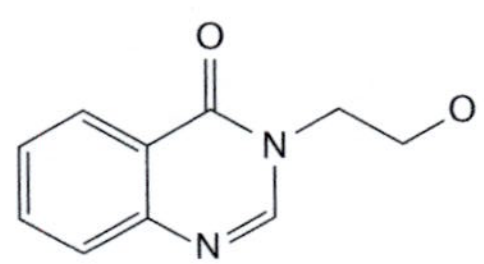	Not identified	

	Dihydroquercetin 4 – methyl ether	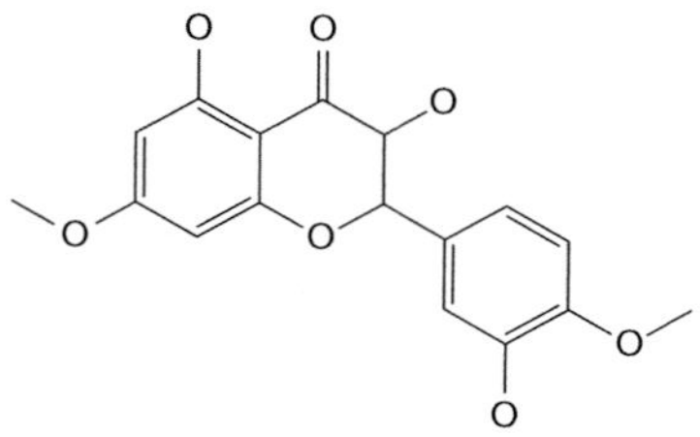	Not identified	
	Lupeol	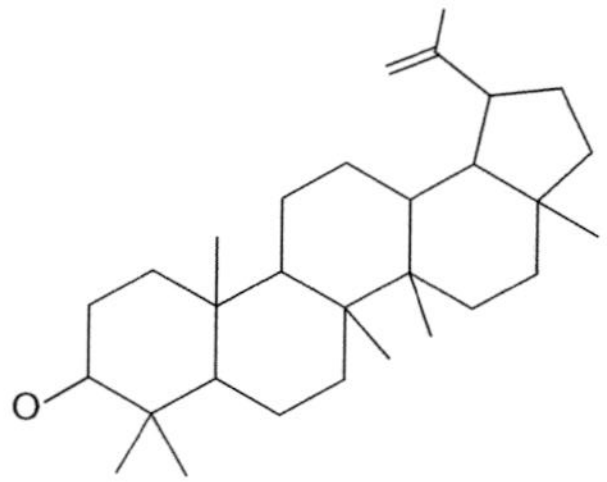	AChE inhibited. Activity of enzyme AChE enhanced. TNF-α, IL-6, and IL-1β rates elevated in hippocampus	[260]
	Echitin	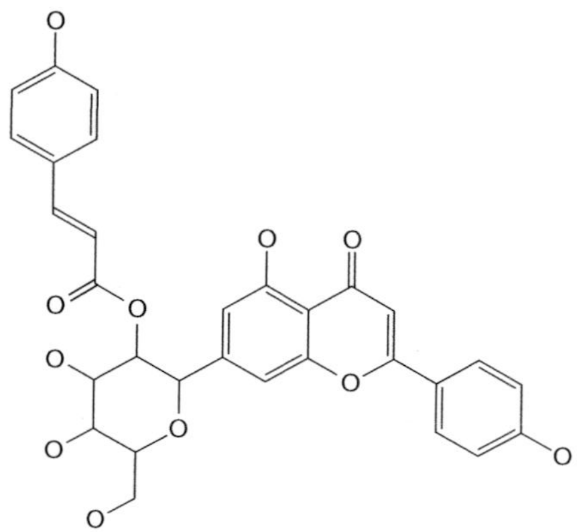	Not identified	
Valeriana wallichii	Valeric acid	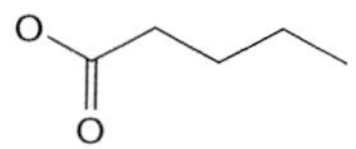	In treatment of experimental dementia, has a substantial GABAergic action	[217]

(*Continued*)

Table 4.1 (Continued)

Plant	Phytochemical	Structure	Target	References
	Sesquiterpene	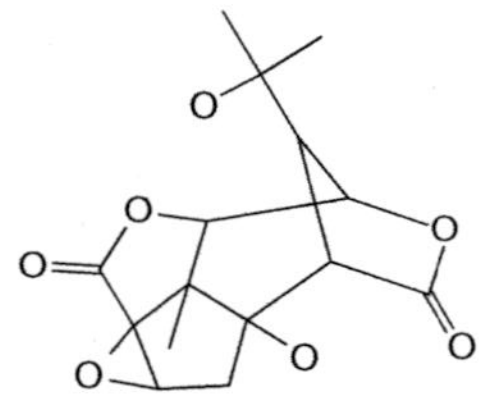 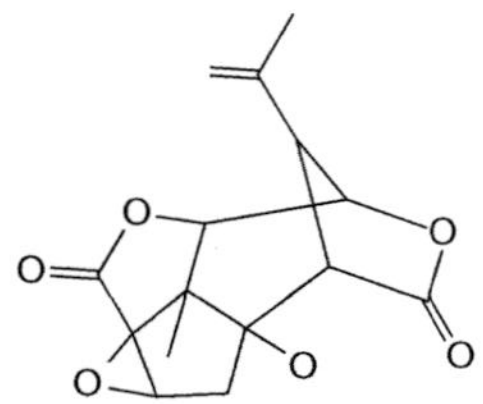	Improves cholinergic transmission by reducing AChE	[258]
	Valeranone	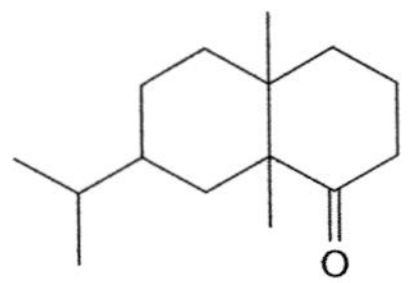	Has been demonstrated to work on brain's amygdaloid body and impede GABA breakdown, resulting in sedation	[261]
	Didrovaltrate	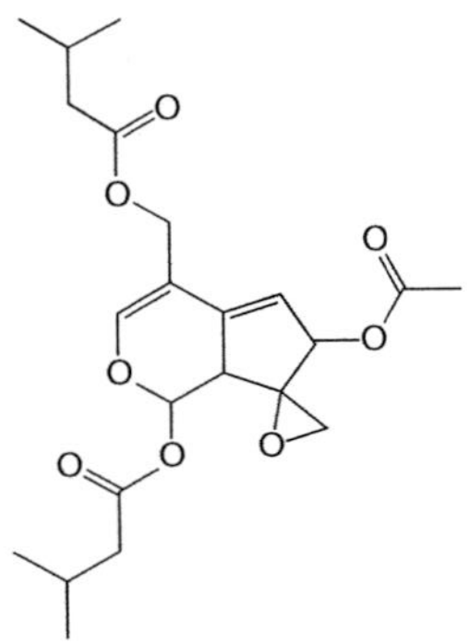	Inhibition of GABA-AT activity; GABA-AT degrades GABA in CNS	[262]

	Valtrate	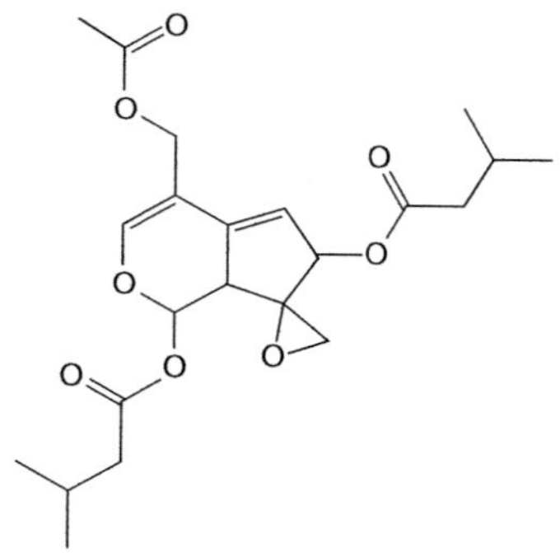	Possesses anxiolytic properties in behavioral studies, which could be mediated by hypothalamus–pituitary adrenal axis	[263]
	Acevaltrate	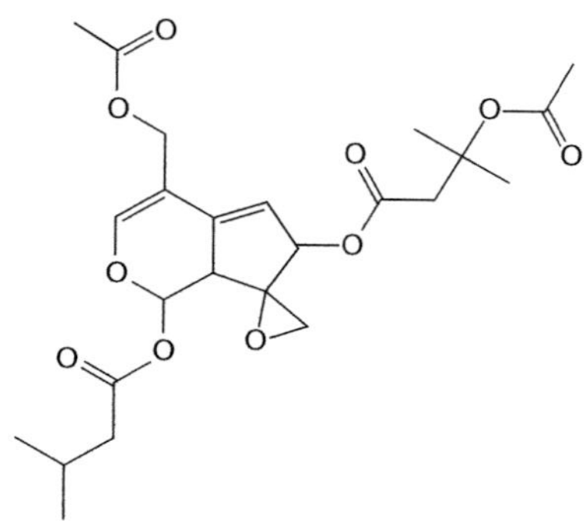	Not identified	[264]
Withania somnifera	Withaferine A	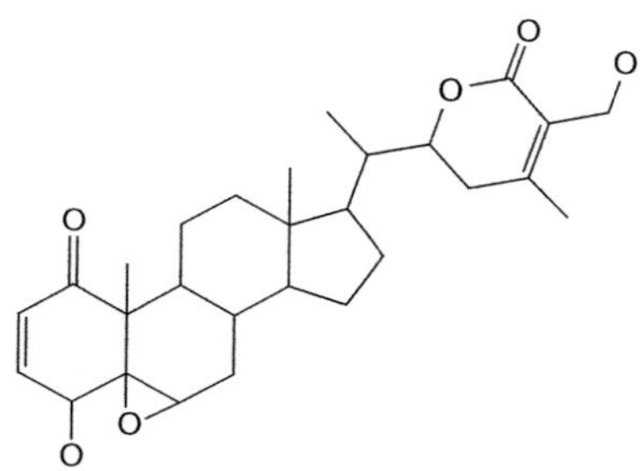	Enhances M1- and M2-muscarinic receptor binding in brain and changes AChE activity in positive way. Increases brain activity of superoxide dismutase, catalase, and glutathione peroxidase	[218]

(*Continued*)

Table 4.1 (Continued)

Plant	Phytochemical	Structure	Target	References
	Somniferine	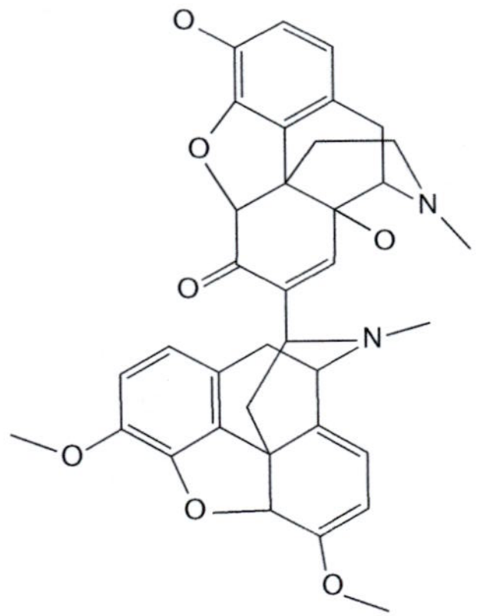	Not identified	
Ziziphus mauritiana	Quercetin 3-O-rutinoside	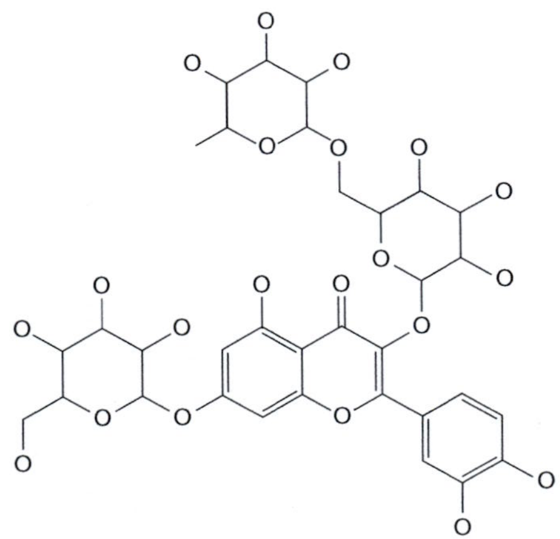	Prevents antienzyme impairment, controls serotonergic and cholinergic neurotransmission, has antianxiety and antidepressant properties, and improves memory	[252]

Puerarin	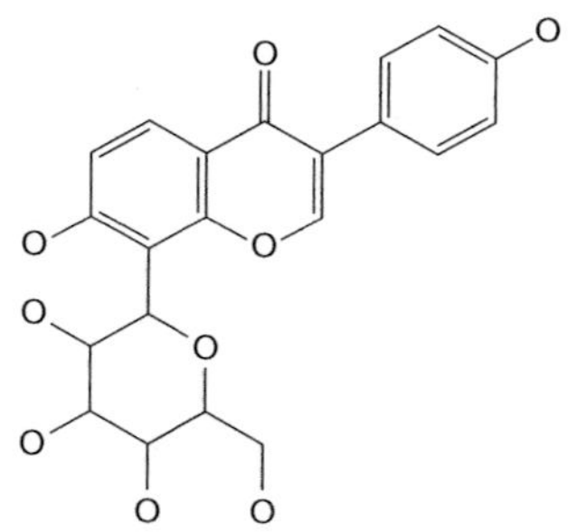	Downregulates transcription of BDNF and mRNA in brain, which has antidepressant-like effects	[265]

Aβ, amyloid beta; AChE, acetylcholinesterase; AMPA, amino-3-hydroxy-5-methyl-4-isoxazole propionic acid; AP-1, activator protein 1; BDNF, brain-derived neurotrophic factor; BuCHe, butyrylcholinesterase; BZD, benzodiazepine; CNS, central nervous system; DA, dopamine; D-DOPA, D-3,4-dihydroxyphenylalanine; ERK, extracellular signal-regulated kinase; GABA, γ-aminobutyric acid; GABA-AT, GABA aminotransferase; IL, interleukin; IP3, inositol trisphosphate; JNK, c-Jun N-terminal kinase; L-DOPA, L-3,4-dihydroxyphenylalanine; MAPK, mitogen-activated protein kinase; MOA, monoamine oxidase; NF-B, nuclear factor B; NFκB, nuclear factor kappa B; NMDA, N-methyl D-aspartate; PI3K, phosphatidylinositol 3-kinase; PNS, peripheral nervous system; THC, tetrahydrocannabinol; TNF, tumor necrosis factor.

required before plant-derived extracts can be widely accepted and used. As summarized in this chapter, ethnobotanical and ethnopharmaceutic (including historical) investigations, as well as other natural active components, account for our present knowledge regarding CNS-active plants of cultural and traditional relevance. The chapter presents a complete overview of the research on these phytochemicals, demonstrating that such substances may be a safe option to approach neurologic illnesses like schizophrenia.

References

1. Tamminga, C.A. and Medoff, D.R. (2000). The biology of schizophrenia. *Dialogues Clin. Neurosci.* 2 (4): 339–348.
2. Praharaj, S., Nizamie, S., Sarkar, S., and Mishra, B. (2012). Add-on effect of Brahmi in the management of schizophrenia. *J. Ayurveda Integr. Med.* 3 (4): 223.
3. Andreasen, N.C. (1995). Symptoms, signs, and diagnosis of schizophrenia. *Lancet* 346 (8973): 477–481.
4. Liddle, P.F. (1987). The symptoms of chronic schizophrenia. A re-examination of the positive-negative dichotomy. *Br. J. Psychiatry* 151: 145–151.
5. van Os, J., Rutten, B.P., and Poulton, R. (2008). Gene-environment interactions in schizophrenia: review of epidemiological findings and future directions. *Schizophr. Bull.* 34 (6): 1066–1082.
6. Correll, C.U. and Schooler, N.R. (2020). Negative symptoms in schizophrenia: A review and clinical guide for recognition, assessment, and treatment. *Neuropsychiatr. Dis. Treat.* 16: 519–534.
7. Sagar, R., Dandona, R., Gururaj, G. et al. (2020). The burden of mental disorders across the states of India: the global burden of disease study 1990–2017. *Lancet Psychiatry* 7 (2): 148–161.
8. Brennand, K., Savas, J.N., Kim, Y. et al. (2015). Phenotypic differences in hiPSC NPCs derived from patients with schizophrenia. *Mol. Psychiatry* 20 (3): 361–368.
9. Stachowiak, E.K., Benson, C.A., Narla, S.T. et al. (2017). Cerebral organoids reveal early cortical maldevelopment in schizophrenia – computational anatomy and genomics, role of FGFR1. *Transl. Psychiatry* 7 (11): 6.
10. Bergdolt, L. and Dunaevsky, A. (2019). Brain changes in a maternal immune activation model of neurodevelopmental brain disorders. *Prog. Neurobiol.* 175: 1–19.
11. Selemon, L.D. and Zecevic, N. (2015). Schizophrenia: a tale of two critical periods for prefrontal cortical development. *Transl. Psychiatry* 5 (8): e623–e623.
12. Park, G.H., Noh, H., Shao, Z. et al. (2020). Activated microglia cause metabolic disruptions in developmental cortical interneurons that persist in interneurons from individuals with schizophrenia. *Nat. Neurosci.* 23 (11): 1352–1364.
13. Page, S.C., Sripathy, S.R., Farinelli, F. et al. (2022). Electrophysiological measures from human iPSC-derived neurons are associated with schizophrenia clinical status and predict individual cognitive performance. *Proc. Natl. Acad. Sci. U. S. A.* 119 (3): e2109395119.
14. Räsänen, N., Tiihonen, J., Koskuvi, M. et al. (2022). The iPSC perspective on schizophrenia. *Trends Neurosci.* 45 (1): 8–26.
15. Sawada, T., Chater, T.E., Sasagawa, Y. et al. (2020). Developmental excitation-inhibition imbalance underlying psychoses revealed by single-cell analyses of discordant twins-derived cerebral organoids. *Mol. Psychiatry* 25 (11): 2695–2711.

16 Volk, D.W. and Lewis, D.A. (2013). Prenatal ontogeny as a susceptibility period for cortical GABA neuron disturbances in schizophrenia. *Neuroscience* 248: 154–164.

17 Lohoff, F.W., Hodge, R., Narasimhan, S. et al. (2014). Functional genetic variants in the vesicular monoamine transporter 1 modulate emotion processing. *Mol. Psychiatry* 19 (1): 129–139.

18 Mulvihill, K.G. (2019). Presynaptic regulation of dopamine release: role of the DAT and VMAT2 transporters. *Neurochem. Int.* 122: 94–105.

19 Multani, P.K., Hodge, R., Estévez, M.A. et al. (2013). VMAT1 deletion causes neuronal loss in the hippocampus and neurocognitive deficits in spatial discrimination. *Neuroscience* 232: 32–44.

20 Huang, M., He, W., Rajagopal, L. et al. (2020). Effects of NBI-98782, a selective vesicular monoamine transporter 2 (VMAT2) inhibitor, on neurotransmitter efflux and phencyclidine-induced locomotor activity: relevance to tardive dyskinesia and antipsychotic action. *Pharmacol. Biochem. Behav.* 190: 172872.

21 Zucker, M. (2002). Increased platelet vesicular monoamine transporter density in adult schizophrenia patients. *Eur. Neuropsychopharmacol.* 12 (4): 343–347.

22 Zubieta, J.K., Taylor, S.F., Huguelet, P. et al. (2001). Vesicular monoamine transporter concentrations in bipolar disorder type I, schizophrenia, and healthy subjects. *Biol. Psychiatry* 49 (2): 110–116.

23 Scorr, L.M. and Factor, S.A. (2018). VMAT2 inhibitors for the treatment of tardive dyskinesia. *J. Neurol. Sci.* 389: 43–47.

24 Gutiérrez, B., Rosa, A., Papiol, S. et al. (2007 Jun 5). Identification of two risk haplotypes for schizophrenia and bipolar disorder in the synaptic vesicle monoamine transporter gene (SVMT). *Am. J. Med. Genet. B Neuropsychiatr. Genet.* 144B (4): 502–507.

25 Lohoff, F.W., Carr, G.V., Brookshire, B. et al. (2019). Deletion of the vesicular monoamine transporter 1 (vmat1/slc18a1) gene affects dopamine signaling. *Brain Res.* 1712: 151–157.

26 Kuepper, R., Skinbjerg, M., and Abi-Dargham, A. (2012). The dopamine dysfunction in schizophrenia revisited: new insights into topography and course. *Handb. Exp. Pharmacol.* 212: 1–26.

27 Metzger, R.R., Brown, J.M., Sandoval, V. et al. (2002). Inhibitory effect of reserpine on dopamine transporter function. *Eur. J. Pharmacol.* 456 (1–3): 39–43.

28 Liedtke, E.I., Zhang, S., Thompson, J.A. et al. (2016). Correlated expression analysis of genes implicated in schizophrenia: identification of putative disease-related pathways. *Eur. J. Mol. Clin. Med.* 3 (5): 224.

29 Brito-Melo, G.E.A., Nicolato, R., de Oliveira, A.C.P. et al. (2012). Increase in dopaminergic, but not serotoninergic, receptors in T-cells as a marker for schizophrenia severity. *J. Psychiatr. Res.* 46 (6): 738–742.

30 Sarkar, C., Basu, B., Chakroborty, D. et al. (2010). The immunoregulatory role of dopamine: an update. *Brain Behav. Immun.* 24 (4): 525–528.

31 Torres, G.E., Gainetdinov, R.R., and Caron, M.G. (2003). Plasma membrane monoamine transporters: structure, regulation and function. *Nat. Rev. Neurosci.* 4 (1): 13–25.

32 Zakharyan, R., Ghazaryan, H., Kocourkova, L. et al. (2020). Association of genetic variants of dopamine and serotonin in schizophrenia. *Arch. Med. Res.* 51 (1): 13–20.

33 Peitl, V., Vidrih, B., Karlović, Z. et al. (2016). Platelet serotonin concentration and depressive symptoms in patients with schizophrenia. *Psychiatry Res.* 239: 105–110.

34 Abi-Dargham, A. (2007). Alterations of serotonin transmission in schizophrenia. In: *Int. Rev. Neurobiol*, vol. 78, 133–164.

35 Lieberman, J.A., Mailman, R.B., Duncan, G. et al. (1998). Serotonergic basis of antipsychotic drug effects in schizophrenia. *Biol. Psychiatry* 44 (11): 1099–1117.

36 Meneses, A. (1999). 5-HT system and cognition. *Neurosci. Biobehav. Rev.* 23 (8): 1111–1125.

37 Quednow, B.B., Geyer, M.A., and Halberstadt, A.L. (2010). Serotonin and schizophrenia. In: *Handbook of Behavioral Neuroscience* (ed. C.P. Müller and B.L. Jacobs), 585–620. Amsterdam: Elsevier https://linkinghub.elsevier.com/retrieve/pii/S1569733910701028.

38 Remington, G. (2008). Alterations of dopamine and serotonin transmission in schizophrenia. In: *Progress in Brain Research*, vol. 172 (ed. G. Di Giovann, V. Di Matteo, and E. Esposito), 117–140. Amsterdam: Elsevier https://linkinghub.elsevier.com/retrieve/pii/S0079612308009060.

39 Jenkins, T., Nguyen, J., Polglaze, K., and Bertrand, P. (2016). Influence of tryptophan and serotonin on mood and cognition with a possible role of the gut-brain Axis. *Nutrients* 8 (1): 56.

40 Gobert, A., Rivet, J.M., Audinot, V. et al. (1998). Simultaneous quantification of serotonin, dopamine and noradrenaline levels in single frontal cortex dialysates of freely-moving rats reveals a complex pattern of reciprocal auto- and heteroreceptor-mediated control of release. *Neuroscience* 84 (2): 413–429.

41 Dean, B. and Hayes, W. (1996). Decreased frontal cortical serotonin2A receptors in schizophrenia. *Schizophr. Res.* 21 (3): 133–139.

42 Rasmussen, H., Erritzoe, D., Andersen, R. et al. (2010). Decreased frontal Serotonin2A receptor binding in antipsychotic-naive patients with first-episode schizophrenia. *Arch. Gen. Psychiatry* 67 (1): 9.

43 Rasmussen, H., Frokjaer, V.G., Hilker, R.W. et al. (2016). Low frontal serotonin 2A receptor binding is a state marker for schizophrenia? *Eur. Neuropsychopharmacol.* 26 (7): 1248–1250.

44 Selvaraj, S., Arnone, D., Cappai, A., and Howes, O. (2014). Alterations in the serotonin system in schizophrenia: a systematic review and meta-analysis of postmortem and molecular imaging studies. *Neurosci. Biobehav. Rev.* 45: 233–245.

45 Burnet, P. (1996). 5-HT1A and 5-HT2A receptor mRNAs and binding site densities are differentially altered in schizophrenia. *Neuropsychopharmacology* 15 (5): 442–455.

46 Cheah, S.Y., Lawford, B., Young, R. et al. (2017). mRNA expression and DNA methylation analysis of serotonin receptor 2A (HTR2A) in the human schizophrenic brain. *Genes* 8 (1): 14.

47 Hernandez, I. and Sokolov, B.P. (1997). Abnormal expression of serotonin transporter mRNA in the frontal and temporal cortex of schizophrenics. *Mol. Psychiatry* 2 (1): 57–64.

48 Lin, C., Tang, W., Hu, J. et al. (2009). Haplotype analysis confirms association of the serotonin transporter (5-HTT) gene with schizophrenia in the Han Chinese population. *Neurosci. Lett.* 453 (3): 210–213.

49 Badner, J.A. and Gershon, E.S. (2002). Meta-analysis of whole-genome linkage scans of bipolar disorder and schizophrenia. *Mol. Psychiatry* 7 (4): 405–411.

50 Liu, J., Juo, S.H., Dewan, A. et al. (2003). Evidence for a putative bipolar disorder locus on 2p13–16 and other potential loci on 4q31, 7q34, 8q13, 9q31, 10q21–24, 13q32, 14q21 and 17q11–12. *Mol. Psychiatry* 8 (3): 333–342.

51 Abdolmaleky, H.M., Nohesara, S., Ghadirivasfi, M. et al. (2014). DNA hypermethylation of serotonin transporter gene promoter in drug naïve patients with schizophrenia. *Schizophr. Res.* 152 (2–3): 373–380.

52 Kim, J.H., Son, Y.D., Kim, J.H. et al. (2015). Serotonin transporter availability in thalamic subregions in schizophrenia: A study using 7.0-T MRI with [11C]DASB high-resolution PET. *Psychiatry Res. Neuroimaging* 231 (1): 50–57.

53 Li, W., Yang, Y., Lin, J. et al. (2013). Association of serotonin transporter gene (SLC6A4) polymorphisms with schizophrenia susceptibility and symptoms in a Chinese-Han population. *Prog. Neuropsychopharmacol. Biol. Psychiatry* 44: 290–295.

54 Bozorgmehr, A., Sadeghi, B., Tabatabaei Zavari, E.S. et al. (2020). An integrative gene network-based approach to uncover the cellular and molecular infrastructures of schizophrenia. *Life Sci.* 260: 118345.

55 Park, M.T.M., Jeon, P., Khan, A.R. et al. (2021). Hippocampal neuroanatomy in first episode psychosis: A putative role for glutamate and serotonin receptors. *Prog. Neuropsychopharmacol. Biol. Psychiatry* 110: 110297.

56 Sanacora, G., Zarate, C.A., Krystal, J.H., and Manji, H.K. (2008). Targeting the glutamatergic system to develop novel, improved therapeutics for mood disorders. *Nat. Rev. Drug Discov.* 7 (5): 426–437.

57 Bojesen, K.B., Broberg, B.V., Fagerlund, B. et al. (2021). Associations between cognitive function and levels of glutamatergic metabolites and gamma-aminobutyric acid in antipsychotic-Naïve patients with schizophrenia or psychosis. *Biol. Psychiatry* 89 (3): 278–287.

58 Schoonover, K.E., Dienel, S.J., and Lewis, D.A. (2020). Prefrontal cortical alterations of glutamate and GABA neurotransmission in schizophrenia: insights for rational biomarker development. *Biomark Neuropsychiatry.* 3: 100015.

59 Ragland, J.D., Maddock, R.J., Hurtado, M.Y. et al. (2020). Disrupted GABAergic facilitation of working memory performance in people with schizophrenia. *Neuroimage Clin.* 25: 102127.

60 Hashimoto, T., Arion, D., Unger, T. et al. (2008). Alterations in GABA-related transcriptome in the dorsolateral prefrontal cortex of subjects with schizophrenia. *Mol. Psychiatry* 13 (2): 147–161.

61 Govindpani, K., Guzmán, B.C.-F., Vinnakota, C. et al. (2017). Towards a better understanding of GABAergic remodeling in Alzheimer's disease. *Int. J. Mol. Sci.* 18 (8): 1813.

62 Rowland, L.M., Summerfelt, A., Wijtenburg, S.A. et al. (2016 Feb 1). Frontal glutamate and γ-aminobutyric acid levels and their associations with mismatch negativity and digit sequencing task performance in schizophrenia. *JAMA Psychiat.* 73 (2): 166.

63 Yoon, J.H., Maddock, R.J., Rokem, A. et al. (2010). GABA concentration is reduced in visual cortex in schizophrenia and correlates with orientation-specific surround suppression. *J. Neurosci.* 30 (10): 3777–3781.

64 Yoon, J.H., Maddock, R.J., DongBo Cui, E. et al. (2020). Reduced in vivo visual cortex GABA in schizophrenia, a replication in a recent onset sample. *Schizophr. Res.* 215: 217–222.

65 Newell, K.A., Zavitsanou, K., Jew, S.K., and Huang, X.F. (2007). Alterations of muscarinic and GABA receptor binding in the posterior cingulate cortex in schizophrenia. *Prog. Neuropsychopharmacol. Biol. Psychiatry* 31 (1): 225–233.

66 Atagün, M.İ., Şıkoğlu, E.M., Soykan, Ç. et al. (2017). Perisylvian GABA levels in schizophrenia and bipolar disorder. *Neurosci. Lett.* 637: 70–74.

67 Förster, A., Model, V., Gos, T. et al. (2022). Reduced GABAergic neuropil and interneuron profiles in schizophrenia: complementary analysis of disease course-related differences. *J. Psychiatr. Res.* 145: 50–59.

68 Steiner, J., Brisch, R., Schiltz, K. et al. (2016). GABAergic system impairment in the hippocampus and superior temporal gyrus of patients with paranoid schizophrenia: A post-mortem study. *Schizophr. Res.* 177 (1–3): 10–17.

69 Folsom, T.D., Thuras, P.D., and Fatemi, S.H. (2015). Protein expression of targets of the FMRP regulon is altered in brains of subjects with schizophrenia and mood disorders. *Schizophr. Res.* 165 (2–3): 201–211.

70 Frajman, A., Maggio, N., Muler, I. et al. (2020). Gene expression meta-analysis reveals the down-regulation of three GABA receptor subunits in the superior temporal gyrus of patients with schizophrenia. *Schizophr. Res.* 220: 29–37.

71 Saleh, A.A., Elhelbawy, N.G., Azmy, R.M. et al. (2022). Evaluation of mRNA expression level of the ATP synthase membrane subunit c locus 1 (ATP5G1) gene in patients with schizophrenia. *Biochem. Biophys. Rep.* 30: 101234.

72 Ji, C.J., Zeng, G.Z., Han, J. et al. (2012). Zizimauritic acids A–C, three novel nortriterpenes from Ziziphus mauritiana. *Bioorg. Med. Chem. Lett.* 22 (20): 6377–6380.

73 Qing, L., Liu, L., Zhou, L. et al. (2020). Sex-dependent association of mineralocorticoid receptor gene (NR3C2) DNA methylation and schizophrenia. *Psychiatry Res.* 292: 113318.

74 Xing, G.Q., Russell, S., Webster, M.J., and Post, R.M. (2004). Decreased expression of mineralocorticoid receptor mRNA in the prefrontal cortex in schizophrenia and bipolar disorder. *Int. J. Neuropsychopharmacol.* 7 (2): 143–153.

75 He, E., Lozano, M.A.G., Stringer, S. et al. (2018). MIR137 schizophrenia-associated locus controls synaptic function by regulating synaptogenesis, synapse maturation and synaptic transmission. *Hum. Mol. Genet.* 27 (11): 1879–1891.

76 Siegert, S., Seo, J., Kwon, E.J. et al. (2015). The schizophrenia risk gene product miR-137 alters presynaptic plasticity. *Nat. Neurosci.* 18 (7): 1008–1016.

77 Yao, Y., Guo, W., Zhang, S. et al. (2021). Cell type-specific and cross-population polygenic risk score analyses of MIR137 gene pathway in schizophrenia. *iScience* 24 (7): 102785.

78 Gong, L., Liu, M., Jen, J., and Yeh, E.T.H. (2000). GNB1L, a gene deleted in the critical region for DiGeorge syndrome on 22q11, encodes a G-protein β-subunit-like polypeptide. *Biochim. Biophys. Acta* 1494 (1–2): 185–188.

79 Bassett, A.S. and Chow, E.W.C. (2008). Schizophrenia and 22q11.2 deletion syndrome. *Curr. Psychiatry Rep.* 10 (2): 148–157.

80 Ishiguro, H., Koga, M., Horiuchi, Y. et al. (2010). Supportive evidence for reduced expression of GNB1L in schizophrenia. *Schizophr. Bull.* 36 (4): 756–765.

81 Vermeesch, J.R. (2022). The hunt for the chromosome 22q11.2 deletion syndrome schizophrenia genes. *Biol. Psychiatry* 91 (8): 692–693.

82 Wang, Y.Y., Hsu, S.H., Tsai, H.Y., and Cheng, M.C. (2021). Sequencing of the coding regions of GNBIL on chromosome 22q11.2 as a risk gene of schizophrenia. *Psychiatry Res.* 300: 113943.

83 Chong, S.Y.C., Rosenberg, S.S., Fancy, S.P.J. et al. (2012). Neurite outgrowth inhibitor Nogo-A establishes spatial segregation and extent of oligodendrocyte myelination. *Proc. Natl. Acad. Sci. U. S. A.* 109 (4): 1299–1304.

84 Sui, Y.P., Zhang, X.X., Lu, J.L., and Sui, F. (2015). New insights into the roles of Nogo-A in CNS biology and diseases. *Neurochem. Res.* 40 (9): 1767–1785.

85 Demirel, Ö.F., Cetin, İ., Turan, Ş. et al. (2017). Decreased expression of α-synuclein, Nogo-A and UCH-L1 in patients with schizophrenia: a preliminary serum study. *Psychiatry Investig.* 14 (3): 344.

86 Tan, E.C., Chong, S.A., Wang, H. et al. (2005). Gender-specific association of insertion/deletion polymorphisms in the nogo gene and chronic schizophrenia. *Brain Res. Mol. Brain Res.* 139 (2): 212–216.

87 Zhu, L., Wu, X., Xu, B. et al. (2021). The machine learning algorithm for the diagnosis of schizophrenia on the basis of gene expression in peripheral blood. *Neurosci. Lett.* 745: 135596.

88 Pvsn, K.K., Mitra, P., Ghosh, R. et al. (2021). Association of the NOTCH4 gene polymorphism with schizophrenia in the Indian population. *Meta Gene.* 29: 100903.

89 Wang, S., Sun, C.E., Walczak, C.A. et al. (1995). Evidence for a susceptibility locus for schizophrenia on chromosome 6pter–p22. *Nat. Genet.* 10 (1): 41–46.

90 Wei, J. and Hemmings, G.P. (2000). The NOTCH4 locus is associated with susceptibility to schizophrenia. *Nat. Genet.* 25 (4): 376–377.

91 Zhang, B., Fan, Q.R., Li, W.H. et al. (2015). Association of the *NOTCH4* gene polymorphism rs204993 with schizophrenia in the Chinese Han population. *Biomed. Res. Int.* 2015: 1–5.

92 Dorrigiv, M., Zareiyan, A., and Hosseinzadeh, H. (2021). Onion (*Allium cepa*) and its Main constituents as antidotes or protective agents against natural or chemical toxicities: a comprehensive review. *Iran J. Pharm. Res.* 20 (1): 3–26.

93 Kadian, R. and Parle, M. (2014). Evaluation of antipsychotic effect of *Allium cepa*. *WJPPS* 3: 1146–1159.

94 Willborn, R.J., Hall, C.P., and Fuller, M.A. (2019). Recycling N-acetylcysteine: a review of evidence for adjunctive therapy in schizophrenia. *Ment. Health Clin.* 9 (3): 116–123.

95 Rossell, S.L., Francis, P.S., Galletly, C. et al. (2016). N-acetylcysteine (NAC) in schizophrenia resistant to clozapine: a double blind randomised placebo controlled trial targeting negative symptoms. *BMC Psychiatry* 16 (1): 320.

96 Mukherjee, P.K., Kumar, V., Mal, M., and Houghton, P.J. (2007). *Acorus calamus*: scientific validation of ayurvedic tradition from natural resources. *Pharm. Biol.* 45 (8): 651–666.

97 Joshi, R.K. (2016). *Acorus calamus* Linn.: phytoconstituents and bactericidal property. *World J. Microbiol. Biotechnol.* 32 (10): 164.

98 Pawar, V.S., Anup, A., Shrikrishna, B., and Shivakumar, H. (2011). Antidepressant–like effects of *Acorus calamus* in forced swimming and tail suspension test in mice. *Asian Pac. J. Trop. Biomed.* 1 (1): S17–S19.

99 Sharma, V., Sharma, R., Gautam, D. et al. (2020). Role of Vacha (*Acorus calamus* Linn.) in neurological and metabolic disorders: evidence from ethnopharmacology, phytochemistry, pharmacology and clinical study. *J. Clin. Med.* 9 (4): 1176.

100 Shougrakpam, P., Bhattacharjee, A., Medhabati, M., and Gunindro, N. (2021). Comparative study of antidepressant-like effect of the leaves of *Sapindus emarginatus* and *Acorus calamus* in experimental animal models. *Int. J. Curr. Pharm. Res.* 13 (1): 28–31.

101 Tripathi, A. and Singh, R. (2010). Experimental evaluation of antidepressant effect of Vacha (*Acorus calamus*) in animal models of depression. *AYU* 31 (2): 153.

102 Yousuf, S., Marifatul Haq, S., Rasool, A. et al. (2020). Evaluation of antidepressant activity of methanolic and hydroalcoholic extracts of *Acorus calamus* L. rhizome through tail suspension test and forced swimming test of mice. *J. Tradit. Chin. Med. Sci.* 7 (3): 301–307.

103 Esfandiari, E., Ghanadian, M., Rashidi, B. et al. (2018). The effects of *Acorus calamus* L. in preventing memory loss, anxiety, and oxidative stress on lipopolysaccharide-induced neuroinflammation rat models. Int. *J. Prev. Med.* 9: 85.

104 Stough, C., Singh, H., and Zangara, A. (2015). Mechanisms, efficacy, and safety of *Bacopa monnieri* (Brahmi) for cognitive and brain enhancement. *Evid. Based Complement. Alternat. Med.* 2015: 717605.

105 Dwivedi, S., Nagarajan, R., Hanif, K. et al. (2013). Standardized extract of *Bacopa monniera* attenuates okadaic acid induced memory dysfunction in rats: effect on Nrf2 pathway. *Evid. Based Complement. Altern. Med.* 2013: 294501.

106 Pase, M.P., Kean, J., Sarris, J. et al. (2012). The cognitive-enhancing effects of *Bacopa monnieri*: a systematic review of randomized, controlled human clinical trials. *J. Altern. Complement. Med.* 18 (7): 647–652.

107 Stough, C., Lloyd, J., Clarke, J. et al. (2001). The chronic effects of an extract of *Bacopa monniera* (Brahmi) on cognitive function in healthy human subjects. *Psychopharmacology* 156 (4): 481–484.

108 Peth-Nui, T., Wattanathorn, J., Muchimapura, S. et al. (2012). Effects of 12-week *Bacopa monnieri* consumption on attention, cognitive processing, working memory, and functions of both cholinergic and monoaminergic Systems in Healthy Elderly Volunteers. *Evid. Based Complement. Altern. Med.* 2012: 606424.

109 Shalini, V.T., Neelakanta, S.J., and Sriranjini, J.S. (2021). Neuroprotection with *Bacopa monnieri*-A review of experimental evidence. *Mol. Biol. Rep.* 48 (3): 2653–2668.

110 Piyabhan, P., Wannasiri, S., and Naowaboot, J. (2016). *Bacopa monnieri* (Brahmi) improved novel object recognition task and increased cerebral vesicular glutamate transporter type 3 in sub-chronic phencyclidine rat model of schizophrenia. *Clin. Exp. Pharmacol. Physiol.* 43 (12): 1234–1242.

111 Rauf, K., Subhan, F., Abbas, M. et al. (2014). Inhibitory effect of bacopasides on spontaneous morphine withdrawal induced depression in mice. *Phytother Res.* 28 (6): 937–939.

112 Sairam, K., Dorababu, M., Goel, R.K., and Bhattacharya, S.K. (2002). Antidepressant activity of standardized extract of *Bacopa monniera* in experimental models of depression in rats. *Phytomedicine* 9 (3): 207–211.

113 Hügel, H.M. and Jackson, N. (2015). Polyphenols for the prevention and treatment of dementia diseases. *Neural Regen. Res.* 10 (11): 1756–1758.

114 Nemetchek, M.D., Stierle, A.A., Stierle, D.B., and Lurie, D.I. (2017). The ayurvedic plant *Bacopa monnieri* inhibits inflammatory pathways in the brain. *J. Ethnopharmacol.* 197: 92–100.

115 Mathur, D., Goyal, K., Koul, V., and Anand, A. (2016). The molecular links of re-emerging therapy: A review of evidence of Brahmi (*Bacopa monniera*). *Front. Pharmacol.* 7: 44.

116 Dowell, A., Davidson, G., and Ghosh, D. (2015). Validation of quantitative HPLC method for bacosides in KeenMind. *Evid. Based Complement. Alternat. Med.* 2015: 696172.

117 Uabundit, N., Wattanathorn, J., Mucimapura, S., and Ingkaninan, K. (2010). Cognitive enhancement and neuroprotective effects of *Bacopa monnieri* in Alzheimer's disease model. *J. Ethnopharmacol.* 127 (1): 26–31.

118 Kumar, N., Abichandani, L.G., Thawani, V. et al. (2016). Efficacy of standardized extract of *Bacopa monnieri* (Bacognize®) on cognitive functions of medical students: A six-week, randomized placebo-controlled trial. *Evid. Based Complement Altern. Med.* 2016: 4103423.

119 Morgan, A. and Stevens, J. (2010). Does *Bacopa monnieri* improve memory performance in older persons? Results of a randomized, placebo-controlled, double-blind trial. *J. Altern. Complement. Med.* 16 (7): 753–759.

120 Raghav, S., Singh, H., Dalal, P.K. et al. (2006). Randomized controlled trial of standardized *Bacopa monniera* extract in age-associated memory impairment. *Indian J. Psychiatry* 48 (4): 238–242.

121 Kapoor, R., Srivastava, S., and Kakkar, P. (2009). *Bacopa monnieri* modulates antioxidant responses in brain and kidney of diabetic rats. *Environ. Toxicol. Pharmacol.* 27 (1): 62–69.

122 Johri, R.K. (2011). *Cuminum cyminum* and *Carum carvi*: An update. *Pharmacogn. Rev.* 5 (9): 63–72.

123 Khan, R., Ahmad, W., Ahmad, M., and Hasan, A. (2016). Phytochemical and pharmacological properties of Carum carvi. *Eur. J. Pharm. Med. Res.* 3: 231–236.

124 Sachan, A.K., Das, D.R., and Kumar, M. (2016). Carum carvi-An important medicinal plant. *J. Chem. Pharm. Res.* 8 (3): 529–533.

125 Mahboubi, M. (2019). Caraway as important medicinal plants in Management of Diseases. *Nat. Prod. Bioprospecting* 9 (1): 1–11.

126 Dhumal, S.R., Shete, R.V., Adak, V.S., and Murthy, K. (2021). A review on *Celastrus paniculatus* wild (Jyotishmati): It's species, geographical sources, phytoconstituents, and therapeutic uses and pharmacological actions. *IJHM* 9 (4): 112–118.

127 Raviraja Shetty, G., Poojitha, K.G., Souravi, K., and Rajasekharan, P.E. (2016). Seed germination studies in *Celastrus paniculatus* Willd: a threated medicinal a plant. *Int. J. Appl. Agric. Res.* 11 (1): 51–56.

128 Kulkarni, Y.A., Agarwal, S., and Garud, M.S. (2015). Effect of Jyotishmati (*Celastrus paniculatus*) seeds in animal models of pain and inflammation. *J. Ayurveda Integr. Med.* 6 (2): 82–88.

129 Sharma, G.N., Kaur, H., Shrivastava, B., and Arora, S.C. (2021). A review from historical to current-*Celastrus paniculatus. Int J Pharm Pharm Sci* 21: 139–144.

130 Maurya, H., Arya, R., Belwal, T. et al. (2021). *Celastrus paniculatus*. In: *Naturally Occurring Chemicals against Alzheimer's Disease* (ed. T. Belwal, S.M. Nabavi, S. Fazel, et al.), 425–435.

131 Raut, S.B., Parekar, R.R., Jadhav, K.S. et al. (2015). Effect of Jyotiṣmatī seed oil on spatial and fear memory using scopolamine induced amnesia in mice. *Anc. Sci. Life* 34 (3): 130–133.

132 Gattu, M., Boss, K.L., Terry, A.V., and Buccafusco, J.J. (1997). Reversal of scopolamine-induced deficits in navigational memory performance by the seed oil of *Celastrus paniculatus. Pharmacol. Biochem. Behav.* 57 (4): 793–799.

133 Borrelli, F., Borbone, N., Capasso, R. et al. (2009). Potent relaxant effect of a *Celastrus paniculatus* extract in the rat and human ileum. *J. Ethnopharmacol.* 122 (3): 434–438.

134 Kumar, M.H.V. and Gupta, Y.K. (2002). Antioxidant property of *Celastrus paniculatus* Willd.: a possible mechanism in enhancing cognition. *Phytomedicine* 9 (4): 302–311.

135 Heywood, V.H. (2014). The socio-economic importance of the Apiales. *J. Fac. Pharm. Istanb Univ.* 44 (2): 113–130.

136 Gohil, K., Patel, J., and Gajjar, A. (2010). Pharmacological review on *Centella asiatica*: A potential herbal cure-all. *Indian J. Pharm. Sci.* 72 (5): 546.

137 Sen, S. and Dutta, S. (2020). A comprehensive review on Thankuni (*Centella asiatica*) as an herbal remedy in diabetes mellitus and wound healing. *J. Pharmacogn. Phytochem.* 9 (5): 1203–1209.

138 Aparna, V., Mallya, S.V., Srikanth, P., and Sunil Kumar, K.N. (2015). Comparative pharmacognosy of two medhya dravyas, Brahmi (*Bacopa monnieri* Linn.) and Mandukaparni (*Centella asiatica* Linn.). *J. Phyto. Pharmacol.* 4 (1): 1–5.

139 James, J. and Dubery, I. (2009). Pentacyclic triterpenoids from the medicinal herb, *Centella asiatica* (L.) urban. *Molecules* 14 (10): 3922–3941.

140 Kulkarni, R., Girish, K.J., and Kumar, A. (2012). Nootropic herbs (medhya rasayana) in ayurveda: an update. *Pharmacogn. Rev.* 6 (12): 147–153.

141 Patel, S., Kumar, S., Kumar, D., and Aggarwal, S. (2022). Importance of medhya rasayana drugsused inthe treatment of mental stress. *Int. Res. J. Ayurveda Yoga* 4: 125–130.

142 Wang, L., Guo, T., Guo, Y., and Xu, Y. (2020). Asiaticoside produces an antidepressant-like effect in a chronic unpredictable mild stress model of depression in mice, involving reversion of inflammation and the PKA/pCREB/BDNF signaling pathway. *Mol. Med. Rep.* 22 (3): 2364–2372.

143 Chen, Y., Han, T., Qin, L. et al. (2003). Effect of total triterpenes from *Centella asiatica* on the depression behavior and concentration of amino acid in forced swimming mice. *Zhong Yao Cai.* 26 (12): 870–873.

144 Devi, P. (2021). An updated review on Shankhpushpi-as medhya rasayana. *J. Ayurvedic. Herb. Med.* 7 (2): 119–123.

145 Agarwa, P., Sharma, B., Fatima, A., and Jain, S.K. (2014). An update on ayurvedic herb *Convolvulus pluricaulis* Choisy. *Asian Pac. J. Trop. Biomed.* 4 (3): 245–252.

146 Malik, J., Karan, M., and Vasisht, K. (2011). Nootropic, anxiolytic and CNS-depressant studies on different plant sources of shankhpushpi. *Pharm. Biol.* 49 (12): 1234–1242.

147 Khan, I., Dubey, W., and Gupta, V. (2014). Taxonomical aspect of coriander (*Coriandrum sativum* L.). *Int. J. Curr. Res.* 6: 9926–9930.

148 Sahib, N.G., Anwar, F., Gilani, A.H. et al. (2013). Coriander (*Coriandrum sativum* L.): a potential source of high-value components for functional foods and nutraceuticals – a review. *Phytother. Res.* 27 (10): 1439–1456.

149 Prachayasittikul, V., Prachayasittikul, S., Ruchirawat, S., and Prachayasittikul, V. (2018). Coriander (Coriandrum sativum): a promising functional food toward the well-being. *Food Res. Int.* 105: 305–323.

150 Gil, A., de la Fuente, E.B., Lenardis, A.E. et al. (2002). Coriander essential oil composition from two genotypes grown in different environmental conditions. *J. Agric. Food Chem.* 50 (10): 2870–2877.

151 Watanabe, Y., Sakurai, J., and Nobuo, I. (2019). Effect of Coriandrum Sativum L. leaf extract on the brain GABA neurons in mice. *J. Nutr. Health Food Sci.* 7: 1–7.

152 Mnif, S. and Aifa, S. (2015). Cumin (*Cuminum cyminum* L.) from traditional uses to potential biomedical applications. *Chem. Biodivers.* 12 (5): 733–742.

153 Pandey, M.M., Rastogi, S., and Rawat, A.K.S. (2013). Indian traditional ayurvedic system of medicine and nutritional supplementation. *Evid. Based Complement. Alternat. Med.* 2013: 376327.

154 Al-Snafi, A. (2016). The pharmacological activities of *Cuminum cyminum* – a review. *IOSR J. Pharm.* 6: 46–65.

155 Kamala, A., Middha, S.K., and Karigar, C.S. (2018). Plants in traditional medicine with special reference to *Cyperus rotundus* L.: a review. *3 Biotech* 8 (7): 309.

156 Pirzada, A.M., Ali, H.H., Naeem, M. et al. (2015). *Cyperus rotundus* L.: traditional uses, phytochemistry, and pharmacological activities. *J. Ethnopharmacol.* 174: 540–560.

157 KK, S. and Soni, S. (2017). *Eclipta alba* (L.). An ethnomedicinal herb plant, traditionally use in Ayurveda. *J. Hortic* 4 (3): 1000208.

158 Yadav, N.K., Arya, R.K., Dev, K. et al. (2017). Alcoholic extract of *Eclipta alba* shows in vitro antioxidant and anticancer activity without exhibiting toxicological effects. *Oxidative Med. Cell. Longev.* 2017: 9094641.

159 Jahan, R., Al-Nahain, A., Majumder, S., and Rahmatullah, M. (2014). Ethnopharmacological significance of *Eclipta alba* (L.) Hassk. (Asteraceae). *Int. Sch. Res. Not.* 2014: 385969.

160 Sathaye, S., Shaikh, M., and Sancheti, J. (2013). Effect of *Eclipta alba* on acute seizure models: a $GABA_A$ –mediated effect. *Indian J. Pharm. Sci.* 75 (3): 380.

161 Chandrasekar, S., Bhanumathy, M., Pawar, A., and Somasundaram, T. (2010). Phytopharmacology of Ficus religiosa. *Pharmacogn. Rev.* 4 (8): 195.

162 Prasad, P.V.V., Subhaktha, P.K.J.P., Narayana, A., and Rao, M.M. (2006). Medico-historical study of "aśvattha" (sacred fig tree). *Bull Indian Inst. Hist. Med. Hyderabad* 36 (1): 1–20.

163 Singh, D., Singh, B., and Goel, R.K. (2011). Traditional uses, phytochemistry and pharmacology of Ficus religiosa: a review. *J. Ethnopharmacol.* 134 (3): 565–583.

164 Bhangale, J.O. and Acharya, S.R. (2016). Anti-Parkinson activity of petroleum ether extract of *Ficus religiosa* (L.) leaves. *Adv. Pharmacol. Sci.* 2016: 9436106.

165 Lim, T.K. (2016). *Edible Medicinal and Non-Medicinal Plants*, vol. 10: Modified Stems, Roots, Bulbs. Dordrecht: Springer.

166 Pastorino, G., Cornara, L., Soares, S. et al. (2018). Liquorice (*Glycyrrhiza glabra*): A phytochemical and pharmacological review. *Phytother Res. PTR* 32 (12): 2323–2339.

167 Sharma, V., Katiyar, A., and Agrawal, R.C. (2018). *Glycyrrhiza glabra*: chemistry and pharmacological activity. In: *Sweeteners* (ed. J.M. Mérillon and K.G. Ramawat), 87–100. Cham: Springer.

168 Wahab, S., Annadurai, S., Abullais, S.S. et al. (2021). *Glycyrrhiza glabra* (licorice): A comprehensive review on its phytochemistry, biological activities, clinical evidence and toxicology. *Plants Basel Switz.* 10 (12): 2751.

169 Cui, Y.M., Ao, M.Z., Li, W., and Yu, L.J. (2008). Effect of glabridin from *Glycyrrhiza glabra* on learning and memory in mice. *Planta Med.* 74 (4): 377–380.

170 Hasanein, P. (2011). Glabridin as a major active isoflavan from *Glycyrrhiza glabra* (licorice) reverses learning and memory deficits in diabetic rats. *Acta Physiol. Hung.* 98 (2): 221–230.

171 Paikra, B.K., Dhongade, H.K.J., and Gidwani, B. (2017). Phytochemistry and pharmacology of Moringa oleifera lam. *J. Pharm.* 20 (3): 194–200.

172 Malage, A., Jadhav, S., Yogekar, T., and Sharma, S. (2020). Phytoassembly and pharmacological activity on moringa oleifera: a review. *Asian J. Pharm. Clin. Res.* 13 (3): 5–8.

173 Abd Rani, N.Z., Husain, K., and Kumolosasi, E. (2018). Moringa genus: A review of phytochemistry and pharmacology. *Front. Pharmacol.* 9: 108.

174 Safaeian, L., Javanmard, S., Heidarinejad, A., and Asghari, G. (2015). The effect of hydroalcoholic extract from the leaves of *Moringa peregrina* (Forssk.) Fiori. on blood pressure and oxidative status in dexamethasone-induced hypertensive rats. *Adv. Biomed. Res.* 4 (1): 101.

175 Ganguly, R. and Guha, D. (2008). Alteration of brain monoamines & EEG wave pattern in rat model of Alzheimer's disease & protection by Moringa oleifera. *Indian J. Med. Res.* 128 (6): 744–751.

176 Ahmad, A., Husain, A., Mujeeb, M. et al. (2013). A review on therapeutic potential of *Nigella sativa*: A miracle herb. *Asian Pac. J. Trop. Biomed.* 3 (5): 337–352.

177 Dzoyem, J.P., McGaw, L.J., Kuete, V., and Bakowsky, U. (2017). Anti-inflammatory and anti-nociceptive activities of African medicinal spices and vegetables. In: *Medicinal Spices and Vegetables from Africa* (ed. V. Kuete), 239–270. Amsterdam: Elsevier https://linkinghub.elsevier.com/retrieve/pii/B9780128092866000091.

178 Yimer, E.M., Tuem, K.B., Karim, A. et al. (2019). *Nigella sativa* L. (black cumin): A promising natural remedy for wide range of illnesses. *Evid. Based Complement Altern. Med* 2019: 1528635.

179 Al-Naggar, T.B., Gómez-Serranillos, M.P., Carretero, M.E., and Villar, A.M. (2003). Neuropharmacological activity of *Nigella sativa* L. extracts. *J. Ethnopharmacol.* 88 (1): 63–68.

180 Perveen, T. (2014). Increased 5-HT levels following repeated Administration of *Nigella sativa* L. (black seed) oil produce antidepressant effects in rats. *Sci. Pharm.* 82 (1): 161–170.

181 Kumar, S., Kamboj, J., and Suman null, Sharma S. (2011). Overview for various aspects of the health benefits of *Piper longum* Linn. fruit. *J. Acupunct. Meridian Stud.* 4 (2): 134–140.

182 Yadav, V., Krishnan, A., and Vohora, D. (2020). A systematic review on *Piper longum* L.: bridging traditional knowledge and pharmacological evidence for future translational research. *J. Ethnopharmacol.* 247: 112255.

183 Taur, D.J. and Patil, R.Y. (2011). Some medicinal plants with antiasthmatic potential: a current status. *Asian Pac. J. Trop. Biomed.* 1 (5): 413–418.

184 Derosa, G., Maffioli, P., and Sahebkar, A. (2016). Piperine and its role in chronic diseases. *Adv. Exp. Med. Biol.* 928: 173–184.

185 Li, S., Wang, C., Li, W. et al. (2007). Antidepressant-like effects of piperine and its derivative, antiepilepsirine. *J. Asian Nat. Prod. Res.* 9 (5): 421–430.

186 Mukherjee, E., Gantait, S., Kundu, S. et al. (2019). Biotechnological interventions on the genus Rauvolfia: recent trends and imminent prospects. *Appl. Microbiol. Biotechnol.* 103 (18): 7325–7354.

187 Dey, A. and De, J.N. (2010). Rauvolfia serpentina (L). Benth. ex Kurz. – a review. *Asian J. Plant Sci.* 9 (6): 285–298.

188 S, R. and Saini, M.K. (2022). Impending threats to the plants with medicinal value in the eastern Himalayas region: an analysis on the alternatives to its non-availability. *Phytomedicine Plus.* 2 (1): 100151.

189 Hood, S.D. and Nutt, D. (2004). Antipsychotics. *Psychiatry* 3 (7): 27–30.

190 Mallick, S.R., Samal, P., Jena, R.C., and Samal, K.C. (2013). Alkaloid profiling of conventionally propagated and in vitro raised plants of Indian Snake plant (*Rauwolfia serpentina* L.). *Asian J. Chem.* 25 (12): 6584–6586.

191 Pathania, S., Ramakrishnan, S.M., Randhawa, V., and Bagler, G. (2015). SerpentinaDB: a database of plant-derived molecules of Rauvolfia serpentina. *BMC Complement. Altern. Med.* 15: 262.

192 Kanba, S. and Richelson, E. (ed.) (2013). *Herbal Medicines for Neuropsychiatric Diseases.* New York: Routledge.

193 Lobay, D. (2015). Rauwolfia in the treatment of hypertension. *Integr. Med. Encinitas Calif.* 14 (3): 40–46.

194 Baumeister, A.A. (2013). The chlorpromazine enigma. *J. Hist. Neurosci.* 22 (1): 14–29.

195 Bein, H.J. (1982). Rauwolfia and biological psychiatry. *Trends Neurosci.* 5: 37–39.

196 Xia, M., Shahane, S.A., Huang, R. et al. (2011). Identification of quaternary ammonium compounds as potent inhibitors of hERG potassium channels. *Toxicol. Appl. Pharmacol.* 252 (3): 250–258.

197 Gupta, S., Khanna, V.K., Maurya, A. et al. (2012). Bioactivity guided isolation of antipsychotic constituents from the leaves of rauwolfia tetraphylla L. *Fitoterapia* 83 (6): 1092–1099.

198 Costa-Campos, L., Elisabetsky, E., Lara, D.R. et al. (1999). Antipsychotic profile of alstonine: ethnopharmacology of a traditional Nigerian botanical remedy. *An. Acad. Bras. Cienc.* 71 (2): 189–201.

199 Elisabetsky, E. and Costa-Campos, L. (2006). The alkaloid alstonine: a review of its pharmacological properties. *Evid. Based Complement. Alternat. Med.* 3 (1): 39–48.

200 Herrmann, A.P., Lunardi, P., Pilz, L.K. et al. (2012). Effects of the putative antipsychotic alstonine on glutamate uptake in acute hippocampal slices. *Neurochem. Int.* 61 (7): 1144–1150.

201 Linck, V.M., Bessa, M.M., Herrmann, A.P. et al. (2012). 5-HT2A/C receptors mediate the antipsychotic-like effects of alstonine. *Prog. Neuro-Psychopharmacol. Biol. Psychiatry* 36 (1): 29–33.

202 Arfan, N., Julie, A., Mohiuddin, A. et al. (2016). Medicinal properties of the sesbania grandiflora leaves. *Ibnosina J. Med. Biomed. Sci.* 8 (6): 271.

203 Ramesh, T. and Begum, V.H. (2008). Protective effect of *Sesbania grandiflora* against cigarette smoke-induced oxidative damage in rats. *J. Med. Food* 11 (2): 369–375.

204 Raja, R.R. and Baba, K.H. (2019). Hummingbird tree-biological review. *Res. J. Pharmacogn. Phytochem.* 11 (3): 150.

205 Hasan, N., Osman, H., Mohamad, S. et al. (2012). The chemical components of Sesbania grandiflora root and their antituberculosis activity. *Pharmaceuticals* 5 (8): 882–889.

206 Mohiuddin, A.K. (2019). Medicinal and therapeutic values of Sesbania grandiflora. *Adv. Res. J. Pharm. Pharmacol.* 4 (2): 87–93.

207 Kasture, V.S., Deshmukh, V.K., and Chopde, C.T. (2002). Anxiolytic and anticonvulsive activity ofSesbania grandiflora leaves in experimental animals. *Phytother. Res.* 16 (5): 455–460.

208 Semwal, B.C., Verma, M., Murti, Y., and Yadav, H.N. (2018). Neuroprotective activity of Sesbania grandifolara seeds extract against celecoxib induced amnesia in mice. *Pharm. J.* 10 (4): 747–752.

209 Galani, V., Patel, B., and Rana, D. (2010). *Sphaeranthus indicus* Linn.: a phytopharmacological review. *Int. J. Ayurveda Res.* 1 (4): 247.

210 Ramachandran, S. (2013). Review on *Sphaeranthus indicus* Linn. (Koṭṭaikkarantai). *Pharmacogn. Rev.* 7 (14): 157–169.

211 Jirovetz, L., Buchbauer, G., Shahabi, M. et al. (2003). Medicinal used plants from India: analysis of the essential oils of *Sphaeranthus indicus* flowers, roots and stems with leaves. *Sci. Pharm.* 71 (3): 251–259.

212 Deepa, B., Babaji, H., Hosmani, J. et al. (2019). Effect of *Tinospora cordifolia*-derived phytocomponents on cancer: a systematic review. *Appl. Sci.* 9 (23): 5147.

213 Saha, S. and Ghosh, S. (2012). Tinospora cordifolia: one plant, many roles. *Anc. Sci. Life* 31 (4): 151–159.

214 Sharma, P., Dwivedee, B.P., Bisht, D. et al. (2019). The chemical constituents and diverse pharmacological importance of *Tinospora cordifolia*. *Heliyon.* 5 (9): e02437.

215 Upadhyay, A., Kumar, K., Kumar, A., and Mishra, H. (2010). *Tinospora cordifolia* (Willd.) Hook. f. and Thoms. (*Guduchi*) – validation of the Ayurvedic pharmacology through experimental and clinical studies. *Int. J. Ayurveda Res.* 1 (2): 112.

216 Asgher, M., Verma, S., Khan, N.A. et al. (2020). Physiological, biochemical and reproductive studies on *Valeriana wallichii*, a critically endangered medicinal plant of the Himalayan region grown under in-situ and ex-situ conditions. *Plan. Theory* 9 (2): 131.

217 Vishwakarma, S., Goyal, R., Gupta, V., and Dhar, K.L. (2016). GABAergic effect of valeric acid from *Valeriana wallichii* in amelioration of ICV STZ induced dementia in rats. *Rev. Bras. Farm.* 26 (4): 484–489.

218 Bhattacharya, S.K., Satyan, K.S., and Ghosal, S. (1997). Antioxidant activity of glycowithanolides from Withania somnifera. *Indian J. Exp. Biol.* 35 (3): 236–239.

219 Becker, A., Felgentreff, F., Schröder, H. et al. (2014). The anxiolytic effects of a valerian extract is based on valerenic acid. *BMC Complement. Altern. Med.* 14: 267.

220 Bharti, V.K., Malik, J.K., and Gupta, R.C. (2016). Ashwagandha. In: *Nutraceuticals* (ed. R.C. Gupta), 717–733. Amsterdam: Elsevier https://linkinghub.elsevier.com/retrieve/pii/B9780128021477000528.

221 Singh, N., Bhalla, M., De Jager, P., and Gilca, M. (2011). An overview on ashwagandha: a rasayana (rejuvenator) of ayurveda. *Afr. J. Tradit. Complement. Altern. Med.* 8 (5S).

222 Mishra, L.C., Singh, B.B., and Dagenais, S. (2000). Scientific basis for the therapeutic use of Withania somnifera (ashwagandha): a review. *Altern. Med. Rev. J. Clin. Ther.* 5 (4): 334–346.

223 Kulkarni, S.K., Sharma, A.C., Verma, A., and Ticku, M.K. (1993). GABA receptor mediated anticonvulsant action of Withania somnifera root extract. 30 (7): 305–312.

224 Mary Princess Sulekha, S. (2022). A comprehensive review on medicinal herb *Withania somnifera* (L.) Dunal in women's health: a rejuvenator from Siddha medicine. *Curr. Pharmacol. Rep.* 8 (1): 72–77.

225 Mirjalili, M., Moyano, E., Bonfill, M. et al. (2009). Steroidal lactones from Withania somnifera, an ancient Plant for Novel Medicine. *Molecules* 14 (7): 2373–2393.

226 Candelario, M., Cuellar, E., Reyes-Ruiz, J.M. et al. (2015). Direct evidence for GABAergic activity of Withania somnifera on mammalian ionotropic GABAA and GABAρ receptors. *J. Ethnopharmacol.* 171: 264–272.

227 Chen, J., Liu, X., Li, Z. et al. (2017). A review of dietary Ziziphus jujuba fruit (jujube): developing health food supplements for brain protection. *Evid. Based Complement Altern. Med.* 2017: 3019568.

228 Patel, G., Murakonda, S., and Dwivedi, M. (2022). Steady and dynamic shear rheology of Indian jujube (Ziziphus mauritiana lam.) fruit pulp with physiochemical, textural and thermal properties of the fruit. *Meas. Food* 5: 100023.

229 Calva, J., Bec, N., Gilardoni, G. et al. (2017). Acorenone B: AChE and BChE inhibitor as a major compound of the essential oil distilled from the Ecuadorian species *Niphogeton dissecta* (Benth.). *J. F. Macbr.Pharmaceuticals* 10 (4): 84.

230 Irie, Y. (2006). Effects of eugenol on the central nervous system: its possible application to treatment of Alzheimers disease, depression, and Parkinsons disease. *Curr. Bioact. Compd.* 2 (1): 57–66.

231 Tiefensee Ribeiro, C., Gasparotto, J., Petiz, L.L. et al. (2019). Oral administration of carvacrol/β-cyclodextrin complex protects against 6-hydroxydopamine-induced dopaminergic denervation. *Neurochem. Int.* 126: 27–35.

232 Sekhar, V.C., Viswanathan, G., and Baby, S. (2019). Insights into the molecular aspects of neuroprotective bacoside A and Bacopaside I. *Curr. Neuropharmacol.* 17 (5): 438–446.

233 Surendran, S., Qassadi, F., Surendran, G. et al. (2021). Myrcene – what are the potential health benefits of this flavouring and aroma agent? *Front. Nutr.* 8: 699666.

234 Sharifi-Rad, M., Lankatillake, C., Dias, D.A. et al. (2020). Impact of natural compounds on neurodegenerative disorders: from preclinical to pharmacotherapeutics. *J. Clin. Med.* 9 (4): 1061.

235 Silva Dos Santos, J., Gonçalves Cirino, J.P., de Oliveira, C.P., and Ortega, M.M. (2020). The pharmacological action of kaempferol in central nervous system diseases: a review. *Front. Pharmacol.* 11: 565700.

236 Gim, S.A. and Koh, P.O. (2014). Ferulic acid prevents the injury-induced decrease of γ-enolase expression in brain tissue and HT22 cells. *Lab. Anim. Res.* 30 (1): 8–13.

237 Trang, A. and Khandhar, P.B. (2022). Physiology, acetylcholinesterase. In: *StatPearls*. Treasure Island, FL: StatPearls Publishing http://www.ncbi.nlm.nih.gov/books/NBK539735.

238 Cho, Y.S., Kim, S.K., Ahn, C.B., and Je, J.Y. (2011). Inhibition of acetylcholinesterase by gallic acid-grafted-chitosans. *Carbohydr. Polym.* 84 (1): 690–693.

239 Tsuji, M., Miyagawa, K., Takeuchi, T., and Takeda, H. (2008). Pharmacological characterization and mechanisms of the novel antidepressive- and/or anxiolytic-like substances identified from Perillae Herba. *Nihon Shinkei Seishin Yakurigaku Zasshi* 28 (4): 159–167.

240 Yugang Gao, Y.Y. (2018). The effect of beta-sitosterol and its derivatives on depression by the modification of 5-HT, DA and GABA-ergic systems in mice. *RSC Adv.* 8: 671–680.

241 Karim, N., Khan, I., Abdelhalim, A. et al. (2021). Stigmasterol can be new steroidal drug for neurological disorders: evidence of the GABAergic mechanism via receptor modulation. *Phytomedicine* 90: 153646.

242 Azimi, A., Ghaffari, S.M., Riazi, G.H. et al. (2016). α-Cyperone of *Cyperus rotundus* is an effective candidate for reduction of inflammation by destabilization of microtubule fibers in brain. *J. Ethnopharmacol.* 194: 219–227.

243 Cai, J., Cai, M., Xia, W. et al. (2022). Explore the mechanism of β-Asarone on improving cognitive dysfunction in rats with diabetic encephalopathy. *J. Alzheimers Dis. Rep.* 6 (1): 195–206.

244 Liang, W., Zhao, X., Feng, J. et al. (2016). Ursolic acid attenuates beta-amyloid-induced memory impairment in mice. *Arq. Neuropsiquiatr.* 74 (6): 482–488.

245 Zhou, F., Chen, S., Xiong, J. et al. (2012). Luteolin reduces zinc-induced tau phosphorylation at Ser262/356 in an ROS-dependent manner in SH-SY5Y cells. *Biol. Trace Elem. Res.* 149 (2): 273–279.

246 Losi, G., Puia, G., Garzon, G. et al. (2004). Apigenin modulates GABAergic and glutamatergic transmission in cultured cortical neurons. *Eur. J. Pharmacol.* 502 (1–2): 41–46.

247 Semaming, Y., Pannengpetch, P., Chattipakorn, S.C., and Chattipakorn, N. (2015). Pharmacological properties of protocatechuic acid and its potential roles as complementary medicine. *Evid. Based Complement. Alternat. Med.* 2015: 593902.

248 Guo, T., Lin, Q., Li, X. et al. (2017). Octacosanol attenuates inflammation in both RAW264.7 macrophages and a mouse model of colitis. *J. Agric. Food Chem.* 65 (18): 3647–3658.

249 Zhang, S., Fritz, N., Ibarra, C., and Uhlén, P. (2011). Inositol 1,4,5-trisphosphate receptor subtype-specific regulation of calcium oscillations. *Neurochem. Res.* 36 (7): 1175–1185.

250 Egbuna, C., Kumar, S., Ifemeje, J.C. et al. (ed.) (2020). *Phytochemicals as Lead Compounds for New Drug Discovery*. Amsterdam: Elsevier.

251 Pathan, H. and Williams, J. (2012). Basic opioid pharmacology: an update. *Br. J. Pain* 6 (1): 11–16.

252 Ay, M., Luo, J., Langley, M. et al. (2017). Molecular mechanisms underlying protective effects of quercetin against mitochondrial dysfunction and progressive dopaminergic neurodegeneration in cell culture and MitoPark transgenic mouse models of Parkinson's disease. *J. Neurochem.* 141 (5): 766–782.

253 Terry-Lorenzo, R.T., Chun, L.E., Brown, S.P. et al. (2014). Novel human D-amino acid oxidase inhibitors stabilize an active-site lid-open conformation. *Biosci. Rep.* 34 (4): e00133.

254 Mori, A., Kabuto, H., and Pei, Y.Q. (1985). Effects of piperine on convulsions and on brain serotonin and catecholamine levels in E1 mice. *Neurochem. Res.* 10 (9): 1269–1275.

255 Han, J., Wu, P., Wang, F., and Chen, J. (2015). S-palmitoylation regulates AMPA receptors trafficking and function: a novel insight into synaptic regulation and therapeutics. *Acta Pharm. Sin. B* 5 (1): 1–7.

256 Mandela, P., Chandley, M., Xu, Y.Y. et al. (2010). Reserpine-induced reduction in norepinephrine transporter function requires catecholamine storage vesicles. *Neurochem. Int.* 56 (6–7): 760–767.

257 Chandwani, A.S. and Pharmaceutically, A. (2021). Therapeutically important genus: Sesbania Adans. *Int. J. Pharm. Res.* doi: 10.31838/ijpr/2021.13.02.412.

258 Arya, A., Chahal, R., Rao, R. et al. (2021). Acetylcholinesterase inhibitory potential of various sesquiterpene analogues for Alzheimer's disease therapy. *Biomolecules.* 11 (3): 350.

259 Manayi, A., Vazirian, M., and Saeidnia, S. (2015). Echinacea purpurea: pharmacology, phytochemistry and analysis methods. *Pharmacogn. Rev.* 9 (17): 63–72.

260 Nejma, A.B., Besbes, M., Guérineau, V. et al. (2017). Isolation and structure elucidation of acetylcholinesterase lipophilic lupeol derivatives inhibitors from the latex of the Tunisian Periploca laevigata. *Arab. J. Chem.* 10: S2767–S2772.

261 Lee, A., Fleisher, M.D., and Roizen, J. (2017). *Essence of Anesthesia Practice*, 4e. Amsterdam: Elsevier.

262 Park, J.Y., Lee, Y., Lee, H.J. et al. (2020). In silico screening of GABA aminotransferase inhibitors from the constituents of Valeriana officinalis by molecular docking and molecular dynamics simulation study. *J. Mol. Model.* 26 (9): 228.

263 Shi, S.N., Shi, J.L., Liu, Y. et al. (2014). The anxiolytic effects of valtrate in rats involves changes of corticosterone levels. *Evid. Based Complement. Alternat. Med.* 2014: 325948.

264 Dong, F.W., Jiang, H.H., Yang, L. et al. (2018). Valepotriates from the roots and rhizomes of Valeriana jatamansi Jones as novel N-type Calcium Channel antagonists. *Front. Pharmacol.* 9: 885.

265 Tantipongpiradet, A., Monthakantirat, O., Vipatpakpaiboon, O. et al. (2019). Effects of Puerarin on the ovariectomy-induced depressive-like behavior in ICR mice and its possible mechanism of action. *Molecules* 24 (24): 4569.

5

Drug Discovery from Medicinal Plants and Phytochemicals against Neuropathic Pain

Santwana Palai[1], Shyam S. Kesh[2], and Mithun Rudrapal[3]

[1] *Department of Veterinary Pharmacology & Toxicology, College of Veterinary Science and Animal Husbandry, Odisha University of Agriculture and Technology, Bhubaneswar, Odisha, India*
[2] *Department of Veterinary Clinical Complex (Veterinary Biochemistry), Faculty of Veterinary and Animal Sciences, West Bengal University of Animal and Fishery Sciences, Kolkata, West Bengal, India*
[3] *Department of Pharmaceutical Sciences, School of Biotechnology and Pharmaceutical Sciences, Vignan's Foundation for Science, Technology & Research (Deemed to be University), Guntur, Andhra Pradesh, India*

Pain is a disagreeable sensation and emotive practice encountered on a regular basis. It serves as a warning sign of tissue damage for preventing more tissue impairment. Acute pain aids in a biologic purpose. It is self-limiting, occurring against a specific injury. Chronic pain, on the other hand, may be considered an illness. If it is accompanied with a sickness or injury, it may linger longer than usual. Depending on the location of the lesion, pain might be categorized as central or peripheral. Metabolic disorders, viral infection, trauma, severe ischemic shocks, and autoimmune diseases are the most common causes of chronic neuropathic pain [1].

Chronic neuropathic pain is the discomfort ascending from abrasion or illness that disturbs the somatosensory structure specifically. Normally, the somatosensory nervous system is critical in the transmission of unpleasant sensations to the central nervous system. Any somatosensory nervous system lesion causes halting of nerve cell innervations, resulting in pain with or without sensory hypersensitivity at the aching location [2].

Neuropathic pain is caused by damage to the outlying or central nervous system. The indications of neuropathic pain are hyperalgesia, allodynia, and enduring pain. Mechanical hyperalgesia, thermal hyperalgesia, and cold allodynia develop significantly. Pain causes an increase in thermal and mechanical hyperalgesia, and biochemical and histologic alterations. Mechanical and thermal hyperalgesia are assessed in aesthesiometer and plantar tests; ultra-high performance liquid chromatography–tandem mass spectrometry (UHPLC-MS), receptors participation, and the paw withdrawal threshold are various tests for the detection of neuropathic pain. The enzyme-linked immunosorbent assay (ELISA) is used to examine inflammatory factors such as interleukin (IL)-6, IL-17, and IL-1 and tumor necrosis factors (TNFs) [3]. Thermal hyperalgesia and mechanical allodynia are measured using the Hargreaves test or the von Frey test, respectively. Histopathology changes along with increased enzymes related to stress are also measured.

Phytochemical Drug Discovery for Central Nervous System Disorders: Biochemistry and Therapeutic Effects, First Edition. Edited by Chukwuebuka Egbuna and Mithun Rudrapal.

Various kinds of medications, such as opiates, tricyclic antidepressants, and antiepileptic compounds, are advantageous in neuropathic pain treatment, but are associated with side effects [4]. Also, because of the varied etiology and complex underlying pathophysiology of neuropathic pain, it is not effectively treated with these current medications.

As conventional therapies have failed to completely regulate the condition, novel approaches of complementary and alternative medicine are advisable for the management of neuropathic pain [5]. In the field of medication research, herbal formulations, the oldest medications from natural items, are receiving greater attention these days as they possess fewer difficulties and adverse actions than synthetic treatments.

This chapter discusses how plants and phytochemicals can aid as a therapeutic strategy in various models of neuropathic pain. This comprehensive assessment of plants and phytochemicals with antineuropathic pain properties will help people involved in phytochemistry, natural product chemistry, and pharmaceutical studies to devise practical uses for these plants and phytochemicals in painful neuropathy management.

5.1 Mechanisms of Neuropathic Pain

Neuropathic pain can be impulsive or induced in response to bodily stimuli, and appear as enhanced pain sensitivity (hyperalgesia) or pain evoked by non-painful stimuli (allodynia). This condition fades as inflammation and healing reduce [6].

If chronic injury or disease causes persistent nociception, the alterations in chief afferent neurons may linger. Peripheral nerve lesions can cause central alterations primarily at the spinal cord or occasionally at supraspinal levels [7].

Microglial stimulation, neuroplasticity, and hyperexcitability due to central sensitization of nociceptive neurons are all examples of changes that might cause pathogenic activation of central nociceptive neurons. The initiation of N-methyl-D-aspartate and glutamate metabotropic receptors may cause intracellular changes that are necessary for central sensitization. Non-neuronal cells, such as microglia found in the brain and macrophages found in the periphery, are stimulated, resulting in various inflammatory cytokines and chemokines release in neuropathic pain development [8].

Ectopic spontaneous action after nerve damage is caused by increased expression of mRNA for voltage-gated Na^+ channels in chief afferent neurons. This event may induce these channels to cluster, lowering the action potential threshold and resulting in hypersensitivity. As a result, sodium channel blockers, such as lidocaine, relieve neuropathic pain through this mechanism. The overexpression of different receptor proteins is also caused by peripheral nerve damage. Upregulation of this channel can be caused by nerve injury, contributing to peripheral C-nociceptor sensitization, resulting in cold hyperalgesia [9].

Also, acid-sensing ion channels contribute to static mechanical hyperalgesia. Both α1- and α2-adrenoceptors present on cutaneous afferent fibers have a critical part in the hypersensitivity of nerve injury [10]. The adrenergic sensitivity is extensive in posttraumatic neuralgias, complex regional pain syndromes II, and postherpetic neuralgias, but no sensitivity is found in primary afferent neurons as in polyneuropathies. As a result, nociceptive neurons can be cured of sympathetically produced and temperature-facilitated pain by blocking the respective receptors.

5.2 Animal Models for Studying Neuropathic Pain

Different animal models of neuropathic pain are crucial for studying pain's molecular mechanisms for the development of new analgesics in chronic pain treatment. The following animal models are used for research in neuropathic pain.

5.2.1 Streptozotocin-Induced Diabetes

This model for diabetic neuropathic pain is generated by streptozotocin that closely resembles diabetic neuropathy. This neuropathy is a common peripheral neuropathy involving cold or hot allodynia, hyperalgesia, and hyperesthesia. Increased blood glucose levels cause oxidative and nitrosative stress, which is a key mechanism in diabetic neuropathy neuronal destruction [11].

5.2.2 High-Fat Diet

Neuronal impairment results from oxidative stress present in lipid metabolism and indicates augmented oxidative–nitrosative stress, sorbitol pathway activity, and proinflammatory variations in the peripheral nervous system, along with hyperinsulinemia, alimentary obesity, and reduced glucose tolerance [12]. This model is employed for research on pathophysiology of diminished glucose tolerance and type 2 diabetes for the advancing novel therapies.

5.2.3 Sciatic Nerve Injury

The symptoms of neuropathy caused by sciatic nerve injury are similar in animals and humans suffering from nerve injuries. So animal models simulating humans can be commonly employed, particularly for behavioral testing. Herc, long-term alterations in neurotransmitter and receptor expression cause central sensitization against various inflammatory and pain intermediaries, which improves peripheral sensory afferent sensitivity at the wound site as well as in the central nervous system [13].

5.2.4 Chemotherapy-Induced Peripheral Neuropathy

Antitubulins like docetaxel, vincristine, ixabepilone, and paclitaxel, platinum analogues like cisplatin, carboplatin, and oxaliplatin, and proteasome inhibitors like thalidomide and bortezomib are among the most prevalent anticancer medications that cause peripheral neuropathy. These medications work directly and also indirectly on sensory neurons to decrease the amplitude of action potentials, slow conduction velocity, and cause aching in cases having nociceptive sensory loss due to cancer treatment [14].

5.3 Medicinal Plants and Phytochemicals against Neuropathic Pain

Various medicinal plants can reduce neuropathy symptoms in animal models of neuropathic pain. The rhizome of *Acorus calamus*; fresh leaves of *Alstonia scholaris*, *Butea monosperma*, and *Ocimum sanctum*; pure red saffron powder of *Crocus sativus*; and whole parts of *Rosmarinus officinalis* have reduced pain in in chronic constriction injury (CCI) of sciatic nerve–brought neuropathic pain in rats. PMI-5011 is the non-toxic ethanolic extract of *Artemisia dracunculus* and was found to be effective against streptozotocin-induced diabetes in high-fat diet–fed C57Bl6/J mice; and the fruit of *Citrullus colocynthis*, rhizome of *Curcuma longa*, and leaves of *Tinospora* were found to be effective against streptozotocin-induced neuropathic pain (Table 5.1).

The leaves of *Salvia officinalis* in vincristine-induced peripheral neuropathy in mice; EGb 761, a *Ginkgo biloba* extract, in lumbar spinal nerve ligation for neuropathic pain in rats; aerial parts of *Phyllanthus amarus,* fruits of *Pterodon pubescens* Benth. in partial ligation of the sciatic nerve; dry fruit powder of *Momordica charantia* in nerve transection-induced neuropathic pain in rats; and oil of *Nigella sativa* in experimentally applied spinal cord injury, all showed their safety and efficacy through clinical trials (Figure 5.1).

Phytochemicals like baicalin, (−)-epicatechin, coumarin, curcumin, diosmin, hesperidin, fisetin, genistein, grape seed proanthocyanidins, hesperidin, icariin, isoorientin, kaempferol, luteolin, matrine, morin, and myricitrin are effective in various models of neuropathic pain (Table 5.2). These polyphenols and terpenoid-like phytochemicals could pave the way for preclinical studies of novel drugs for neuropathic pain.

With the application of medicinal plants and phytochemicals against neuropathic pain, mechanical allodynia and thermal hyperalgesia reduced in all behavioral as well as physiologic events with oxidative stress. As oxidative damage to neural cells is the main mechanism of painful neuropathy induction, plants or phytochemicals with antioxidant and neuroprotective actions can manage neuropathy. They decreased proinflammatory cytokines, IL-1 and -6, and TNF-α in injured neurons and reduced phosphoactivation of extracellular signal regulated kinase (pERK) and transient receptor potential vanilloid 1 (TRPV1) expression. Bcl-2 levels rise while caspase-3 and Bax levels fall, indicating an antiapoptotic action (Figure 5.2).

The antinociceptive action on thermal hyperalgesia is linked to the spinal serotonergic system (5-HT_7). Medicinal plants and phytochemicals against neuropathic pain work by activating the 5-$HT_{1A/1B/1D/5A}$ serotonergic and opioid receptors, and the nitric oxide (NO)–cyclic guanosine monophosphate (cGMP)–potassium (K) channels pathway for antinociception. They lowered nuclear factor-kappa-B, proinflammatory cytokine, and the NO system activated through immunomodulatory and anti-inflammatory properties. They raised total superoxide dismutase, total antioxidant capacity, and catalase levels, while decreasing malondialdehyde (MDA) levels. Glial cell and matrix metalloproteinase (MMP)-9 activation was inhibited, while TNF, IL-6, and IL-1 expression was reduced.

Furthermore, biochemical and histologic alterations in the tissue were reduced in mechanical and cold allodynia with the application of medicinal plants and phytochemicals in a rodent model of neuropathic pain.

These plants and phytochemicals can be considered for further neuropathic pain studies in the long path from laboratory to clinic.

5.4 Role of Plants and Phytochemicals in Different Neuropathic Pain Models

5.4.1 Diabetic Neuropathy

Diabetic neuropathy, the most common neuropathy, affects millions of individuals worldwide. The most common type of diabetic neuropathy is peripheral neuropathy, affecting the outer nerves of the limbs, specifically the feet. It primarily affects sensory function, resulting

Table 5.1 Medicinal plants effective against neuropathic pain.

Image no. in Figure 5.1	Plant name (common name)	Family	Part used	Neuropathic pain model used	Effect seen	Probable mechanism	References
1	*Acorus calamus* (sweet flag)	Acoraceae	Rhizome	Chronic constriction injury of sciatic nerve induced neuropathic pain in rats	Meaningfully enhanced nociceptive pain threshold, functional and electrophysiologic alterations in sciatic nerve	Ameliorative effect owing to neuroprotective, anti-inflammatory, and antioxidative actions	[15]
2	*Alstonia scholaris* (white cheesewood)	Apocynaceae	Fresh leaves	Chronic constriction injury of sciatic nerve induced neuropathic pain in rats	Reduced nociceptive threshold and led to metabolic alterations nearly the same as pregabalin	Kaempferol in *A. scholar* has antioxidative and anti-inflammatory characteristics responsible for amelioratory action in neuropathic pain	[16]
3	*Artemisia dracunculus* (tarragon)	Asteraceae	PMI-5011 (ethanolic extract)	Streptozotocin-induced diabetes in high-fat diet–fed C57Bl6/J mice	Relieves diabetes-induced nerve conduction sluggishness, small sensory nerve fiber dysfunction, amplified intra-epidermal nerve fiber density	Inhibited pain in prediabetes and obesity model through lipoxygenase activation and oxidative–nitrosative stress of sciatic nerve and spinal cord	[17]
4	*Butea monosperma* (bastard teak)	Fabaceae	Leaves	Chronic constriction damage of sciatic nerve led to neuropathic pain in rats	Pretreatment reduced behavioral, biochemical, and histologic changes similar to pregabalin	Antioxidant, neuroprotective, and calcium channel modulatory properties aiding in neuropathic pain	[18]
5	*Citrullus colocynthis* (bitter apple)	Cucurbitaceae	Fruit	Streptozotocin (STZ)-induced diabetic rats with peripheral neuropathy	Fruit has beneficial effect in treatment of diabetic neuropathy	Antioxidant effect and hypoglycemic properties, which reduced amount of demyelinated and degenerative nerve fibers	[19]

(*Continued*)

Table 5.1 (Continued)

Image no. in Figure 5.1	Plant name (common name)	Family	Part used	Neuropathic pain model used	Effect seen	Probable mechanism	References
6	*Curcuma longa* (turmeric)	Zingiberaceae	Rhizome	STZ-induced diabetic peripheral neuropathic pain	Therapy prevents weight loss and reduces mechanical allodynia in STZ-induced diabetic mice	Antinociceptive impact of curcumin has novel therapeutic effect in diabetic neuropathic pain mediated by opioid system	[20]
7	*Crocus sativus* (saffron)	Iridaceae	Pure red saffron powder	Chronic constriction injury in rats	Reduced thermal hyperalgesia and mechanical allodynia	Possesses antinociceptive and anti-inflammatory properties and effective in lowering pain behaviors	[21]
8	*Ginkgo biloba* (maidenhair tree)	Ginkgoaceae	EGb 761 (extract)	Left L5 and L6 spinal nerve ligation for neuropathic pain in rat	Reduces mechanical and cold allodynia, suggesting it as effective in neuropathic pain treatment	Oriental herbal medicine that has various pharmacologic actions like antinociceptive and anti-inflammatory effects	[22]
9	*Momordica charantia* (bitter-melon)	Cucurbitaceae	Dry fruit powder	Tibial and sural nerve transection-induced neuropathic pain in rats	Reduced cold allodynia, dynamic and mechanical allodynia, mechanical and thermal hyperalgesia, behavioral and biochemical alterations	Antinociceptive impact in neuropathic pain due to PPAR-γ agonistic action, antioxidative and anti-inflammatory potential	[23]
10	*Nigella sativa* (black cumin)	Ranunculaceae	Seed oil	Experimentally applied spinal cord injury	Decreased oxidative stress-induced cell damage and pain	Total antioxidant and oxidant status, nitric oxide, malondialdehyde levels, and tumor necrosis factor are all decreased, proving decreased oxidative stress-induced cell damage	[24]

11	*Ocimum sanctum* (holy basil)	Lamiaceae	Fresh leaves	Chronic constriction injury-induced neuropathic pain	Effectively reduced chronic constriction injury-induced neuropathic pain and oxidative stress	Alleviates neuropathic pain due to reduction in oxidative stress and levels of calcium	[25]
12	*Phyllanthus amarus* (Indian gooseberry)	Phyllanthaceae	Aerial part	Partial ligation of sciatic nerve	Decreased mechanical allodynia, paw edema development, and myeloperoxidase activity, both in paw and in damaged sciatic nerve	Hexanic extract of *P. amarus* therapy generates significant antiallodynia and anti-inflammatory effect by reducing increased myeloperoxidase activity	[26]
13	*Pterodon pubescens Benth.* (faveira)	Fabaceae	Fruits	Partial sciatic nerve ligation	Reduces neuropathic pain and causes reduction in thermal and mechanical hyperalgesia	Has substantial antinociceptive impact due to limiting of glutamatergic receptors, TRPA1, TRPV1 channels, and partly limiting proinflammatory cytokines	[27]
14	*Rosmarinus officinalis* (rosemary)	Lamiaceae	Whole plant	Chronic constriction injury-induced sciatic nerve rat model of neuropathic pain	Helps in treating moderating neuroinflammation of neuropathic pain and similar neurologic illnesses related to inflammation	Reduced levels of inflammatory markers (MMP2, COX-2, FGE-2, IL-1b, and NO)	[28]

(*Continued*)

Table 5.1 (Continued)

Image no. in Figure 5.1	Plant name (common name)	Family	Part used	Neuropathic pain model used	Effect seen	Probable mechanism	References
15	*Salvia officinalis* (common garden sage)	Lamiaceae	Leaves	Vincristine-induced peripheral neuropathy in mice	Caused dose-dependent analgesic effects and decreased vincristine-induced pain	Remarkable anti-inflammatory effect, antineuropathic effect probably due to significant analgesic, antipyretic, antiepileptic, anti-inflammatory properties	[29]
16	*Tinospora cordifolia* (guduchi)	Rubiaceae	Leaves	Streptozotocin induced neuropathic pain	In diabetic rat sciatic nerves, *T. cordifolia* leaf extract dramatically reduced thermal hyperalgesia and allodynia by raising antioxidant enzyme levels, lowering lipid peroxidation, and enhancing NGF expression	Extract may help to relieve diabetic neuropathic pain by acting as antioxidant and inducing NGF mRNA in sciatic nerves	[30]

COX-2, cyclooxygenase-2; IL-1b, interleukin 1b; MMP2, matrix metalloproteinase-2; NGF, nerve growth factor; NO, nitric oxide; PGE-2, prostaglandin E2; PPAR, peroxisome proliferator-activated receptor; TRPA1, transient receptor potential ankyrin 1; TRPV1, transient receptor potential cation channel subfamily V member. 1.

Figure 5.1 Medicinal plants effective in neuropathic pain. *Sources:* (1) Volodymyr/Adobe stock; (2) Rupesh/Adobe stock; (3) Fineblick/Adobe stock; (7) goldika/Adobe stock; (8) Radila/Adobe stock; (9) Bob Peterson / Flickr / CC BY 2.0; (11) gv image/Adobe stock; (12) tuaindeed/Adobe stock; (13) João Medeiros / Flickr; (14) Visharo/Adobe Stock; (15) Kurt Stüber/Wikipedia commons; (16) Tmd/Wikipedia.

in abnormal sensations and progressive numbness, which aids in the development of ulcers. Diabetes causes excessive free radical creation and, ultimately, oxidative stress due to activated glucose metabolic pathways. Diabetic neuropathy develops in the context of hyperglycemia and a complex metabolic imbalance, primarily oxidative stress [55].

The streptozocin (STZ)-induced diabetic rat is a popular animal model for studying the mechanisms of painful diabetic neuropathy and evaluating potential treatments [30]. The high-fat diet–fed mouse has a typical condition of obesity and prediabetes. It advances the oxidative stress along with proinflammatory variations in peripheral nerve organization [56]. Thermal and mechanical hypoalgesia, obesity, nerve conduction deficit, moderate non-fasting hyperglycemia, and tactile allodynia were detected in mice fed a high-fat diet for four months. They had nitrosative stress, buildup of 12(S)-hydroxyeicosatetraenoic acid, and 12/15-lipoxygenase overexpression in the outlying nerve and spinal cord.

PMI-5011 is a non-toxic botanical extract that is used to treat neuropathic changes in the early phases of disease. It normalized glycemia, reduced upregulation of 12/15-lipoxygenase, and nitrated protein expression peripherally in the nervous system [56].

Some flavonoids, such as genistein, catechin, rutin, luteolin, and pelargonidin, lower MDA levels in diabetes animal models. MDA is also a critical indicator of lipid damage and oxidative stress, both of which can be generated by free radicals.

Genistein, naringenin, luteolin, hesperidin, catechin, kaempferol, fisetin, rutin, and morin can reduce reactive oxygen species (ROS) levels by growing the levels of antioxidative enzymes such as glutathione peroxidase, superoxide dismutase, and so on [57].

Rutin, luteolin, and morin can increase Nrf2 expression and its downstream effecter heme oxygenase-1 (HO-1) in the nervous tissues of diabetic animals. This Nrf-2/HO-1 can protect diabetic animal models from oxidative stress-induced nerve damage and neuroinflammation.

Table 5.2 Phytochemicals effective against neuropathic pain.

Name of phytochemical	Structure	Model used	Tests done	Effect seen	Probable mechanism	References
Baicalin	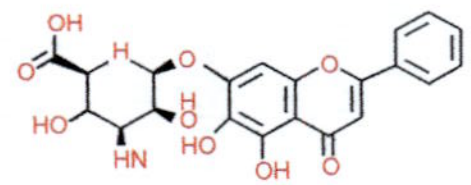	L5/L6 spinal nerve ligation in Sprague-Dawley rats	Paw withdrawal limit, ELISA to examine inflammatory factors such as IL-6, IL-17, and IL-1 and TNF, histopathology change	TNF, IL-6, IL-17, and IL-1 levels increased, histologic damage in spinal cord reduced	Antiallodynic effect related to modulation of A2AR expression	[31]
(−)-epicatechin	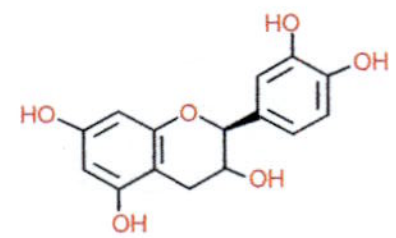	L5/L6 spinal nerve ligation for neuropathic pain model	Formalin and carrageenan tests Plethysmometer for inflammation measurement	Carrageenan-induced inflammation and nociception decreased	Works by activating 5-HT1A/1B/1D/5A serotonergic and opioid receptors, NO–cGMP–K channels pathway for antinociception	[32]
Coumarin	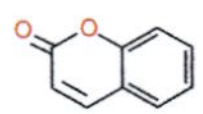	SNI model of neuropathy	Western blotting, ELISA, and immunofluorescence	Triggered development of dose-dependent mechanical hypersensitivity	Decreased proinflammatory cytokines, TNF-α, IL-1, and IL-6 in injured neurons and reduced pERK and TRPV1 expression	[33]
Curcumin	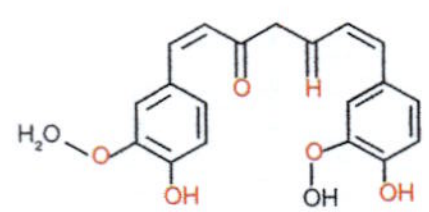	CCI Sprague-Dawley rat model of neuropathic pain	Behavioral measurements, double-labeling immunofluorescence, chromatin immunoprecipitation assay, qRT-PCR Western immunoblot analysis	CCI-induced thermal hyperalgesia and mechanical allodynia greatly reduced	Acted as antinociceptive in neuropathic pain by inhibiting BDNF and COX-2 gene expression	[34]

Diosmin	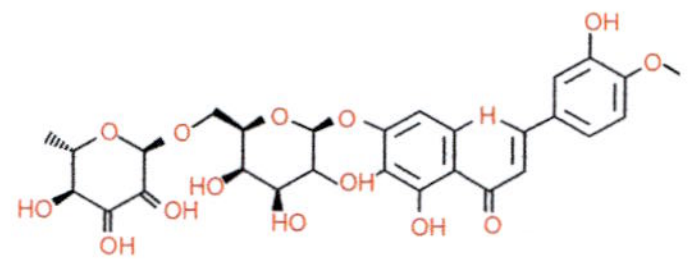	CCI of sciatic nerve pain male Swiss mice model	Thermal (hot-plate) hyperalgesia, mechanical (electronic) analgesimeter	CCI-induced thermal and mechanical hyperalgesia reduced in dose-dependent manner	Activates glial cells (astrocytes) while inhibiting spinal cytokine (TNF-α, Il-1, and Il-33/St2) and stimulating NO–cGMP–PKG–KATP channel signaling path	[35]
Hesperidin	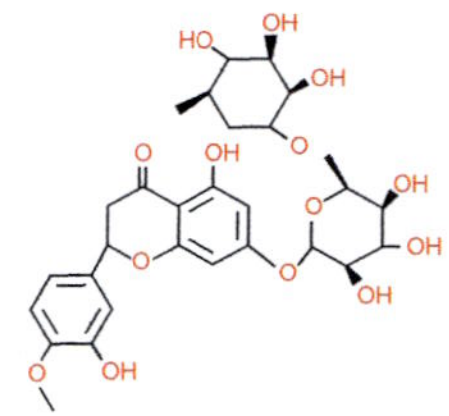	CCI rat model	Aesthesiometer and plantar tests, UHPLC-MS, receptors participation	Mechanical and thermal hyperalgesia dramatically reduced	Central activity influenced in part by GABAA, D2, and opioids, rather than by 5-HT_{1A} receptors	[36]
Fisetin	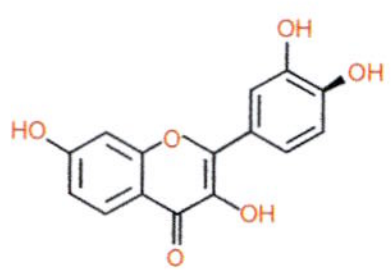	Loose ligation of sciatic nerves causing CCI in mouse	Novelty suppressed feeding test, forced swim test, light–dark test, Hargreaves test, von Frey test	Comorbid behavioral symptoms of depression, anxiety, and thermal hyperalgesia reduced	Antinociceptive effect on thermal hyperalgesia linked with spinal serotonergic system (5-HT7)	[37]
Genistein	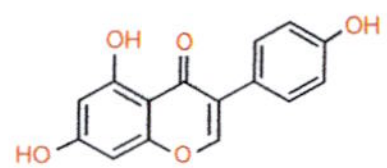	Chronic sciatic nerve constriction in mice	Thermal hyperalgesia and mechanical allodynia IL-1b content in sciatic nerve, Western blot analysis for NOS determination	Reversed pain hypersensitivity Ameliorated painful neuropathy with antinociceptive dose	Lowered nuclear factor-kappa-B, proinflammatory cytokine, and NO system activated through immunomodulatory and anti-inflammatory properties	[38]

(*Continued*)

Table 5.2 (Continued)

Name of phytochemical	Structure	Model used	Tests done	Effect seen	Probable mechanism	References
Grape seed proanthocyanidins	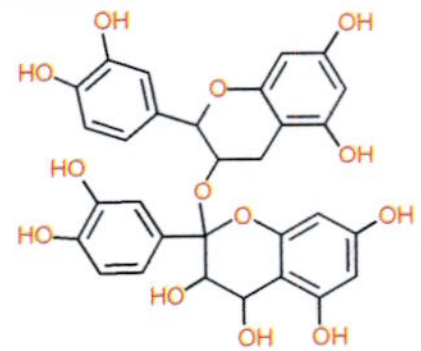	CCI in sciatic nerve causing neuropathic pain in rats	Behavioral and biochemical parameters	Antihyperalgesic and antinociceptive effects greatly improved	Mechanical allodynia and thermal hyperalgesia reduced in all behavioral and physiologic measures, and oxidative stress produced	[39]
Hesperidin	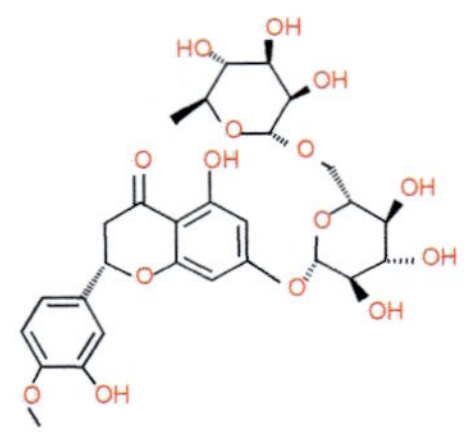	PTX-induced sciatic nerve damage inducing peripheral neuropathy in Sprague-Dawley rats	Behavioral tests SOD, CAT, and GPx activities	Alleviates neuropathic pain caused by PTX-induced peripheral neuropathy with anti-inflammatory, antiapoptotic action	Endoplasmic reticulum stress alleviated by reducing GRP78, pERK, IRE1, ATF-6, and MDA, NF-B, IL-1, TNF, COX-2, and nNOS levels produced by PTX Bcl-2 levels rise while Caspase-3 and Bax fall	[40]
Icariin	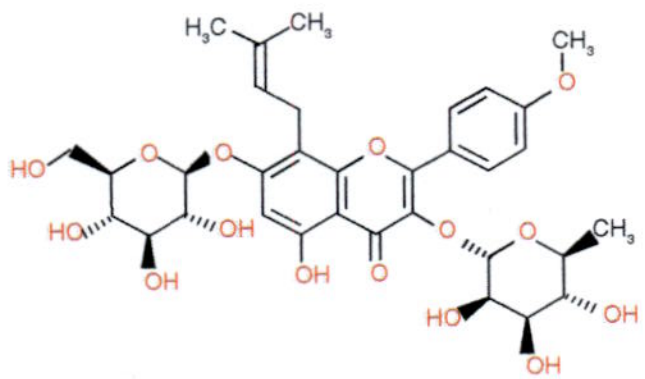	Partial sciatic nerve ligation caused neuropathic pain in rats	in vivo and in silico research studies	Antiapoptotic and anti-inflammatory action for treatment of neuropathic pain	TNF, IL-6, and Bax expression reduced, whereas Bcl-2 expression elevated Reduces inflammation via decreasing NR2B, TRPV1 receptors, and Cav2.2 channel action	[41]

Isoorientin	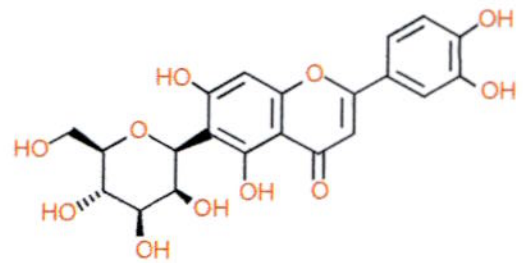	CCI of sciatic nerve in mice	Electrophysiologic, histopathologic, and biochemical indices	Allodynia and hyperalgesia reduced, sensory nerve conduction velocities enhanced, and CCI-induced sciatic nerve injury repaired	Raised total SOD, total antioxidant capacity, and CAT levels while lowering MDA levels Glial cell and MMP-9 activation inhibited, while TNF, IL-6, and IL-1 expression reduced	[42]
Kaempferol	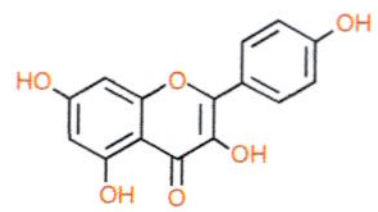	CCI surgery of sciatic nerve in vivo NP model LPS-induced injury in microglial BV2 cells in vitro	Proinflammatory cytokine levels, H&E staining and immunohistochemistry, Western blotting and immunofluorescence	Can greatly lower synthesis of NP and proinflammatory cytokines	Reduced TLR4/NF-B pathway stimulation in LPS-activated BV2 cells, inhibited microglia activation, and switched M1 to M2 phenotype	[43]
Luteolin	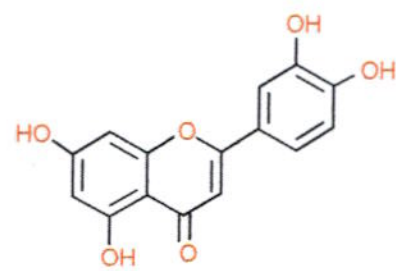	Acute and chronic pain with sciatic nerve ligated-induced neuropathy in mice	Hot-plate test, formalin test	Pain-killing effect of luteolin potentiated with morphine coadministration	Antinociceptive effects of luteolin comparable to imipramine that substantially decreased neuropathic pain with or without morphine	[44]
Matrine	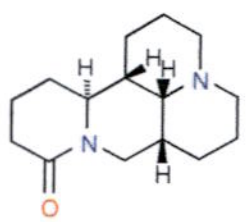	Vincristine-induced neuropathic pain in mouse model	Hot-plate, cold-plate Von Frey hair, and paw pressure tests, thermal and mechanical hyperalgesia, mechanical and cold allodynia	Neuroprotective benefits linked to reduced painful neuropathy, behavioral, electrophysiologic, and histologic alterations	Induced oxidative stress (MDA, GSH-Px, T-AOC, SOD, and TCA) and inflammation (TNF-α, MPO, IL-6, IL-10) reduced	[45]

(*Continued*)

Table 5.2 (Continued)

Name of phytochemical	Structure	Model used	Tests done	Effect seen	Probable mechanism	References
Morin	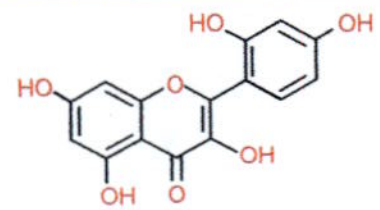	Vincristine-induced NP	Paw withdrawal latency and paw withdrawal threshold	Neuronal hyperexcitability and rise in IL-6, NF-B, and pNF-B expression eliminated Histologic and functional impairments reversed	Reduced neuropathic pain caused by vincristine by preserving sciatic nerve and suppressing inflammation via NF-B pathway	[46]
Mitragynine	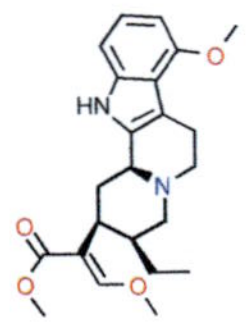	Oxaliplatin injected sore peripheral neuropathy	Mechanical allodynia with Von Frey filaments, paw withdrawal threshold, and other neuropathic pain experiments Experiments with locomotion	Antiallodynic effects, reduced mechanical sensitivity, and increased locomotor activity α-adrenoceptor and opioid antagonists decreased mitragynine's antiallodynic effectiveness	Reduced NP via mechanism necessitating active adrenoceptors proposes it includes stimulation of both adrenergic and opioid systems	[47]
Myricitrin	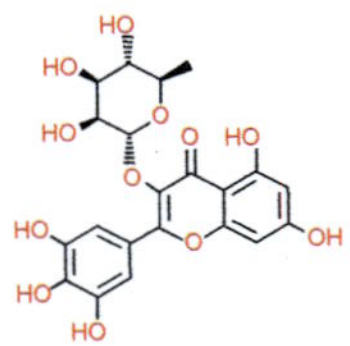	Spinal nerve ligation in pain model of rats	Whole-cell patch-clamp test of isolated rat DRG neurons	Myricetin lowers neuropathic pain behavior temporarily	Analgesic effect could be due to reduction in I(Ca[V]) in DRG neurons caused by PKC	[48]
Naringin	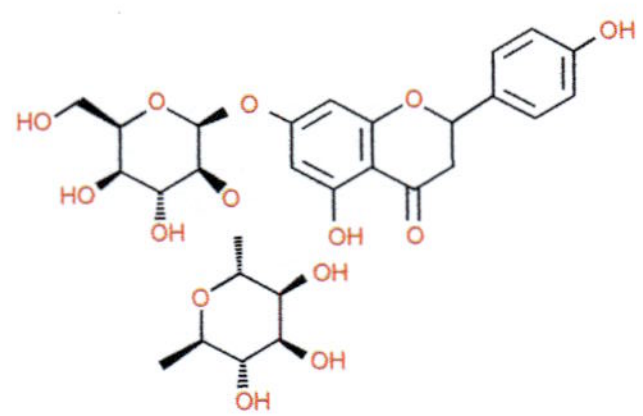	Streptozotocin-induced diabetic NP in Wistar rat	Hyperalgesia and allodynia linked with enhanced oxidative–nitrosative stress, inflammatory mediators	Behavioral, biochemical, and molecular modifications, motor nerve conduction velocity changed	Reversal of NP, as well as oxidative–nitrosative stress; inflammatory cytokine production reduced with MMP-9 modulation	[49]

Paeoniflorin	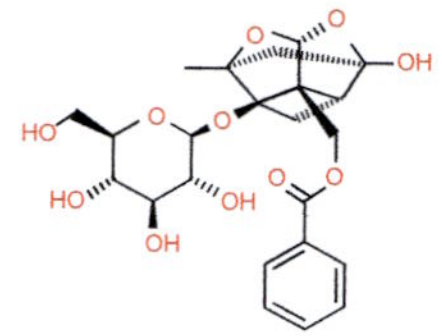	CCI surgery in rats	Von Frey test Cell signaling assay by immunohistochemistry and Western blotting	Inhibition of phosphorylation of ASK1 treats NP	Decreased p-p38, p-JNK expression, late NP, reduced astrocyte and microglia damage responses, decreased TNF-α, IL-1 expression	[50]
Pelargonidin	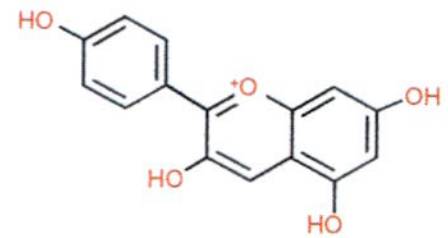	STZ-diabetic neuropathic rat	Hyperalgesia with formalin and hot tail immersion tests, measurements of markers of oxidative stress in brain	Reduction in oxidative damage and thereby avoids diabetic neuropathic hyperalgesia	Reduced synthesis of thiobarbituric acid reactive compounds caused by diabetes and corrected nitrite and SOD elevations	[51]
Puerarin	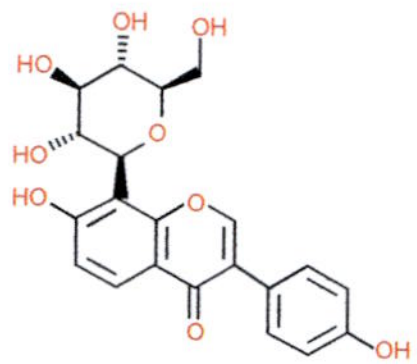	NP induced by chemotherapy paclitaxel	Behavioral tests, mechanical allodynia, thermal hyperalgesia Whole-cell patch-clamp recordings	Paclitaxel induced NP reduced	DRG neurons of rat excitability lowered and their voltage-gated sodium (Nav) channels inhibited	[52]
Quercetin	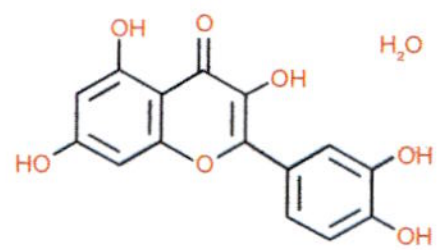	Rat basophilic leukemia (RBL-2H3) cells enthused with paclitaxel	Heat hyperalgesia and mechanical allodynia PKCε and TRPV1 expression levels and histamine release analysis	Paclitaxel-induced peripheral neuropathy reduced in both acute and chronic cases	By stabilizing mast cell membrane and activating downstream protein PKC, reduced paclitaxel-induced NP	[53]

(*Continued*)

Table 5.2 (Continued)

Name of phytochemical	Structure	Model used	Tests done	Effect seen	Probable mechanism	References
Rutin	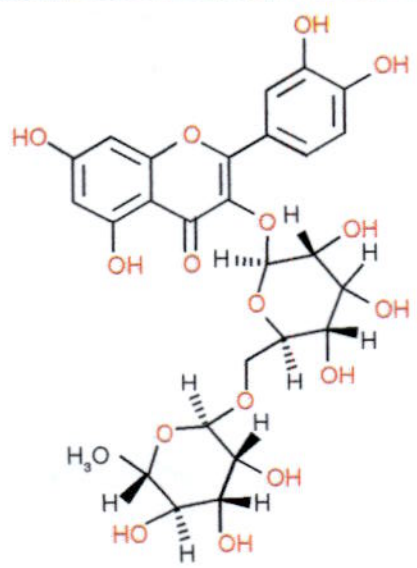	Oxaliplatin brought sore peripheral neuropathy	Tail immersion test, von Frey test Thermal and mechanical allodynia, and immunohistochemical histopathologic and malondialdehyde assay	Reduced oxaliplatin-induced painful peripheral neuropathy	Antioxidant effects, due to lipid peroxidation and protein nitrosylation that partially inhibit oxidative stress-induced injury in dorsal horn neurons	[54]

ASK1, apoptosis signal-regulating kinase 1; ATF-6, activating transcription factor 6; ATP, adenosine triphosphate; BDNF, brain-derived neurotrophic factor; CAT, catalase; CCI, chronic constriction injury; cGMP, cyclic guanosine monophosphate; COX-2, cyclooxygenase-2; DRG, dorsal root ganglia; ELISA, enzyme-linked immunosorbent assay; GABAA, γ-aminobutyric acid type A; GPx, glutathione peroxidase; GRP78, 78-kDa glucose-regulated protein; GSH-Px, glutathione peroxidase; H&E, hematoxylin and eosin; IL, interleukin; IRE1, inositol-requiring enzyme type 1; K, potassium; LPS, lipopolysaccharide; MDA, malondialdehyde; MMP-9, matrix metallopeptidase 9; MPO, myeloperoxidase; NF-B, nuclear factor B; nNOS, neuronal nitric oxide synthase; NO, nitric oxide; NOS, nitric oxide synthase; NP, neuropathic pain; NR2B, N-methyl D-aspartate receptor subtype 2B; pERK, phosphoactivation of extracellular signal regulated kinase; PKC, protein kinase C; PKG, protein kinase G; pNF-B, phospho-nuclear factor B; PTX, paclitaxel; qRT-PCR, quantitative reverse transcription–polymerase chain reaction; SNI, spared nerve injury; SOD, superoxide dismutase; T-AOC, total antioxidant capacity; TCA, trichloroacetic acid; TLR4, Toll-like receptor 4; TNF, tumor necrosis factor; TRPV1, transient receptor potential cation channel subfamily V member; UHPLC-MS, ultra-high performance liquid chromatography–tandem mass spectrometry.

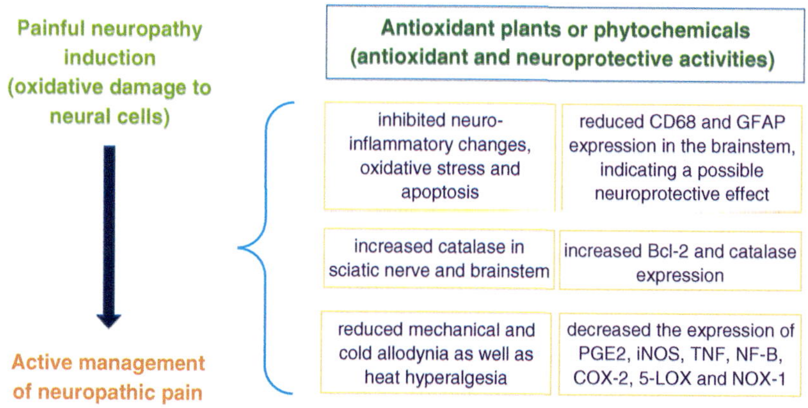

Figure 5.2 Diagrammatic representation of the key mechanisms by which curative plants and their bioactive constituents can avert neuropathic pain. 5-LOX, 5-lipoxygenase; COX-2, cyclooxygenase-2; GFAP, glial fibrillary acidic protein; iNOS, inducible nitric oxide synthase; NF-B, nuclear factor B; NOX-1, nicotinamide adenine dinucleotide phosphate oxidase-1; PGE2, prostaglandin E2; TNF, tumor necrosis factor.

Genistein and naringenin also increase nerve growth factor (NGF) in the sciatic nerves in a diabetic animal model. NGF is essential for the survival and welfare of neurons [58].

Although hyperglycemic conditions can quicken diabetic neuropathy, multiple complex pathways contribute to nerve toxicity. The unifying mechanism in many of these pathways is oxidative stress. Even in long-term diabetes, where strict glycemic control is difficult to maintain, amelioration of this oxidative stress could halt the progression of diabetic neuropathy and its symptoms.

Seven weeks of treatment with *T. cordifolia* reversed thermal hyperalgesia, allodynia, motor deficits, and pathologic changes in the sciatic nerves. Thus, *T. cordifolia* may be useful in preventing early diabetic neuropathic symptoms. Isolation, characterization of the active ingredient(s) responsible for the neuroprotective effect of *T. cordifolia*, and evaluation of functional biomarkers such as motor nerve and sensory nerve conduction velocity measurements in future research could demonstrate the herb's potential in the prophylaxis and cure of diabetic neuropathy [30].

5.4.2 Chemotherapy-Induced Peripheral Neuropathy

Chemotherapy-induced peripheral neuropathy (CIPN) is a type of neuropathy produced by antineoplastic medications. Peripheral nerve damage is caused by various chemotherapeutic agents and other medications in cancer. Platinum compounds are widely utilized to treat various solid cancers. Oxaliplatin, a third-generation platinum drug, is important in reducing antitumoral resistance while also causing significant cytotoxicity. The adverse effects of CIPN reduce quality of life and lead to discontinuation of chemotherapy. Several medications like duloxetine and gabapentin are suggested to treat CIPN, with restricted use due to less efficacy or adverse effects like nausea, vomiting, somnolence, and dizziness [59].

The most important accepted products with healing effects in CIPN are curcumin, rutin, quercetin, matrine, and cannabinoids [60]. Furthermore, medicinal plants such as *Matricaria chamomilla*, *Xylopia aethiopica*, and *Snedrella nodiflora* have protective or therapeutic activities in neuropathic pain.

Medicinal plants and their phytochemicals benefit CIPN through a variety of pharmacologic mechanisms, like reduction of total calcium levels, regulation of neural cell apoptosis, inflammatory cytokines transduction signaling (TNF-, ILs, cyclooxygenase-2 [COX-2]), inhibition of inflammatory response and axonal degeneration, nitrergic and nuclear factor B (NF-B) pathways, and reinforcement of enzymatic antioxidant agents such as superoxide dismutase (SOD) and catalase (CAT) [61].

Rutin and quercetin, a flavonoid found in plants, were found to lower mechanical and cold nociceptive thresholds induced by oxaliplatin. Rutin and a nanoemulsion of quercetin were found to alleviate oxaliplatin-induced mechanical allodynia.

Moreover, 6-methoxyflavone possesses antinociceptive efficacy. It significantly decreased cisplatin-induced mechanical allodynia and thermal hypoalgesia by rising the paw withdrawal threshold and improving the paw thermal threshold in a rodent model of CIPN. The pain-relieving potency of plants and their phytochemicals on CIPN is mainly facilitated by glial modulation [59].

5.4.3 Sciatic Nerve Chronic Constriction Injury

The CCI model of chronic neuropathic pain is very well studied. Hyperalgesia, paresthesia, allodynia, dysesthesia, and impulsive pain are some of the symptoms of CCI-induced pain. Neuropathy and neuroinflammation found in the nervous systems both centrally and peripherally are frequently associated with sciatic nerve injury [62].

Both mechanical and thermal hyperalgesia were reduced by flavonoids such as hesperidin, diosmin, and grape seed proanthocyanidins. Other flavonoids, such as genistein, epigallocatechin gallate (EGCG), EGCG-derived compounds, morin, and isoorientin, only reduced thermal hyperalgesia. Flavonoids such as genistein, EGCG, puerarin, morin, and isoorientin reduce CCI-mediated mechanical allodynia. Morin and isoorientin reduced cold allodynia. Puerarin condensed the increased immunoreactivity of glial fibrillary acidic protein and ionized calcium-binding adaptor protein-1, both being astroglia and microglial activation markers [63].

The plant extract pretreatment reduced heat hyperalgesia and cold and mechanical allodynia more effectively than pregabalin. Furthermore, CCI increased the expression of TNF-α, nicotinamide adenine dinucleotide phosphate (NADPH) oxidase-1 (NOX-1), COX-2, NF-B, 5-lipoxygenase (5-LOX), inducible nitric oxide synthase (iNOS), and prostaglandin E2 (PGE2) and diminished catalase in the brainstem and sciatic nerve [64].

The neuroinflammatory variations linked with amplified apoptosis (increased Bax), glial cell activation (glial fibrillary acidic protein [GFAP]- and CD68-positive cells) occurs along with organizational variations in both the brainstem and the sciatic nerve. The plant extract inhibited CCI-persuaded oxidative stress, neuroinflammatory variations, and apoptosis during increasing Bcl-2 and CAT expression, depending on dose. It reduced GFAP and CD68 expression in the brainstem, indicating a possible neuroprotective effect [62].

5.4.4 Other Neuropathic Pain Signaling Pathways

Because of their activity on γ-aminobutyric acid type A ($GABA_A$) receptors, flavonoids have anti-inflammatory and antioxidant properties. The majority of metabolic diseases are caused by oxidative stress. Aside from foreign influences, normal oxygen metabolism in tissues and cells generates ROS and free radicals that are constantly threatening them [65].

Rutin and epicatechin are capable of being oxidized by free radicals, resulting in less reactive and more stable compounds. Similarly, quercetin protects cells from NO-mediated cell damage [66].

Various flavonoids, including myricetin, EGCG, and baicalein, were reported to reduce mechanical allodynia caused by spinal nerve ligation.

Quercetin, silibinin, and luteolin are antioxidants that function by inhibiting the enzyme xanthine oxidase [67].

Reperfusion triggered by leukocyte mobility results in inflammatory mediators and cytotoxic oxidants release, activating the complement system. Numerous flavonoids contribute to the immobility of leukocytes, decreasing the serum complement system and both inflammation of neuropathic pain of peripheral origin and inflammation linked by similar pathophysiologic pathways. Hyperalgesia and allodynia are symptoms of both types of diseases [68].

5.5 Future Perspectives

The absence of a thorough valuation of the meticulous cellular and extracellular signaling paths found in pharmacologic action of plants and phytochemicals necessitates studies that can directly measure inflammatory signaling mediators as biomarkers of oxidative stress such as Nrf2.

Additional investigations are needed to evaluate mechanisms of action in comprehensive animal models for improved understanding of the molecular targets implicated in neuropathic pain treatment. Well-constructed clinical trials and mechanistic studies could measure the protection and effectiveness of medicinal plants and phytochemicals in neuropathic pain.

5.6 Conclusion

Remedial plants and phytochemicals are appreciated causes of antineuropathic pain effects. Herbal medicines, according to this review, may be an alternative option for relieving and managing neuropathic pain. Numerous medicinal plants and phytochemicals are beneficial in animal models of neuropathic pain through their antioxidant, anti-inflammatory, and antinociceptive properties. Their broad mechanisms of action can aid in pathologies involving inflammation and oxidative stress of neuropathic pain.

References

1 Morris, G., Berk, M., Galecki, P. et al. (2016). The neuro-immune pathophysiology of central and peripheral fatigue in systemic immune-inflammatory and neuro-immune diseases. *Mol. Neurobiol.* 53 (2): 1195–1219.

2 Dym, H., Lin, S., and Thakkar, J. (2020). Neuropathic pain and burning mouth syndrome: an overview and current update. *Dent. Clin.* 64 (2): 379–399.

3 Steiner, G.E., Newman, M.E., Paikl, D. et al. (2003). Expression and function of pro-inflammatory interleukin IL-17 and IL-17 receptor in normal, benign hyperplastic, and malignant prostate. *Prostate* 56 (3): 171–182.

4 Sindrup, S.H. and Jensen, T.S. (1999). Efficacy of pharmacological treatments of neuropathic pain: an update and effect related to mechanism of drug action. *Pain* 83 (3): 389–400.

5 Brunelli, B. and Gorson, K.C. (2004). The use of complementary and alternative medicines by patients with peripheral neuropathy. *J. Neurol. Sci.* 218 (1–2): 59–66.

6 Covington, E.C. (2000). The biological basis of pain. *Int. Rev. Psychiatry* 12 (2): 128.

7 Pascuzzi, R.M. (2003). Peripheral neuropathies in clinical practice. *Med. Clin.* 87 (3): 697–724.

8 Mense, S. and Hoheisel, U. (2004). Central nervous sequelae of local muscle pain. *J. Musculoskelet. Pain* 12 (3–4): 101–109.

9 Miesenböck, G. and Kevrekidis, I.G. (2005). Optical imaging and control of genetically designated neurons in functioning circuits. *Annu. Rev. Neurosci.* 28: 533.

10 Meisner, J.G. (2008). Effect of nerve injury on noradrenaline augmentation of $P2X_3$-mediated peripheral pain signaling: role of injured fibres, strain and receptor subtype. Ottawa: Library and Archives Canada/Bibliothèque et Archives Canada.

11 Ismail, C.A. and Long, I. (2022). Models of diabetes in rats: a focus on diabetic neuropathy and biomarkers. In: *Biomarkers in Diabetes* (ed. V.B. Patel and V.R. Preddy), 1–23. Cham: Springer.

12 Lupachyk, S., Watcho, P., Hasanova, N. et al. (2012). Triglyceride, nonesterified fatty acids, and prediabetic neuropathy: role for oxidative–nitrosative stress. *Free Radic. Biol. Med.* 52 (8): 1255–1263.

13 Greening, J.B. (2000). Changed vibration threshold and loss of nerve movement in patients with repetitive strain injury: the peripheral neuropathology of RSI. PhD thesis. University College London.

14 Argyriou, A.A., Bruna, J., Marmiroli, P., and Cavaletti, G. (2012). Chemotherapy-induced peripheral neurotoxicity (CIPN): an update. *Crit. Rev. Oncol. Hematol.* 82 (1): 51–77.

15 Muthuraman, A. and Singh, N. (2012). Neuroprotective effect of saponin rich extract of Acorus calamus L. in rat model of chronic constriction injury (CCI) of sciatic nerve-induced neuropathic pain. *J. Ethnopharmacol.* 142 (3): 723–731.

16 Singh, H., Arora, R., Arora, S., and Singh, B. (2017). Ameliorative potential of *Alstonia scholaris* (Linn.) R. Br. against chronic constriction injury-induced neuropathic pain in rats. *BMC Complement. Altern. Med.* 17 (1): 1–9.

17 Watcho, P., Stavniichuk, R., Tane, P. et al. (2011). Evaluation of PMI-5011, an ethanolic extract of *Artemisia dracunculus* L., on peripheral neuropathy in streptozotocin-diabetic mice. *Int. J. Mol. Med.* 27 (3): 299–307.

18 Thiagarajan, V.R., Shanmugam, P., Krishnan, U.M. et al. (2012). Ameliorative potential of *Butea monosperma* on chronic constriction injury of sciatic nerve induced neuropathic pain in rats. *An. Acad. Bras. Cienc.* 84 (4): 1091–1104.

19 Ostovar, M., Akbari, A., Anbardar, M.H. et al. (2020). Effects of *Citrullus colocynthis* L. in a rat model of diabetic neuropathy. *J. Integr. Med.* 18 (1): 59–67.

20 Banafshe, H.R., Hamidi, G.A., Noureddini, M. et al. (2014). Effect of curcumin on diabetic peripheral neuropathic pain: possible involvement of opioid system. *Eur. J. Pharmacol.* 723: 202–206.

21 Safakhah, H.A., Taghavi, T., Rashidy-Pour, A. et al. (2016). Effects of saffron (*Crocus sativus* L.) stigma extract and its active constituent crocin on neuropathic pain responses in a rat model of chronic constriction injury. Iran. *J. Pharm. Res.* 15 (1): 253.

22 Kim, Y.S., Park, H.J., Kim, T.K. et al. (2009). The effects of *Ginkgo biloba* extract EGb 761 on mechanical and cold allodynia in a rat model of neuropathic pain. *Anesth. Analg.* 108 (6): 1958–1963.

23 Jain, V., Pareek, A., Paliwal, N. et al. (2014). Antinociceptive and antiallodynic effects of *Momordica charantia* L. in tibial and sural nerve transection-induced neuropathic pain in rats. *Nutr. Neurosci.* 17 (2): 88–96.

24 Çelik, F., Göçmez, C., Karaman, H. et al. (2014). Therapeutic effects of thymoquinone in a model of neuropathic pain. *Curr. Ther. Res. Clin. Exp.* 76: 11–16.

25 Kaur, G., Bali, A., Singh, N., and Jaggi, A.S. (2015). Ameliorative potential of *Ocimum sanctum* in chronic constriction injury-induced neuropathic pain in rats. *An. Acad. Bras. Cienc.* 87: 417–429.

26 Kassuya, C.A., Silvestre, A.A., Rehder, V.L., and Calixto, J.B. (2003). Anti-allodynic and anti-oedematogenic properties of the extract and lignans from *Phyllanthus amarus* in models of persistent inflammatory and neuropathic pain. *Eur. J. Pharmacol.* 478 (2–3): 145–153.

27 Nucci-Martins, C., Martins, D.F., Nascimento, L.F. et al. (2015). Ameliorative potential of standardized fruit extract of *Pterodon pubescens* Benth on neuropathic pain in mice: evidence for the mechanisms of action. *J. Ethnopharmacol.* 175: 273–286.

28 Rahbardar, M.G., Amin, B., Mehri, S. et al. (2017). Anti-inflammatory effects of ethanolic extract of *Rosmarinus officinalis* L. and rosmarinic acid in a rat model of neuropathic pain. *Biomed. Pharmacother.* 86: 441–449.

29 Abad, A.N., Nouri, M.H., and Tavakkoli, F. (2011). Effect of *Salvia officinalis* hydroalcoholic extract on vincristine-induced neuropathy in mice. *Chin. J. Nat. Med.* 9 (5): 354–358.

30 Archana, J., Annapurna, A., and Devayani, P. (2022). Neuroprotective role of *Tinospora cordifolia* extract in streptozotocin induced neuropathic pain. *Braz. J. Pharm. Sci.* 58: e18501.

31 Huang, L.J., Jia, S.S., Sun, X.H. et al. (2020). Baicalin relieves neuropathic pain by regulating α2 adrenoceptor levels in rats following spinal nerve injury. *Exp. Ther. Med.* 20 (3): 2684–2690.

32 Quiñonez-Bastidas, G.N., Pineda-Farias, J.B., Flores-Murrieta, F.J. et al. (2018). Antinociceptive effect of (−)-epicatechin in inflammatory and neuropathic pain in rats. *Behav. Pharmacol.* (2 and 3-Spec Issue): 270–279.

33 Li, R., Zhao, C., Yao, M. et al. (2017). Analgesic effect of coumarins from Radix angelicae pubescentis is mediated by inflammatory factors and TRPV1 in a spared nerve injury model of neuropathic pain. *J. Ethnopharmacol.* 195: 81–88.

34 Zhu, X., Li, Q., Chang, R. et al. (2014). Curcumin alleviates neuropathic pain by inhibiting p300/CBP histone acetyltransferase activity-regulated expression of BDNF and cox-2 in a rat model. *PLoS One* 9 (3): e91303.

35 Bertozzi, M.M., Rossaneis, A.C., Fattori, V. et al. (2017). Diosmin reduces chronic constriction injury-induced neuropathic pain in mice. *Chem. Biol. Interact.* 273: 180–189.

36 Carballo-Villalobos, A.I., González-Trujano, M.E., Pellicer, F., and López-Muñoz, F.J. (2016). Antihyperalgesic effect of hesperidin improves with Diosmin in experimental neuropathic pain. *Biomed. Res. Int.* 2016: 8263463.

37 Zhao, X., Wang, C., Cui, W.G. et al. (2015). Fisetin exerts antihyperalgesic effect in a mouse model of neuropathic pain: engagement of spinal serotonergic system. *Sci. Rep.* 5 (1): 1–2.

38 Valsecchi, A.E., Franchi, S., Panerai, A.E. et al. (2008). Genistein, a natural phytoestrogen from soy, relieves neuropathic pain following chronic constriction sciatic nerve injury in mice: anti-inflammatory and antioxidant activity. *J. Neurochem.* 107 (1): 230–240.

39 Kaur, G., Bedi, O., Sharma, N. et al. (2016). Anti-hyperalgesic and anti-nociceptive potentials of standardized grape seed proanthocyanidin extract against CCI-induced neuropathic pain in rats. *J. Basic Clin. Physiol. Pharmacol.* 27 (1): 9–17.

40 Semis, H.S., Kandemir, F.M., Kaynar, O. et al. (2021). The protective effects of hesperidin against paclitaxel-induced peripheral neuropathy in rats. *Life Sci.* 287: 120104.

41 Pokkula, S. and Thakur, S.R. (2021). Icariin ameliorates partial sciatic nerve ligation induced neuropathic pain in rats: evidence of in silico and in vivo studies. *J. Pharm. Pharmacol.* 73 (7): 874–880.

42 Zhang, G., Liu, N., Zhu, C. et al. (2019). Antinociceptive effect of isoorientin against neuropathic pain induced by the chronic constriction injury of the sciatic nerve in mice. *Int. Immunopharmacol.* 75: 105753.

43 Chang, S., Li, X., Zheng, Y. et al. (2022). Kaempferol exerts a neuroprotective effect to reduce neuropathic pain through TLR4/NF-κB signaling pathway. *Int. J. Phytother. Res.* 36 (4): 1678–1691.

44 Hashemzaei, M., Abdollahzadeh, M., Iranshahi, M. et al. (2017). Effects of luteolin and luteolin-morphine co-administration on acute and chronic pain and sciatic nerve ligated-induced neuropathy in mice. *J. Complement. Integr. Med.* 14 (1).

45 Gong, S.S., Li, Y.X., Zhang, M.T. et al. (2016). Neuroprotective effect of matrine in mouse model of vincristine-induced neuropathic pain. *Neurochem. Res.* 41 (11): 3147–3159.

46 Jiang, K., Shi, J., and Shi, J. (2019). Morin alleviates vincristine-induced neuropathic pain via nerve protective effect and inhibition of NF-κB pathway in rats. *Mol. Neurobiol.* 39 (6): 799–808.

47 Foss, J.D., Nayak, S.U., Tallarida, C.S. et al. (2020). Mitragynine, bioactive alkaloid of kratom, reduces chemotherapy-induced neuropathic pain in rats through α-adrenoceptor mechanism. *Drug Alcohol Depend.* 209: 107946.

48 Hagenacker, T., Hillebrand, I., Wissmann, A. et al. (2010). Anti-allodynic effect of the flavonoid myricetin in a rat model of neuropathic pain: involvement of p38 and protein kinase C mediated modulation of Ca^2+ channels. *Eur. J. Pain* 14 (10): 992–998.

49 Singh, P., Bansal, S., Kuhad, A. et al. (2020). Naringenin ameliorates diabetic neuropathic pain by modulation of oxidative-nitrosative stress, cytokines and MMP-9 levels. *J. Funct. Foods* 11 (5): 4548–4560.

50 Zhou, D., Zhang, S., Hu, L. et al. (2019). Inhibition of apoptosis signal-regulating kinase by paeoniflorin attenuates neuroinflammation and ameliorates neuropathic pain. *J. Neuroinflammation* 16 (1): 83.

51 Mirshekar, M., Roghani, M., Khalili, M. et al. (2010). Chronic oral pelargonidin alleviates streptozotocin-induced diabetic neuropathic hyperalgesia in rat: involvement of oxidative stress. *Iran. Biomed. J.* 14 (1–2): 33.

52 Zhang, X.L., Cao, X.Y., Lai, R.C. et al. (2019). Puerarin relieves paclitaxel-induced neuropathic pain: the role of Nav1. 8 β1 subunit of sensory neurons. *Front. Pharmacol.* 9: 1510.

53 Gao, W., Zan, Y., Wang, Z.J. et al. (2016). Quercetin ameliorates paclitaxel-induced neuropathic pain by stabilizing mast cells, and subsequently blocking PKCε-dependent activation of TRPV1. *Acta Pharmacol. Sin.* 37 (9): 1166–1177.

54 Azevedo, M.I., Pereira, A.F., Nogueira, R.B. et al. (2013). The antioxidant effects of the flavonoids rutin and quercetin inhibit oxaliplatin-induced chronic painful peripheral neuropathy. *Mol. Pain* 9: 1744–8069.

55 Hosseini, A. and Abdollahi, M. (2013). Diabetic neuropathy and oxidative stress: therapeutic perspectives. *Oxid. Med. Cell. Longev.* 2013: 168039.

56 Obrosova, I.G., Ilnytska, O., Lyzogubov, V.V. et al. (2007). High-fat diet–induced neuropathy of pre-diabetes and obesity: effects of "healthy" diet and aldose reductase inhibition. *J. Diabetes* 56 (10): 2598–2608.

57 Khan, H., Ullah, H., Tundis, R. et al. (2020). Dietary flavonoids in the management of Huntington's disease: mechanism and clinical perspective. *EFood* 1 (1): 38–52.

58 Basu, P. and Basu, A. (2020). in vitro and in vivo effects of flavonoids on peripheral neuropathic pain. *Molecules* 25 (5): 1171.

59 Lee, J.H., Kim, N., Park, S., and Kim, S.K. (2021). Analgesic effects of medicinal plants and phytochemicals on chemotherapy-induced neuropathic pain through glial modulation. *Pharmacol. Res. Perspect.* 9 (6): e00819.

60 Oveissi, V., Ram, M., Bahramsoltani, R. et al. (2019). Medicinal plants and their isolated phytochemicals for the management of chemotherapy-induced neuropathy: therapeutic targets and clinical perspective. *DARU J. Pharm. Sci.* 27 (1): 389–406.

61 Liu, Y.W., Liu, C.T., Su, Y.L., and Tsai, M.Y. (2020). A narrative review of complementary nutritional supplements for chemotherapy-induced peripheral neuropathy. *Altern. Ther. Health Med.* 26 (4): 43–49.

62 Mahmoud, M.F., Rezq, S., Alsemeh, A.E. et al. (2021). Extract attenuates neuroinflammation and neuropathic pain in sciatic nerve chronic constriction injury-induced peripheral neuropathy in rats. *Front. Pharmacol.* 12: 799444.

63 Uddin, M., Mamun, A.A., Rahman, M. et al. (2020). Exploring the promise of flavonoids to combat neuropathic pain: from molecular mechanisms to therapeutic implications. *Front. Neurosci.* 14: 478.

64 Sobeh, M., Mahmoud, M.F., Rezq, S. et al. (2019). Salix tetrasperma Roxb. extract alleviates neuropathic pain in rats via modulation of the NF-κB/TNF-α/NOX/iNOS pathway. *Antioxidants* 8 (10): 482.

65 Singh, M., Kaur, M., and Silakari, O. (2014). Flavones: an important scaffold for medicinal chemistry. *Eur. J. Med. Chem.* 84: 206–239.

66 de Groot, H.D. and Rauen, U. (1998). Tissue injury by reactive oxygen species and the protective effects of flavonoids. *Fundam. Clin. Pharmacol.* 12 (3): 249–255.

67 Pauff, J.M. and Hille, R. (2009). Inhibition studies of bovine xanthine oxidase by luteolin, silibinin, quercetin, and curcumin. *J. Nat. Prod.* 72 (4): 725–731.

68 Nijveldt, R.J., Van Nood, E.L., Van Hoorn, D.E. et al. (2001). Flavonoids: a review of probable mechanisms of action and potential applications. *Am. J. Clin. Nutr.* 74 (4): 418–425.

6

Brain Function, Stroke, and Medicinal Herbs

Shahira M. Ezzat[1,2], Rana Marghany[3], Nehal El Mahdi[4], and Mohamed A. Salem[5]

[1] Department of Pharmacognosy, Faculty of Pharmacy, Cairo University, Cairo, Egypt
[2] Department of Pharmacognosy, Faculty of Pharmacy, October University for Modern Sciences and Arts (MSA), Giza, Egypt
[3] Department of Pharmacognosy, National Research Centre, Giza, Egypt
[4] Department of Pharmaceutics and Industrial Pharmacy, Faculty of Pharmacy, October University for Modern Sciences and Arts (MSA), Giza, Egypt
[5] Department of Pharmacognosy and Natural Products, Faculty of Pharmacy, Menoufia University, Menoufia, Egypt

6.1 Brain Function and Stroke

The mammalian central nervous system (CNS) is the most complex system in the body, coordinating several organ systems. Anatomically, the CNS is mainly composed of the brain, brainstem, and spinal cord [1]. The brain comprises two hemispheres, and the brainstem connects the spinal cord to the brain [2]. The brain controls the processing of sensory information, the regulation of involuntary physiologic processes such as digestion, heart rate, blood pressure, and breathing, as well as the release of hormones and neurotransmitters [3, 4]. The human brain is vulnerable to damage from neurodegeneration, injury, or pathogen invasions, leading to cognitive and neurobehavioral problems [5]. The brain contains blood vessels and nerves, including neurons and glial cells. Blood carries oxygen and essential nutrients to brain cells and this blood flow is controlled by neurons and astrocytes [6]. The disruption of cerebral blood flow is a driving factor for the progression of several disorders such as ischemic stroke [7].

Stroke is the second major cause of death and the main cause of dementia and depression worldwide [8]. Stroke is defined as a lack of oxygen, with the death of some brain cells due to the stopping of blood flow to such cells as a result of blockage of a cerebral artery [9]. The burden of stroke is higher in low- and middle-income countries than in high-income countries and is mainly of the ischemic type (only 34% are of the hemorrhagic type) [9]. For this reason, correct diagnosis of stroke and finding effective drugs for this disease are essential.

The mechanism behind ischemic stroke may involve the presence of toxic amino acids, free radicals, and high intracellular calcium [10]. The lack of blood supply to the brain causes a reduction in glucose, oxygen, and energy required for maintaining the ionic gradient. This leads to the activation of calcium channels and accumulation of excitatory amino acids, especially glutamate, occurs in the extracellular space [11]. The accumulation of free

Phytochemical Drug Discovery for Central Nervous System Disorders: Biochemistry and Therapeutic Effects, First Edition. Edited by Chukwuebuka Egbuna and Mithun Rudrapal.

radicals, particularly reactive oxygen species (ROS), results from the activation of phospholipase A and cyclooxygenase, causing an inflammatory response that is accompanied by apoptosis caused by the release of cytochrome C from mitochondria [12].

6.2 Strategies for Treatment of Ischemic Stroke

There are two main strategies: neuroprotective and neurorestorative.

6.2.1 Neuroprotective Strategy

The neuroprotective strategy means inhibition of the neuronal cell death cascade, which cannot be fully achieved so it is limited to reduction of ischemic cerebral damage through targeting the following pathways [13].

6.2.1.1 Oxidative Stress Targeting

This target is very important as brain ischemia is mainly caused by imbalance between the ROS produced and the antioxidant enzyme [14].

6.2.1.2 Excitotoxicity Targeting

Regulating the activity of excitatory amino acids through using drugs that inhibit glutamate release or regulate the glutamate and 5-HT1 receptor modulators and thus exert an antibrain ischemic effect is a very effective strategy [15].

It has been found that hypoxia, glucose depletion, and energy failure caused by ischemia leads to accumulation of glutamate (due to reuptake inhibition), which activates the N-methyl-D-aspartate (NMDA) receptors of the excitatory amino acids, resulting in opening of calcium channels, increased intracellular calcium concentration, necrosis of neurons [16], and mitochondrial destruction that leads to apoptosis [17].

6.2.1.3 Apoptosis Targeting

Activation of enzymes such as Bcl-2 can inhibit brain ischemia–perfusion injury-induced neuronal apoptosis [18]. Other genes may be involved in apoptosis, such as the *P53* gene, and the caspase family can be also a target for inhibition of apoptosis [19]. In addition, inflammation and autophagy can also be effective targets for treatment.

6.2.2 Neurorestorative Strategy

The neurorestorative strategy may take longer than the neuroprotective strategy, but has still proved its efficacy.

6.2.2.1 Angiogenesis Targeting

Angiogenesis is the process of production of new blood vessels through proliferation and migration of endothelial cells from mature or original blood vessels. This mechanism is crucial in wound healing [20]. It was found that patients with an increased number of blood vessels after stroke have higher survival rates [20]. Stimulation of angiogenesis after stroke can be achieved through various angiogenesis-promoting factors, among which the vascular endothelial growth factor (VEGF) and nitric oxide (NO) are the most effective targets [21].

6.2.2.2 Neurogenesis Targeting

Neurogenesis is the production of new functional neurons from neuronal stem/progenitor cells through proliferation and migration then differentiation, which mainly occurs in the subventricular zone (SVZ) in cases of ischemic stroke [22].

Brain-derived neurotrophic factor (BDNF) binds to tyrosine kinase B (TrkB) receptors and its activation promotes neurogenesis [23]. The phosphatidylinositol-3-kinase (PI3K)/Akt signaling pathway plays a role in axonal sprouting, which controls neurogenesis and neurologic function repair after stroke [24]. Moreover, cyclic adenosine monophosphate (cAMP)-response element-binding protein (CREB) is involved in neurogenesis and synaptic development [25]. Thus, BDNF–TrkB, PI3K/Akt, and CREB are the major therapeutic targets that promote neurogenesis and repair neurologic function after ischemic stroke.

6.3 Medicinal Plants for the Treatment of Stroke

Currently, after 1000 years of practice, many studies have confirmed that medicinal plants can afford a potent therapeutic effect against brain stroke. Medicinal plants and their different constituents can perform their antiischemic action by inhibiting oxidative stress, cell apoptosis, and different inflammatory reactions, as well as promoting angiogenesis and neurogenesis through regulating different signaling pathways [26].

For instance, EGb (a common natural herbal product extracted from *Ginkgo biloba* leaves) was evaluated for its anti-ischemic role in vivo. EGb (15 mg/kg body weight [BW] intragastrically [IG]/d in middle cerebral artery occlusion [MCAO]-induced stroke rats) showed a significant increase in the activity of glutathione (GSH) and superoxide dismutase (SOD), while it decreased the levels of malondialdehyde (MDA) and nitric oxide (NO) on day 7. This antioxidative mechanism was mediated by offsetting extracellular signal-regulated kinase (ERK) activation and stimulating protein phosphatase 2A (PP2A) expression [27].

Moreover, EGb (60 mg/kg BW IG/d in MCAO-induced stroke rats) for 15 days showed a significant reduction in infarct volume, decrease in damage to gray and white matter, and enhancement in collateral circulation, cerebral perfusion, as well as axonal remodeling. It also improved behavioral recovery and endogenous neurogenesis. In addition, it helped in increasing the levels of phosphorylated Akt (p-Akt) and glycogen synthase kinase-3 beta (p-GSK-3β) and decreasing the phosphorylated collapsin response mediator protein-2 (p-CRMP2) level [28].

Furthermore, EGb (15 mg/kg BW IG/d in MCAO-induced stroke rats) showed a significant decrease in cerebral infarcts and neurologic deficiency. Moreover, it showed a reduction in signal transducer and activator of transcription 3 (STAT3) and Janus kinase 2 (JAK2) phosphorylation as well as lipocalin-2 (LCN2) expression. It also inhibited astrocyte activation and neuronal apoptosis by inhibiting caspase-3 cleavage and hindering the extrinsic apoptotic pathway [29, 30].

In an *in vitro* study conducted by Yang et al., EGb (0.1 mg/ml in oxygen-glucose deprivation/reoxygenation [OGD/R]-induced stroke in the astrocyte-neuron cells network) showed a significant increase in neuron viability and electrical resistance. It also showed a significant decrease in cell injury and apoptosis, as well as endothelial permeability for sodium fluorescein. Moreover, it increases the levels of zonula occludens-1 (ZO-1) and occludin. Further, it helped in enhancing Bcl-2 mRNA and protein levels, while reduced caspase-3 and Bax mRNA and protein levels [31].

Bu Yang Huan Wu (BYHW) decoction (a combination of astragalus, angelica tail, red peony root, earthworm, Chuanxiong, safflower, and peach kernel) and Houshiheisan (a combination of Flos Chrysanthemi, Radix Saposhnikoviae, Ramulus Cinnamomi, Rhizoma Chuanxiong, Radix et Rhizoma Asari, Radix Platycodonis, Rhizoma Atractylodis macrocephalae, Poria, Rhizoma Zingiberis, Radix Angelicae sinensis, Radix et Rhizoma Ginseng, Radix Scutellariae, and Concha Ostreae) are traditional Chinese medicine formulas that are commonly used to treat different forms of stroke [32–35].

BYHW (7, 14, 28 g/kg/d IG in MCAO/R-induced ischemia in rats) showed significant suppression of the hypoxia inducible factor-1 alpha (HIF-1α)/VEGF pathway [36].

Houshiheisan (5.25 or 10.5 g/kg/d IG in MCAO/R-induced ischemia in rats) revealed a significant reduction in edema and damage to blood vessels and neurons, and reduced the damage to blood vessels and neurons in the ischemic areas. Immunostaining and quantitative real-time polymerase chain reaction (PCR) analysis showed that Houshiheisan upregulated CD31, VEGF, VEGFA, VEGFR2, and angiopoietin 1 (Ang-1) expression, while it downregulated Ang-2, stromal cell-derived factor 1 (SDF-1), and CXC chemokine receptor 4 (CXCR4) expression in vivo. As well, Houshiheisan showed a significant enhancement of proliferation and migration after persistent hypoxia in immortalized human umbilical vein cells (Im-HUVECs) in vitro. It also revealed significant upregulation in the expression of HIF-1α, VEGFA, Ang-1, Ang-2, and CXCR4, and downregulation of Ang-2 [37].

Angelica sinensis extract also proved to have potential neuroprotective activity on cerebral ischemia [38, 39]. For instance, Cheng et al. proved that *A. sinensis* extract (0.5 g and 1 g/kg BW intraperitoneally [IP] in cortical ischemic penumbra) revealed potent effects on the angiogenesis and antiapoptotic effects of neurons through the activation of p38/HIF-1/VEGFA/von Willebrand factor (vWF) and p38/HIF-1/VEGFA/p-Bad–mediated regulation of cytochrome c/caspase-3 signaling. Docking studies revealed that compounds with a high lipid-water partition coefficient and ring structures with a molecular weight under 500 g/mol and fewer than 10 hydrogen bond acceptors are related to the regulation of HIF-1-related pathways [40].

Panax ginseng and its saponin glycosides proved to have potent efficiency in preventing and treating experimental strokes [41]. For instance, Korean red ginseng (100–360 mg/kg/d IG in MCAO-induced ischemia in mice) showed a significant decrease in infarct volume and brain edema, as well as an increase in the protein levels of SOD2, glutathione peroxidase 1 (GPx1), heme oxygenase 1 (HO-1), and nicotinamide adenine dinucleotide phosphate quinone dehydrogenase 1 (NQO-1). It also revealed significant upregulation of Nrf2 pathway-mediated neuroprotection [42].

Moreover, Korean red ginseng (100 mg/kg/d IG in MCAO-induced ischemia in rats) showed a significant reduction in the neurologic deficits score and the infarct volume. It also attenuated the oxidative damage by reducing the levels of 8-hydroxyguanosine (8-OHG) and lipid peroxidation [43].

Further, ginseng total saponins (ginsenosides 25 mg/kg/d IG twice in MCAO-induced ischemia in rats) showed a significant decrease in neurologic deficits score. It also enhanced neurogenesis by increasing the number of neuronal precursors and promoting the proliferation of endogenous neural stem cells [44].

In addition, black ginseng is a recent type of processed ginseng with a unique component profile, which revealed a significant improvement in learning and memory of rats after MCAO at a dose of 100 or 400 mg/kg orally (PO), indicated by the Morris water maze [45].

6.4 Natural Products for the Treatment of Stroke

Several studies proved the efficacy of natural compounds in treating and preventing brain stroke [46]. These natural compounds were divided into flavonoids (baicalin, icariin, puerarin, and breviscapine), phenols (resveratrol, curcumin, and salvianolic acid B), terpenes (ginkgolide B and catalpol), and glycosides (ginsenoside Rg1). They displayed antibrain ischemia activities in different in vitro and in vivo experiments. These compounds are summarized in Table 6.1 with their possible mechanisms of action. The structures of these compounds are illustrated in Figure 6.1.

Moreover, computational studies have achieved great attention for the development of new active compounds and relating to their structure–activity relationships. For instance, in a recent in silico docking study conducted by Lee et al., the authors proved that asernestioside B, a saponin glycoside (one of the compounds constituting HGWT, a traditional Chinese herbal formulation used for controlling poststroke symptoms) revealed high potency and stability against mitogen-activated protein kinase 3 (MAPK3), a target protein for stroke-mediated numbness and weakness [92].

6.5 Recent Applications of Nanomedicine for Treatment of Stroke

Nanomedicine has offered great advancement in treatment of stroke. Nanomedicine refers to the manipulation of materials at the nano range from 1 to 1000 nm for encapsulation or delivery of drugs to the target site. Nanotechnology offers great advantages in the treatment of brain disorders and specifically stroke owing to its direct therapeutic effect, or its capacity to target therapeutic drugs or phytochemicals toward the brain owing to their smaller size and larger surface area, which enable them to cross the blood–brain barrier (BBB), reach the brain directly, and achieve rapid therapeutic effect with maximal drug concentration at the target site [93]. Nanoparticles can cross the BBB either through passive transport due to the "leaky" nature of the BBB following a stroke, or through receptor-mediated transport through functionalization of the nanoparticles with receptor targets to facilitate delivery across the BBB. Several types of nanoparticles have been utilized in stroke treatment either due to their intrinsic neuroprotective effects or through loading of drugs onto them [94].

One type of nanoparticles are lipid-based nanoparticles such as liposomes, which are nanospheres composed of a lipid bilayer allowing them to be used as carriers for several therapeutic agents. Liposomes have been extensively studied in treatment of stroke owing to their capability to cross the BBB and their long circulation time in the blood. However, they suffer from low loading capacities, which can hinder their use as drug carriers [95]. Imaizumi et al. studied liposomes in the treatment of focal ischemic stroke through loading of SOD, a free radical scavenger, to increase their BBB permeability and circulation time [96]. Liposomes can also be functionalized with moieties that allow specific targeting or active transport through the BBB. One attempt was conducted by Agulla et al., who studied the effect of cytosine-loaded PEGylated liposome functionalized with heat shock protein 72 (HSP72) antibody to enhance aggregation in the ischemic parts and increase the drug's efficacy through selective transport through the BBB. Polyethylene glycol (PEG) has also served to increase liposomes' circulation time [97]. The liposomal surface was modified with a transferrin receptor by Zhao et al., which improved the permeability of the liposomes across the BBB, as well as with a stroke-homing peptide (SHp) to actively target the ischemic areas,

Table 6.1 Reported in vitro and in vivo studies on the effects of different flavonoids on ischemic brain strokes with their possible mechanisms.

Natural compound	Class of compound	In vitro/ in vivo	Concentration/dose	Model	Result	References
Baicalin	Flavonoid	In vitro	34.38, 17.19, 8.59, 4.30, 2.15, and 1.08 μg/ml	OGD/R-induced stroke in astrocytes	Increasing outgrowth of neurites and formation of neuronal networks, decreasing activity of Bax, caspase-3, and caspase-9, as well as reducing levels of NO, MDA, TNF-α, IL-1β, and IL-6, while increasing activity of Bcl-2 and SOD. Modulating BDNF–TrkB/PI3K/Akt as well as MAPK/ERK1/2 signaling pathways	[47]
			0.1, 1, and 10 μM	OGD/R-induced stroke in PC12 cells	Inhibiting activity of Drp-1 and so mitochondrial fission through modulating AMPK pathway as well as increasing generation of MFN2. Reducing oxidative stress	[48]
			1 μM	OGD/R-induced stroke in SY5Y cells	Decreasing activity of caspase-3 and Bax, while increasing activity of Bcl-2. Inhibiting neuronal apoptosis	[49]
			1 and 10 μmol/l	OGD/R-induced stroke in primary astrocytes	Reducing ROS production and increasing activity of glutamine synthetase	[50]
		In vivo	50 mg/kg IP	MCAO/R-induced ischemic stroke in rats	Decreasing SDH activation and ROS production, so protecting glutamine synthetase from degradation	[50]
			100 mg/kg IP	Cerebral ischemia-induced memory impairment in gerbils	Improving learning and memory and inhibiting phosphorylation of CaMKII	[49]
			120 mg/kg IP	Hypoxic ischemia-induced encephalopathy in rats	Decreasing infarct volume and so neuronal loss by inhibiting neuronal apoptosis. Enhancing activity of glutamate transporter 1 through PI3K/Akt signaling pathway	[51]
			100 mg/kg PO	Hyperglycemia-induced ischemia in rats	Decreasing blood glucose level, neurologic deficit, as well as infarct volume	[48]

Icariin	Flavonoid	In vitro	0.25, 0.5, and 1 mg/l	OGD/R-induced ischemia in microglia	Decreasing levels of IL-1β, IL-6, and TNF-α. Reducing TUNEL-positive cells. Downregulating IRE1α–XBP1 signaling pathway	[52]
		In vivo	60 mg/kg IG	MCAO/R-induced ischemic stroke in rats	Reducing infarction volume and neurologic deficits. Increasing VEGF and BDNF levels and so enhancing angiogenesis and neurogenesis via activating PI3K and ERK1/2 signaling pathways	[53]
			60 mg/kg IG	MCAO/R-induced ischemic stroke in rats	Decreasing infarct volume and neurologic deficit. Reducing levels of TNF-α, IL-6, activity of caspase 3 and Bax, as well as inhibiting NF-κB through PPARs/Nrf2/NF-κB pathway	[54]
			10 and 30 mg/kg IG	MCAO/R-induced ischemic stroke in rats	Decreasing neurologic deficit score and infarct volume. Reducing level of IL-1β. Suppressing degradation of IκB-α and activation of NF-κB. Inhibiting ↓ inflammatory response induced by PPARα, PPARγ, and NF-κB pathways	[55]
Puerarin	Flavonoid	In vivo	4.2 mg/kg IP	MCAO/R-induced ischemic stroke in rats	Reducing neurologic deficit score, water content, and infarct volume as well as autophagy in neurons	[56]
			50 and 100 mg/kg IP	MCAO/R-induced ischemic stroke in rats	Decreasing infarct volume, Beclin-1, LC3-II/LC3-I ratio, and autophagy via AMPK–mTOR–ULK1 pathway	[57]
			50 and 100 mg/kg IP	4-vessel occlusion-induced ischemia in rats	Enhancing motor symptoms by increasing traveling distance and number of crossings. Activating PI3K/Akt1/GSK-3β/MCL-1 pathway	[58]
Breviscapine	Flavonoid	In vivo	20, 50, and 100 mg/kg IP	MCAO/R-induced ischemic stroke in rats	Decreasing neurologic score, infarct volume, and cerebral water content. Reducing levels of IL-6, IL-1β, TNF-α, ROS, MDA, and PARP-1. Increasing activity of SOD, catalase, and GSH. Suppressing neuronal apoptosis and NF-κB signaling pathway	[59]
			0.33 mg/kg IP	MCAO/R-induced ischemic stroke in rats	Reducing infarct volume, brain water content, LC3-II/LC3-I ratio, and neuronal autophagy	[60]

(*Continued*)

Table 6.1 (Continued)

Natural compound	Class of compound	In vitro/ in vivo	Concentration/dose	Model	Result	References
Resveratrol	Phenol	In vitro	40 μmol/l	OGD/R-induced stroke in cortical neurons	Decreasing positive TUNEL cells. Increasing Shh, Ptc-1, Smo, and Gli-1	[61]
			10 μM	OGD/R-induced stroke in SY5Y cells	Reducing activity of caspase 3 and 9. Increasing activity of Bcl-2 and CREB. Decreasing mitochondrial deficiency. Enhancing PGC1, Nrf1, Tfam, and AMPK	[62]
			10, 20, 40, 60, and 80 mmol/l	OGD/R-induced stroke in neurons	Suppressing activity of LDH and caspase-3. Increasing activity of SOD, Nrf2, NQO-1, HO-1, and Bcl-2	[63]
			1, 5, and 20 μmol/l	OGD/R-induced stroke in primary cortical neurons	Increasing GAP-43, synaptophysin, Shh, Ptc-1, Smo, Gli-1, and Sirt1. Enhancing neurite outgrowth and synaptogenesis through sonic Hedgehog signaling	[64]
			1, 5, 10, 20, 50, and 100 μmol/l	OGD/R-induced stroke in neurons	Increasing neuron survival and proliferation. Suppressing neuron apoptosis and MDA activity. Increasing activity of SOD and GSH	[65]
			5 μmol/l	OGD/R-induced stroke in neurons	Suppressing activation of astrocytes and microglia. Enhancing synaptophysin	[66]
		In vivo	30 mg/kg IP	MCAO/R-induced ischemic stroke in rats	Enhancing neurologic function. Decreasing infarct volume and neuronal damage as well as activity of cleaved caspase-3 and Bax. Upregulating PI3K/AKT/mTOR pathway by activating JAK2/STAT3 pathway	[67]
			30 mg/kg IP	MCAO/R-induced ischemic stroke in mice	Decreasing levels of CD32, IL-1β, IL-1, and IL-6. Increasing levels of CD206, arginase-1, IL-4, and IL-10. Increasing M2 polarization of microglia and decreasing neuroinflammation by inhibiting miR-155 pathway	[68]
			30 mg/kg IP	MCAO/R-induced ischemic stroke in rats	Reducing infarct volume. Enhancing neurologic function and upregulating Shh signaling pathway	[61]
			20 mg/kg IP	MCAO/R-induced ischemic stroke in rats	Reducing neurologic deficits and infarct size. Increasing activity of ATP, p-AMPK, SIRT1, and cAMP	[69]
			30 mg/kg IP	MCAO/R-induced cerebral ischemic stroke in rats	Decreasing neurologic deficit scores, brain water content, and MPO activity as well as = levels of IL-1β, TNF-α, and COX-2. Upregulating PI3K/Akt pathway	[70]

			30 mg/kg IP	MCAO/R-induced focal cerebral ischemic stroke in rats	Suppressing neuronal death. Enhancing cognitive and memory skills. Increasing phosphorylation of ERK1/2 and CREB	[71]
			30 mg/kg IP	MCAO/R-induced ischemic stroke in rats	Decreasing neurologic deficit scores. Enhancing cell proliferation, differentiation, and migration	[66]
Curcumin	Phenol	In vitro	12.5 and 25 μmol/l	LPS and IFN-γ-induced stroke in BV2 microglia	Decreasing M1 microglia/macrophage markers. Increasing M2 microglia/macrophage markers. Decreasing levels of TNF-α, IL-6, and IL-12	[72]
			0.25–10 μM	OGD/R-induced stroke in primary cortical neurons	Suppressing neuronal death. Decreasing LDH release and caspase-3 activity. Regulating flotillin-1 and ERK1/2 pathway	[73]
		In vivo	200 mg/kg IP	MCAO/R-induced ischemic stroke in rats	Reducing brain infarction. Increasing neurologic function. Upregulating p-Akt and p-mTOR pathways. Downregulating LC3-II/LC3-I, IL-1, TLR4, p-38, and p-p38 pathways	[74]
			300 mg/kg IP	MCAO/R-induced ischemic stroke in rats	Reducing neurobehavioral deficits. Increasing number of BrdU/DCX-positive cells, NICD level. Upregulating Notch signaling pathway	[75]
			300 mg/kg IP	MCAO/R-induced ischemic stroke in rats	Reducing brain infarction, edema, and dysfunction. Suppressing activity of ICAM-1, MMP-9, and caspase-3	[76]
			100, 200, 300, and 400 mg/kg IP	MCAO/R-induced ischemic stroke in rats	Decreasing infarct volume. Increasing activity of Bcl-2, while decreasing Bax and caspase-3. Controlling mitochondrial membrane integrity and suppressing its apoptosis through suppressing Bax activation	[77]
			100 and 300 mg/kg IP	MCAO/R-induced ischemic stroke in rats	Decreasing cerebral infarction. Increasing cognitive function. Reducing levels of TNF-α, IL-6, IL-1β, ROS, and MDA. Increasing activity of SOD and catalase. Regulating miR-7/RELA p65 axis1	[78]
			25 mg/kg IP	MCAO/R-induced ischemic stroke in rats	Reducing brain water content and so edema. Decreasing levels of TNF-α and IL-6. Decreasing activity of MMP, p53, and Bax. Enhancing activity of Bcl-2 and Sirt1	[79]
			50 mg/kg IP	MCAO/R-induced ischemic stroke in rats	Reducing infarct volume, brain water content, and neurologic scores. Decreasing levels of TNF-α and IL-6. Enhancing potential of mitochondrial membrane, activity of mitochondrial complex I, and level of mitochondrial cytochrome c. Increasing activity of SIRT1 and Bcl-2. Suppressing activity of Ac-p53 and Bax	[80]
			300 mg/kg IP	MCAO/R-induced ischemic stroke in rats	Decreasing water content and infarct volume. Reducing levels of MDA. Downregulating NF-kB pathway, while upregulating Nrf2 pathway	[81]

(*Continued*)

Table 6.1 (Continued)

Natural compound	Class of compound	In vitro/ in vivo	Concentration/dose	Model	Result	References
Salvianolic acid B	Phenol	In vitro	200, 400, and 800 ng/ml	OGD/R-induced in primary cortical neurons	Suppressing NF-kB activity. Decreasing levels of IL-1β, IL-6, and TNF-α	[82]
		In vivo	30 mg/kg IG	MCAO/R-induced ischemic stroke in rats	Increasing NeuN release. Downregulating TLR4/MyD88/TRAF6 pathway	[82]
			3, 6, and 12 mg/kg IP	MCAO/R-induced ischemic stroke in rats	Suppressing damage to DNA and neurons. Decreasing levels of ICAM-1, IL-1β, IL-6, IL-8, and MCP-1. Downregulating CD40/NF-κB pathway	[83]
Ginkgolide B	Terpene	In vitro	0.1, 1, and 10 uM	LPS, IFN-γ, and IL-4-induced stroke in BV2 microglia and primary bone marrow–derived macrophage	Decreasing levels of CCL3, iNOS, and TNF-α. Increasing levels of CD206, Arg1, Ym1, IL-10, and TGF-β. Modifying microglia/ macrophage polarization	[84]
			25 mg/l	OGD/R-induced stroke in SY5Y cells	Increasing activity of SOD, HO-1, and NQO-1. Upregulating Akt/Nrf2 pathway	[85]
		In vivo	1, 2, and 4 mg/kg IP =	MCAO/R-induced ischemic stroke in mice	Decreasing neurologic deficits, infarct volume, and water content. Decreasing activity of caspase-3, while increasing level of BDNF	[86]
			1.75, 3.5, and 7.0 mg/kg IP	MCAO/R-induced ischemic stroke in mice	Suppressing cerebral ischemic damage and neurologic deficits	[84]
Catalpol	Terpene	In vitro	0.1, 1, 10, and 100 μg/ml	OGD/R-induced stroke in primary neurons	Enhancing GAP-43 and p-S6 expression. Activating axonal regeneration through PI3K/Akt/mTOR pathway	[87]
			25, 50, and 100 μM	OGD/R-induced stroke in BMEC cells	Increasing proliferation, migration, and tube formation. Upregulating HIF-1α/VEGF pathway	[88]
			10 μg/ml	OGD/R-induced stroke in primary neurons	Increasing GAP-43 expression. Enhancing axonal growth through regulating miR-124-mediated PI3K/Akt/mTOR pathway	[89]

In vivo	2.5, 5, and 7.5 mg/kg IP	Electrocoagulation-induced stroke in rats	Decreasing neurologic score, while increasing neuronal cell activity	[87]
	5 and 10 mg/kg	MCAO/R-induced ischemic stroke in rats	Reducing neurologic deficits. Increasing BrdU + -Nestin+ and BrdU+ -Tuj-1+. Suppressing activity of caspase 3. Increasing levels of Tuj-1 and MAP2 as well as BDNF, but not BDNF/TrkB pathway	[90]
	2.5, 5, and 10 mg/kg IV	MCAO/R-induced ischemic stroke in rats	Reducing neurologic deficit scores and infarct volume. Increasing levels of CD31+/EDU+. Enhancing proliferation, migration, and tube formation. Upregulating HIF-1α/VEGF pathway	[88]
	5 mg/kg IP	Electrocoagulation-induced stroke in rats	Decreasing neurologic deficits. Increasing cerebral blood flow. Increasing levels of EPO and EPOR. Enhancing STAT3 activation. Improving VEGF production through JAK2/STAT3 pathway	[91]

AMPK, 5′ adenosine monophosphate-activated protein kinase; ATP, adenosine triphosphate; BDNF, brain-derived neurotrophic factor; BMEC, bone marrow microvascular endothelial cells; BrdU, bromodeoxyuridine; CaMKII, calmodulin-dependent protein kinase II; cAMP, cyclic adenosine monophosphate; CCL3, chemokine (C-C motif) ligand 3; COX-2, cyclooxygenase 2; CREB, cAMP response element-binding protein; DCX, doublecortin; Drp-1, dynamin-related protein 1; EPO, erythropoietin; EPOR, erythropoietin receptor; ERK, extracellular signal-regulated kinase; GAP-43, growth associated protein 43; GSH, glutathione; HIF-1α, hypoxia inducible factor-1 alpha; HO-1, heme oxygenase 1; ICAM-1, intercellular adhesion molecule 1; IFN, interferon; IG, intragastric; IL, interleukin; iNOS, inducible nitric oxide synthase; IP, intraperitoneal; IV, intravenous; JAK2, Janus kinase 2; LC3, light chain 3; LDH, lactate dehydrogenase; LPS, lipopolysaccharide; MAP2, microtubule-associated protein 2; MAPK, mitogen-activated protein kinase; MCAO/R, middle cerebral artery occlusion/reperfusion; MCP-1, monocyte chemoattractant protein 1; MDA, malondialdehyde; MFN2, mitofusin 2; MMP, matrix metalloproteinase; MPO, myeloperoxidase; mTOR, mammalian target of rapamycin; NF-κB, nuclear factor kappa B; NICD, Notch intracellular domain; NO, nitric oxide; NQO-1, nicotinamide adenine dinucleotide phosphate quinone dehydrogenase 1; OGD/R, oxygen-glucose deprivation/reoxygenation; PARP-1, Poly (ADP-ribose) polymerase 1; PI3K, phosphatidylinositol-3-kinase; PO, orally; PPAR, peroxisome proliferator-activated receptor; ROS, reactive oxygen species; SDH, succinate dehydrogenase; SOD, superoxide dismutase; STAT3, signal transducer and activator of transcription 3; TGF, transforming growth factor; TLR4, Toll-like receptor 4; TNF, tumor necrosis factor; TRAF6, TNF receptor-associated factor 6; TrkB, tyrosine kinase B; TUNEL, terminal deoxynucleotidyl transferase dUTP nick end labeling; VEGF, vascular endothelial growth factor.

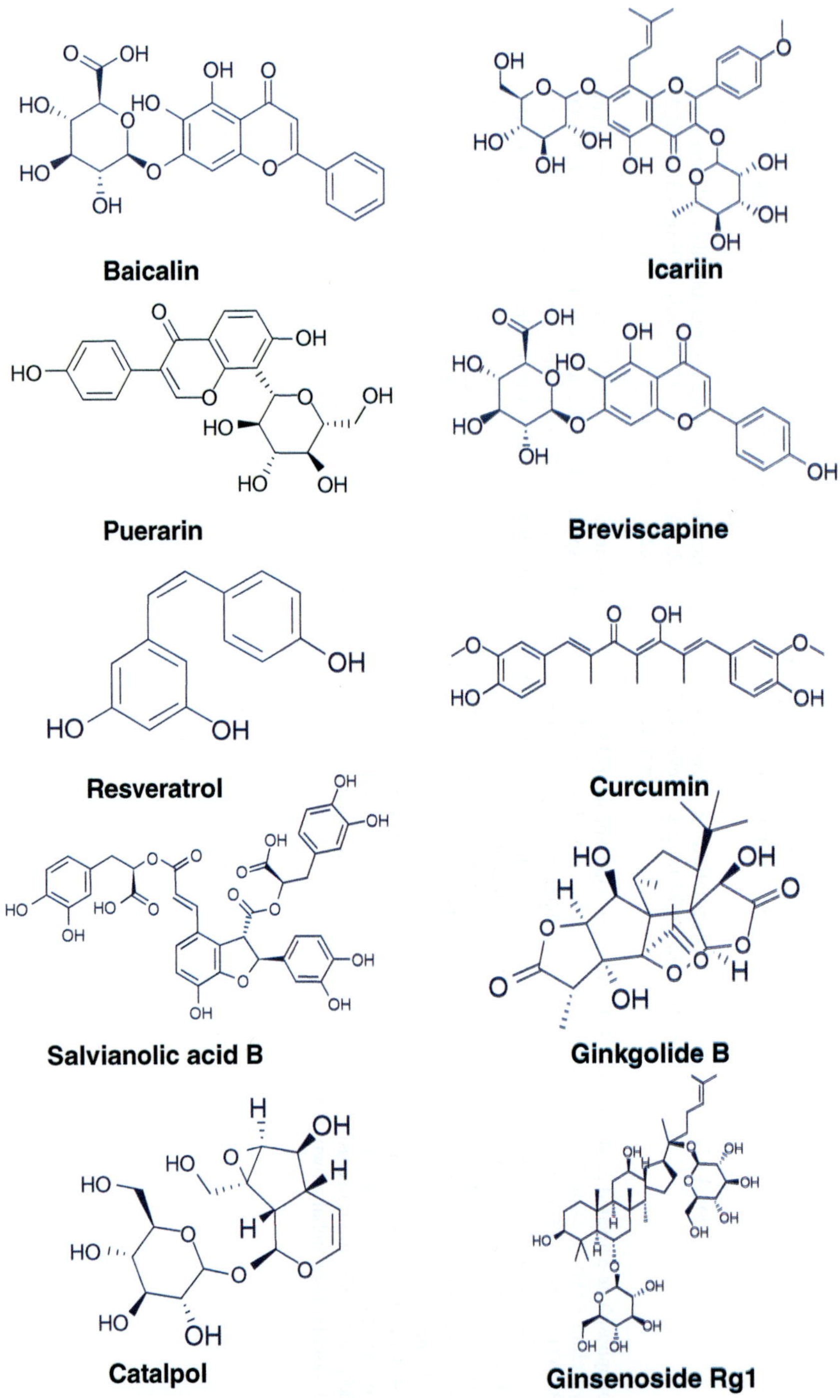

Figure 6.1 The structures of active compounds against brain strokes.

which decreased the neurologic deficit, infarction size, and histopathologic changes in the MCAO rat model [98]. Nanostructured lipid carriers were utilized to load ferulic acid (FA), which showed a controlled release profile, enhanced bioavailability, and improved the pharmacologic profile of FA through activating the PI3K pathway, which can therapeutically be significant in cerebral stroke [99]. Solid lipid nanoparticles were exploited by Kakkar et al. to incorporate curcumin, leading to higher oral absorption and longer circulation time as well as enhanced mitochondrial complex enzyme activity and declined lipid peroxidation, nitrite, and acetylcholinesterase rates [100].

Inorganic nanoparticles have also shown promise as ROS-scavenging agents in case of ischemic stroke. Bao et al. developed monodisperse cerium nanoparticles loaded with edaravone and modified with Angiopep-2 and PEG on their surface (E-A/P-CeO_2), which were highly effective in BBB crossing and showed synergistic scavenging of ROS by the ceria nanoparticles and edaravone, which was effective in protection of the BBB and in management of stroke [101]. Platinum nanoparticles have also displayed antioxidant effect by scavenging superoxide anions and hydrogen peroxide in vivo [102]. Amani et al. studied the effects of selenium (Se) nanoparticles on ischemic stroke in a murine model. The prepared nanoparticles were successful in resolving brain edema, protecting axons in the hippocampus region, and myelination of the hippocampal area following a cerebral ischemic stroke [103]. Zhao et al. prepared *Calendula officinalis* L. synthesized gold nanoparticles, which were successful in treatment of cerebral ischemia stroke–reperfusion injury in rats [104].

Polymeric nanoparticles have been applied as potential drug delivery systems for the treatment of stroke. Poly(D,L-lactide-co-glycolide) acid (PLGA) is a commonly used, biodegradable non-toxic polymer that was utilized to deliver a variety of antioxidants and neuroprotective agents. Rajkovic et al. prepared ROS-scavenging PEGylated polymeric nanoparticles (NPs) composed of poly(propylene sulfide) (PPS) for ischemic stroke therapy, which demonstrated antioxidant and anti-inflammatory properties with insignificant cytotoxicity [105]. Mdzinarishvili et al. exploited thyroxine (T3) hormone-loaded PLGA-PEG nanoparticles coated with GSH, which showed a significant decrease in brain edema and infarction in a MCAO stroke model [106]. Another study by Shahani et al. also utilized PLGA-PEG through loading with reelin and embryonic neural crest stem cells (NSCs), which promoted neurologic outcome and functional recovery in a photothrombotic stroke model of mice through improved NSC dynamic growth, neuronal differentiation, and local angiogenesis [107].

Nanomicelles have been utilized, which comprise amphiphilic molecules that self-assemble above their critical micelle concentration to form colloidal spheres in the nonorange. Lu et al. studied rapamycin-loaded nanomicelles that consist of ROS-responsive and fibrin-binding polymers with controlled drug release within the ischemic lesion, which showed improved neuron survival and remodeling through micelle-polarized M2 microglia repair and BBB preservation, thereby achieving increased neuroprotection and blood perfusion in the affected areas [108].

Dendrimers, which are supramolecular moieties, were exploited as drug delivery systems for treatment of stroke, specifically polyamidoamine (PAMAM) dendrimers, which have been studied as an efficient gene carrier and showed promising success in treatment of ischemic stroke due to their low toxicity and high delivery efficiency [109].

Carbon-based nanoparticles and most prominently carbon nanotubes (CNTs) have shown great promise in stroke treatments, as they exert a neuroprotective effect against oxidative stress and have been shown to reduce the volume of cerebral infarction by 50% [110]. A study conducted by Lee et al. showed that pretreating rats with amine-modified single-walled CNTs can protect neurons and enhance the recovery of behavioral functions in rats with induced stroke [111]. Another study concluded that hydrophobic CNTs loaded with SVZ neural progenitor cells have been successful in repairing damaged neural tissue through improved stem cell differentiation to cure damage induced by strokes [112].

6.6 Conclusion

Developing multitargeted drugs from natural sources is a very effective way for improving general health with cheap and safe remedies. This chapter has hopefully highlighted the causes and the mechanisms of ischemic strokes that can help in developing promising therapeutic strategies from different natural sources. Either on the plant extract level or the single compound level, natural remedies have been proved to control and treat ischemic stroke in several in vitro and in vivo studies discussed in this chapter. More attention in the future should be targeted at implicating these active natural remedies in clinical studies for improving general health and reducing death rates.

References

1 Maynard, R.L. and Downes, N. (2019). The brain and spinal cord. In: *Anatomy and Histology of the Laboratory Rat in Toxicology and Biomedical Research* (ed. R.L. Maynard and N. Downes), 231–260. Cambridge, MA: Academic Press.

2 Donnelly, L. (2020). The brain: functional divisions. *Anaesth. Intensive Care Med.* 21 (6): 322–327.

3 Zavitz, E. and NSC, P. (2019). Understanding sensory information processing through simultaneous multi-area population recordings. *Front. Neural Circuits* 12: 115.

4 Moini, J., Koenitzer, J., and LoGalbo, A. (2021). Brain neurotransmitters. In: *Global Emergency of Mental Disorders* (ed. J. Moini, J. Koenitzer, and A. LoGalbo), 31–40. Cambridge, MA: Academic Press.

5 Heneka, M.T., McManus, R.M., and Latz, E. (2018). Inflammasome signalling in brain function and neurodegenerative disease. *Nat. Rev. Neurosci.* 19 (10): 610–621.

6 Attwell, D., Buchan, A.M., Charpak, S. et al. (2010). Glial and neuronal control of brain blood flow. *Nature* 468 (7321): 232–243.

7 Cui, P., McCullough, L.D., and Hao, J. (2021). Brain to periphery in acute ischemic stroke: mechanisms and clinical significance. *Front. Neuroendocrinol.* 63: 100932.

8 World Health Organization (2018). *World Health Statistics 2018: Monitoring Health for the SDGs, Sustainable Development Goals*. Geneva: World Health Organization.

9 Owolabi, M.O., Arulogun, O., Melikam, S. et al. (2015). The burden of stroke in Africa: a glance at the present and a glimpse into the future. *Cardiovasc. J. Afr.* 26(2 H3Africa Suppl): S27.

10 Sasaki, T., Hamada, J., Shibata, M. et al. (2001). Inhibition of nitric oxide production during global ischemia ameliorates ischemic damage of pyramidal neurons in the hippocampus. *Keio J. Med.* 50 (3): 182–187.

11 Coyle Joseph, T. and Puttfarcken, P. (1993). Oxidative stress, glutamate, and neurodegenerative disorders. *Science* 262 (5134): 689–695.

12 Kroemer, G. and Reed, J.C. (2000). Mitochondrial control of cell death. *Nat. Med.* 6 (5): 513–519.

13 Zhou, Z., Lu, J., Liu, W.-W. et al. (2018). Advances in stroke pharmacology. *Pharmacol. Ther.* 191: 23–42.

14 Shirley, R., Ord, E.N., and Work, L.M. (2014). Oxidative stress and the use of antioxidants in stroke. *Antioxidants.* 3 (3): 472–501.

15 de Aguiar, R.P., Newman-Tancredi, A., Prickaerts, J., and de Oliveira, R.M.W. (2021). The 5-HT1A receptor as a serotonergic target for neuroprotection in cerebral ischemia. *Prog. Neuro-Psychopharmacol. Biol. Psychiatry* 109: 110210.

16 Lai, T.W., Zhang, S., and Wang, Y.T. (2014). Excitotoxicity and stroke: identifying novel targets for neuroprotection. *Prog. Neurobiol.* 115: 157–188.

17 Szydlowska, K. and Tymianski, M. (2010). Calcium, ischemia and excitotoxicity. *Cell Calcium* 47 (2): 122–129.

18 Shehadah, A., Chen, J., Kramer, B. et al. (2013). Efficacy of single and multiple injections of human umbilical tissue-derived cells following experimental stroke in rats. *PLoS One* 8 (1): e54083.

19 Broughton, B.R., Reutens, D.C., and Sobey, C.G. (2009). Apoptotic mechanisms after cerebral ischemia. *Stroke* 40 (5): e331–e339.

20 Fan, Y. and Yang, G.-Y. (2007). Therapeutic angiogenesis for brain ischemia: a brief review. *J. Neuroimmune Pharmacol.* 2 (3): 284–289.

21 Beck, H. and Plate, K.H. (2009). Angiogenesis after cerebral ischemia. *Acta Neuropathol.* 117 (5): 481–496.

22 Libert, S., Cohen, D., and Guarente, L. (2008). Neurogenesis directed by Sirt1. *Nat. Cell Biol.* 10 (4): 373–374.

23 Edelbrock, A.N., Àlvarez, Z., Simkin, D. et al. (2018). Supramolecular nanostructure activates TrkB receptor signaling of neuronal cells by mimicking brain-derived neurotrophic factor. *Nano Lett.* 18 (10): 6237–6247.

24 Li, S., Overman, J.J., Katsman, D. et al. (2010). An age-related sprouting transcriptome provides molecular control of axonal sprouting after stroke. *Nat. Neurosci.* 13 (12): 1496–1504.

25 Shang, Y., Wang, X., Li, F. et al. (2019). rTMS ameliorates prenatal stress–induced cognitive deficits in male-offspring rats associated with BDNF/TrkB signaling pathway. *Neurorehabil. Neural Repair* 33 (4): 271–283.

26 Zhu, T., Wang, L., Feng, Y. et al. (2021). Classical active ingredients and extracts of Chinese herbal medicines: pharmacokinetics, pharmacodynamics, and molecular mechanisms for ischemic stroke. *Oxidative Med. Cell. Longev.* 2021: 8868941.

27 Zhou, X., Qi, Y., and Chen, T. (2017). Long-term pre-treatment of antioxidant Ginkgo biloba extract EGb-761 attenuates cerebral-ischemia-induced neuronal damage in aged mice. *Biomed. Pharmacother.* 85: 256–263.

28 Li, M.-Z., Zhang, Y., Zou, H.-Y. et al. (2018). Investigation of Ginkgo biloba extract (EGb 761) promotes neurovascular restoration and axonal remodeling after embolic stroke in rat using magnetic resonance imaging and histopathological analysis. *Biomed. Pharmacother.* 103: 989–1001.

29 Zhang, Y., Liu, J., Yang, B. et al. (2018). Ginkgo biloba extract inhibits astrocytic lipocalin-2 expression and alleviates neuroinflammatory injury via the JAK2/STAT3 pathway after ischemic brain stroke. *Front. Pharmacol.* 9: 518.

30 Wu, C., Zhao, X., Zhang, X. et al. (2015). Effect of Ginkgo biloba extract on apoptosis of brain tissues in rats with acute cerebral infarction and related gene expression. *Genet. Mol. Res.* 14 (2): 6387–6394.

31 Yang, X., Zheng, T., Hong, H. et al. (2018). Neuroprotective effects of Ginkgo biloba extract and ginkgolide B against oxygen–glucose deprivation/reoxygenation and glucose injury in a new in vitro multicellular network model. *Front. Med.* 12 (3): 307–318.

32 Chen, K.-Y., Wu, K.-C., Hueng, D.-Y. et al. (2020). Anti-inflammatory effects of powdered product of Bu Yang Huan Wu decoction: possible role in protecting against transient focal cerebral ischemia. *Int. J. Med. Sci.* 17 (12): 1854.

33 Lu, Y., Hsiang, F., Chang, J.-H. et al. (2018). Houshiheisan and its components promote axon regeneration after ischemic brain injury. *Neural Regen. Res.* 13 (7): 1195.

34 Weng, S.-W., Chang, C.-C., Chen, T.-L. et al. (2021). Risk of diabetes in stroke patients who used Bu Yang Huan Wu Tang: a nationwide propensity-score matched study. *Phytomedicine* 80: 153376.

35 Chang, J., Yao, X., Zou, H. et al. (2016). BDNF/PI3K/Akt and Nogo-a/RhoA/ROCK signaling pathways contribute to neurorestorative effect of Houshiheisan against cerebral ischemia injury in rats. *J. Ethnopharmacol.* 194: 1032–1042.

36 Chen, Z.-Z., Gong, X., Guo, Q. et al. (2019). Bu Yang Huan Wu decoction prevents reperfusion injury following ischemic stroke in rats via inhibition of HIF-1 α, VEGF and promotion β-ENaC expression. *J. Ethnopharmacol.* 228: 70–81.

37 Xiang, Y., Yao, X., Wang, X. et al. (2019). Houshiheisan promotes angiogenesis via HIF-1α/VEGF and SDF-1/CXCR4 pathways: in vivo and in vitro. *Biosci. Rep.* 39 (10): BSR20191006.

38 Hu, J., Zeng, C., Wei, J. et al. (2020). The combination of *Panax ginseng* and *Angelica sinensis* alleviates ischemia brain injury by suppressing NLRP3 inflammasome activation and microglial pyroptosis. *Phytomedicine* 76: 153251.

39 Cheng, C.-Y., Kao, S.-T., and Lee, Y.-C. (2020). *Angelica sinensis* extract protects against ischemia-reperfusion injury in the hippocampus by activating p38 MAPK-mediated p90RSK/p-bad and p90RSK/CREB/BDNF signaling after transient global cerebral ischemia in rats. *J. Ethnopharmacol.* 252: 112612.

40 Cheng, C.-Y., Ho, T.-Y., Hsiang, C.-Y. et al. (2017). *Angelica sinensis* exerts angiogenic and anti-apoptotic effects against cerebral ischemia–reperfusion injury by activating p38MAPK/HIF-1 α/VEGF-A signaling in rats. *Am. J. Chin. Med.* 45 (08): 1683–1708.

41 Liu, L., Anderson, G.A., Fernandez, T.G., and Doré, S. (2019). Efficacy and mechanism of *Panax ginseng* in experimental stroke. *Front. Neurosci.* 13: 294.

42 Liu, L., Vollmer, M.K., Fernandez, V.M. et al. (2018). Korean red ginseng pretreatment protects against long-term sensorimotor deficits after ischemic stroke likely through Nrf2. *Front. Cell. Neurosci.* 12: 74.

43 Ban, J.Y., Kang, S.W., Lee, J.S. et al. (2012). Korean red ginseng protects against neuronal damage induced by transient focal ischemia in rats. *Exp. Ther. Med.* 3 (4): 693–698.

44 Zheng, G.-q., Cheng, W., Wang, Y. et al. (2011). Ginseng total saponins enhance neurogenesis after focal cerebral ischemia. *J. Ethnopharmacol.* 133 (2): 724–728.

45 Park, H.-J., Shim, H.S., Kim, K.S., and Shim, I. (2011). The protective effect of black ginseng against transient focal ischemia-induced neuronal damage in rats. *Korean J. Physiol. Pharmacol.* 15 (6): 333–338.

46 Zhu, T., Wang, L., Wang, L.-p., and Wan, Q. (2022). Therapeutic targets of neuroprotection and neurorestoration in ischemic stroke: applications for natural compounds from medicinal herbs. *Biomed. Pharmacother.* 148: 112719.

47 Li, C., Sui, C., Wang, W. et al. (2021). Baicalin attenuates oxygen–glucose deprivation/reoxygenation–induced injury by modulating the BDNF-TrkB/PI3K/Akt and MAPK/Erk1/2 signaling axes in neuron–astrocyte cocultures. *Front. Pharmacol.* 12: 599543.

48 Li, S., Sun, X., Xu, L. et al. (2017). Baicalin attenuates in vivo and in vitro hyperglycemia-exacerbated ischemia/reperfusion injury by regulating mitochondrial function in a manner dependent on AMPK. *Eur. J. Pharmacol.* 815: 118–126.

49 Wang, P., Cao, Y., Yu, J. et al. (2016). Baicalin alleviates ischemia-induced memory impairment by inhibiting the phosphorylation of CaMKII in hippocampus. *Brain Res.* 1642: 95–103.

50 Song, X., Gong, Z., Liu, K. et al. (2020). Baicalin combats glutamate excitotoxicity via protecting glutamine synthetase from ROS-induced 20S proteasomal degradation. *Redox Biol.* 34: 101559.

51 Zhou, Z.-q., Li, Y.-l., Ao, Z.-b. et al. (2017). Baicalin protects neonatal rat brains against hypoxic-ischemic injury by upregulating glutamate transporter 1 via the phosphoinositide 3-kinase/protein kinase B signaling pathway. *Neural Regen. Res.* 12 (10): 1625.

52 Mo, Z.-t., Liao, Y.-l., Zheng, J., and Li, W.-n. (2020). Icariin protects neurons from endoplasmic reticulum stress-induced apoptosis after OGD/R injury via suppressing IRE1α-XBP1 signaling pathway. *Life Sci.* 255: 117847.

53 Liu, D., Ye, Y., Xu, L. et al. (2018). Icariin and mesenchymal stem cells synergistically promote angiogenesis and neurogenesis after cerebral ischemia via PI3K and ERK1/2 pathways. *Biomed. Pharmacother.* 108: 663–669.

54 Dai, M., Chen, B., Wang, X. et al. (2021). Icariin enhance mild hypothermia-induced neuroprotection via inhibiting the activation of NF-κB in experimental ischemic stroke. *Metab. Brain Dis.* 36 (7): 1779–1790.

55 Xiong, D., Deng, Y., Huang, B. et al. (2016). Icariin attenuates cerebral ischemia–reperfusion injury through inhibition of inflammatory response mediated by NF-κB, PPARα and PPARγ in rats. *Int. Immunopharmacol.* 30: 157–162.

56 Hongyun, H., Tao, G., Pengyue, Z. et al. (2017). Puerarin provides a neuroprotection against transient cerebral ischemia by attenuating autophagy at the ischemic penumbra in neurons but not in astrocytes. *Neurosci. Lett.* 643: 45–51.

57 Wang, J.-F., Mei, Z.-G., Fu, Y. et al. (2018). Puerarin protects rat brain against ischemia/reperfusion injury by suppressing autophagy via the AMPK-mTOR-ULK1 signaling pathway. *Neural Regen. Res.* 13 (6): 989.

58 Tao, J., Cui, Y., Duan, Y. et al. (2017). Puerarin attenuates locomotor and cognitive deficits as well as hippocampal neuronal injury through the PI3K/Akt1/GSK-3β signaling pathway in an in vivo model of cerebral ischemia. *Oncotarget* 8 (63): 106283.

59 Li, Y., Li, S., and Li, D. (2020). Breviscapine alleviates cognitive impairments induced by transient cerebral ischemia/reperfusion through its anti-inflammatory and anti-oxidant properties in a rat model. *ACS Chem. Neurosci.* 11 (24): 4489–4498.

60 Pengyue, Z., Tao, G., Hongyun, H. et al. (2017). Breviscapine confers a neuroprotective efficacy against transient focal cerebral ischemia by attenuating neuronal and astrocytic autophagy in the penumbra. *Biomed. Pharmacother.* 90: 69–76.

61 Yu, P., Wang, L., Tang, F. et al. (2017). Resveratrol pretreatment decreases ischemic injury and improves neurological function via sonic hedgehog signaling after stroke in rats. *Mol. Neurobiol.* 54 (1): 212–226.

62 Lin, C.-H., Nicol, C.J., Cheng, Y.-C. et al. (2020). Neuroprotective effects of resveratrol against oxygen glucose deprivation induced mitochondrial dysfunction by activation of AMPK in SH-SY5Y cells with 3D gelatin scaffold. *Brain Res.* 1726: 146492.

63 Yang, J., Huang, J., Shen, C. et al. (2018). Resveratrol treatment in different time-attenuated neuronal apoptosis after oxygen and glucose deprivation/reoxygenation via enhancing the activation of Nrf-2 signaling pathway in vitro. *Cell Transplant.* 27 (12): 1789–1797.

64 Tang, F., Guo, S., Liao, H. et al. (2017). Resveratrol enhances neurite outgrowth and synaptogenesis via sonic hedgehog signaling following oxygen-glucose deprivation/reoxygenation injury. *Cell. Physiol. Biochem.* 43 (2): 852–869.

65 Shen, C., Cheng, W., Yu, P. et al. (2016). Resveratrol pretreatment attenuates injury and promotes proliferation of neural stem cells following oxygen-glucose deprivation/reoxygenation by upregulating the expression of Nrf2, HO-1 and NQO1 in vitro. *Mol. Med. Rep.* 14 (4): 3646–3654.

66 Yu, P., Wang, L., Tang, F. et al. (2021). Resveratrol-mediated neurorestoration after cerebral ischemic injury-sonic hedgehog signaling pathway. *Life Sci.* 280: 119715.

67 Hou, Y., Wang, K., Wan, W. et al. (2018). Resveratrol provides neuroprotection by regulating the JAK2/STAT3/PI3K/AKT/mTOR pathway after stroke in rats. *Genes Dis.* 5 (3): 245–255.

68 Ma, S., Fan, L., Li, J. et al. (2020). Resveratrol promoted the M2 polarization of microglia and reduced neuroinflammation after cerebral ischemia by inhibiting miR-155. *Int. J. Neurosci.* 130 (8): 817–825.

69 Wan, D., Zhou, Y., Wang, K. et al. (2016). Resveratrol provides neuroprotection by inhibiting phosphodiesterases and regulating the cAMP/AMPK/SIRT1 pathway after stroke in rats. *Brain Res. Bull.* 121: 255–262.

70 Lei, J. and Chen, Q. (2018). Resveratrol attenuates brain damage in permanent focal cerebral ischemia via activation of PI3K/Akt signaling pathway in rats. *Neurol. Res.* 40 (12): 1014–1020.

71 Li, Z., Fang, F., Wang, Y., and Wang, L. (2016). Resveratrol protects CA1 neurons against focal cerebral ischemic reperfusion-induced damage via the ERK-CREB signaling pathway in rats. *Pharmacol. Biochem. Behav.* 146: 21–27.

72 Liu, Z., Ran, Y., Huang, S. et al. (2017). Curcumin protects against ischemic stroke by titrating microglia/macrophage polarization. *Front. Aging Neurosci.* 9: 233.

73 Lu, Z., Liu, Y., Shi, Y. et al. (2018). Curcumin protects cortical neurons against oxygen and glucose deprivation/reoxygenation injury through flotillin-1 and extracellular signal-regulated kinase1/2 pathway. *Biochem. Biophys. Res. Commun.* 496 (2): 515–522.

74 Huang, L., Chen, C., Zhang, X. et al. (2018). Neuroprotective effect of curcumin against cerebral ischemia-reperfusion via mediating autophagy and inflammation. *J. Mol. Neurosci.* 64 (1): 129–139.

75 Liu, S., Cao, Y., Qu, M. et al. (2016). Curcumin protects against stroke and increases levels of notch intracellular domain. *Neurol. Res.* 38 (6): 553–559.

76 Li, W., Suwanwela, N.C., and Patumraj, S. (2017). Curcumin prevents reperfusion injury following ischemic stroke in rats via inhibition of NF-κB, ICAM-1, MMP-9 and caspase-3 expression. *Mol. Med. Rep.* 16 (4): 4710–4720.

77 Xie, C.J., Gu, A.P., Cai, J. et al. (2018). Curcumin protects neural cells against ischemic injury in N2a cells and mouse brain with ischemic stroke. *Brain Behav.* 8 (2): e00921.

78 Xu, H., Nie, B., Liu, L. et al. (2019). Curcumin prevents brain damage and cognitive dysfunction during ischemic-reperfusion through the regulation of miR-7-5p. *Curr. Neurovasc. Res.* 16 (5): 441–454.

79 Zhang, Y., Yan, Y., Cao, Y. et al. (2017). Potential therapeutic and protective effect of curcumin against stroke in the male albino stroke-induced model rats. *Life Sci.* 183: 45–49.

80 Miao, Y., Zhao, S., Gao, Y. et al. (2016). Curcumin pretreatment attenuates inflammation and mitochondrial dysfunction in experimental stroke: the possible role of Sirt1 signaling. *Brain Res. Bull.* 121: 9–15.

81 Li, W., Suwanwela, N.C., and Patumraj, S. (2016). Curcumin by down-regulating NF-kB and elevating Nrf2, reduces brain edema and neurological dysfunction after cerebral I/R. *Microvasc. Res.* 106: 117–127.

82 Wang, Y., Chen, G., Yu, X. et al. (2016). Salvianolic acid B ameliorates cerebral ischemia/reperfusion injury through inhibiting TLR4/MyD88 signaling pathway. *Inflammation* 39 (4): 1503–1513.

83 Xu, S., Zhong, A., Ma, H. et al. (2017). Neuroprotective effect of salvianolic acid B against cerebral ischemic injury in rats via the CD40/NF-κB pathway associated with suppression of platelets activation and neuroinflammation. *Brain Res.* 1661: 37–48.

84 Shu, Z.M., Shu, X.D., Li, H.Q. et al. (2016). Ginkgolide B protects against ischemic stroke via modulating microglia polarization in mice. *CNS Neurosci. Ther.* 22 (9): 729–739.

85 Liu, Q., Jin, Z., Xu, Z. et al. (2019). Antioxidant effects of ginkgolides and bilobalide against cerebral ischemia injury by activating the Akt/Nrf2 pathway in vitro and in vivo. *Cell Stress Chaperones* 24 (2): 441–452.

86 Wei, H., Sun, T., Tian, Y., and Wang, K. (2017). Ginkgolide B modulates BDNF expression in acute ischemic stroke. *J. Korean Neurosurg. Soc.* 60 (4): 391.

87 Wang, J., Wan, D., Wan, G. et al. (2019). Catalpol induces cell activity to promote axonal regeneration via the PI3K/AKT/mTOR pathway in vivo and in vitro stroke model. *Ann. Transl. Med.* 7 (23): 756.

88 Wang, H., Xu, X., Yin, Y. et al. (2020). Catalpol protects vascular structure and promotes angiogenesis in cerebral ischemic rats by targeting HIF-1α/VEGF. *Phytomedicine* 78: 153300.

89 Zhu, H., Wang, J., Shao, Y., and Wan, D. (2019). Catalpol may improve axonal growth via regulating miR-124 regulated PI3K/AKT/mTOR pathway in neurons after ischemia. *Ann. Transl. Med.* 7 (14): 306.

90 Zhu, H.-F., Shao, Y., Qin, L. et al. (2019). Catalpol enhances neurogenesis and inhibits apoptosis of new neurons via BDNF, but not the BDNF/Trkb pathway. *Drug Des. Devel. Ther.* 13: 4145.

91 Dong, W., Xian, Y., Yuan, W. et al. (2016). Catalpol stimulates VEGF production via the JAK2/STAT3 pathway to improve angiogenesis in rats' stroke model. *J. Ethnopharmacol.* 191: 169–179.

92 Lee, S., Hung, A., Li, H., and Yang, A.W.H. (2022). Mechanisms of action of a herbal formula Huangqi Guizhi Wuwu Tang for the Management of Post-Stroke Related Numbness and Weakness: a computational molecular docking study. *J. Evid.-Based Integr. Med.* 27: 2515690X221082989.

93 Sarmah, D., Banerjee, M., Datta, A. et al. (2021). Nanotechnology in the diagnosis and treatment of stroke. *Drug Discov. Today* 26 (2): 585–592.

94 Tian, X., Fan, T., Zhao, W. et al. (2021). Recent advances in the development of nanomedicines for the treatment of ischemic stroke. *Bioact. Mater.* 6 (9): 2854–2869.

95 Kawaguchi, A.T., Haida, M., Ohba, H. et al. (2013). Liposome-encapsulated hemoglobin ameliorates ischemic stroke in nonhuman primates: longitudinal observation. *Artif. Organs* 37 (10): 904–912.

96 Imaizumi, S., Woolworth, V., Fishman, R.A., and Chan, P.H. (1990). Liposome-entrapped superoxide dismutase reduces cerebral infarction in cerebral ischemia in rats. *Stroke* 21 (9): 1312–1317.

97 Agulla, J., Brea, D., Campos, F. et al. (2013). In vivo theranostics at the peri-infarct region in cerebral ischemia. *Theranostics.* 4 (1): 90–105.

98 Zhao, Y., Jiang, Y., Lv, W. et al. (2016). Dual targeted nanocarrier for brain ischemic stroke treatment. *J. Control. Release* 233: 64–71.

99 Hassanzadeh, P., Arbabi, E., Atyabi, F., and Dinarvand, R. (2018). Ferulic acid-loaded nanostructured lipid carriers: a promising nanoformulation against the ischemic neural injuries. *Life Sci.* 193: 64–76.

100 Kakkar, V., Muppu, S.K., Chopra, K., and Kaur, I.P. (2013). Curcumin loaded solid lipid nanoparticles: an efficient formulation approach for cerebral ischemic reperfusion injury in rats. *Eur. J. Pharm. Biopharm.* 85(3 Pt A): 339–345.

101 Bao, Q., Hu, P., Xu, Y. et al. (2018). Simultaneous blood–brain barrier crossing and protection for stroke treatment based on Edaravone-loaded ceria nanoparticles. *ACS Nano* 12 (7): 6794–6805.

102 Takamiya, M., Miyamoto, Y., Yamashita, T. et al. (2012). Strong neuroprotection with a novel platinum nanoparticle against ischemic stroke – and tissue plasminogen activator-related brain damages in mice. *Neuroscience* 221: 47–55.

103 Amani, H., Habibey, R., Shokri, F. et al. (2019). Selenium nanoparticles for targeted stroke therapy through modulation of inflammatory and metabolic signaling. *Sci. Rep.* 9 (1): 6044.

104 Zhao, L., Lan, T., Jiang, G., and Yan, B. (2022). Protective effect of the gold nanoparticles green synthesized by *Calendula officinalis* L. extract on cerebral ischemia stroke-reperfusion injury in rats: a preclinical trial study. *Inorg. Chem. Commun.* 141: 109486.

105 Rajkovic, O., Gourmel, C., d'Arcy, R. et al. (2019). Reactive oxygen species-responsive nanoparticles for the treatment of ischemic stroke. *Adv. Ther.* 2 (7): 1900038.

106 Mdzinarishvili, A., Sutariya, V., Talasila, P.K. et al. (2013). Engineering triiodothyronine (T3) nanoparticle for use in ischemic brain stroke. *Drug Deliv. Transl. Res.* 3 (4): 309–317.

107 Shabani, Z., Rahbarghazi, R., Karimipour, M. et al. (2022). Transplantation of bioengineered reelin-loaded PLGA/PEG micelles can accelerate neural tissue regeneration in photothrombotic stroke model of mouse. *Bioeng. Transl. Med.* 7 (1): e10264.

108 Lu, Y., Li, C., Chen, Q. et al. (2019). Microthrombus-targeting micelles for neurovascular remodeling and enhanced microcirculatory perfusion in acute ischemic stroke. *Adv. Mater.* 31 (21): e1808361.

109 Lee, Y., Lee, J., Kim, M. et al. (2021). Brain gene delivery using histidine and arginine-modified dendrimers for ischemic stroke therapy. *J. Control. Release* 330: 907–919.

110 Song, G., Zhao, M., Chen, H. et al. (2021). The role of nanomaterials in stroke treatment: targeting oxidative stress. *Oxidative Med. Cell. Longev.* 2021: 8857486.

111 Lee, H.J., Park, J., Yoon, O.J. et al. (2011). Amine-modified single-walled carbon nanotubes protect neurons from injury in a rat stroke model. *Nat. Nanotechnol.* 6 (2): 121–125.

112 Moon, S.U., Kim, J., Bokara, K.K. et al. (2012). Carbon nanotubes impregnated with subventricular zone neural progenitor cells promotes recovery from stroke. *Int. J. Nanomedicine* 7: 2751–2765.

7

Plant-Based Analgesics

Gabriel O. Anyanwu[1] *and Dorathy O. Anzaku*[2]

[1] *Department of Biochemistry, Faculty of Science and Technology, Bingham University, Karu, Nasarawa State, Nigeria*
[2] *Department of Biochemistry, College of Science and Technology, Covenant University, Ota, Ogun State, Nigeria*

Pain is an unpleasant sensation caused by any form of damage to tissue, which could be real or potential, that activates certain nerve fibers that transmit the signal to the brain, where it is consciously perceived [1]. Pain is vital for the preservation of the body, as it signals to the body the presence of injury, trauma, and danger [2, 3]. Many disorders, including tumors, surgical procedures, physical damage, and unpleasant chemical stimuli, include pain as a common and uncomfortable component [4]. Excessive pain can cause perspiration, nervousness, nausea, and palpitation, although it is largely a warning signal and a protective mechanism. Damage to nociceptors is induced by chemical, mechanical, and thermal stimuli, and when cells are injured it results in the release of chemical mediators, which in turn activate and sensitize the sensory receptors to other pain mediators. Bradykinin, histamine, serotonin, and prostaglandins are all important pain mediators [5].

Although numbers vary depending on the approach used to quantify pain, chronic pain is thought to affect 50 million individuals in the United States, with 19.6 million of those adults experiencing high-impact chronic pain that interferes with daily life activities [6]. In 2019, the Centers for Disease Control and Prevention (CDC) clearly stated that the 2016 Guideline for Prescribing Opioids for Chronic Pain had been misconstrued or misapplied in some cases [7]. The CDC reported that about 30% of adults aged 65 years and above report suffering from chronic pain, 1 in 5 adults in the US population experiences chronic pain, while between 11% and 40% of US adults are living with chronic pain [6]. The prevalence of lower back pain is estimated at 39% in Africa [8]. In 2013, out of the 170 million people in Nigeria, 162 257 died of cancer or HIV with untreated severe or moderate pain [9]. The prevalence rate of pain for those 65 years and above ranged from 13% to 49% worldwide, and 25% specifically in Nigeria for the same age group [10].

Pain therapy differs significantly between industrialized and developing countries. Internationally prescribed drugs are limited in developing countries, making it unnecessarily difficult for patients to obtain them. Australia, the United States, the United Kingdom, and Canada combined consume around 68% of all opioids, whereas middle- and low-income countries account for only 7% of global consumption [11]. Africa has the average lowest narcotics

Phytochemical Drug Discovery for Central Nervous System Disorders: Biochemistry and Therapeutic Effects, First Edition. Edited by Chukwuebuka Egbuna and Mithun Rudrapal.

consumption of about 50 daily doses per million per day, which is significantly lower than the United States, which consumes approximately 14000 daily doses per million [12].

Pain management has been classified into two main categories: pharmacologic management and non-pharmacologic management. Pharmacologic pain management mostly involves the use of medications widely known as analgesics, which are chosen primarily on the severity of the pain. Basic pain relievers, such as non-steroidal anti-inflammatory drugs (NSAIDs), are the first-line pharmacologic agents suggested by the World Health Organization (WHO) analgesic ladder for the symptomatic management/treatment of mild or moderate pain [13–15]. They are usually widely available without a prescription. Strong opioids, such as morphine, are necessary for moderate to severe pain, according to the WHO [16].

Analgesics are pain-relieving medications that are generally referred to as painkillers [17]. Treatment should always begin with minimal dosages of mild medicines, which can then be gradually increased or switched to stronger drugs until the best pain relief possible is achieved. The correct dose of any medication is usually the smallest amount that gives the most benefit. Painkillers are rarely effective in totally removing pain. NSAIDs such as paracetamol, celecoxib, aspirin, diclofenac, naproxen, and ibuprofen; light opioids like codeine, tramadol, and dihydrocodeine; and strong opioids like morphine, oxycodone, buprenorphine, and fentanyl are all available for pain management [18]. Beyond synthetic drugs, natural products are also available for the treatment or management of pain issues [19]. This chapter focuses on the review of plants and/or their bioactive compounds that possess analgesic effects.

7.1 Current Analgesic Drugs and Their Mechanisms of Action

Analgesics are primarily classified into two categories: anti-inflammatory drugs and opioids; the former ease pain by reducing local inflammatory responses, while the latter act on the brain. NSAIDs inhibit prostaglandin biosynthesis, which is responsible for the pain caused by mediators of inflammation [20–22]. They are frequently associated with serious adverse effects such as stomach ulcers, renal damage, cardiac problems, gastrointestinal ulceration, and bronchospasm [22–24]. NSAIDs include ibuprofen, diclofenac, fenbufen, ketoprofen, piroxicam, naproxen, aspirin, indomethacin, celecoxib, and entericoxib. Table 7.1 gives information about some analgesics, their mechanisms of action, and side effects.

Opioids produce an analgesic effect by operating both presynaptically and postsynaptically. They presynaptically block calcium channels on nociceptive afferent neurons, inhibiting the release of nociceptive neurotransmitters such as substance P (which is a neuropeptide and acts as a neurotransmitter and as a neuromodulator) and glutamate [48, 49]. Opioids may postsynaptically activate potassium channels, thereby raising the necessary action potential for nociceptive transmission when cell membranes are hyperpolarized. Opioids' analgesic effect is mediated by their receptors (mu, kappa, and delta) in the spine and supraspinally [48, 49]. Their side effects may include nausea/vomiting, constipation, and blurred vision [50, 51]. Opioids include morphine, meperidine, tramadol, remifentanil, methadone, oxycodone, hydrocodone, pethidine, and codeine.

7.2 Plant-Derived Lead Compounds with Analgesic Activities

Flavonoids, alkaloids, cardenolides, and saponins have been found in plant products with analgesic activities [52]. The analgesic activities of plants may be due to these active substances [53], and some phytochemicals with analgesic properties have been highlighted in the following.

Table 7.1 Some analgesics, their mechanisms of action and side effects.

Analgesics	Mechanism of action	Side effects	References
Acetylsalicylic acid (ASA)	Inhibits platelet thromboxane A2 (TXA-2) formation via irreversible inhibition of cyclooxygenase 1 (COX-1)	Nausea, vomiting, indigestion, watery diarrhea, mild allergic reactions with rash, itchiness, and fever	[17, 25]
Diclofenac	Inhibits lipoxygenase pathways and phospholipase A2	Hepatotoxicity, gastrointestinal ulceration and bleeding, renal damage	[26, 27]
Indomethacin	Prevents prostaglandin production via COX inhibition	Depression, psychiatric disturbances, Parkinsonism, epilepsy	[28, 29]
Paracetamol	Inhibits COX pathways Lowers pain threshold and suppresses central prostaglandin synthesis	Allergic reaction, including rash or swelling, blood disorders; higher dosages could lead to liver and kidney damage	[17]
Ziconotide	Selective blockage of N-type calcium channels that prevents release of pronociceptive neurochemicals (glutamate, material substance P, and calcitonin gene-connected peptide)	Blurred vision, crawling feelings, dizziness or lightheadedness, excessive muscle tone, tension or tightness, headache, loss of memory, pain in arms or legs	[30, 31]
Nefopam	Inhibits synaptic uptake of dopamine, noradrenaline, and serotonin	Nausea, nervousness, confusion or hallucinations, numbness and tingling of hands or feet, dry mouth, lightheadedness	[32, 33]
Naloxone (opioid antagonist)	Acts as competitive antagonist of opioid receptor	Elevated blood pressure, irregular heart rate, nausea, vomiting, withdrawal symptoms, rapid breathing, dry cough, seizures, fainting, headaches, anxiety, tinnitus, and confusion	[34, 35]
Propofol (hypnotic analgesic)	Increase γ-aminobutyric acid (GABA) channel activity.	Decreased or increased heart rate, dizziness, headache, dry cough, low blood pressure, nausea, rash, vomiting	[36–38]
Methadone (opioid agonist)	μ-receptor agonist and mimics endogenous opioids like endorphins and enkephalins	Respiratory depression, constipation of biliary spasm	[39, 40]
Meperidine	Inhibits adenylate cyclase and subsequently release of GABA, substance P, dopamine, noradrenaline, and acetylcholine is inhibited	Agitation, disorientation, or hallucinations. Shakiness, seizures, abnormal muscle movements, bradycardia or tachycardia, hypotension or hypertension, fainting	[35]
Fentanyl	Potent μ-opiate receptor agonist	Respiratory depression	[41, 42]

(*Continued*)

Table 7.1 (Continued)

Analgesics	Mechanism of action	Side effects	References
Morphine	μ, δ, and κ-opioid receptor agonist in brain and spinal cord	Decreased peristalsis, decreased respiratory rate, overdose leads to coma	[35, 43]
Pentazocine	Activates κ- and μ-opioid receptors	Sedation, euphoria, agitation, itching, nausea, sweating, abdominal bloating, vomiting and constipation, confusion, clammy skin, changes in heart rate or blood pressure, sweating or lightheadedness, respiratory depression	[44, 45]
Tramadol	μ-opioid receptor agonist Serotonin and norepinephrine reuptake inhibitor	Drowsiness, dry mouth, unusual cold, heartburn, itching or skin rash, irritability, joint and muscle pain, loss of interest, weakness, drowsiness, sore throat, stomach pain	[46, 47]

7.2.1 Saponins and Terpenoids

Terpenoids inhibit tumor necrosis factor (TNF) production, phospholipase A2 (PLA2) activity, cyclooxygenase-2 (COX-2) and inducible nitric oxide synthase (iNOS) expression, and also inhibit nuclear factor kappa B (NF-κB) activation, whereas saponins interfere with iNOS expression [54]. Terpenoids have anti-inflammatory effect by inhibiting PLA2 activity [55]. Ginsenosides and other ginseng saponins from *Panax ginseng* possess analgesic properties [56]. Sesquiterpene lactones isolated from *Tanacetum parthenium*, the feverfew plant, displayed antinociceptive action, and inhibited iNOS expression and NF-κB activation [54].

7.2.2 Flavonoids

Certain flavonoids and related compounds such as kaempferol, genistein, quercetin, resveratrol, and resorcinol have exhibited analgesic and anti-inflammatory effects via COX-2 inhibition and dose-dependent reduction in TNF-α, an inflammatory-mediating cytokine [57]. Unlike NSAIDs, plant-based COX-2 inhibitors relieve pain very slowly.

The flavonoids exhibit potent anti-inflammatory activity by inhibiting prostaglandin synthesis in the same way as the inflammatory response is mediated by NSAIDs. This effect might be due to inhibition of key enzymes in prostaglandin biosynthesis by flavonoids; these enzymes include lipoxygenase, phospholipase, and cyclooxygenase [58, 59]. Several flavonoids with analgesic potential isolated from *Waltheria indica* include (−)-epicatechin, kaempferol-3-O-β-D-(6″-E-p-coumaroyl)-glucopyranoside, tiliroside, quercetin, and kaempferol [60, 61]. Also, the active constituents that are responsible for the analgesic, anti-inflammatory, and antipyretic activities of *Drymaria odorata* are mainly flavonoid compounds [62].

7.2.3 Alkaloids

Alkaloids, like morphine, either isolated from plants or their synthetic derivatives, are used as analgesic agents [63]. Alkaloids exhibit anti-inflammatory activity by interfering with

COX expression, prostaglandin E2 (PGE2) production, and inhibiting proinflammatory cytokines like TNF-α, interleukin (IL)-1, and IL-6 [55]. Also, alkaloids exert analgesic activity by interfering with neurotransmitters that facilitate the sensation of pain in the central nervous system [24]. Most studies on analgesia have reported the presence of alkaloids from phytochemical screening of the plant parts used for studies [64–67].

Some alkaloids from Aconitum species like aconitine, heteratistine, and lappacodine have shown analgesic effects in mechanical, thermal, and chemical nociceptive experiments in animals [68]. As a result of the growing popularity of *Papaver somniferum* as a pain reliever, German scientist Sertüner isolated the plant's principal analgesic element, the alkaloid morphine, in 1905 [54]. The opium latex contains a variety of alkaloids, including codeine, thebaine, and papaverine, in addition to morphine; however, the mechanism behind most of the alkaloids with analgesic effects remains unknown [54]. *Papaver somniferum*, *Cannabis sativa*, Capsicum species, *Panax ginseng*, feverfew, Aconitum species, *Drymis winteri*, *Siphocampylus verticillatus*, *Hedyosinum brasiliense*, and *Phyllanthus* species all have biologic activity that has contributed to the development of current analgesics [43].

7.2.4 Glycosides

A wide range of structurally diverse glycosides that have been isolated from several useful plants have shown in vivo and/or in vitro analgesic activities. Luteolin, quercetin, rutin, and triterpenoid glycosides, for example, have anti-inflammatory properties by inhibiting the COX and lipoxygenase (LOX) pathways, which play significant roles in the formation of the inflammatory mediator PGE2 from arachidonic acid [3]. COX is an enzyme that converts arachidonic acid to prostaglandin H2, then transforms it to a variety of active metabolites [3].

Examples of glycoside compounds from plant sources with analgesic effects include pseudopterosin from *Pseudopterogorgia elisabethae* [69], harpagoside from *Harpagophytum procumbens* [70], and kaempferol and kaempferol-3-O-glucoside extracted from *Thesium chinense* [61]. Several studies have also reported plant extracts rich in glycosides that displayed analgesic activities, such as a triterpene glycoside-rich extract of *Cussonia paniculata* stem bark [71] and *Buddleja globose* extract [72], among others.

7.3 Analgesic Effects of Medicinal Plants Found in Nigeria

Different plant parts, which include leaves, stem, pulp, fruit, root, and even whole plants found across different areas of Nigeria, have been screened for analgesic activity by different test methods, which include the formalin test, tail-flick test, hot-plate test, and acetic acid–induced writhing test. Most of these plants were also screened for phytochemicals, of which flavonoids, glycosides, tannins, and saponins were the most reported. These phytochemicals are said to be the active ingredients hypothesized to be responsible for the mechanism by which the plant extracts alleviate pain. Plants with analgesic activities found in Nigeria are outlined in Table 7.2. For example, the methanol leaf extract *Ficus ingens* was found to have attenuated acetic acid–induced abdominal writhing in mice and the mechanism by which it exerts this effect was suggested as being linked to LOX or COX pathways. In the same way, evaluation of the ethanolic leaf extract of *Loranthus micranthus*, ethanolic extract of *Spathodea campanulate* and *Syzygium guineense*, and methanol extract of *Alafia barteri oliver* in Wistar rats, Sprague-Dawley rats, and mice, respectively, reported that they exhibited analgesic properties in dose-dependent studies.

Table 7.2 Plants with analgesic activity found in Nigeria.

Scientific names	Common names	Native names	Part of crude extract	Active ingredient (fractions or compounds)	Reversal of pain	Pharmacological evaluation	References
Acalypha wilkesiana Müll. Arg	Copperleaf and Jacob's coat	Aworoso (Y)	Leaf (E)	Flavonoids and triterpenoids	ASA	AAIWT, FT, HIP	[73]
Adansonia digitata L.	Boabab tree	Kuka (H)	Fruit pulp	Steriods, saponins, tritepenes	ASA	FT, HPT	[74]
Alafia barteri Oliv.	Alafia chewing stick, guinea-fowl's crest	Ota nza (I), Agbárí ẹtù (Y)	Root extract	Not reported	Morphine, diclofenac, naloxone	AAIWT, FT, TFT	[75]
Allium cepa L.	Onions	Albasa (H)	Bulb			FT, HPT, carrageenan-induced edema	[76]
Alstonia boonei De Wild	Cheese wood; alstonia	Egbu (I), Ahun (Y)	Stem bark; leaves	ASA indomethacin	Not reported	AAIWT, FT, HPT, TFT	[77, 78]
Anacardium occidentale L.	Cashew	Fisa (H), kashu (I), kajú (Y)	Leaves	Pentazotocine, indomethacin	Alkaloids, tannins, saponins, and cardenolides	AAIWT, FT, TFT	[53]
Anthocleista djalonensis A. Chev.	Cabbage tree	Sapo (Y), kwarii (H), okpokolo (I)	Stem bark	Indomethacin	Flavonoids and tannins	Analgesiometer, mechanically induced pain	[79]
Anthocleista nobilis G. Don	Forest fever	Wudifokete/ Bontodee (Ghana)	Stem bark	Diclofenac sodium, indomethacin	N-hexane, ethyl-acetate, butanol	AAIWT	[80]
Beta vulgaris L.	Beets, beetroot		Root	ASA	Bioflavonoid (rutin)	AAIWT, FT, HPT	[81]
Byrsocarpus coccineus Schumach. & Thonn.	Huntsman's pepper or short pod	Kasa, tsaamiyar-kasa (H), oke abolo (I), ado (Y)	Leaves	ASA, morphine	Alkaloids, tannins, saponins, reducing sugars, glycosides, anthraquinones	AAIWT, FT, TFT	[64]

Calliandra portoricensis (Jacq.) Benth.	Powder puff (USA, Bates) Corpse awakener, snowflake acacia		Leaves and roots	Indomethacin		AAIWT, FT	[82]
Capparis thonningii Schumach.	Humbug	Awara (Y)	Root	Naloxone		FT, HPT	[75]
Carissa edulis (Forssk.) Vahl	Carandas plum; simple-spined num-num	Cizaki (H)	Root bark, stem bark, leaves, fruits, seeds	ASA, metamizol	Not reported	AAIWT, tail-clip method	[83]
Manihot esculenta Crantz	Cassava, manioc, mandioca, or yuca	Rogo, karaza (H), akpu, abacha (I), ege (Y)	Leaves	Paracetamol	Flavonoids	AAIWT	[84]
Carpolobia lutea G. Don	Cattle stick	Angalagala (I), egbo oshunshun (Y)	Root	Acetylsalicylic acid	Flavonoid	AAIWT, FT, TFT	[85]
Chenopodium ambrosioides L.	Wormseed, Mexican tea	Ewe imí (I)	Leaves	Indomethacin	Not reported	AAIWT, FT, HPT	[86]
Chromolaena odorata L.	Siam weed, Christmas bush, devil weed	Abani di egwu or nsiibilibe (I)	Fractions	Indomethacin	Flavonoids	FT, HPT	[87]
Combretum mucronatum Schumach. & Thonn.	Bushwillows or combretums	Ọ̀kàn, ọ̀gàn (Y)	Root	Diclofenac sodium, diazapem, quercetin, morphine	Flavonoids, saponins, glycosides	AAIWT, FT, HPT	[75]

(*Continued*)

Table 7.2 (Continued)

Scientific names	Common names	Native names	Part of crude extract	Active ingredient (fractions or compounds)	Reversal of pain	Pharmacological evaluation	References
Crinum glaucum A. Chev.		Édè-chúkwú, òlòdì, édẹ ọ̀bàsị̀ (I), ẹdẹ́suku (Y)	Bulbs	ASA, morphine, indomethacine	Alkaloids, reducing sugars, flavonoids, phenols	Cold-water TFT, AAIWT, FT	[54]
Crossopteryx febrifuga (Afzel. ex G. Don) Benth.	African bark; ordeal tree		Stem bark	ASA	Saponins, terpenes, carbohydrates, monossacharides, reducing sugars, tannins, anthraquinones, flavonoids, phenols	AAIWT, analgesiometer	[88]
Culcasia Scandens P. Beauv.	Climbing arum, climbing plant		Leaves	ASA, pentazocine	Flavonoids and terpenoids	Formalin-induced paw licking, AAIWT	[89]
Cussonia paniculate Eckl. & Zeyh.	Mountain cabbage tree		Stem bark	Indomethacin and cyproheptadine	Triterpene glycosides	AAIWT, FT	[71]
Cyathula prostrata (Linn.) Blume	Pasture weed	Dánká dáfi, tsattsarar ßera (H), àgbí, rì, gbá (I), çawere pèpè (Y)	Whole plant	ASA, morphine	Flavonoids	AAIWT, HPT	[90]
Cynodon dactylon (L.) Pers.	Bermuda grass; Bahama grass	Tsárkíyàr zóómóó, jáàjàà mazá (H)	Whole plant	Aspirin, pentazocin	Flavonoids and glycosides	AAIWT, HPT	[91]
Dennettia tripetala Baker f.	Pepper fruit	Mmị̀mị̀ (I), àtà (I)	Fruit	Aspirin, morphine diclofenac, indomethacin		AAIWT, HPT, FT	[92]
Dialium guineense Wild	Black velvet or velvet tamarind	Icheku (I), awin (Y)	Stem bark	Aspirin	Not reported	AAIWT, HPT, TFT	[93]
Diodia scandens Sw.			Leaves	Aspirin, indo methacin	Not reported	Analgesiometer	[94]

Drymaria cordata (L.) Willd. ex Schult.	Chick weed		Whole plant	ASA, morphine, amphentamine	Glycosides, saponins, alkaloids, flavonoids, phenols, tannins	AAIWT, FT, tail-clip test	[62]
Erythrina senegalensis DC.	Coral tree; coral flower	Minjirya (H)	Stem bark	ASA	Alkaloid	AAIWT	[95]
Ficus ingens (Miq.) Miq.	Red-leaved fig		Leaves	Morphine sulfate, piroxicam	Flavonoids, saponins, tannins	AAIWT, HPT	[96]
Gongronema latifolium Benth		Utazi (I), arọ̀kẹ̀kẹ̀ (Y)	Leaves	ASA, paracetamol	Tannins and flavonoids	AAIWT, FT	[97, 98]
Garcinia Kola Heckel	Bitter kola, male kola, or false kola	Cida goro (H), àdù, ugugolu, akụ ilū (I), orógbó (Y)	Seeds	Aspirin		AAIWT, TFT	[99]
Hibiscus sabdariffa Linn.	Sorrel; Florida cranberry (Blench); roselle	Zobo (H)	Seeds	Indomethacin, diclofenac, sodium	Not reported	AAIWT	[100]
Hunteria umbellata K. Schum		Erin (Yoruba), nkpokiri (Ibo)	Fruit pulp	ASA	Flavonoids, simple sugars, alkaloids, saponins, steroidal compounds	AAIWT	[101]
Ipomoea asarifoliai (Desr.) Roem. & Schult.	Ginger-leaf morning glory	Dúmán kadaá (H), gbọọrọ ayaba (Y)	Aerial part	Ketoprofen	Carbohydrates, saponins, flavonoids, tannins	AAIWT	[102]
Ipomoea involucrata	Morning glory weed or railway creeper	Duman-kwaadi (H), alukerese or atewagba (Y), eriri ala or fifilori (I)	Leaves	Aspirin	Flavonoids, alkaloids	AAIWT	[103]
Irvingia gabonensis (Aubry-Lecomte ex O'Rorke) Baill.	African mango, wild mango, bush mango	Ogbono (I)	Stem bark	Morphine, metamizole sodium	Not reported	HPT, mechanical method applying pressure on tail of mouse using analgesiometer	[104]

(*Continued*)

Table 7.2 (Continued)

Scientific names	Common names	Native names	Part of crude extract	Active ingredient (fractions or compounds)	Reversal of pain	Pharmacological evaluation	References
Jatropha curcas L.	Physic nut; fig nut; pig nut	Bíí ní dà zúgúú (H), bulu olu, ugbolo, okwata, ugbolu (I), bòtújẹ̀ (Y)	Leaves	Paracetamol, piroxicam	Alkaloids, flavonoids	AAIWT, HPT, TFT	[105, 106]
Kigelia Africana (Lam.) Benth.	Cucumber or sausage tree	Hantsar giíwaá, noónòn giíwaá (H), alambọrọgoda, amọ-ibi, izhi (I), orora (Y)	Stem bark	Morphine, indomethacin		AAIWT, HPT	[107]
Lagenaria breviflora (Benth.) Roberty	Wild colocynth	Gojin jima, gunar jiímaà, jiímaà (H), ányụ́ḿmụ̄ọ́ (I), eso gbegbe (Y)	Leaves	Indomethacin	Tannins, saponins, alkaloids, flavonoids	AAIWT, FT	[108]
Landolphia owariensis P. beauv.	White rubber vine; rubber – white ball rubber	Nwalikali, mba (I), Ibo alaitipa (Y), ciwo (H)	Leaves	ASA	Flavonoid	AAIWT, TFT	[109]
Leptadenia hastata Vatke		Adizindir (H), iran-aji igbó, isanaje-igbó (Y)	Leaves, stem	Piroxicam, pentazocine, paracetamol	Flavonoids and tannins, terpenoids, saponin, alkaloids, steroidal ring, resins	AAIWT, HPT, FT, carrageenan-induced edema	[66, 110]
Lippia multiflora Moldenke	Bush tea, healer herb	Bunsurun fadama (H)	Leaves	ASA	Alkaloids, tannins, flavonoids, saponins, volatile oils	AAIWT	[63]
Loranthus micranthus Linn.	African mistletoe		Leaves	Pentazocine, aspirin	Alkaloids and flavonoids	AAIWT, HPT	[67]

Margaritaria discoidea (Baill.) G.L. Webster		Isi m°kpị (I), àçàçà (Y)	Stem bark	Indomethacin	Not reported	AAIWT, FT	[111]
Melanthera scandens (Schumach. & Thonn.) Roberty		Anwuliriwani, azuzo (I), abo yúnyún (Y)	Leaves	ASA	Flavonoids	Formalin-induced paw licking, AAIWT, thermal-induced pain tests	[112]
Moringa stenopetala Baker f.	Cabbage tree		Leaves	Aspirin, morphine	Flavonoids, terpenoids, steroids	AAIWT, FT	[24]
Nelsonia canescens (Lam.) Spreng.	Blue pussy leaf	Dambun makiyaayaa (H)	Leaves	Indomethacin	Flavonoids and saponins	HPT, FT	[59]
Nothospondias staudtii Engl.			Leaves	Aspirin	Flavonoids	AAIWT, TFT	[113]
Ocimum gratissimum L.	Clove basil, African basil		Leaves		ASA	HPT	[114]
Paullinia pinnata L.	Tietie		Leaves	Saponins, tannins, glycosides, flavonoids	Aspirin, pentazocine, piroxicam	Acetic acid–induced abdominal constriction test, HPT	[58]
Peperomia pellucida (L.) Kunth	Pepper elder, shining bush plant	Erínrín (Y)	Aerial parts	Anthraquinones, cardiac glycosides, tannins	Paracetamol, ASA	AAIWT	[115]
Physalis angulate L.	Cut-leaf ground cherry; angular winter cherry	Hausa language (Nigeria) "Saadi Birii"	Leaves	Ibuprofen		AAIWT	[116]
Piper guineense Schumach. & Thonn.	Benin pepper, Ashanti pepper, West African pepper	Uziza (I)	Leaves	Morphine, ASA, diclofenac, dexamethasone	Not reported	AAIWT	[117, 118]

(*Continued*)

Table 7.2 (Continued)

Scientific names	Common names	Native names	Part of crude extract	Active ingredient (fractions or compounds)	Reversal of pain	Pharmacological evaluation	References
Pseudocedrela kotschyi Harms	Dry-zone cedar, hard cedar-mahogany	Tuna or tunas (H), emibegbe or emigbegeri (Y)	Leaves, n-butanol	Piroxicam	Flavonoids, saponins, terpenoids, tannins	AAIWT, FT	[119]
Rothmannia Longiflora Salisb.		Katambiri (H), uli, alankitauku, mbembe (I), edo-pata, kerebuje (Y)	Leaves	Piroxicam, morphine		AAIWT, FT, HPT	[120]
Securinega virosa Baill.		Gussu (H), njisi ntà (I), ìranjé (Y)	Leaves	Piroxicam, morphine	Flavonoids, tannins, alkaloid, saponins	AAIWT, FT	[121]
Sida acuta Burn f.	Broom weed		Whole plant	Paracetamol	Flavonoid	AAIWT, FT	[122]
Sida cordifolia L.	Flannel weed, heart-leaf sida		Whole plant	Paracetamol	Flavonoid	AAIWT, FT	[122]
Spathodea campanulate P. Beauv.	English tulip tree; African tulip tree	Ímí éwū, akpoti, oruru (I), mójútòrò (Y)	Leaf		Morphine, ASA	AAIWT, TFT, HPT; carrageenan-induced edema	[123]
Syzygium guineense (Willd.) DC.	Waterberry, Bambara; woodland waterberry, waterpear	Malmo (H), adere (Y)	Leaves	Alkaloids, saponins, tannins, glycosides	Aspirin	AAIWT, HPT	[124]
Tithonia diversifolia (Hemsl.) A. Gray	Mexican sunflower, tree marigold		Leaves	Not reported	Indomethacin	HPT, FT	[113]

Vernonia amygdalina Delile	Bitter leaf	Shiwáákáá, (H), òlúgbù (I); ewúrò, ewúrò jíję (Y)	Leaves	Saponins, tannins, phenols, steroids, flavonoids, alkaloids	ASA, pethidine	AAIWT, FT, TFT	[125]
Vernonia cinerea (L.) Less.		Bójúrę́, ęlęgbę-oju, ewe jèdíjèdí (Y)	Leaves	Saponin, steroids, terpenoids, alkaloids, flavonoids	ASA, morphine, paracetamol	AAIWT, HPT, analgesiometer	[126]
Vernonia hymenolepis Vatke	Sweet bitterleaf, bitterleaf (En).		Leaves		Acetylsalicylate	FT	[127]
Waltheria americana L.	Velvetleaf, marshmallow, monkey bush, boater bush		Leaves	Alkaloids	Prazosin, atropine, indomethacin, propranolol, nifedipine, glibenclamide	AAIWT, FT, TFT	[67]
Waltheria indica L.	Sleepy morning	Hankufah or hankubah (H), korikodi (Y)	Root and leaves	Flavonoids, tannins	Aspirin	FT	[128]
Zea mays L.	Maize or corn		Husk	Tannins, polyphenols	Indomethacin	HPT, FT	[129]
Ziziphus spina-christi (L.) Desf.	Christ's thorn jujube		Root, bark		ASA	AAIWT, FT, HPT	[130]

AAIWT, acetic acid–induced writhing; ASA, acetylsalicylic acid; FT, formalin test; HIP, heat-induced pains, HPT, hot-plate test, TFT, tail-flick test.

7.4 Limitations of Plant-Based Analgesics

Limitations to the use of plant analgesics include the fact that a lot of vital information about many plants, especially on traditional plants in Africa, is yet to be documented, as opposed to plant information in certain countries of the world such as China and India [131]. Also, the mode of consumption and dosage of plant analgesics differ greatly from one another; the method of preparation also could have an impact on the effect of the analgesics in the body [132]. All compounds are poisonous; the only variation is the amount and dosage used. Although they have been verified to be useful in the treatment of pain, most of the plants that have been reported to be active analgesics have not undergone any type of clinical research. The majority of acute toxicity tests on these plants demonstrate that they are non-toxic in laboratory animals [53, 66, 67]; however, side effects have not been widely recorded. Because dose-dependent studies have primarily been conducted in mice and rats, the dosage that will be effective in people has yet to be defined or reported. Plants also contain many antinutrients [132] in addition to active components and some of these antinutrients could lead to other illnesses or exacerbate a patient's diseased condition unknowingly, which limits the general use of plant-based analgesics for pain treatment.

7.5 Future Directions and Perspective for Plant-Based Analgesics

Despite numerous previous achievements in the production of plant-derived therapeutic products, the majority of plants are frequently simply tested for one or a few types of biologic features [133]. Because numerous secondary metabolites from plants are genus or species specific, many more plant elements with potentially valuable biologic features are likely to remain unexplored, uninvestigated, and undeveloped. It would be expected that virtual screening for potential analgesic compounds from plant sources would facilitate plant drug discovery and development.

Aside from efficacy and potency, a major goal in the development of analgesic medications is the avoidance of adverse effects on the digestive system owing to the significant risk of producing gastric ulcers with the persistent use of some analgesics [27, 43]. As a result, developing new synthetic analgesics that are less ulcer inducing is a key problem [3]. The quest for novel bioactive plant-based molecules with a better safety profile, on the other hand, is a relevant topic of research around the world. Indeed, in recent years computational tools used in drug discovery have aided in the identification of ulcer-free analgesic medicines derived from medicinal plants. Various types of innovative pharmaceutical techniques and formulations are also used to discover answers to the difficulties that exist. However, these strategies have not been very successful in alleviating the side effects associated with clinically utilized analgesics [3].

7.6 Conclusion

A great deal of research has already been done on plant extracts and their potential analgesic effects in vivo and in vitro, but most of it has not gone any further. However, it is recommended that most of these plants be studied further to discover their active components and elucidate their reaction mechanisms. Screening medicinal plants with well-known analgesic potential could lead to the identification of safe and effective analgesic drugs. The existing analgesic drugs have various adverse effects, thus the need to discover more plant-based analgesic bioactive compounds with less toxicity or fewer side effects is the target of researchers, whether in the education system or the pharmaceutical industry.

References

1 Andersen, L.P.H., Gogenur, I., Fenger, A.Q. et al. (2015). Analgesic and antihyperalgesic effects of melatonin in a human inflammatory pain model: a randomized, doubleblind, placebo-controlled, three-arm crossover study. *Pain* 156 (11): 2286–2294.

2 Koech, S.C., Ouko, R.O., Michael, N.M. et al. (2017). Analgesic activity of dichloromethanolic root extract of *Clutia abyssinica* in Swiss albino mice. *Nat. Prod. Chem. Res.* 5 (2): 255.

3 Khan, H., Pervaiz, A., Intagliata, S. et al. (2020). The analgesic potential of glycosides derived from medicinal plants. *DAR U J. Pharm. Sci.* 28 (1): 387–401.

4 Aliu, Y.O. (2007). *Veterinary Pharmacology*, 111–131. Kaduna: Tamaza Publishing.

5 Craig, C.R. and Stitzel, R.E. (2003). *Modern Pharmacology with Clinical Applications*, 5e. Philadelphia, PA: Lippincott Williams & Wilkins.

6 Centers for Disease Control and Prevention (2018). Prevalence of chronic pain and high-impact chronic pain among adults — United States, 2016. *MMWR Morb. Mortal. Wkly. Rep.* 67 (36): 1–2. https://www.cdc.gov/mmwr/volumes/67/wr/pdfs/mm6736a2-H.pdf.

7 Dahlhamer, J., Lucas, J., Zelaya, C. et al. (2018). Prevalence of chronic pain and high-impact chronic pain among adults-United States. *MMWR Morb. Mortal. Wkly. Rep.* 67 (36): 1001–1006.

8 Morris, L.D., Daniels, K.J., Ganguli, B., and Louw, Q.A. (2018). An update on the prevalence of low back pain in Africa: a systematic review and meta-analyses. *BMC Musculoskelet. Disord.* 19 (1): 1–5.

9 Federal Ministry of Health (2018). Guidelines for the management of pain in Nigeria. http://www.health.gov.ng (accessed 8 March 2023).

10 Ayanniyi, O., Mbada, C.E., and Muolokwu, C.A. (2011). Prevalence and profile of back pain in Nigerian adolescents. *Med. Princ. Pract.* 20: 368–373.

11 Global Alliance to Pain Relief Initiative (GAPRI) (2010). Access to essential pain medicines brief. http://gapri.org/understand-problem (accessed 8 March 2023).

12 Nchakoa, E., Bussella, S., Nesbetha, C., and Odoh, C. (2018). Barriers to the availability and accessibility of controlled medicines for chronic pain in Africa. *Int. Health* 10 (2): 71–77.

13 Schug, S.A. and Auret, K. (2008). Clinical pharmacology: principles of analgesic drug management. In: *Clinical Pain Management: Cancer Pain* (ed. N. Sykes, M. Bennet, and C.-S. Yuan), 104–122. London: Hodder Arnold.

14 World Health Organization (2018). *WHO Guidelines for the Pharmacological and Radiotherapeutic Management of Cancer Pain in Adults and Adolescents*. Geneva: WHO.

15 Yang, J., Bauer, B.A., Wahner-Roedler, D.L. et al. (2020). The modified WHO analgesic ladder: is it appropriate for chronic non-cancer pain? *J. Pain Res.* 13: 411–417.

16 World Health Organization (2000). Narcotic and psychotropic drugs: achieving balance in national opioids control policy: guidelines for assessment. https://apps.who.int/iris/handle/10665/66496 (accessed 8 March 2023).

17 Aloyo, V.J. (2007). Analgesics: Opioid and non-opioid. In: *Rang & Dale's Pharmacology*, 6e (ed. J. Ritter, R. Flower, G. Henderson, et al.), 226–237.

18 British Pain Society (2010). Understanding and managing pain: information for patients. https://www.britishpainsociety.org/static/uploads/resources/files/book_understanding_pain.pdf (accessed 8 March 2023).

19 Tick, H., Nielsen, A., Pelletier, K.R. et al. (2018). Evidence-based nonpharmacologic strategies for comprehensive pain care: the consortium pain task force white paper. *Explore* 14 (3): 177–211.

20 World Health Organization (2005). *WHO Consultation on Selected Medicinal Plants*, 4e. Geneva: WHO https://apps.who.int/iris/handle/10665/42052.

21 Rauf, A., Uddin, G., Siddiqui, B.S. et al. (2014). In-vivo antinociceptive, anti-inflammatory and antipyretic activity of pistagremic acid isolated from *Pistacia integerrima*. *Phytomedicine* 21 (12): 1509–1515.

22 Burke, A., Smyth, E., and FitzGerald, G.A. (2005). Analgesic-antipyretic agents; pharmacotherapy of gout. In: *Goodman & Gilman's: The Pharmacological Basis of Therapeutics* (ed. L.B. Brunton, J.S. Lazo, and K.L. Parker), 671–715. New York: McGraw-Hill.

23 Geremew, H., Shibeshi, W., Tamiru, W., and Engdawork, E. (2015). Experimental evaluation of analgesic and anti-inflammatory activity of 80% methanolic leaf extract of *Moringa stenopetala* Bak. F. in mice. *Ethiop. Pharm. J.* 31 (1): 15–26.

24 Tamrat, Y., Nedi, T., Assefa, S. et al. (2017). Anti-inflammatory and analgesic activities of solvent fractions of the leaves of *Moringa stenopetala* Bak. (Moringaceae) in mice models. *BMC Complement. Altern. Med.* 17 (1): 1–10.

25 Rauf, A., Jehan, N., Ahmad, Z., and Mubarak, M.S. (2017). Analgesic potential of extracts and derived natural products from medicinal plants. In: *Pain Relief – from Analgesics to Alternative Therapies* (ed. C. Maldonado), 339–351. Rijeka: InTech.

26 Saxena, A., Kumari, R., Mukherjee, U. et al. (2014). Draft genome sequence of the rifamycin producer *Amycolatopsis rifamycinica* DSM 46095. *Genome Announc.* 2 (4): e00662-14.

27 Smith, H.S. and Peppin, J.F. (2014). Toward a systematic approach to opioid rotation. *J. Pain Res.* 7: 589–608.

28 Brayfield, A. (ed.) *Indometacin. Martindale: The Complete Drug Reference*. London: Pharmaceutical Press.

29 Lucas, S. (2016). The pharmacology of indomethacin. *Headache* 56 (2): 436–446.

30 McGivern, J.G. (2007). Ziconotide: a review of its pharmacology and use in the treatment of pain. *Neuropsychiatr. Dis. Treat.* 3 (1): 69–85.

31 Nair, A.S., Poornachand, A., and Kodisharapu, P.K. (2018). Ziconotide: indications, adverse effects, and limitations in managing refractory chronic pain. *Indian J. Palliat. Care* 24 (1): 118–119.

32 Girard, P., Chauvin, M., and Verleye, M. (2016). Nefopam analgesia and its role in multimodal analgesia: a review of preclinical and clinical studies. *Clin. Exp. Pharmacol. Physiol.* 43 (1): 3–12.

33 Kim, K.H. and Abdi, S. (2014). Rediscovery of nefopam for the treatment of neuropathic pain. *Kor. J. Pain* 27 (2): 103–111.

34 Substance Abuse and Mental Health Services Administration (2017). National Survey on Drug Use and Health. https://www.samhsa.gov/data/nsduh/reports-detailed-tables-2017-NSDUH (accessed 8 March 2023).

35 Salah, S., Awadallah, H., and Karaman, R. (2015). Pain killers. In: *Commonly Used Drugs* (ed. R. Karaman), 75–129. New York: Nova Science Publishers.

36 Sahinovic, M.M., Struys, M.M.R.F., and Absalom, A.R. (2018). Clinical pharmacokinetics and pharmacodynamics of propofol. *Clin. Pharmacokinet.* 57: 1539–1558.

37 Vasile, B., Rasulo, F., Candiani, A., and Latronico, N. (2003). The pathophysiology of propofol infusion syndrome: a simple name for a complex syndrome. *Intensive Care Med.* 29 (9): 1417–1425.

38 Ke, J.J., Zhan, J., Feng, X.B. et al. (2008). A comparison of the effect of total intravenous anaesthesia with propofol and remifentanil and inhalational anaesthesia with isoflurane on the release of pro- and anti-inflammatory cytokines in patients undergoing open cholecystectomy. *Anaesth. Intensive Care* 36 (1): 74–78.

39 World Health Organization (2012). *Health Statistics*. Geneva: WHO.

40 United Nations Office on Drugs and Crime (2012). World drug report 2012. https://www.unodc.org/unodc/en/data-and-analysis/WDR-2012.html (accessed 8 March 2023).

41 Drug Enforcement Administration (2020). Drug fact sheets. https://www.getsmartaboutdrugs.gov/publication/drug-fact-sheets (accessed 8 March 2023).

42 Stanley, T.H. (2014). The fentanyl story. *J. Pain* 15 (12): 215–1226.

43 Hallett, M., Albanese, A., Dressler, D. et al. (2013). Evidence-based review and assessment of botulinum neurotoxin for the treatment of movement disorders. *Toxicon* 67: 94–114.

44 Dewey, W. (2008). Morphine. In: *xPharm: The Comprehensive Pharmacology Reference* (ed. S.J. Enna and D.B. Bylund), 1–6. Amsterdam: Elsevier https://www.sciencedirect.com/science/article/pii/B9780080552323622081.

45 Kimura, S., Ohi, Y., and Haji, A. (2016). Mechanisms of pentazocine-induced ventilatory depression and antinociception in anesthetized rats. *J. Pharmacol. Sci.* 130 (3): 181–184.

46 Driessen, B. and Reimann, W. (1992). Interaction of the central analgesic, tramadol, with the uptake and release of 5-hydroxytryptamine in the rat brain *in vitro*. *Br. J. Pharmacol.* 105 (1): 147–151.

47 Subedi, M., Bajaj, S., Kumar, M.S., and Mayur, Y.C. (2019). An overview of tramadol and its usage in pain management and future perspective. *Biomed. Pharmacother.* 111: 443–451.

48 Soin, A., Cheng, J., Brown, L. et al. (2008). Functional outcomes in patients with chronic nonmalignant pain on long-term opioid therapy. *Pain Pract. Off. J. World Inst. Pain* 8 (5): 379–384.

49 Schmid, C.L., Kennedy, N.M., Ross, N.C. et al. (2017). Bias factor and therapeutic window correlate to predict safer opioid analgesics. *Cell* 171 (5): 1165–1175.

50 Schug, S.A., Garrett, W.R., and Gillespie, G. (2003). Opioid and non-opioid analgesics. *Best Pract. Res. Clin. Anaesthesiol.* 17 (1): 91–110.

51 Munir, M.A., Enany, N., and Zhang, J.M. (2007). Nonopioid analgesics. *Anesthesiol. Clin.* 25: 761–774.

52 Fernanda, L.B., Victor, A.K., Amelia, T.H., and Elisabestsky, E. (2002). Analgesic properties of Umbellatine, *Psychotria umbellata*. *Pharm. Biol.* 44: 54–56.

53 Onasanwo, S.A., Fabiyi, T.D., Oluwole, F.S., and Olaley, S.B. (2012). Analgesic and anti-inflammatory properties of the leaf extracts of *Anacardium occidentalis* in the laboratory rodents. *Niger. J. Physiol. Sci.* 27: 65–71.

54 Calixto, J.B., Scheidt, C., Otuki, M., and Santos, A.R.S. (2001). Biological activity of plant extracts: novel analgesic drugs. *Expert Opin. Emerg. Drugs* 6 (2): 261–279.

55 Yimer, T., Birru, E.M., Adugna, M. et al. (2020). Evaluation of analgesic and anti-inflammatory activities of 80% methanol root extract of *Echinops kebericho* M. (Asteraceae). *J. Inflamm. Res.* 13: 647–658.

56 Yoon, S.R., Nah, J.J., Shin, Y.H. et al. (1998). Ginsenosides induce differential antinociception and inhibit substance P induced-nociceptive response in mice. *Life Sci.* 62 (21): PL319–PL325.

57 Mutoh, M., Takahashi, M., Fukuda, K. et al. (2000). Suppression of cyclooxygenase-2 promoter-dependent transcriptional activity in colon cancer cells by chemopreventive agents with a resorcin-type structure. *Carcinogenesis* 21 (5): 959–963.

58 Ior, L.D., Uguru, M.O., Olotu, P.N. et al. (2011). Evaluation of analgesic and anti-inflammatory activities and phytochemical screening of the leaves extract of *Paullinia pinnata* (Sapindaceae). *J. Chem. Pharm. Sci.* 3 (4): 351–356.

59 Owoyele, V.B., Oloriegbe, Y.Y., Balogun, E.A., and Soladoye, A.O. (2005). Analgesic and anti-inflammatory properties of *Nelsonia canescens* leaf extract. *J. Ethnopharmacol.* 99: 153–156.

60 Owemidu, I.O., Ajayi, A.M., and Onasanwo, S.A. (2022). Anti-neuroinflammatory properties of *Waltheria americana* L. leaf in experimental animals. *Phytomed. Plus* 2 (1): 100217.

61 Parveen, Z., Deng, Y., Saeed, M.K. et al. (2007). Anti-inflammatory and analgesic activities of *Thesium chinense* Turcz extracts and its major flavonoids, kaempferol and kaempferol −3-O-glucoside. *Yakugaku Zas shi* 127 (8): 1275–1279.

62 Akindele, A.J., Ibe, I.F., and Adeyemi, O.O. (2012). Analgesic and antipyretic activities of *Drymaria cordata* (Linn.) Willd (caryophyllaceae) extract. *Afr. J. Tradit. Complement. Altern. Med.* 9 (1): 25–35.

63 Jigam, A.A., Akanya, H.O., Ogbadoyi, E.O. et al. (2009). In vivo antiplasmodial, analgesic and anti-inflammatory activities of the leaf extract of *Lippia multiflora* mold. *J. Med. Plant Res.* 3 (3): 148–154.

64 Akindele, A.J. and Adeyemi, O.O. (2006). Analgesic activity of the aqueous leaf extract of *Byrsocarpus coccineus*. *Niger. J. Health Sci.* 5 (1): 43–46.

65 Okpo, S.O., Fatokun, F., and Adeyemi, O.O. (2001). Analgesic and anti-inflammatory activity *of Crinum glaucum* aqueous extract. *J. Ethnopharmacol.* 78: 207–211.

66 Mailafiya, M.M., Pateh, U.U., Iliya, I., and Olorukooba, A.B. (2018). Phytochemical, toxicological, antinociceptive and anti-inflammatory effect of methanol leaf extract of *Leptadenia hastata* (PERS). *Niger. J. Pharm. Biomed. Res.* 3 (2): 1–5.

67 Akefe, I., Yusuf, I., Adamu, A. et al. (2017). Phytochemical investigation and evaluation of analgesic effect of ethanolic leaves extract of *Loranthus micranthus* Linn (Nigerian mistletoe). *Am. J. Phytomed. Clin. Ther.* 5 (3): 23.

68 Ameri, A. (1998). The effects of Aconitum alkaloids on the central nervous system. *Prog. Neurobiol.* 56 (2): 211–235.

69 Rodríguez, I.I., Shi, Y.-P., García, O.J. et al. (2004). New pseudopterosin and seco-pseudopterosin diterpene glycosides from two colombian isolates of *Pseudopterogorgia elisabethae* and their diverse biological activities. *J. Nat. Prod.* 67 (10): 1672–1680.

70 Lanher, M.C., Fleurentin, J., Mortier, F. et al. (2007). Anti-inflammatory and analgesic effects of an aqueous extract of *Harpagophytum procumbens*. *Planta Med.* 58: 117–123.

71 Adedapo, A.A., Sofidiya, M.O., Maphosa, V. et al. (2008). Anti-inflammatory and analgesic activities of the aqueous extract of *Cussonia paniculata* stem Bark. *Rec. Nat. Prod.* 2 (2): 46–53.

72 Backhouse, N., Rosales, L., Apablaza, C. et al. (2008). Analgesic, anti-inflammatory and antioxidant properties of *Buddleja globose,* Buddlejaceae. *J. Ethnopharmacol.* 116 (2): 263–269.

73 Udobang, J.A., Nwafor, P.A., and Okokon, J.E. (2010). Analgesic and antimalarial activities of crude leaf extract and fractions of *Acalypha wilkensiana*. *J. Ethnopharmacol.* 127: 373–378.

74 Ramadan, A., Harraz, F.M., and El-Mougy, S.A. (1994). Anti-inflammatory, analgesic and antipyretic effects of fruit pulp of *Adansonia digitata*. *Fitoterapia* 65 (5): 418–422.

75 Ishola, I.O., Agbaje, E.O., Adeyemi, O.O., and Shukla, R. (2014). Analgesic and anti-inflammatory effects of the methanol root extracts of some selected Nigerian medicinal plants. *Pharm. Biol.* 52 (9): 1208–1216.

76 Nasri, S., Anoush, M., and Khatami, N. (2012). Evaluation of analgesic and anti-inflammatory effects of fresh onion juice in experimental animals. *Afr. J. Pharm. Pharmacol.* 6 (23): 1679–1684.

77 Olajide, O.A., Awe, S.O., Makinde, J.M. et al. (2000). Studies on the anti-inflammatory, antipyretic and analgesic properties of *Alstonia boonei* stem bark. *J. Ethnopharmacol.* 71: 179–186.

78 Iniaghe, L.O., Okpo, S.O., Olung, J.E., and Eguae, A.A. (2012). Analgesic effect of methanol leaf extract of Alstonia boonei De Wild (Apocynaceae). *Trop. J. Pharm. Res.* 11 (5): 793–798.

79 Shorinwa, O.A., Enemuoh, A.O., and Uche, F.I. (2015). Anti-inflammatory and analgesic activities of methanol extracts of stem bark of *Anthocleista djalonensis* in wistar rats. *J. Appl. Pharm Sci.* 5 (11): 117–120.

80 Ngwoke, K.G., Akwagbulam, A.G., Erhirhie, E.O. et al. (2018). Antioxidant, anti-inflammatory, analgesic properties, and phytochemical characterization of stem bark extract and fractions of *Anthocleista nobilis*. *Pharmacognosy Res.* 10 (1): 81–87.

81 Sarfaraz, S. and Ikram, R. (2019). Anti-nociceptive potential of lyophilized *Beta vulgaris* L. (Beet root) powder. *Pak. J. Pharm. Sci.* 32 (2): 529–534.

82 Agunua, A., Abdurahmana, E.M., Shoka, M., and Yusuf, S.A. (2005). Analgesic activity of the roots and leaves extracts of *Calliandra portoricensis*. *Fitoterapia* 76: 442–445.

83 Ibrahim, H., Abdulrahman, E.M., Shok, M. et al. (2007). Comparative analgesic activity of the root bark, stem bark, leaves, fruits, and seeds of *Carissa edulis* VAHL (Apocynaceae). *Afr. J. Biotechnol.* 6 (10): 1233–1235.

84 Miladiyah, I., Dayi, F., and Desrini, S. (2011). Analgesic activity of ethanolic extract of *Manihot esculenta* Crantz leaves in mice. *Univ. Med.* 30: 3–10.

85 Jackson, C., Mbagwu, H., Jackson, I. et al. (2011). Analgesic activities of ethanolic extract of the root of Carpolobia lutea. *Afr. J. Pharm. Pharmacol.* 5 (3): 367–370.

86 Ibironke, G.F. and Ajiboye, K.I. (2007). Studies on the anti-inflammatory and analgesic properties of *Chenopodium ambrosioides* leaf extract in rats. *Int. J. Pharmacol.* 3 (1): 111–115.

87 Owoyele, B.V., Oguntoye, S.O., Dare, K. et al. (2008). Analgesic, anti-inflammatory and antipyretic activities from flavonoid fractions of *Chromolaena odorata*. *J. Med. Plant Res.* 2 (9): 219–225.

88 Salawu, O.A., Chindo, B.A., Tijani, A.Y., and Adzu, B. (2008). Analgesic, anti-inflammatory, antipyretic and antiplasmodial effects of the methanolic extract of *Crossopteryx febrifugea*. *J. Med. Plant Res.* 2 (8): 213–218.

89 Okoli, C.O., Akah, P.A., and Egbuniwe, O.N. (2006). Analgesic activity of leaf extracts of *Culcasia scandens* P. Beauv. *Indian J. Exp. Biol.* 44: 422–424.

90 Ibrahima, B., Sowemimoa, A., van Rooyenb, A., and Van de Venter, M. (2012). Anti-inflammatory, analgesic and antioxidant activities of *Cyathula prostrata* (Linn.) Blume (Amaranthaceae). *J. Ethnopharmacol.* 141: 282–289.

91 Garg, V.K. and Khosa, R.L. (2006). Analgesic and anti-pyretic activity of aqueous extract of *Cynodon dactylon*. *Pharmacologyonline* 3: 12–18.

92 Oyemitan, I.A., Iwalewa, E.O., Akanmu, M.A., and Olugbade, T.A. (2008). Antinociceptive and anti-inflammatory effects of essential oil of *Dennettia tripetala* G.baker (Annonaceae) in rodents. *Afr. J. Trad.* 5 (4): 355–362.

93 Ezeja, M.I., Omeh, Y.S., Ezeigbo, I.I., and Ekechukwu, A. (2011). Evaluation of the analgesic activity of the methanolic stem bark extract of *Dialium guineense* (Wild). *Ann. Med. Res.* 1 (1): 55–62.

94 Akah, P.A., Okogun, J.I., and Ekpendu TO (1993). Antioedema and analgesic actions of *Diodia scandens* extract in rats and mice. *Phytother. Res.* 7: 317–319.

95 Saidu, K., Onah, J., Orisadipe, A. et al. (2000). Antiplasmodial, analgesic, and anti-inflammatory activities of the aqueous extract of the stem bark of *Erythrina senegalensis*. *J. Ethnopharmacol.* 71: 275–280.

96 Aiyelero, O.M., Ibrahim, Z.G., and Yaro, A.H. (2009). Analgesic and anti-inflammatory properties of the methanol leaf extract of *Ficus ingens* (Moraceae) in rodents. *Niger. J. Pharm. Sci.* 8 (2): 79–86.

97 Akuodor, G.C., Idris-Usman, M.S., Mbah, C.C. et al. (2010). Studies on anti-ulcer, analgesic and antipyretic properties of the ethanolic leaf extract of *Gongronema latifolium* in rodents. *Afr. J. Biotechnol.* 9 (5): 2316–2321.

98 Ukwuani, A.N., Abubakar, M.G., Warra, S.H., and Agaie, B.M. (2012). Antipyretic activity of some Nigerian medicinal plants in rats. *Int. J. Pharm. Clin. Res.* 4 (4): 48–51.

99 Olaleye, S.B., Farombi, E.O., Adewoye, E.A. et al. (2000). Analgesic and anti-inflammatory effects of kolaviron (a garcinia kola seed extract). *Afr. J. Biomed. Res.* 3: 171–174.

100 Ali, S.A.E., Mohamed, A.H., and Mohammed, G.E.E. (2014). Fatty acid composition, anti-inflammatory, and analgesic activities of *Hibiscus sabdariffa* Linn. seeds. *J. Adv. Vet. Anim. Res.* 1 (2): 50–57.

101 Igbe, I., Ozolua, R.I., Okpo, S.O., and Obasuyi, O. (2009). Antipyretic and analgesic effects of the aqueous extract of the fruit pulp of *Hunteria umbellata* K Schum (Apocynaceae). *Trop. J. Pharm. Res.* 8 (4): 331–336.

102 Lawal, U., Ibrahim, H., Agunu, A., and Abdulahi, Y. (2010). Anti-inflammatory and analgesic activity of water extract from *Ipomoea asarifolia* Desr (Convolvulaceae). *Afr. J. Biotechnol.* 9 (51): 8877–8880.

103 Uche, F.I., Shorinwa, O.A., Okorie, O. et al. (2011). Antinociceptive and anti-inflammatory activities of crude extracts of *Ipomoea involucrata* leaves in mice and rats. *Asian Pac. J. Trop. Med.* 4 (2): 121–124.

104 Okolo, C.O., Johnson, P.B., Abdurahman, E.M. et al. (1995). Analgesic effect of *Irvingia gabonensis* stem bark extract. *J. Ethnopharmacol.* 45: 125–129.

105 Uche, F.I. and Aprioku, J.S. (2008). The phytochemical constituents, analgesic and anti-inflammatory effects of methanol extract of *Jatropha curcas* leaves in mice and Wistar albino rats. *J. Appl. Sci. Environ. Manage.* 12 (4): 99–102.

106 Omeh, Y.S. and Ezeja, M.I. (2010). Analgesic activity of the methanolic leaf extract of *Jatropha Curcas* (Linn). *Afr. J. Biomed. Res.* 13: 149–152.

107 Owolabi, O.J. and Omogbai, E.K.I. (2007). Analgesic and anti-inflammatory activities of the ethanolic stem bark extract of *Kigelia africana* (Bignoniaceae). *Afr. J. Biotechnol.* 6 (5): 582–585.

108 Adedapo, A., Adewuyi, T., and Sofidiya, M. (2013). Phytochemistry, anti-inflammatory and analgesic activities of the aqueous leaf extract of *Lagenaria breviflora* (Cucurbitaceae) in laboratory animals. *Rev. Biol. Trop.* 61 (1): 281–290.

109 Owoyele, B.V., Olaleye, S.B., Oke, J.M., and Elegbe, R.A. (2001). Anti-inflammatory and analgesic activities of leaf extracts of *Landolphia owariensis*. *Afr. J. Biomed.* 4: 131–133.

110 Sarkiyayi, S., Umaru, H.A., and Onche, H.O. (2015). Anti-inflammatory and analgesic effect of *Leptadenia hastata* on albino rats. *Am. J. Biochem.* 5 (2): 35–41.

111 Adedapo, A.A., Sofidiya, M.O., Maphosa, V. et al. (2009). Anti-inflammatory and analgesic activities of the aqueous extracts of *Margaritaria discoidea* (Euphorbiaceae) stem bark in experimental animal models. *Rev. Biol. Trop.* 57 (4): 1193–1200.

112 Okokon, J.E., Udoh, A.E., Frank, S.G., and Amazu, L.U. (2012). Anti-inflammatory and analgesic activities of *Melanthera scandens*. *Asian Pac. J. Trop. Biomed.* 2 (2): 144–148.

113 Owoyele, B.V., Olaleye, S.B., Oke, J.M., and Elegbe, R.A. (2004). Anti - inflammatory and analgesic activities of *Nothospondias staudtii*. *Niger. J. Physiol. Sci.* 19 (1–2, 102): -105.

114 Paula-Freire, L.I., Andersen, M.L., Molska, G.R. et al. (2013). Evaluation of the antinociceptive activity of *Ocimum gratissimum* L.(Lamiaceae) essential oil and its isolated active principles in mice. *Phytother. Res.* 27 (8): 1220–1224.

115 Aziba, P.I., Adedejia, U.A., Ekor, M., and Adeyemib, O. (2001). Analgesic activity of *Peperomia pellucida* aerial parts in mice. *Fitoterapia* 72: 57–58.

116 Ukwubile, C.A. and Oise, I.E. (2016). Analgesic and anti-inflammatory activity of *Physalis angulata* Linn. (solanaceae) leaf methanolic extract in Swiss albino mice. *Int. J. Biomed. Res.* 2 (4): 167–170.

117 Kabiru, A.Y., Ibikunle, G.F., Innalegwu, D.A. et al. (2016). In vivo anti-plasmodial and analgesic effect of crude ethanol extract of *Piper guineense* leaf extract in albino mice. *Scientifica* 2016: 8687313.

118 Idris, A.O., Fatmat, K., and Adebola, O.O. (2014). Acute toxicity, antinociceptive and anti-inflammatory activity of the essential oil of fresh fruits of *Piper guineense* Schum & Thonn (Piperaceae) in rodents. *J. Med. Plant Res.* 8 (40): 1191–1197.

119 Musa, Y.M., Haruna, A.K., Ilyas, M. et al. (2007). Analgesic and anti-inflammatory effects of the leaf extracts of *Pseudocedrella kotschyii* harms (Meliaceae). *J. Pharmacol. Toxicol.* 2: 542–550.

120 Mallam, D., Anuka, J.A., Zezi, A.U. et al. (2016). Analgesic and anti-inflammatory activities of *Rothmannia longiflora* Salisb in mice and rats. *IOSR J. Pharm.* 6 (8): 1–7.

121 Yerima, M., Magaji, M.G., Yaro, A.H. et al. (2009). Analgesic and anti-inflammatory activities of the methanolic leaves extract of *Securinega Virosa* (Euphorbiaceae). *Niger. J. Pharm. Sci.* 8 (1): 47–53.

122 Konaté, K., Bassolé, I.H.N., Hilou, A. et al. (2012). Toxicity assessment and analgesic activity investigation of aqueous acetone extracts of *Sida acuta* Burn f. and *Sida cordifolia* L. (Malvaceae), medicinal plants of Burkina Faso. *BMC Complement. Altern. Med.* 12 (1): 1–1.

123 Ilodigwe, E.E. and Akah, P.A. (2009). *Spathodea campanulata*: an experimental evaluation of the analgesic and anti-inflammatory properties of a traditional remedy. *Asian J. Med. Sci.* 1 (2): 35–38.

124 Ior, L.D., Otimenyin, S.O., and Umar, M. (2012). Anti-inflammatory and analgesic activities of the ethanolic extract of the leaf of *Syzygium guineense* in rats and mice. *IOSR J. Pharm.* 2 (4): 33–36.

125 Njan, A.A., Adzu, B., Agaba, A.G. et al. (2007). The analgesic and antiplasmodial activities and toxicology of *Vernonia amygdalina*. *J. Med. Food* 11 (3): 574–581.

126 Iwalewa, E.O., Iwalewa, O.J., and Adeboye, J.O. (2003). Analgesic, antipyretic, anti-inflammatory effects of methanol, chloroform, and ether extracts of *Vernonia cinerea* less leaf. *J. Ethnopharmacol.* 86: 229–234.

127 Onzago, R.O., Kiama, S.G., Gakuya, D.W. et al. (2013). Analgesic activity of aqueous extract of *Vernonia hymenolepis* (A. Rich) a traditional medicine plant used in Kenya for toothache. *J. Pharmacol.* 2 (6): 41–45.

128 Mohammed, Z., Shok, M., Ilyas, N. et al. (2007). Analgesic activity of *Waltheria indica* Linn. *Eur. J. Sci. Res.* 16 (1): 6–9.

129 Owoyele, B.V., Negedu, M.N., Olaniran, S.O. et al. (2010). Analgesic and anti-inflammatory effects of aqueous extract of Zea mays husk in male wistar rats. *J. Med. Food* 13 (2): 343–347.

130 Adzu, B., Amos, S., Wambebe, C., and Gamaniel, K. (2001). Antinociceptive activity of *Zizyphus spina-christi* root bark extract. *Fitoterapia* 72: 344–350.

131 Sofowora, A., Ogunbodede, E., and Onayade, A. (2013). The role and place of medicinal plants in the strategies for disease prevention. *Afr. J. Tradit. Complement. Altern. Med.* 10 (5): 210–229.

132 Sofowora, A. (2008). *Medicinal Plants and Traditional Medicine Practice in Africa*. Ibadan: Spectrum Books.

133 Owemidu, I.O., Olubori, M.A., Faborode, O.S. et al. (2018). Anti-nociceptive and anti-inflammatory activities of the methanol extract of *Waltheria americana* Linn. Leaf in experimental animals. *J. Intercult. Ethnopharmacol.* 9 (2): 47–54.

8

Medicinal Plants and Phytochemicals against Depression

Neelma Munir, Ayesha Qamar, Maria Hasnain, Huma Waqif, Maria Hanif, Zirwa Sarwar, and Shagufta Naz

Department of Biotechnology, Lahore College for Women University, Lahore, Punjab, Pakistan

Depression is a typical and severe medical health issue that has harmful effects on how you feel, how you think, and the way you act. It is classified as a mood disorder and is a curable disease. Depression may lead to feelings of sadness or a loss of interest in desired activities. It can cause a variety of emotional, mental, and physical problems and can reduce the ability of a person to function properly in daily activities [1]. A report by the World Health Organization (WHO) suggested that depression leads to the heaviest burden of disability among all mental disorders. The average occurrence of major depression is 4.7% or 1 out of 20 people worldwide. The occurrence is 5.9% in women and 3.8% in men. Depression may affect people of all ages and compromises quality of the life in all areas, from work to social relationships [2].

There is a need to understand that being sad is different from being in depression. The ending of a relationship, loss of a job, or death of a loved one are difficult experiences for a person and it is normal to feel sadness; this is a natural process and it varies from person to person [3]. Depression and grief both involve extreme sadness and are different in important ways. For example, in major depression, interest, mood, or pleasure is reduced for most of the time; in grief, the painful feelings are often intermixed with positive memories of the deceased. In major depression, feelings of self-loathing and worthlessness are common; in grief, self-esteem is commonly maintained. For some people being a victim of physical assault, a major disaster, or losing a job can lead to depression and the death of a loved one can cause major depression for some people [4].

8.1 Causes of Depression

There are several attainable causes of depression, together with faulty mood regulation by the brain, genetic vulnerability, nerve-wracking life events, medications, and medical issues [5].

Phytochemical Drug Discovery for Central Nervous System Disorders: Biochemistry and Therapeutic Effects, First Edition. Edited by Chukwuebuka Egbuna and Mithun Rudrapal.

8.1.1 Genetic Factors

There are different possible causes of depression, which can range from biologic to circumstantial. One common cause is family history: if someone in the family has a history of depression or another related disorder, then the person is at a higher risk for developing depression [6].

8.1.2 Environmental Factors

Early childhood trauma and other events may traumatize a person, which can lead to stress. It also affects the brain structure; that is, if a person's frontal lobe of the brain is less active, then there is a higher risk of depression. However, scientists do not know whether this occurs before or after the beginning of depressive symptoms. Medical conditions like insomnia, attention-deficit hyperactivity disorder, chronic pain, or chronic illness may put a person at higher risk for developing depression, and misuse of alcohol or a history of drug misuse can cause a risk of depression [7].

8.1.3 Risk Factors for Depression

A person who appears to live in ideal circumstances can be affected by depression. Various factors can play a role in depression, such as personality, which includes people with less self-confidence, easily overcome by stress, or feel hopeless and have negative thoughts. Genetic factors can run in families, for instance in the case of identical twins, if one has depression then the other one has a 70% chance of having depression or a related disorder. Biochemical differences in the brain may lead to depressive symptoms. Environmental factors also play a significant role in causing depression, for example poverty, abuse, neglect, or continuous exposure to violence may make some people susceptible to depression [8]. Other factors for depression may include certain medications, a personal history of mental illness, being self-critical or having low self-esteem, and stressful events such as economic problems, loss of a loved one, or divorce [9].

8.2 Symptoms of Depression

There are various symptoms of major depression. Some affect the body, while others affect a person's mood. Depression can have multiple effects on men, women, and children.

8.2.1 Symptoms of Depression in Men

Depression symptoms in men include mood (aggressiveness, anger, irritability, restlessness, anxiousness), behavioral (loss of interest in daily activities, feeling tired easily, drinking excessively, thoughts of suicide, engaging in high-risk activities, using drugs), emotional (feeling sad, empty, hopeless), cognitive (difficulty completing tasks, inability to concentrate, late responses during conversations), physical (headache, fatigue, digestive problems, pains), and sleep related (restless sleep, insomnia, not sleeping through the night, excessive sleepiness) [10].

8.2.2 Symptoms of Depression in Women

Depression symptoms in women may include mood (irritability), behavioral (withdrawing from social engagements, loss of interest in activities, thoughts of suicide), emotional (feeling empty or sad, anxious or hopeless), cognitive (talking or thinking more slowly), physical (greater fatigue, decreased energy, changes in appetite, pain, weight change, increased cramps, aches, headaches), and sleep related (waking early, difficulty sleeping through the night, sleeping too much) [11].

8.2.3 Symptoms of Depression in Children

Depression symptoms in children may include mood (anger, irritability, crying, mood swings), behavioral (refusal to go to school or getting into trouble at school, avoiding siblings or friends, thoughts of suicide or death), emotional (feelings of lack of skill, for example "I can't do anything right", intense sadness, crying), cognitive (decline in school performance, difficulty concentrating, changes in grades), physical (digestive problems, loss of energy, changes in appetite, weight gain or loss), and sleep related (sleeping too much or difficulty sleeping) [12].

8.3 Diagnosis of Depression

A person has depression when they have five or more of the following symptoms for at least two weeks:

- Depressed mood throughout the day, especially in the morning.
- Feeling worthless or guilty almost daily.
- Feeling tired or having a lack of energy almost daily.
- Having a hard time focusing, making decisions, and remembering details.
- Taking almost no pleasure or interest in many activities almost every day.
- No sleep or sleeping too much almost daily.
- Often thinking about suicide or death (no fear of death).
- Feeling restless.
- Weight gain or weight loss [13].

The diagnosis may also include irritability, losing pleasure in life, not feeling hungry or overeating, have pains or aches, cramps, headaches, or digestive problems, or having anxious, sad, or "empty" feelings.

Not everyone with = depression will have the same signs, as these symptoms are common. How often they happen, how severe they are, and how long they last can vary [14].

8.4 Types of Depression

Depression can be divided into classifications depending on the severity of the symptoms, as some people have temporary and mild episodes, whereas others have severe and ongoing depressive episodes. Depression is divided into two main types: major depressive disorder and persistent depressive disorder (PDD) [15]. The other types are unipolar major depression, disruptive mood dysregulation disorder (usually teens and children get extremely angry or irritable), premenstrual dysphoric disorder (usually a woman has mood problems before her period and more intense than typical premenstrual syndrome [PMS]), substance-induced mood disorder (SIMD; symptoms happen when someone is taking a drug or after they stop), depressive disorder (which may be due to another medical condition), and other depressive disorders (such as minor depression) [16].

Depression can also have other particular features such as anxious distress (a person worries about losing control or about things that may happen), mixed features (having both depression and manic periods of high energy, high self-esteem, and talking too much), atypical features (a person feels good after happy occasions, but also feels hungrier, more sleepy, and is sensitive to rejection), psychotic features (believing things that are not true, or hearing and seeing things that are not there), catatonia (being unable to move body normally, unresponsive or uncontrolled movements), peripartum depression (symptoms begin during the pregnancy or after birth), or a seasonal pattern (symptoms become worse with variations in the seasons, especially in the colder, darker months) [17].

8.4.1 Major Depressive Disorder

Major depressive disorder is a very severe kind of depression. It is described by persistent feelings of hopelessness, worthlessness, and sadness. To be diagnosed with clinical depression, one must experience five or more of the following signs over a two-week period: feeling depressed most of the day; significant weight gain or loss; loss of interest in most regular activities; not being able to sleep or sleeping too much; low energy or fatigue most days; slowed thinking or movement; feelings of guilt or worthlessness; recurring thoughts of suicide or death; loss of indecisiveness or concentration. There are also different subcategories of major depressive disorder and what the American Psychiatric Association states as "specifiers," which include atypical features, mixed features, anxious distress, peripartum onset, during pregnancy or right after birth, melancholic features, seasonal patterns, catatonia, and psychotic features [18].

8.4.2 Persistent Depressive Disorder

PDD is also called dysthymia. It is a milder but chronic type of depression. To diagnose PDD, the symptoms must last for at least two years. As it lasts for a longer period, so PDD can affect a person more than major depression. In someone with PDD it is common to lose interest in normal daily activities, feel hopeless, have low self-esteem, and lack productivity. Depression can be successfully treated, but it is important to stick to the treatment plan [19].

8.4.3 Depression with Psychosis

Some people who have been diagnosed with major depression can also have signs of another mental disorder, psychosis. When these two conditions occur together, it is known as depressive psychosis. This disorder causes people to believe, see, smell, or hear things that are not real; people with depressive psychosis can also experience feelings of irritability, hopelessness, and sadness [20]. Someone with depressive psychosis can experience delusions that lead them to take unusual risks or to be suicidal, which is why the combination of these two situations is particularly dangerous. It is unclear why they occur together or what the cause of these two conditions is. Treatment can successfully cure the symptoms. Treatments include electroconvulsive therapy and medications. Understanding the possible causes and risk factors may help to be aware of the early symptoms [21].

8.4.4 Depression in Pregnancy

Pregnancy is usually an exciting time for people, but experiencing depression is common for an expectant mother. The symptoms of depression during pregnancy include altered eating habits or changes in appetite, feeling hopeless, losing interest in activities and things that were previously enjoyed, anxiety, difficulty remembering or concentrating, persistent sadness, sleep problems including sleeping too much or insomnia, and thoughts of suicide or death. The treatment for depression during pregnancy may focus on talk therapy or other natural treatments. Although some women also take antidepressants during pregnancy, it is not clear which are the safest. The doctor can encourage the woman to try an alternative option until after the baby is delivered, but the risks for depression do not end when the baby arrives. Major depressive disorder is a major concern for new mothers and recognizing the signs can help to spot a problem and seek help before it becomes irresistible [22].

8.5 Treatment of Depression

Depression is one of the most treatable psychologic disorders. Between 80% and 90% of depressed people ultimately respond well to treatment and almost all patients get some relief from their symptoms (Figure 8.1). Before diagnosis or treatment, a health professional must perform a complete diagnostic evaluation, including an interview and also a physical

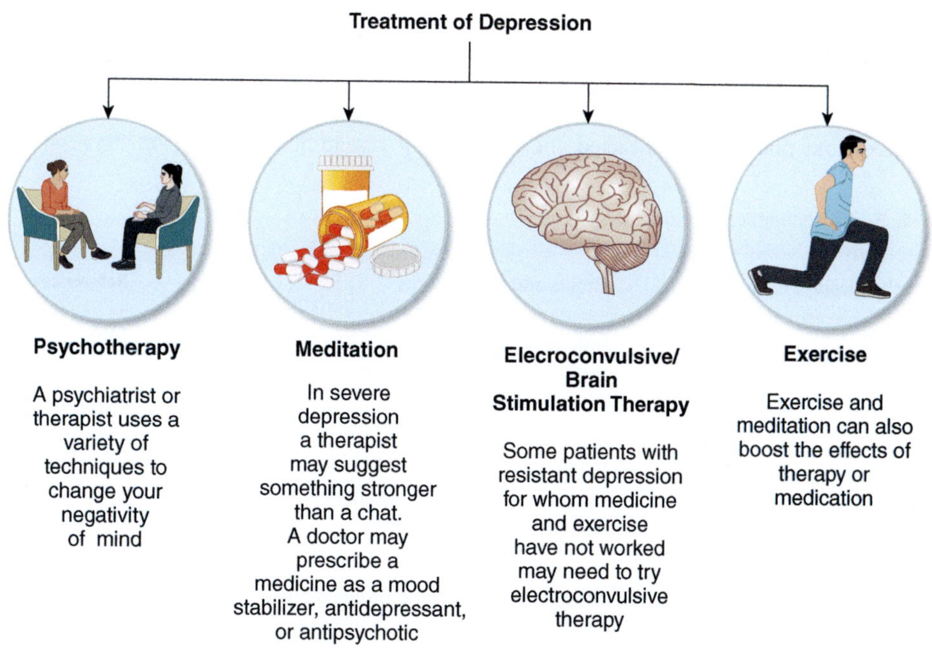

Figure 8.1 Treatment of depression. *Source:* Nastasic/DigitalVision Vectors/Getty Images.

examination. Sometimes a blood test can be done to make sure that the depression is not due to a medical condition such as a thyroid problem. Evaluation consists of identifying specific symptoms, family and medical history, environmental factors, and cultural factors to reach a diagnosis and plot a course of action. Living with depression can be difficult, but the quality of life can be improved by treatment.

The traditional treatment of depression uses a combination of prescription drugs and counseling. The leading medical treatment for depression is antidepressants. There is a lot of information on antidepressant drugs and, while there is no clear clarification on how they work, they can be useful in the treatment of moderate to severe depression and some anxiety disorders.

8.5.1 Natural Treatment for Depression

There are also complementary or alternative treatments that can be tried. Table 8.1 describes the effects and mechanisms of different phytochemicals against depression. The doctor can discuss the possible options with the patient. A person can control symptoms with just one form of treatment, or may find that a combination of treatments works better. It is very common to combine medical treatments and lifestyle therapies [29].

Different plants and formulations have been used to cure depression for thousands of years. It is important to remember that several natural treatments have been studied that show their effects on depression, good or bad. The US Food and Drug Administration (FDA) does not approve many of the food supplements on the market in the United States, so it is important to make sure that products are from a reliable brand. The person should talk to a doctor before taking supplements as part of a treatment plan [30].

8.5.2 Medicinal Plants as Antidepressants

Even though a variety of synthetic medicines are being used as treatment for the clinically depressed patient, they have side effects that may compromise satisfactory treatment [30]. Currently researchers are searching for more specific drugs with greater security and lower cost. Medicinal plants have attracted the attention of researchers working on depression

Table 8.1 Phytochemicals against depression.

Phytochemicals	Dose	Effects and mechanisms	References
Carvacrol	12.5 mg/kg	Induces antidepressant effects Influences neuronal activity through modulation of neurotransmitters	[23]
Curcumin	1.25–10 mg/kg	Reverses bilateral olfactory bulbectomy-induced hyperactivity in the open field	[24]
	20–40 mg/kg	Restores biochemical and behavioral changes induced by chronic stress	[25]
Ferulic acid	100–250 mg/kg	Attenuates stress-induced behavior	[26]
L-Theanine	1–20 mg/kg	Antagonizes reserpine-induced ptosis and hypothermia	[27]
Proanthocyanidin	25–50 mg/kg	Reduces immobility period	[28]

because these plants have long been used to cure various diseases, including psychiatric disorders, and cause fewer side effects than chemical and synthetic drugs [31]. A variety of medicinal plants and the medicines derived from them have exhibited antidepressant properties by virtue of the combined impact of their medicinal constituents. It is valuable to search for antidepressants from plants with proven advantages and a positive benefit-to-risk ratio [32].

The authors conducted a search of PubMed articles in English from 1995 to December 2015 on plants and essential oils that have antidepressant activity. The plants have been identified with the chemical name and the name of their bioactive complexes. Table 8.2 demonstrates the antidepressant activity of herbal plants and their parts. The reason for selecting some plant species was to estimate the antidepressant activity and identify the mechanism of action of the oils or particular components of oils. The plants that proved to have antidepressant properties are shown in Figure 8.2.

8.5.2.1 *Lavandula officinalis*

Lavandula officinalis is a herb. It has 39 different species and exhibits a wide distribution. The flower and the oil of lavender are used to make medicine. The inhaled *L. officinalis* extract causes a rise in plasma levels of the monoamine neurotransmitter and its metabolites. Its smell causes calmness and sleep. Lavender oil is available in an oral form, collected into microscopic bubbles and enclosed in a capsule that allows it to cross the viscus barrier. Once it does, it causes calm and decreases anxiety. It is marketed as Lavela and it is not dangerous. Lavender oil has been in use as sprays, oils, lotions, and as an inhalant for many years [39].

Among all psychiatric disorders, the most common is anxiety disorders, which have debilitating effects on individual function. Available pharmacotherapies for anxiety disorders are limited by delay to therapeutic effect, withdrawal, tolerance, dependence, and abuse potential [32].

Silexan is an orally administered lavender oil preparation that is effective in the cure of subsyndromal anxiety disorder. The anxiolytic efficacy of lavender oil has been examined, and orally administered Silexan has been studied in generalized anxiety disorder (GAD) compared to placebo and paroxetine. Silexan indicated a pronounced antidepressant

Table 8.2 Antidepressant activity of herbal plants and their parts.

Herbal plant	Part used	Uses	References
Glycyrrhiza uralensis (Mulethi)	Root	Anti-inflammatory, antiviral, and antiulcer	[33]
Lafoensia pacari (Didal)	Leaves	Antibacterial, anti-inflammatory, and antinociceptive	[34]
Schinus molle L. (Peruvian pepper tree)	Leaves	Antioxidant and anticancer activity	[35]
Tabebuia Avellaneda (Lapacho)	Bark	Antitumor, anti-inflammatory, and antimicrobial	[36]
Bupleurum falcatum (Chai Hu)	Root	Anti-inflammatory and antitussive	[37]
Piper laetispicum (Piper)	Stem and root	Antinociceptive activity; analgesic, insecticidal, anti-inflammatory, and antitumor	[38]

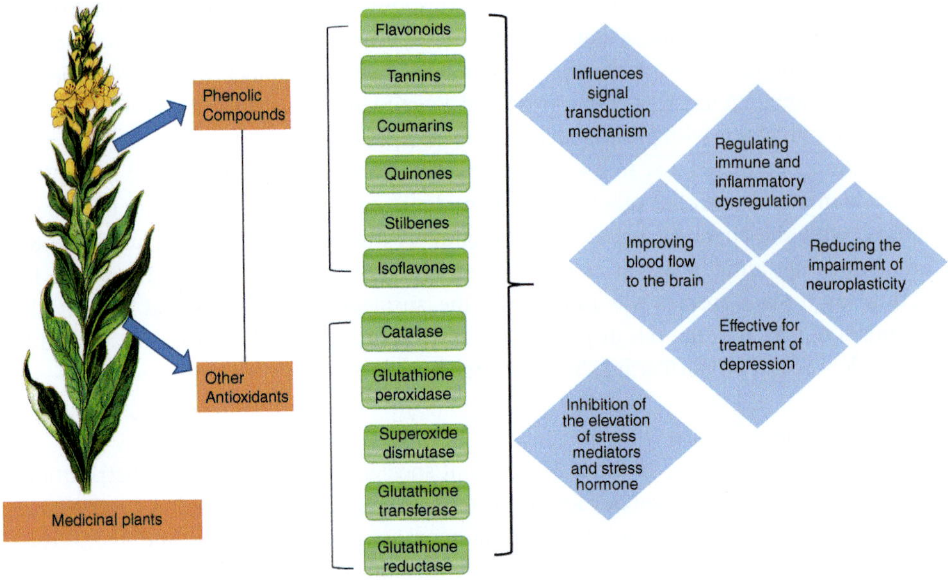

Figure 8.2 Importance of medicinal plants in depression.

outcome and an improvement in general mental health and health-related quality of life. In GAD, Silexan was more effective than placebo. The treatment exposure rates for Silexan were comparable to those of placebo [40].

Lavender essential oil showed a significant reduction in anxiety, tobacco cravings, systolic blood pressure, and heart rate compared to placebo. Much research aims to evaluate the acute effects of inhaled administration of lavender in adult smokers with withdrawal syndrome. The experimental substances demonstrated an efficacy similar to conventional tobacco treatment in relation to the control of anxiety and desire [41].

The two primary terpenoid ingredients of lavender essential oil, linalyl acetate and linalool, might produce an anxiolytic effect in combination via inhibition of voltage-dependent calcium channels, reduction of 5-HT1 (serotonin), and activity of receptors. It has been observed that lavender oil has a significant use in aromatherapy. It will be useful to explore variability in the ingredients of lavender oil and its safety profile and alongside that to describe that body of research that has been conducted for anxiety [42].

8.5.2.2 *Kaempferia parviflora*

Kaempferia parviflora is a herbal plant. Its extracts are commonly used as alternative medicines. It is widely used against inflammation and peptic ulcers, and as an antimicrobial. Its common names are Thai black ginger, Thai ginseng, or krachai dum and it is a herbaceous plant in the family Zingiberaceae, native to Thailand. *K. parviflora* has been the topic of increased scientific interest in recent years [43].

It was observed that *K. parviflora*, a health-promoting herb, has traditionally been used to treat a variety of diseases. Pharmacologic studies have affirmed the various benefits of *K. parviflora* and its effective methoxyflavones, including regulatory activity of the cellular metabolism, antitumor activity, vascular relaxation and cardioprotective activity, sexual activity, neuroprotective activity, antiallergic activity, anti-inflammatory and antioxidant activity, and antimicroorganism activity. The molecular mechanisms underlying *K. parviflora* and its methoxyflavones are still under investigation. The clinical applications of *K. parviflora* and its methoxyflavones might be limited due to their low bioavailability. A review exhaustively discussed the biologic activities of *K. parviflora* and its methoxyflavones [44].

8.5.2.3 *Asparagus racemosus*

Asparagus racemosus is a medicinal plant well known for its varied application in Ayurveda. It contains different types of secondary metabolites of natural active products (or ingredients), such as essential oils, steroids, flavonoids, alkaloids, and furan derivatives. In the current scenario due to the increasing catastrophic effect of diseases on human life, researchers are working to replace synthetic drugs with herbal medicines. In general, herbal drug treatments are used to provide public and front-line health services, both for people living in remote areas where they are the only social assistance service available, and for people living in areas where only financial aid is offered. From this perspective, *A. racemosus* can exhibit a new herbal treatment pathway [45].

A. racemosus belongs to the Liliaceae family, is generally known as Satamuli, Satawar, and Shatavari, and is found at low altitudes throughout India. The dried root part of the plant is used as a drug. The plant has a wide range of secondary metabolites, including steroids, alkaloids, flavonoids, dihydrophenanthrene derivatives, furan derivatives, and essential oils [46].

A. racemosus is an Ayurvedic rasayana that is used as an adaptogen. Adaptogenic medications are those that are helpful as antidepressant agents by stimulating non-specific opposition of the body. Therefore, investigations have evaluated the antistress impact of methanolic extract of roots (MAR) of *A. racemosus* standardized to saponins, and the results indicated that MAR reduces immobility in the forced swim test (FST) and enhances the dodging response in the learned helplessness test (LH). Thus, MAR has important medicinal drug activity, a result that may be mediated through the serotonergic and noradrenergic systems and augmentation of the antioxidant defenses [47].

So far there has been no convincing evidence to justify these uses of *A. racemosus* root extract. However, recently a few reports have become available that demonstrated the beneficial impacts of the aqueous and alcoholic extract of *A. racemosus* roots in a few clinical conditions and experimentally induced diseases, for example antihepatotoxic, immunomodulatory effects, immunoadjuvant effect, antilithiatic impact, and teratogenicity. The plant is known for its many applications in Ayurveda, such as hypertension, hyperlipidemia, angina, dysmenorrhea, benign prostatic hyperplasia (BPH), anxiety disorders, leukorrhea, and urinary tract infections [48].

8.5.2.4 *Passiflora foetida*

Passiflora foetida (family Passifloraceae) is a species of passionflower also called stinking passionflower, bush passion fruit, maracuja, or wild water lemon. A herbal climbing plant, it is native to Central, North, and South America and has been wide employed in Mexican ancient medication for the management of various disorders of the central system (CNS). Researchers have examined the mood-elevating impacts of *P. foetida* in animals [49]. However, the scientific information available on this species of plant is rare and there have been no reports of its possible impacts on the CNS [50].

8.5.2.5 *Momordica charantia*

Momordica charantia Linn. (Family Cucurbitaceae) is also known as bitter melon, balsam pear, and pare, and grows in tropical areas of the Amazon, East Africa, Asia, India, South Africa, and the Caribbean. It has been employed historically as a food and a medicine. It is known as karela in Bengali and Hindi, bitter gourd in English, Karke in Marathi, pavakay in Tamil, and kaippa or kaippa-valli in Malayalam [51].

8.5.2.6 *Bacopa monniera*

Bacopa monniera is an ordinarily used Ayurvedic drug for mental disorders. It has been studied as a traditional memory enhancer and also has testified antidepressant properties. The standardized extract has been reported to have important antioxidant impact and anxiolytic activity [52].

8.5.2.7 *Valeriana officinalis*

Valerian root works as a gentle antidepressant drug and has been exhibited a sedative impact in testing. The extract has been shown to possess sleep-enhancing qualities. Valerian root is the best herb for sleep, according to analysis. It is conjointly a good antianxiety herb, but, in contrast to lavender or L-theanine, it will cause tiredness. It is a good possibility for severe anxiety spells and for evening use, although after taking it the person should not be driving or operating machinery. Valerian is often taken throughout the day in very small doses, 20 mg or less. In fact, it is widely available on the market in "sleep teas" at a dose of 20 mg [53].

8.5.2.8 *Rhodiola rosea*

Rhodiola rosea, generally referred to as Arctic root or golden root, is associated with being an adaptogenic herb, which means that it acts in non-specific ways to extend resistance to anxiety. It grows at high altitudes within the Arctic areas of Europe and Asia, and its root has been utilized in ancient drugs in Russia and the Scandinavian countries for hundreds of years. Studies of *R. rosea*'s medicative applications have appeared within the scientific literature in Scandinavia, France and Germany. *R. rosea* is widely utilized in Russia as a tonic and remedy for fatigue, poor attention span, and reduced memory.

Rhodiola is one of the essential plants studied for its therapeutic properties in ancient times and some of the most famous species of *Rhodiola* are *R. rosea*, *R. heterodonta*, *R. imbricata*, and *R. quadrifida*. These species were known to possess potent pharmacologic and biologic activities against leprosy, as well as being antioxidant, antistress, and anti-inflammatory. The plants grow at a height of about 4000–5000 m above sea level at about 10 °C, therefore they survive in harsh conditions, which is due to their adaptation to that environment as well as the kind of compounds these plants produce as part of their biologic mechanism [54].

8.5.2.9 *Withania somnifera*

Withania somnifera is also known as Ashwagandha and has proven effectiveness in lowering stress hormones, promoting intellect and reminiscence, and being defensive in opposition to neuronal injury. It has been observed that Ashwagandha is powerful as an anxiolytic, as a mood stabilizer, and in the treatment of obsessive compulsive disease [55].

The steroidal lactones known as withanolides obtained from the plant roots have been associated with a wide variety of beneficial therapeutic events and sustaining general vigor: as a memory enhancer, to promote good physical and mental health, immunomodulation, fighting infectious causes, anticancer and antiepileptic, as a general tonic and rejuvenator, mood elevator, diuretic, stress reliever, cardiorespiratory endurance enhancer, antioxidant, antiaging, hypoglycemic, hypocholesterolemic, and as an active adaptogen. Withanolides and lactones are the active elements of the plant. Its antistress and radiosensitizing activities, valuable properties in relation to the cardiovascular system and sexual behavior, therapeutic activities against neurodegenerative syndromes and injury due to venoms and other substances (including snake poison) make this plant a gem of Mother Nature [56].

8.5.2.10 *Matricaria recutita*

Matricaria recutita is popularly known as chamomile and is specifically advantageous for those suffering with both depression and anxiety. There are two main species of chamomile used medicinally: German or wild chamomile (*M. recutita*) and Roman or English chamomile (*Chamaemelum nobile*). German chamomile is more often used, but both of them offer comparable health benefits.

8.5.2.11 *Ginkgo biloba*

The extract of *Ginkgo biloba* is widely used in Europe for multiple medicinal purposes. In the United States, ginkgo is offered as a nutritional complement, commonly as the standardized extract of *G. biloba* crafted from the dried leaves of the plant [57].

Currently, the standardized extracts are used to treat a huge variety of situations, including memory loss, attention problems, psychological confusion, melancholy, anxiety, dizziness, tinnitus, and headache. Ginkgo helps to increase the blood supply, decreasing blood viscosity, boosting neurotransmitters, and lowering harmful free radicals. Ginkgo extract can also lessen signs of anxiety and stress by reducing levels of the stress hormone cortisol [58].

8.6 Conclusion

Depression is a popular disease closely related to stress, in particular mood disorder and psychiatric anxiety disorders. Several medications are available for the treatment of depression, but the clinical evaluation of these medications has revealed drug interactions, incidence of recurrence, and side effects. This was the reason for the development of new antidepressants in which herbal drugs are included. Herbal plants have been reported to be effective in treating various health problems. Efforts are directed toward the characterization and isolation of active ingredients. Moreover, detailed investigation of the active components of natural medicines should be oriented to clinical relevance. The study of plants with potential properties for the CNS has increased, especially in preclinical in vivo and in vitro studies to validate plant protection products, in particular to treat anxiety and depression. These studies have delivered evidence of advantageous therapeutic activity in recent decades. Plant-based originations can be efficiently useful for the treatment of different cases of depression from mild to moderate and with very few side effects. So it can be concluded that medicinal plants are a very rich source of constituents that are responsible for enhancing antidepressant activity.

References

1 Rahnavard, F., Modaresi, M., and Farhadi, H. (2018). The comparative effects of chamomile's hydro alcoholic extract and imipramine on decreasing depression in mice. *World Fam. Med.* 16 (1): 198–203.

2 Bakhshaei, S. (2017). Phyto-pharmacological effect of nine medicinal plants as a traditional treatment on depression. *J. Appl. Pharm.* 9 (3): 1–5.

3 Nutt, D.J. (2008). Relationship of neurotransmitters to the symptoms of major depressive disorder. *J. Clin. Psychiatry* 69 (Suppl E1): 4–7.

4 Patocka, J. and Jakl, J. (2010). Biomedically relevant chemical constituents of *Valeriana officinalis*. *J. Appl. Biomed.* 8 (5): 11–18.

5 Becker, A., Felgentreff, F., Schroder, H. et al. (2014). The anxiolytic effects of a valerian extract is based on valernic acid. *BMC Complement. Altern. Med.* 14 (5): 1–5.

6 Bangratz, M., Abdellah, S.A., Berlin, A. et al. (2018). A preliminary assessment of a combination of rhodiola and saffron in the management of mild–moderate depression. *Neuropsychiatr. Dis. Treat.* 14 (3): 1821–1829.

7 Dey, A., Chatterjee, S.S., and Kumar, V. (2016). Low dose effects of a *Withania somnifera* extract on altered marble burying behavior in stressed mice. *J. Intercult. Ethnopharmacol.* 5 (3): 274–277.

8 Amsterdam, J.D., Shults, J., Soeller, I. et al. (2012). Chamomile (*Matricaria recutita*) may have antidepressant activity in anxious depressed humans – an exploratory study. *Altern. Ther. Health Med.* 18 (5): 44–49.

9 Dhingra, D. and Sharma, A. (2006). A review on antidepressant plants. *Nat. Prod. Radian.* 5 (2): 144–152.

10 Hamidpour, R., Shahlari, M., Sohraby, M. et al. (2015). *Rhodiola rosea* from the selection of traditional applications to the novel phytotherapy for the prevention and treatment of serious diseases. *Glob. J. Med. Res.* 15 (3): 1–11.

11 Benke, D., Barberis, A., Kopp, S. et al. (2009). GABA (A) receptors as in vivo substrate for the anxiolytic action of valerenic acid, a major constituent of valerian root extracts. *Neuropharmacology* 56 (1): 174–181.

12 Singh, G., Sharma, P.K., Dudhe, R., and Singh, S. (2010). Biological activities of *Withania somnifera*. *Ann. Biol. Res.* 1 (3): 56–63.

13 Beji, R.S., Khemir, S., Wannes, W.A. et al. (2018). Antidiabetic, antihyperlipidemic and antioxidant influences of the spice cinnamon (*Cinnamomum zeylanicumon*) in experimental rats. *Braz. J. Pharm. Sci.* 54 (2): 1–8.

14 Amsterdam, J.D., Li, Y., Soeller, I. et al. (2009). A randomized, double-blind, placebo-controlled trial of oral *Matricaria recutita* (chamomile) extract therapy of generalized anxiety disorder. *J. Clin. Psychopharmacol.* 29 (4): 378–382.

15 Anghelescua, I.-G., Edwardsb, D., Seifritzc, E., and Kasperd, S. (2018). Stress management and the role of *Rhodiola rosea*: a review. *Int. J. Psychiatry Clin. Pract.* 22 (4): 242–252.

16 Bystritsky, A., Kerwin, L., and Feusner, J.D. (2008). A pilot study of *Rhodiola rosea* (rhodax) for generalized anxiety disorder (GAD). *Altern. Complem. Med.* 14 (2): 175–180.

17 Auddy, B., Hazra, J., Mitra, A. et al. (2008). A standardized *Withania somnifera* extract significantly reduces stress-related parameters in chronically stressed humans: a double-blind, randomized, placebo-controlled study. *JANA* 11 (1): 50–56.

18 Chiang, H.-M., Chen, H.-C., Wu, C.-S. et al. (2015). Rhodiola plants: chemistry and biological activity. *J. Food Drug Anal.* 23 (10): 359–369.

19 Bhattacharya, S.K. and Muruganandam, A.V. (2003). Adaptogenic activity of *Withania somnifera*: an experimental study using a rat model of chronic stress. *Pharmacol. Biochem. Behav.* 7 (5): 547–555.

20 Crosta, P. (2018). What you need to know about Lyrica (pregabalin). *Med. News Today* 3 (7): 1–14.

21 Dai, C.-X., Hu, C.-C., Shang, Y.-S., and Xie, J. (2018). Role of *Ginkgo biloba* extract as an adjunctive treatment of elderly patients with depression and on the expression of serum S100B. *Medicine* 97 (39): 1–8.

22 Gutierreza, S.L.G., Chilpab, R.R., and Jaimea, H.B. (2014). Medicinal plants for the treatment of "nervios", anxiety, and depression in Mexican traditional medicine. *Braz. J. Pharmacogn.* 24 (9): 591–608.

23 Holzmann, I., Filho, V.C., Mora, T.C. et al. (2011). Evaluation of behavioral and pharmacological effects of hydroalcoholic extract of valerian a prionophylla Standl. from Guatemala. *Evid. Based Complement. Altern. Med.* 4 (9): 1–10.

24 Xu, Y., Ku, B.-S., Yao, H.-Y. et al. (2005). Antidepressant effects of curcumin in the forced swim test and olfactory bulbectomy models of depression in rats. *Pharmacol. Biochem. Behav.* 82 (1): 200–206.

25 Galdino, P., Nascimento, M., Sampaio, B. et al. (2009). Antidepressant-like effect of *Lafoensia pacari* A. St.-Hil. ethanolic extract and fractions in mice. *J. Ethnopharmacol.* 124 (3): 581–585.

26 Machado, D.G., Bettio, L.E., Cunha, M.P. et al. (2008). Antidepressant-like effect of rutin isolated from the ethanolic extract from *Schinus molle* L. in mice: evidence for the involvement of the serotonergic and noradrenergic systems. *Eur. J. Pharmacol.* 587 (1–3): 163–168.

27 Freitas, A.E., Budni, J., Lobato, K.R. et al. (2010). Antidepressant-like action of the ethanolic extract from *Tabebuia avellanedae* in mice: evidence for the involvement of the monoaminergic system. *Prog. Neuro-Psychopharmacol. Biol. Psychiatry* 34 (2): 335–343.

28 Kwon, S., Lee, B., Kim, M. et al. (2010). Antidepressant-like effect of the methanolic extract from *Bupleurum falcatum* in the tail suspension test. *Prog. Neuro-Psychopharmacol. Biol. Psychiatry* 34 (2): 265–270.

29 Cropley, M., Banks, A.P., and Boyle, J. (2015). The effects of *Rhodiola rosea* L. extract on anxiety, stress, cognition and other mood symptoms. *Phytother. Res.* 4 (9): 1–6.

30 Yao, C., Wang, J., Dong, D. et al. (2009). Laetispicine, an amide alkaloid from *Piper laetispicum*, presents antidepressant and antinociceptive effects in mice. *Phytomedicine* 16 (9): 823–829.

31 Murphy, K., Kubin, Z.J., Shepherd, J.N., and Ettinger, R.H. (2010). *Valeriana officinalis* root extracts have potent anxiolytic effects in laboratory rats. *Phytomedicine* 17 (5): 674–678.

32 Namjou, A., Yazdani, N., Abbasi, E., and Rafieian-Kopaei, M. (2018). The antidepressant activity of *Matricaria chamomilla* and *Melissa officinalis* ethanolic extracts in non-reserpinized and reserpinized balb/c mice. *Jundishapur J. Nat. Pharm. Prod.* 13 (4): 1–6.

33 Perfumi, M. and Mattioli, L. (2007). Adaptogenic and central nervous system effects of single doses of 3% rosavin and 1% salidroside *Rhodiola rosea* L. extract in mice. *Phytother. Res.* 21 (5): 37–43.

34 Mirjalili, M.H., Moyano, E., Bonfill, M. et al. (2009). Steroidal lactones from *Withania somnifera*, an ancient plant for novel medicine. *Molecules* 14 (5): 2373–2393.

35 Kaur, P., Mathur, S., Sharma, M. et al. (2001). A biologically active constituent of *Withania somnifera* (ashwagandha) with anti-stress activity. *Indian J. Clin. Biochem.* 16 (2): 195–198.

36 Hattesohl, M., Feistel, B., Sievers, H. et al. (2008). Extracts of *Valeriana officinalis* L. s.l. show anxiolytic and antidepressant effect but neither sedative nor myorelaxant properties. *Phytomedicine* 15 (7): 2–15.

37 Hazra, S., Banerjee, R., Das, B.K. et al. (2012). Evaluation of antidepressant activity of *Bacopa monnieri* in rat: a study in animal model of depression. *Drug Des. Discov.* 2 (4): 8–13.

38 Ali, M.S., Mukherjee, S., Roy, D. et al. (2018). *Asparagus racemosus*, a climbing ayurvedic medicinal plant: review on its cultivation, morphology and medicinal significance. *Pharma Tutor.* 6 (12): 46–54.

39 Ismail, H., Amanat, M.A., Iqbal, A., and Mirza, B. (2018). Medicinal plants: a complementary and alternative antidepressant therapy. *Curr. Pharm. Des.* 24 (22): 2609–2624.

40 Jayanthi, M.K., Prathima, C., Huralikuppi, J.C. et al. (2012). Anti-depressant effects of *Withania somnifera* fat (ashwagandha ghrutha) extract in experimental mice. *Int. J. Pharma Bio Sci.* 3 (1): 1–11.

41 Jawaid, T., Gupta, R., and Siddiqui, Z.A. (2011). A review on herbal plants showing antidepressant activity. *Pharm. Sci. Res.* 2 (12): 3051–3060.

42 Barrington, R. (2019). *Passiflora foetida* (passionflower): antidepressant effects. *J. Nutr.* 3 (4): 1–3.

43 Bhutani, M.K., Bishnoi, M., and Kulkarni, S.K. (2009). Anti-depressant like effect of curcumin and its combination with piperine in unpredictable chronic stress-induced behavioral, biochemical and neurochemical changes. *Pharmacol. Biochem. Behav.* 92 (1): 39–43.

44 Jain, S., Shukla, S.D., Sharma, K., and Bhatnagar, M. (2001). Neuroprotective effects of *Withania somnifera* Dunn. in hippocampal sub-regions of female albino rat. *Phytother. Res.* 15 (3): 544–548.

45 Jowko, E., Sadowski, J., Dlugolecka, B. et al. (2018). Effects of *Rhodiola rosea* supplementation on mental performance, physical capacity, and oxidative stress biomarkers in healthy men. *J. Sport Health Sci.* 7 (16): 473–480.

46 Kamble, S. and Rambhimaiah, S. (2013). Antidiabetic activity of aqueous extract of *Cinnamomum cassia* in alloxan-induced diabetic rats. *Biomed. Pharmacol. J.* 6 (1): 83–88.

47 Khanum, F., Bawa, A.S., and Singh, B. (2005). *Rhodiola rosea*: a versatile adaptogen. *Compreh. Rev. Food Sci. Food Saf.* 4 (3): 1–8.

48 Khurshid, F., Govindasamy, J., Khalilullah, H. et al. (2017). Effect of herb-drug interactions of *Bacopa monnieri* Linn. (Brahmi) formulation on the pharmacokinetics of amitriptyline in rats. *Pharm. Sci.* 53 (4): 1–10.

49 Kong, Y., Wang, T., Wang, R. et al. (2017). Inhalation of roman chamomile essential oil attenuates depressive-like behaviors in wistar Kyoto rats. *Sci. China Life Sci.* 60 (6): 647–655.

50 Kizilaslan, N. and Erdem, N.Z. (2019). The effect of different amounts of cinnamon consumption on blood glucose in healthy adult individuals. *Int. J. Food Sci.* 9 (2): 1–10.

51 Leach, M.J. and Kumar, S. (2013). Cinnamon for diabetes mellitus. *Cochrane Database Syst. Rev.* 4 (9): 1–46.

52 Lee, G. and Bae, H. (2017). Therapeutic effects of phytochemicals and medicinal herbs on depression. *Bio Med Res. Int.* 2 (7): 1–11.

53 MacGill, M. (2017). What is depression and what can I do about it? *Med. News Today* 2 (4): 1–10.

54 Ranasinghe, P., Galappaththy, P., Constantine, G.R. et al. (2017). *Cinnamomum zeylanicum* (Ceylon cinnamon) as a potential pharmaceutical agent for type-2 diabetes mellitus: study protocol for a randomized controlled trial. *Trials* 18 (7): 1–8.

55 Mao, J.J., Xie, S.X., Keefe, J.R. et al. (2016). Long-term chamomile (*Matricaria chamomilla L.*) treatment for generalized anxiety disorder: a randomized clinical trial. *Phytomedicine* 23 (14): 1735–1742.

56 Reis, L.S.L.S., Pardo, P.E., Oba, E. et al. (2006). *Matricaria chamomilla* CH_{12} decreases handling stress in nelore calves. *J. Vet. Sci.* 7 (2): 189–192.

57 Maity, T., Adhikari, A., Bhattacharya, K. et al. (2011). A study on evalution of antidepressant effect of imipramine adjunct with aswagandha and bramhi. *Nepal Med. Coll. J.* 13 (4): 250–253.

58 Sharma, L., Sharma, A., Gupta, G.L., and Bisht, G.S. (2018). Pharmacological evaluation of *Bacopa monnieri* extract against depressive like behavior induced by ethanol withdrawal in rats. *Pharm. J.* 10 (6): 48–53.

9

Anti-inflammatory Agents from Medicinal Plants

Bui T. Tung, Tran V. Linh, Trinh P. Thao, and Nguyen D. Thuan

VNU University of Medicine and Pharmacy, Vietnam National University, Ha Noi, Vietnam

Inflammation is a localized protective reaction of cells and tissues of the body to physical trauma, noxious chemicals, or microbiologic agents [1]. The typical manifestations of inflammation are tissue damage; metabolic disorders; microcirculation disorders, causing leukocytes to drift; and increased venule permeability and capillaries leading to fluid drainage, inflammatory edema, and leukocyte leakage. Mediators of the inflammatory response are released as plasma proteins, or come from cells such as macrophages, mast cells, neutrophils, and platelets. These mediators play an important role in inflammation [1]. Some mediators of the inflammatory response include nitric oxide (NO), prostaglandins, leukotrienes (LTs), histamine, serotonin, and cytokines. These substances bind to specific target receptors on cells and can increase vascular permeability, stimulate smooth muscle contraction, and increase direct enzyme activity, causing pain and mediating oxidative damage [1]. The control of the inflammatory response is a necessary and useful process for the removal of harmful stimuli and the restoration of normal physiology.

There are three stages in the inflammatory response:

- Stage 1 – circulatory disturbances: vasomotor disorders, inflammatory exudate, transvascular leukocytes, and phagocytosis.
- Stage 2 – metabolic disorders (glucid, protid, lipid) and tissue damage.
- Stage 3 – cell proliferation to heal and repair inflammatory tissue.

The inflammatory process is a combination of multiple pathways such as the synthesis of prostaglandins, interleukins, or other chemotherapeutic toxins, adhesion protein receptor activity, and platelet activating factors [1]. When the body is injured, infected, or allergic, cell membranes release membrane phospholipids. Membrane phospholipid hydrolysis is stimulated by phospholipase A2 to form arachidonic acid, which is a subsequent substrate for the enzymes cyclooxygenase and lipoxygenase and their byproducts, prostaglandins PGE2, PGH2, PGI2, and LTs LTC4 and LTB4 [1]. PGE2 is a major chemical mediator of inflammation and pain. The release of PGE2 due to mechanical, chemical, thermal, and bacterial stimulation works to increase blood flow as well as increase vascular permeability by aiding the release of NO from weak endothelial-derived releasing factor, which causes vasodilatation and helps with platelet adhesion [1]. Besides, LTB4 is an LT

Phytochemical Drug Discovery for Central Nervous System Disorders: Biochemistry and Therapeutic Effects, First Edition. Edited by Chukwuebuka Egbuna and Mithun Rudrapal.

produced from neutrophils, eosinophils, and monocytes in response to inflammatory mediators. LTB4 promotes the adhesion of neutrophils to vascular endothelial cells and their transendothelial migration, and stimulates the synthesis of inflammatory cytokines from macrophages and lymphocytes [1]. Several cytokines also play an essential role in the regulation of inflammation, particularly interleukin-1 (IL-1) and tumor necrosis factor-α (TNF-α). IL-1 and TNF-α are considered to be major chemical mediators of bacterial lipopolysaccharide (LPS) biologic reactions. They are secreted by monocytes and macrophages and other cells [1].

There are two main types of inflammation: acute and chronic. Acute inflammation usually appears within minutes or at most hours after injury, and is associated with increased blood vessel osmosis, capillary penetration, and migration of leukocytes, especially neutrophils [2]. This is a short-term process and is characterized by classic symptoms such as swelling, pain, heat, and redness. This acute inflammatory response is useful to the body to destroy bacteria, viruses, and parasites while still facilitating wound repair. If acute inflammation fails to control proinflammatory stimulus, this leads to chronic inflammation, autoimmunity, and excessive tissue damage [3]. Chronic inflammation is a disease that persists for a long time, ranging from three months to more than one year. It is associated with infiltration of immune cells, macrophages, monocytes, neutrophils, fibroblast activation, proliferation, and fibrosis [2]. Chronic inflammation cannot be prevented with vaccines or cured with drugs, nor does it go away on its own. It is necessary to control symptoms on a daily basis as directed by your doctor.

Inflammation is the body's protective response to eliminate microorganisms, antigens, dead cells, foreign bodies, and repair to restore damaged tissue. But sometimes its action can have disastrous consequences, such as deformity and stiffness in osteoarthritis, or neck skin keloid scars that cause cosmetic loss. Some overinflammatory conditions affect quality of life, such as rheumatoid arthritis, osteoarthritis, inflammatory bowel disease, retinitis, multiple sclerosis, psoriasis, and atherosclerosis [1]. To overcome this problem, non-steroidal anti-inflammatory drugs (NSAIDs) are among the most popularly used medicines because of their efficacy in reducing pain and inflammation [4]. However, there have been many studies showing adverse effects of NSAIDs in gastrointestinal, cardiovascular, liver, kidney, brain, and lung complications. Typically, long-term use of NSAIDs will cause gastric ulcers, gastrointestinal bleeding, hepatotoxicity, and cardiovascular disease [4]. Therefore, there is a need for safe and effective solutions that both improve inflammation and reduce side effects on other organs.

9.1 Role of Neuroinflammation in Neurodegenerative Diseases

We must first define inflammation in the context of the central nervous system (CNS) and how the term "neuroinflammation" is used in comparison before we can consider the role of neuroinflammation in inflammation (Figure 9.1). Because the term "inflammation" has sometimes been misused, this is necessary [5]. The four cardinal signs of inflammation were first identified historically by the Roman physician Celsus: *rubor et tumor cum calore et dolore* (redness and swelling with heat and pain); Virchow added the fifth cardinal sign – *functio laesa*, loss of function – in 1871 [6]. The conventional view of inflammation

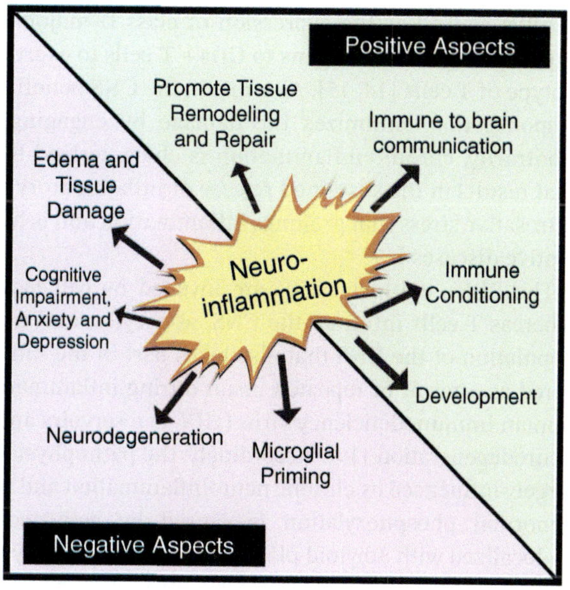

Figure 9.1 Role of neuroinflammation in neurodegenerative diseases.

is still dominated by this descriptive and clinical definition. The only aspect of the cardinal symptoms that is directly applicable to CNS pathology is the tissue correlate, cellular inflammatory changes [7]. Furthermore, rather than being based on a cellular understanding of inflammation, new models of inflammation have been created in recent years that are based on molecular phenomena [8]. Although the indications and anticipated benefits of some of these molecular models are still far from being supported by evidence, experimental findings are occasionally blindly applied to clinical medicine and may have significant effects on anti-inflammatory treatments and the use of anti-inflammatory drugs [9]. Around 120 billion aspirin tablets are consumed annually, and they are thought to be responsible for 100 000 hospital admissions and 16.5 million older people dying from NSAID-related gastrointestinal complications [9]. Inflammation is thought to play a role in almost every CNS disorder, but the exact nature of inflammation in the CNS is frequently unclear, and CNS-resident cells may not play as important a role in inflammation as cells that invade the tissue from the systemic immune compartment [10]. Thus, neuroinflammation represents a concept.

Innate and acquired immune responses are tightly regulated in relation to the periphery, making the CNS an immune-privileged organ [11]. Evidence suggests that systemic LPS or viral infections cause a strong inflammatory response in the periphery, which then causes leukocytes from the periphery to infiltrate the CNS, leading to neuroinflammation and neurodegeneration [12]. An offense is followed by the initial activation of microglia, which causes the release of proinflammatory mediators that encourage permeabilization of the blood–brain barrier (BBB) [13]. Following this, peripheral leukocytes infiltrate the CNS, including T cells and macrophages that are similar to microglia in that they express Toll-like receptors (TLRs), which allows them to be activated by pathogen-associated molecular patterns or aggregated proteins; inflammatory M1 and anti-inflammatory M2 phenotypes, which can be influenced by inflammatory T cells and lymphocyte regulatory

T cells; as well as the expression of class II major histocompatibility complex and the capacity to present antigens to CD4+ T cells to exert an influence on the functional phenotype of T cells [14, 15]. As a result, the CNS benefits from an acute neuroinflammatory response that minimizes the damage by engaging the innate immune system [16]. Contrarily, chronic inflammation is characterized by persistent activation of microglia that results in the sustained release of inflammatory mediators, increasing oxidative and nitrosative stress that prolongs inflammation and is harmful for a number of neurodegenerative diseases [17].

The BBB's epithelial cells are invaded by cell factors that determine microglial fate, whereas T cells infiltrate the CNS, astrocytes, and neurons, the most prevalent glial cell population of the CNS that also takes part in the innate immune response, which is triggered as a result of repeated insult during inflammation or infection [18]. Astrocytes are human immunodeficiency virus (HIV-1) reservoirs and are important in the virus-induced neurodegeneration [19]. Accordingly, the pathophysiology of neurodegenerative disease is largely influenced by chronic neuroinflammation and activated microglia [20]. For instance, abnormal phosphorylation is caused by activated microglia that are IL-1-positive, co-localized with amyloid plaques and neurofibrillary tangles in Alzheimer's disease (AD), or present in degenerative motor neuron regions in amyotrophic lateral sclerosis (ALS) patients [21]. Similar to how they affect viral neurodegenerative diseases, neurotropic viruses cause long-term neuroimmune activation. Additionally, it has been discovered that neuroinflammation is either the origin or result of chronic oxidative stress, a crucial component of all neurodegenerative diseases that alters the genetic makeup of proteins and lipids and causes neurodegeneration [22]. Reactive oxygen and nitrogen species, tumor necrosis, and glutamate are primarily produced by microglial cells and are all neurotoxic when released in high doses following microglial activation, which is most likely brought on by the stimulation of TLRs via aggregated proteins, as is the case in AD, Parkinson's disease (PD), and ALS patients [23].

What are known as neuroinflammatory disorders affect the part of the nervous system that is damaged by the immune system. In the CNS, inflammatory effectors from the innate and acquired immune systems as well as glial cells, in particular microglia, act as sensors for disturbed brain tissue homeostasis and accumulate locally in response to injury to neuronal cells or foreign material entry into the brain [24]. The differential activation of microglia cells is the key factor that regulates neuroinflammation, which has two outcomes: neurotoxicity or neuroprotection. The crucial factor determining whether neurons will degenerate or be protected is therefore environmental exposure [25]. Since activated microglia can also produce anti-inflammatory mediators and neurotrophic factors like insulin-like growth factor-1, glial cell-derived neurotrophic factor, brain-derived neurotrophic factor, and other factors, and obtain beneficial effects, more research must be done to take advantage of the versatility of microglia [26].

9.2 Neuroinflammatory Drugs

Some neuroinflammatory drugs for neurodegenerative diseases are summarized in Tables 9.1, 9.2, 9.3, 9.4, 9.5, and 9.6.

Table 9.1 Neuroinflammatory drugs for stroke.

Drug name	Method of action	Status	Treatment for	Outcome
Pexelizumab	Blocks C5 cleavage and generation of C5a and MAC	Phase III	Coronary artery bypass graft	Reduces neurologic injury and deficits
Pioglitazone (thiazolidones)	PPAR-γ agonist and alters transcription of genes implicated in fatty acid oxidation, which results in metabolic changes	Phase III	Ischemic stroke	Neuroprotection: induction of genes involved in oxidative stress defense, induction of anti-inflammatory responses, and stimulation of mitochondrial biogenesis
Minocycline	Inhibits microglial activation	Phase I and II	Stroke and neuroinflammation	Anti-inflammatory and neuroprotective properties
Verapamil	Minimizes carotid intima-media thickening	Phase I	Acute ischemic stroke	Anticerebral vasospasm secondary to subarachnoid hemorrhage
Neu2000	Combines both NR2B subtype-selective blockade of NMDA receptor and ROS scavenging	Phase II	Acute ischemic stroke	Better neuroprotective effect than either NMDA receptor antagonist or antioxidants alone
Citicoline (CDP-choline)	Promotes membrane stability and inhibits glutamate excitotoxicity, apoptosis, and oxidative stress	Phase II/ III	Acute ischemic stroke and other neurologic disorders	Combines neurovascular protection and repair-promoting effects
Nerinetide (NA-1)	Disrupts interaction of the scaffolding protein, PSD-95, with NMDA receptor, thus preventing receptor signaling and protecting neurons from excitotoxicity	Phase III	Acute stroke patients with small infarct core identified for abrupt EVT	Did not show benefit in majority of patients who had successful clinical outcomes after EVT when compared with those receiving placebo
Darbepoetin-α (DARB)	Stimulates bone marrow to produce red blood cells	Phase II	Prevention of ischemic complications in surgery	Harmful effects may occur
Enoxaparin	Inhibits adhesion molecule function and heparinase activity, thus inhibiting leukocyte accumulation in ischemic tissue	Phase I/II	Hematologic cancer patients afflicted with acute ischemic stroke	Failure in patient enrollment

CDP, cytidine diphosphate; EVT, endovascular therapy; MAC, membrane attack complex; NMDA, N-methyl-D-aspartate; PPAR-γ, peroxisome proliferator-activated receptor gamma; ROS, reactive oxygen species.

Table 9.2 Neuroinflammatory drugs for traumatic brain injury.

Drug name	Method of action	Status	Treatment for	Outcome
Memantine	Inhibits overactivation of NMDA receptors, thus decreasing accumulation of phosphorylated tau proteins in cortical tissue at early stage	Phase IV	Traumatic brain injury	Failure in patient enrollment
Magnesium	Blocks calcium channel within NMDA receptors in regulatory manner, prevents overactivation of NMDA receptors, and glutamatergic excitatory action	Phase III	Traumatic brain injury	Not neuroprotective and increased mortality in high doses
Minocycline	Inhibits microglial activation, caspase-mediated apoptosis, and excitotoxic NMDA pathway	Phase I/II	Traumatic brain injury	Reduced lesion volume and improved neurologic outcome Reduced level of microglial activation but increased neurodegeneration
Cyclosporine	Inhibits opening of MPTP, resulting in oxidative phosphorylation and rupturing of mitochondrial outer membrane, leading to caspase cascade activation and neurodegeneration	Phase II	Severe traumatic brain injury	Decreased cerebrospinal fluid levels of several traumatic brain injury biomarkers, including GFAP (astroglial injury marker), NFL, tau, and UCHL-1 (neuronal cell body injury marker)
Atorvastatin	Reduces glial activation and increases blood–brain barrier integrity	Phase II	Mild traumatic brain injury	Safe, but had no impact on neurologic recovery
Docosahexaenoic acid (DHA)	Reduces level of endoplasmic reticulum stress and inhibits abnormal protein accumulation	Phase III	Mild traumatic brain injury	Reduced hippocampal and cortical damage, decreased oxidative stress and microglial-proinflammatory activation, and improved short-term cognitive function
Estrogen Progesterone	Promoted neuronal survival in hippocampus, decreased neuronal degeneration in hippocampus and cortex, and decreased brain edema	Phase II/III	Severe and moderate traumatic brain injury	Did not show improvement for severe patients and had moderate to good outcome compared to placebo-treated group in moderate patients

GFAP, glial fibrillary acidic protein; MPTP, mitochondrial permeability transition pore; NFL, neurofilament light chain; NMDA, N-methyl-D-aspartate; UCHL-1, ubiquitin carboxyl-terminal hydrolase isozyme L1.

Table 9.3 Neuroinflammatory drugs for neuromyelitis optica (NMO).

Drug name	Method of action	Status	Treatment for	Outcome
Cinryze (C1-esterase inhibitor)	Inhibits C1 complex of the antibody-initiated classical complement pathway	Phase Ib	NMO	Safe in humans, with no adverse events, but the clinical efficacy poor
Eculizumab	Anti-C5 monoclonal antibody	Phase I/II	NMO and Devic's disease	Well tolerated, reduces attack frequency, and improves disability measures
		Phase III	NMO in patients who test positive for anti-aquaporin-4 antibodies	Reduces risk of relapse in combination with standard immunosuppression compared to placebo and immunosuppression
		Phase II/III	Pediatric NMO patients	Reduces annualized relapse rate
Ravulizumab	Second-generation anti-C5 monoclonal antibody	Phase III	NMO spectrum disorder	Ongoing
Alpha-1-antitrypsin (A1AT)	Inhibits plasminogen activators, chymotrypsin, as well as elastases, and can function as anti-inflammatory and tissue repair molecule	Phase I	NMO relapses	Trial withdrawn and no further updates
Rituximab	Humanized glycosylated IgG antibody specific for CD20, transmembrane protein expressed on both healthy and malignant B cells	Phase I	NMO	Safe and well tolerated
		Phase II/III	NMO with anti-aquaporin 4 antibody	Prevented relapses from occurring for up to 72 wk in NMO patients
Ublituximab	Another monoclonal antibody that targets CD20, but with increased toxicity against malignant B cells compared to rituximab	Phase I	NMO and NMO spectrum disorder	Safe, no serious adverse events, and no opportunistic infections
Inebilizumab	Monoclonal antibody that targets CD19, another protein expressed on B cells that is viable therapeutic target independently of or in addition to CD20-targeting antibodies	Phase II/III	NMO and NMO spectrum disorder	Reduces risk of relapse

(Continued)

Table 9.3 (Continued)

Drug name	Method of action	Status	Treatment for	Outcome
Mycophenolate mofetil (MMF)	Reversible inhibitor of inosine monophosphate dehydrogenase, enzyme essential for synthesis of DNA nucleotides, which are important for cell proliferation	Phase IV		Reduces clinical relapse and stabilizes or reduces disability
Plasma exchange	Eliminates pathogenic antibodies, complement components, and cytokines from blood of patients suffering from NMO		Acute attacks in NMO patients	Used early after first attack resulted in a better outcome as assessed by percentage improvement in EDSS score

EDSS, Expanded Disability Status Scale; IgG, immunoglobulin G.

Table 9.4 Neuroinflammatory drugs for amyotrophic lateral sclerosis (ALS).

Drug name	Method of action	Status	Treatment for	Outcome
Riluzole	Inhibits glutamate release presynaptically, and activates receptors and channels postsynaptically		ALS	Exerts neuroprotective effect in spinal cord, where glutamate used as transmitter
Edaravone	Increases prostacyclin production, decreases lipoxygenase metabolism of arachidonic acid by trapping hydroxyl radicals, inhibits alloxan-induced lipid peroxidation, and quenches active oxygen species [27]	Phase III	ALS	Slows the rate of physical decline in ALS patients through neuroprotective function
Deferiprone	Iron-chelating activity – binds to ferric ions and forms 3 : 1 (deferiprone : iron) stable complex [28]	Phase III	ALS	Decreases iron accumulation with manageable side effects
Olesoxime	Binds to two outer mitochondrial membrane proteins associated with mitochondrial response to oxidative stress	Phase II–III	ALS	Failed two clinical trials as add-on therapy for riluzole for ALS (did not show survival benefit)

Table 9.4 (Continued)

Drug name	Method of action	Status	Treatment for	Outcome
Tauroursodeoxycholic acid	Inhibits apoptosis by stabilizing mitochondrial membrane and inhibiting translocation of proapoptotic protein, Bax, from cell to mitochondria	Phase III	ALS	Reduction in functional decline
Ibudilast	Enhances autophagy of protein aggregates through inhibiting mTORC1 activity, and protects motor neuron-like cells from TDP-43-induced cytotoxicity	Phase IIb/III	ALS	As add-on for riluzole reduces ALS disease progression relative to riluzole alone, but only in patients with short (<600 d) history of ALS
		Phase I/II		Pending
Ravulizumab	Complement inhibition at level of C5	Phase III	ALS	Ongoing
Zilucoplan		Phase II/III		

mTORC1, mammalian target of rapamycin complex 1; TDP-43, transactive response DNA binding protein 43 kDa.

Table 9.5 Neuroinflammatory drugs for multiple sclerosis (MS).

Drug name	Method of action	Status	Treatment for	Outcome
Polyphenon E	Mitochondrial antioxidant	Phase II	MS	Terminated early due to hepatotoxicity in large number of participants
Dimethyl fumarate	Activation of the Nrf2 pathway	Phase II/III	MS	Reduction in relapse episodes, as well as lesion number and size in MS patients
Vitamin D	Modulation of T-helper 17 cell differentiation and interleukin-17a expression	Phase IV	MS	Failed to provide conclusive evidence that vitamin D slowed disease progression
Amiloride	Sodium channel blocker	Phase II	MS	No evidence of neuroprotection in any patient group compared to placebo
Fluoxetine	Selective serotonin reuptake inhibitor			
Riluzole	Glutamate receptor antagonist			

(Continued)

Table 9.5 (Continued)

Drug name	Method of action	Status	Treatment for	Outcome
Memantine	Blockade of current flow through channels of NMDA-type glutamate receptor [29]	Phase III	RRMS patients with cognitive impairment	No effect on memory or cognition in patients with MS-related dementia. Also significant neurologic side effects
Rivastigmine	Acetylcholinesterase inhibitor	Phase IV	MS patients with cognitive impairment	No effect of rivastigmine relative to placebo
D-aspartate	Localized to multiple brain regions and produces various hormones like gonadotropin-releasing hormones, luteinizing hormone, testosterone, and melatonin	Early phase I	MS	Demonstrated positive effects of D-aspartate on neuroplasticity measured by TMS, suggesting further clinical studies warranted
Lamotrigine	Selectively binds and inhibits voltage-gated sodium channels, stabilizing presynaptic neuronal membranes and inhibiting presynaptic glutamate and aspartate release [30]	Phase II	Secondary progressive MS	Did not find evidence of neuroprotection or clinical improvement
Carbamazepine	Enhancement of sodium channel inactivation by reducing high-frequency repetitive firing of action potentials and action on synaptic transmission [31]	N/A	Trigeminal neuralgia	Higher doses shown to have low efficacy, but with significant side effects
Phenytoin	Blockade of voltage-dependent membrane sodium channels responsible for increasing action potential [32]	Phase II	MS	No clinical benefit found
Interferon beta (IFN-β)	Skews both T-cell activity and cytokine profiles toward anti-inflammatory phenotype	N/A	RRMS	Alters fundamental course of relapsing MS
Methylprednisolone	Exert multiple anti-inflammatory effects, including reducing infiltration of inflammatory cells into CNS and possibly inducing apoptosis	Phase IV	Acute exacerbation of MS	Reduces severity of episodes

Table 9.5 (Continued)

Drug name	Method of action	Status	Treatment for	Outcome
Adrenocorticotrophic hormone	Regulates anti-inflammatory and immunomodulatory functions involving lymphocytes, macrophages, sympathetic nervous system involved in inflammatory processes, and reduction of pro-inflammatory cytokines [33]	N/A	MS relapse	Stimulates endogenous production of corticosteroids
Glatiramer acetate	Due to similar structure to myelin basic protein, competes with myelin binding to antigen-presenting cells, suppresses autoimmunity, and induces anti-inflammatory response	Early phase III	RRMS	Decreased relapse rate over 2 yr period, coupled with improved disability
Methotrexate	Inhibits nucleic acid synthesis, causing preferential toxicity in rapidly dividing cells such as lymphocytes	Phase IV	RRMS	Has efficacy similar to that of IFN-1α and safe in treating progressive forms of MS
Minocycline	Crosses blood–brain barrier and modulates T-cell behavior, reduces microglial activation, and prevents neuronal apoptosis	Phase III	RRMS	Suggested trend toward lower relapse rates or lower risk of conversion of CIS to MS in short term, but more studies needed
Ocrelizumab	Anti-CD20 antibodies complement-dependent cytotoxicity and antibody-dependent cellular cytotoxicity [34]	Phase III	Primary-progressive MS	Effect in treatment compared to IFN-β1α with regard to improved outcomes in areas such as disability progression and suppression of new inflammatory lesions in brain detected by MRI
Rituximab	Directed at CD20-positive B lymphocytes resulting in cell-mediated apoptosis [35]	Phase II	MS	Reduced inflammatory brain lesions along with decrease in clinical relapses for 48 wk
Alemtuzumab	Anti-CD52 antibody that depletes B and T cells	Phase II	RRMS	Significant clinical benefit when compared to IFN-β1α

(Continued)

Table 9.5 (Continued)

Drug name	Method of action	Status	Treatment for	Outcome
Natalizumab	Targets α4 integrin expressed on surface of activated lymphocytes	Phase III	Relapsing multiple sclerosis	Visual improvement in relapsing MS patients, reduces relapse rate and severity

CIS, clinically isolated syndrome; CNS, central nervous system; MRI, magnetic resonance imaging; N/A, not applicable; NMDA, N-methyl-D-aspartate; Nrf2, nuclear factor (erythroid-derived 2)-2; RRMS, relapsing–remitting multiple sclerosis; TMS, transmagnetic stimulation.

Table 9.6 Neuroinflammatory drugs for Parkinson's disease (PD).

Drug name	Method of action	Status	Treatment for	Outcome
Levodopa Bromocriptine Pramipexole R-apomorphine	Decrease nigrostriatal cell loss following dopaminergic autoreceptor stimulation	Phase III Phase IV	PD	Slows progression of PD or has prolonged effect on symptoms of disease
Selegiline	Binds to brain regions with a high MAO-B content, such as thalamus, striatum, cortex, and brainstem [36]		PD	Modest symptom delay when using selegiline compared to placebo, although unclear if due to neuroprotective mechanism
Rasagiline	Inhibiting oxidative breakdown of dopamine in striatum [37]	Phase III	PD	1 mg/day dose provided benefits of possible disease-modifying effect using UPDRS
Amantadine	Nicotine and NMDA antagonist	Phase III	PD	Temporary improvement in motor complications and significantly reduced dyskinesia with more consistent daily effect
Memantine	Non-competitive NMDA receptor antagonist	Phase II	PD	Improves PD symptoms, specifically PD-associated dementia
		Phase IV	PD, dementia, dementia with Lewy bodies	No significant improvement in PD-dementia group and possible clinical benefit in dementia with Lewy bodies group
Riluzole	Sodium channel blocker with indirect blockade effect on NMDA receptors		PD	Failed as treatment modality for PD

Table 9.6 (Continued)

Drug name	Method of action	Status	Treatment for	Outcome
Deferoxamine	Hexadentate iron chelator [38]		PD	Positive results in preclinical models, but clinical studies have not indicated any neuroprotective properties
Phytic acid	Chelated iron required for radical formation via Fenton reaction [38]			
Deferiprone	Iron-chelating activity – binds to ferric ions and forms 3 : 1 (deferiprone : iron) stable complex [28]	Phase II	PD	Decreased iron content in dentate and caudate nuclei
Israpidine	LTCC blockers restore balance and reduce degenerative signaling to maintain protective microenvironment	Phase III	Early PD	No benefit relative to placebo
Glial-derived neurotrophic factor (GDNF)	Binds to receptor complex consisting of member of GDNF family receptor-α and Ret tyrosine kinase [39]		PD	Intraventricular or intraputaminal administration did not result in significant clinical improvement
Coenzyme Q10	Antioxidant present within mitochondrial respiratory chain and functions as ROS scavenger	Phase III	PD	Failed to show any benefit in patients treated with coenzyme Q10, although some studies did suggest mild symptomatic relief due to alternative effect as antidepressive agent
PD01A PD03A	Peptides mimicking c-terminus portions of α-synuclein protein	Phase I	PD	Ongoing
PRX002	Monoclonal immunoglobulin G1 antibody preferentially targeting soluble and insoluble aggregated forms of α-synuclein [40]	Phase II	PD	Ongoing

LTCC, L-type calcium channel; MAO-B, monoamine oxidase B; NMDA, N-methyl-D-aspartate; ROS, reactive oxygen species; UPDRS, Unified Parkinson's Disease Rating Scale.

9.3 Medicinal Plants as Sources of Anti-inflammatory Agents

Plants play a very important role in the discovery of new drugs [41]. The use of medicinal plants as anti-inflammatory agents has been known for a long time. Many scientific studies in vitro, in vivo, and clinical have proven the anti-inflammatory ability of medicinal plants [41]. Many medicinal plant species have been shown to have significant anti-inflammatory and immunomodulatory effects, including suppression of both cellular and humoral immunity, lymphocyte activation, and propagation of apoptosis [3]. In addition, anti-inflammatory agents from medicinal plants have been shown to reduce adverse effects on other organs. Science is developing more and more, so it can be applied in isolating and identifying many medicinal herbs and biologically active substances [42] that inhibit the inflammatory response. The use of anti-inflammatory agents from medicinal plants will be the target of new drug development in the future [42].

Since ancient times, plants have played an important role in human healthcare. Traditional plants are extremely important in the development of new drugs. Inflammation has become a major problem for humanity in recent years. Many anti-inflammatory drugs are available, but it is believed that some of these drugs, such as opioids and analgesic-inducing drugs like NSAIDs, are not effective in all cases, and that these drugs also have side effects. To address this issue, new drugs are needed, and plants contain many phytoconstituents that are effective in inflammation and have fewer side effects. In this chapter we are going to introduce some anti-inflammatory agents from plants, such as *Allium sativum*, *Cassia tora*, *Euphorbia hirta*, and *Punica granatum*.

9.3.1 *Allium sativum*

A. sativum, commonly known as garlic, is an aromatic herbaceous plant extensively used worldwide for centuries as both flavoring and a well-trusted medicine, from Ayurvedic medicine to traditional Chinese medicine (TCM) and even to traditional European medicine [43]. In TCM, garlic, also called *Da suan*, was included in the Yang herbs, which are naturally characterized as hot and warm [44], so it has been used as a remedy against not only common diseases but also diabetes and cardiovascular diseases, immune system problems, and so on. [45]. Garlic contains more than 200 compounds [46], mostly sulfur-containing phytoconstituents such as alliin, allicin, ajoene, diallyl sulfide, 2-vinyl-4H-1,3-dithiin, diallyl trisulfide, and flavonoids such as quercetin [45].

A significant number of studies have shown that *A. sativum* and its bioactive constituents exhibit various biologic activities, such as being antioxidant, anti-inflammatory, antibacterial, antifungal, immunomodulatory, cardiovascular protective, and anticancer [47]. The secretion of IL-6 and monocyte chemoattractant protein-1 (MCP-1) in human preadipocytes treated with macrophage factor, both of which are associated with inflammation, decreased via 1,2-vinyldithiin from garlic. Alliin, the main cysteine sulfoxide, showed its effect on preventing the genes and proteins related to the proinflammatory state induced by LPS in 3T3-L1 adipocytes from increasing [48]. A study focusing on water-soluble garlic polysaccharide (WSGP), an agent extracted from garlic, demonstrated that WSGP is able to downregulate tissue inflammation and the expression of proinflammatory cytokines (TNF-α, IL-6, IL-1β, and their related mRNA) in mice with colitis [49]. Moreover, *A. sativum* inhibits the activation of nuclear factor kappa B (NF-κB) in enzymes, namely cyclooxygenase 2 (COX-2) and inducible nitric oxide synthase (iNOS). Also, *A. sativum* has been demonstrated to reduce the expression of pro-inflammatory cytokines such as IL-1, IL-6, IL-17, and TNF-α, which results in the attenuation in inflammation [50]. Moreover, garlic and its supplements showed their potential in osteoarthritis treatment with obese or overweight patients by reducing resistin [51]. Overall, *A. sativum* has potential to treat inflammatory diseases with its anti-inflammatory activity, but needs more preclinical and clinical studies in the future.

9.3.2 *Azadirachta indica*

The neem tree, *Az. indica* L., is well known for its wide range of biologically and pharmacologically interesting compounds [52]. *Az. indica* L. is a Meliaceae family multipurpose medicinal tree. It can be found in tropical and semitropical areas around the world. Hypolipidemic, antifertility, microbicidal, antidiabetic, anti-inflammatory, hepatoprotective, antipyretic, hypoglycemic, insecticidal, nematicidal, antiulcer, antioxidant, neuroprotective, cardioprotective, and antileishmaniasis are some of the pharmacologic properties of the tree [53].

Az. indica is a plant species that has been studied for its anti-inflammatory properties. Carrageenan-induced paw edema and excision and incision wound models were the most common in vivo techniques used for the anti-inflammatory and wound healing assays, respectively [54]. Because of their anti-inflammatory, antimalarial, antifungal, antibacterial, antiviral, antioxidant, and anticarcinogenic properties, *Az. Indica* leaves are renowned for their medicinal applications [54]. *Az. Indica* can produce silver nanoparticles, and the anti-inflammatory (69.77%) and antidiabetic (73.5%) activities of these kernel fabricated silver nanoparticles at 100 μg/ml were partially comparable to standards (anti-inflammatory 81.15%; antidiabetic 87.9%) [55]. On LPS, biomimetization of hydroxyapatite (HA) with *Az. indica* was proposed, and its antioxidant, antibacterial, and anti-inflammatory properties were established [56]. The acetone-water extract of the *Az. indica* leaf has been shown to prevent malaria parasitized red blood cells (pRBCs) from adhering to the vascular endothelium, cancer cell cytoadherence during metastasis, and HIV invading target T lymphocytes [57].

9.3.3 *Cassia tora*

C. tora L. is a member of the Fabaceae family that grows primarily in tropical African countries such as Nigeria, Niger, and Ghana. It is also grown in other countries such as China and India [58]. In Korea, *C. tora* has been used as a folk medicine [59]. It has been used in traditional medicine for its anticonstipation, anti-inflammatory, visual acuity, and hepatoprotective properties [60]. Tapan Kumar et al. investigated the anti-inflammatory effects of a methanolic extract of *C. tora* leaves in rat hind paw edema induced by carrageenin, histamine, serotonin, and dextran. *C. tora* was found to have significant anti-inflammatory activity against all of these agents in the study. A dose of 400 mg/kg was shown to be the most effective in inhibiting edema [61]. Natural plant extracts from *Ulmus pumila* L., *C. tora*, and *Taraxacum officinale* make up AF-343. In HS68 cells and NC/Nga mouse models of atopic dermatitis, AF-343 has been shown to have anti-inflammatory and skin hydration properties [62]. *C. tora* leaves were used to isolate ononitol monohydrate (OM). Male Wistar rats and mice were used to test OM's anti-inflammatory and analgesic properties. Carrageenan-induced paw edema, croton oil-induced ear edema, acetic acid-induced vascular permeability, cotton pellet-induced granuloma, and adjuvant-induced arthritis were used to test the efficacy of OM against inflammation [63].

9.3.4 *Euphorbia hirta*

E. hirta, which belongs to the family Euphorbiaceae and genus Euphorbia, is a traditional medicinal plant of the Ayurvedic system of medicine. It is slender stemmed and reddish or purplish in color [64]. *E. hirta*, often found in wasteland along the roadside, is distributed throughout the hotter parts of India and Australia [65]. *E. hirta* possesses anti-inflammatory, antibacterial, antifungal, antispasmodic, antimalarial, anthelmintic, and sedative properties [66]. The aqueous extract exhibits anti-inflammatory, analgesic, and antipyretic activities [64]. Flavonoids, triterpenoids, alkanes, amino acids, and alkaloids are also found in *E. hirta* [66]. Specifically, quercitrin, afzelin, and myricitrin were compounds isolated from the methanolic extract of *E. hirta* [67]. *E. hirta* also contains euphorbin A, euphorbin B, euphorbin C, euphorbin D, rutin, kaempferol, gallic acid, 2,4,6-tri-O-galloyl-β-D-glucose, 1,3,4,6-tetra-O-galloyl-β-D-glucose, and protocatechuic acid [64].

The compounds extracted from *E. hirta* have many biologic effects, especially anti-inflammatory. The anti-inflammatory activity of *E. hirta* extract may be due to compounds such as phytol, hydroxymethylfurfural (HMF), fatty acids, and others [68]. Shih et al. demonstrated that the extracts in *E. hirta* such as flavones, glucosides, and tannins have the ability to inhibit NO, thereby increasing anti-inflammatory activity [69]. The n-hexane extract of *E. hirta* showed anti-inflammatory effects in the model of phorbol acetate-induced ear inflammation in mice [70]. The study by Ahmad et al. showed that *E. hirta* ethanol extract has important antiarthritic properties against adjuvant-induced arthritis in mice [71]. In vitro, the aqueous extract of *E. hirta* inhibited the production of PGE2 on rabbit synovial fibroblasts [72]. Treatment with *E. hirta* extract in neonatal asthmatic rats significantly reduced the eosinophil and decreased the gene expression of TNF-α, IL-6, iNOS, and COX-2 [73]. Using extract of *E. hirta* is a new direction in the preparation of anti-inflammatory substances from medicinal plants.

9.3.5 *Garcinia mangostana*

G. mangostana L. is a tropical plant native to Southeast Asia, commonly known as mangosteen, which is famous for its white, soft, and juicy edible pulp inside a tough, dark purple cover. Its sweet flavor and pleasant aroma gave it the name of "queen of fruits" [74]. Mangosteen contains diverse bioactive compounds such as xanthones (α-mangostin, γ-mangostin, 8-deoxygartanin, garcinone E, etc.), terpenes, anthocyanins, tannins, phenols, and some vitamins (C, B1, B2, B5) [75]. At least 68 xanthone derivatives has been found in mangosteen, which makes xanthones the major bioactive substances [76].

Many biologic effects of *G. mangostana* have been recorded, including being antioxidant, anticancer, antinociceptive, anti-inflammatory, neuroprotective, antiobesity, and antihyperglycemic [77]. *G. mangostana* was highly effectual in scavenging free radicals and was capable of terminating the production of proinflammatory cytokines [78]. α-Mangostin, the most abundant xanthone in the pericarp of mangosteen fruit, inhibits the LPS-induced production of NO, IL-8 secretion by activated macrophage-like THP-1 and colonic HT-29, and also TNF-α via phorbol 12-myristate 13-acetate (PMA)-activated HepG2 cells [79]. In addition, stem bark of *G. mangostana*-isolated β-mangostin was reported to attenuate NO production in LPS-stimulated RAW264.7 macrophage [80]. An in vivo β-mangostin test resulted in the decrease of carrageenan-induced peritonitis and TNF-α and IL-β concentrations in peritoneal fluid [81]. Other previous studies revealed that 1 μg/ml mangosteen extract reduced IL-6 and IL-8 expression in *Porphyromonas gingivalis* KCOM 2804 LPS-treated hTERT-hNOF cells ($p < 0.05$) [82]. Another study on methicillin-resistant *Staphylococcus aureus* (MRSA)-infected tape-stripping model mice showed that *G. mangostana* pericarp ethanolic extract significantly downregulated proinflammatory cytokine expression, including TNF-α, IL-6, and IL-1β via the TLR-2 pathway, and also maintained the wound-healing process, as it completely healed the MRSA-infected wound by the 10th day [83].

9.3.6 *Punica granatum*

P. granatum L. (pomegranate), a member of the Punicaceae family, is now distributed all around the world from West to East. Almost all parts of the pomegranate are high in potential bioactive compounds, including seed, juice, peel, and leaves [84]. Products derived from *P. granatum* appear in the prevention and treatment of different diseases such as acquired immune deficiency syndrome (AIDS), cancer, allergic symptoms, and cardiovascular conditions [85]. Recent studies have shown that pomegranate has various potential bioactive activities, including being anti-inflammatory, antibacterial, antidiarrheal, immune modulatory, antitumor, wound healing, and antifungal [86]. Significant numbers of studies demonstrated pomegranate's effect on chronic inflammatory disorders, for example rheumatoid arthritis, inflammatory bowel disease, or chronic obstructive pulmonary disease [87].

Pomegranate peel extract used on bovine mammary epithelial cells (BME-UV1) reduced proinflammatory cytokines TNF and IL-1β, especially IL-10, which could be positively correlated with reduced transcription levels of proinflammatory cytokines during the inflammatory response [88]. In vitro, pomegranate juice together with total pomegranate tannins and punicalagin inhibited Akt activity, NF-κB activation, and COX-2 expression induced by TNFα. Moreover, pomegranate peel significantly inhibited secretion of the chemokine IL-8, also PGE2 by COX-1 as well as COX-2 in response to IL-1β stimulation, and the production of NO on stimulation by IL-1β or a mixture of proinflammatory stimuli [89]. Punicalagin, a hydrolysable tannin of pomegranate juice, exhibited an inhibitory effect in macrophage inflammation by inhibiting NF-κB, the mitogen-activated protein kinase (MAPK) signaling pathway, and FoxO3a/autophagy signaling because of suppressed ability of LPS through the FoxO3a/autophagy signaling pathway [90]. Not only punicalagin, Lamees et al. showed that ellagic acid and gallic acid also prevented NO, PGE-2, and IL-6 from production in LPS-induced RAW264.7 cells [91]. Moreover, the mechanism of the anti-inflammatory effect of *P. granatum* has been found to be tightly connected to the gut microbiota – a large and diverse group of microorganisms that live in the gastrointestinal tract [87].

9.3.7 *Ricinus communis*

The castor plant (*R. communis* L.) has been used in traditional medicine in Mediterranean and Eastern ancient cultures since time immemorial. Furthermore, it is still used in folk medicine all over the world. The castor bean has traditionally been used as an anti-inflammatory, anthelmintic, antibacterial, laxative, and abortifacient, as well as for wounds, ulcers, and a variety of other ailments [92]. Phytochemicals such as alkaloids, flavonoids, terpenes, saponins, and phenolic compounds such as kaempferol, gallic acid, ricin, rutin, lupeol, ricinoleic acid, pinene, thujone, and gentisic acid have been discovered in various studies. Anticancer, antimicrobial, insecticidal, antioxidant, antidiabetic, antinociceptive, anti-inflammatory, bone-regenerative, analgesic, and anticonvulsant activities are among the pharmacologic and therapeutic effects of these phytochemicals [93]. Castor oilseed bioactive compounds such as polyphenols, phytosterols, and tocopherols have anti-inflammatory and antioxidant properties, which may help to extend the shelf life of the oil [94]. Antiviral activity was found in the leaves of *R. communis*. All extracts had high antireplicative activity against hepatitis A virus (HAV), especially the methanol and methylene chloride fractions, and moderate antireplicative activity against COX-B4, butanol > methylene chloride and ethyl acetate > methanol. All extracts showed antiviral activity against HAV, particularly butanol extract, while methanol extracts had higher non-significant antiviral protective activity against herpes simplex virus (HSV1) compared to acyclovir [95].

R. communis extract (RCE) reduced paw swelling and arthritic score, as well as restoring body weight and organ indices. RCE treatment significantly improved arthritic rats' hematologic parameters, serum inflammatory markers like C-reactive protein (CRP) and rheumatoid factor (RF), and liver function markers. The plant convincingly reduced IL-1β, IL-6, IL-17a, TNF-α, and receptor activator of nuclear factor kappa-B ligand (RANKL) relative expression, while increasing IL-4, interferon (INF)-γ, and osteoprotegerin (OPG) relative expression and alleviating hepatic oxidative stress parameters. In the extracts-treated groups, histopathologic and radiologic findings revealed a significant reduction in tissue inflammation and bone erosion [96]. The findings suggest that by modulating pro-/anti-inflammatory cytokine expression and reducing oxidative stress, *R. communis* leaves and roots significantly reduced inflammation and improved health [96].

9.3.8 *Scutellaria baicalensis*

S. baicalensis, which belongs to the family Lamiaceae and genus Lamiaceae, is important in TCM with the functions of clearing away heat and dampness, purging fire, and detoxification [97]. It is widely distributed in China, Russia, Mongolia, North Korea, and Japan. *S. baicalensis* is used to treat respiratory tract infections, pneumonia, colitis, hepatitis, and allergic diseases, because it has important pharmacologic actions like anti-inflammatory, antiviral, antitumor, antioxidant, and antibacterial effects [98]. Root extracts of *S. baicalensis* have multiple biologic activities, thus that is the main part used as a medicine. The chemical constituents of *S. baicalensis* are mainly divided into five categories: flavonoids, volatile oils, terpenoids, polysaccharides, and other chemical components [99]. Specially, flavonoids, which have potent anti-inflammatory activity [100], are the main active substances in *S. baicalensis*. More than 40 different polyphenols have been isolated and identified from *S. baicalensis*, including flavonoids and their flavanols, dihydroflavones and their dihydroflavonols, chalcones, and biflavonoids [97]. Among them, the compounds with many biologic effects are baicalein, baicalin, wogonin, wogonoside, and oroxylin A. These biologic compounds can act directly on immune cells like lymphocytes, macrophages, monocytes, neutrophils, mast cells, and dendritic cells. They also inhibit the expression of the inflammatory cytokines IL-1β, IL-6, IL-8, and TNF-α, and other inflammatory mediators like prostaglandins, NO, LTs, and reactive oxygen species (ROS) [99]. The molecular mechanisms of the anti-inflammatory effects of the active compounds of *S. baicalensis* include downregulation of TLRs, activation of the nuclear factor erythroid 2-related factor 2 (Nrf2) and peroxisome proliferator-activated receptor (PPAR) signaling pathways, and inhibition of the nuclear thioredoxin system and inflammation-associated pathways such as those of MAPK, Akt, and NF-κB [99].

Many studies have demonstrated that chemical compounds in *S. baicalensis* exert anti-inflammatory effects. Baicalin ameliorates cognitive impairment and protects microglia from LPS-induced neuroinflammation via the SIRT1/HMGB1 pathway [101]. Baicalein ameliorates pristane-induced lupus nephritis via activating Nrf2/heme oxygenase 1 (HO-1) in myeloid-derived suppressor cells [102]. Wogonin exerts potent anti-inflammatory and chondroprotective effects through the activation of ROS/extracellular signal-regulated kinase (ERK)/Nrf2 signaling pathways in human osteoarthritis chondrocytes [103].

Wogonoside exerts anti-inflammatory action through downregulating ubiquitin-specific protease 14 (USP14) to promote the degradation of activating transcription factor 2 [104]. Oroxylin A exerts anti-inflammatory activity on LPS-induced mouse macrophage via Nrf2/antioxidant-responsive element (ARE) activation [105]. *S. baicalensis* shows potential in the preparation of anti-inflammatory agents from medicinal plants.

9.3.9 *Solanum melongena*

So. melongena, which belongs to the family Solanaceae, is a traditional Chinese medicinal plant with anti-inflammatory, antioxidant, and anticancer properties [106]. It is commonly known as eggplant and is widely distributed in the subtropics and tropics, with the greatest concentration in China and India. *So. melongena* is a herbaceous plant; the fruit is a berry and the seeds have a large endosperm and are grown mainly for medicinal and food purposes [107]. Various parts of *So. melongena* are used in the treatment of inflammatory conditions, neuralgia, cardiac debility, bronchitis, asthma, and cholera. It also has important pharmacologic actions like analgesic, antipyretic, anti-inflammatory, anticonvulsant, and hypolipidemic activity [108]. The compounds extracted from *So. melongena* mainly contain flavonoids, alkaloids, phenolic acids, and steroids [109]. In addition, the aqueous extract of *So. melongena* possesses glycosides, saponins, tannins, sterols, carbohydrates (1.5%), fixed oils, and resin [110].

Flavonoids, alkaloids, and glycosides were previously reported as having potent anti-inflammatory activity [100, 111]. There were 30 compounds that were isolated from the ethanol extract of the roots of *So. melongena*. The compounds were evaluated for their potential anti-inflammatory activity by virtue of a RAW264.7 cells model induced by LPS in vitro inhibitory activities on NO production. Pseudo-protodioscin and 3,5-dimethoxy-4-hydroxybenzaldehyde, which were two compounds isolated from the ethanol extract of the roots of *So. melongena*, showed moderate inhibition of NO production [112]. Melongenamide B and cannabisin G also exhibited inhibition of NO production in LPS-induced RAW264.7 macrophages [113]. A study by Umamageswari and Maniyar showed the anti-inflammatory activity of the aqueous leaf extract of *So. melongena* in Wistar albino rats by the carrageenan-induced paw edema method [110]. The aqueous extract of *So. melongena* inhibits the inflammatory response in mouse paw edema by inhibiting protease-activated receptor 2 (PAR2) agonist activation [114], because PAR2 plays an important role in the process of inflammation, which is caused by increase of vascular permeability, infiltration of neutrophils, and secretion of proinflammatory cytokines. *So. melongena* has potential in the preparation of anti-inflammatory substances from medicinal plants, but more preclinical and clinical studies are needed in the future.

9.4 Bioactive Compounds as Anti-inflammatory Agents

Many herbal medicines for inflammation and rheumatism have yet to be investigated scientifically. As a result, it is imperative that all herbal medicines be evaluated for their pharmacologic activities, the isolation of a single entity responsible for anti-inflammatory activity, and the development of appropriate formulations to combat inflammatory disorders. This part of the chapter focuses on substances that are anti-inflammatory agents from plants, including curcumin, eugenol, and epigallocatechin 3-gallate.

9.4.1 Curcumin

Curcumin has a wide range of biologic effects, indicating that it is important in both health and disease [115]. Curcumin is known to have antioxidant, anti-inflammatory, and anticancer properties, and to play an important role in the prevention and treatment of a variety of illnesses, including cancer, autoimmune, neurologic, cardiovascular, and diabetic diseases [116]. Curcumin is a phenolic compound with anti-inflammatory and immunomodulatory properties [117]. It has a wide range of biologic activities, among which the anti-inflammatory effect is a significant feature [118].

Curcumin's antioxidant, anti-inflammatory, and antiangiogenic properties have been highlighted in several studies [119]. Curcumin has strong pharmacologic activity in reducing the inflammatory response in macrophages and has therapeutic effects on arthritis in collagen-induced arthritis (CIA) rats. Its mechanism could be linked to the NF-κB signaling pathway being inhibited and macrophage apoptosis being promoted [102]. Curcumin is an antioxidant and anti-inflammatory compound that can help to prevent the formation of advanced glycation end products (AGEs) [120]. In the Wistar rat model, curcumin had a relatively weak effect on arthritis progression; however, microencapsulated curcumin effectively prevented arthritis progression – the disease stabilized after 10 days of supplementation. Curcumin also reduced the levels of immune cells (neutrophils and leukocytes) as well as proinflammatory cytokines – TNFα, IL-1, and IL-6 – bringing them closer to the levels seen in arthritis-free people [121]. Consumption of the curcumin supplement, along with drug therapy, is associated with significant improvement of the clinical outcomes, quality of life, high-sensitivity CRP, and erythrocyte sedimentation rate (ESR) in patients with mild to moderate ulcerative colitis [122].

9.4.2 Eugenol

Eugenol (4-allyl-2-methoxyphenol) is a pale yellow oil with a spicy aroma from the class of phenylpropanoids, obtained mostly from clove (*Eugenia caryophyllata*) buds and leaves, and is also in soybean (*Glycine max* [L.] Merr.), beans, coffee, cinnamon (*Cinnamomum verum* J. Presl), basil (*Ocimum basilicum* L.), and other foods, could be used both as flavoring agents for food and cosmetics [123]. For centuries, the ancients used clove oil as an antimicrobial, antiseptic, and antispasmodic in TCM [124]. Nowadays, many properties from eugenol have been demonstrated such as being antioxidant, analgesic, antimutagenic, antiplatelet, antiallergic, antiswelling, and anti-inflammatory [125].

Significant numbers of stories have shown that clove essential oil and eugenol – its main bioactive compound – both exhibited anti-inflammatory activities in murine macrophages and also inhibited the production of proinflammatory cytokines [126]. Previous studies reported that eugenol is able to suppress the expressions of COX-2. Moreover, eugenol prevented cell proliferation through inhibition of NF-kB and also regulated the expression of NF-kB target genes, which are responsible for the regulation of cell proliferation and cell survival [124]. Estevao-Silva et al. demonstrated that in vitro eugenol similarly inhibited leukocyte migration induced by N-formyl-methionyl-leucyl-phenylalanine (fMLP) and by rolling and adherent cells [127]. In addition, a study using eugenol for an experimental model of allergic asthma induced by ovalbumin (OVA) showed that eugenol distribution inhibited the OVA-induced eosinophilia in the lung tissue, decreased the IL-4 and IL-5 levels, and reduced the NF-κB signalizing pathways [123]. In LPS-induced acute lung injury, pretreatment with eugenol exhibit a significantly downregulation in the neutrophils in lung tissues via decreasing myeloperoxidase (MPO) activity. Contemporaneously, the levels of superoxide dismutase (SOD1), catalase (CAT), glutathione peroxidase (GPx1), and glutathione S-transferase (GST) in lung improved remarkably in eugenol pretreatment [128]. Furthermore, eugenol treatment not only reduced TNF-α levels in lung homogenate and NF-κB activation, but also decreased IL-1β and IL-6 levels in lung homogenate to values similar to those observed in control mice [129].

In vitro anti-inflammatory effects were observed in an ethanol extract of *Senna septemtrionalis* (SSE) aerial parts, with IC50 values calculated as follows: 163.3 µg/ml (IL-6), 154.7 µg/ml (H_2O_2), and >200 µg/ml (IL-1β, TNF-α, and NO). SSE showed also in vivo anti-inflammatory actions in the tissue polypeptide antigen (TPA) test (40% of inhibition of ear edema) and the carrageenan test (ED50 = 137.8 mg/kg orally). Overall, SSE had anti-inflammatory and antinociceptive effects in vitro and in vivo, possibly due to the involvement of adenosine triphosphate (ATP)-sensitive K+ channels [130]. In vitro and in vivo studies show that the ethanol extract from *Clerodendrum cyrthophyllum* Turcz (EE-CC) leaves has potent anti-inflammatory properties and could be used to treat a variety of inflammatory diseases [131]. TNFα, IL-6, and IL-8 inflammatory cytokine production was strongly inhibited by an ethanol extract of *Vernonia amygdalina* leaves plus a known flavonoid, cynaroside, at a concentration of 30 g/ml. These results showed that it inhibited proinflammatory cytokine production in RAW264.7 cells stimulated with LPS via ROS [131]. In experimental models, the ethanolic extract from *Vatairea macrocarpa* (EEVM) leaves showed anti-inflammatory properties, indicating that EEVM could be used as a natural anti-inflammatory agent [132]. In all studied models, the ethanol extract from *Uapaca staudtii* stem bark fraction showed good anti-inflammatory, analgesic, and antioxidant capacity, which was comparable to the standard drugs used. The findings back up the medicinal use of *U. staudtii* stem bark for pain and inflammatory disease management [133]. Due to the involvement of NO and ATP-sensitive K+ channels, an ethanol extract from the leaves and stem of *Cuphea aequipetala* (CAE) induced antinociceptive effects without affecting motor coordination. CAE also increases the release of IL-10, which has anti-inflammatory properties in vitro and in vivo [134].

9.4.3 Epigallocatechin 3-Gallate

Epigallocatechin 3-gallate is a natural flavonoid derived from the leaves of *Camellia sinensis*, commonly known as green tea [135]. It has important pharmacologic effects such as being antimicrobial, anti-inflammatory, antioxidative, and anticarcinogenic [136]. Through cellular, animal, and human experiments, epigallocatechin-3-gallate has been demonstrated to have anti-inflammatory effects [137]. Epigallocatechin 3-gallate is known to inhibit COX-1 and COX-2, which are enzymes involved in the inflammatory response. Epigallocatechin 3-gallate exhibits antioxidant properties by altering several proinflammatory probes involving multiple pathways, such as the NF-κB, Janus kinase/signal transducer and activator of transcription (JAK/STAT), and phosphatidylinositol 3-kinase/protein kinase B (PI3K/Akt) pathways, and ameliorates the effects of COX-1 and arachidonate 5-lipoxygenase (5-LOX) by affecting the arachidonic acid (AA) pathway [3]. Epigallocatechin 3-gallate can downregulate inflammatory cytokines (COX) and reduce translocation of NF-κB to the nucleus [3]. Therefore, epigallocatechin 3-gallate has been shown to be effective in chronic inflammatory diseases such as atherosclerosis, colitis, pancreatitis, ulceration, and prevention of precancerous lesions [3]. In addition, epigallocatechin 3-gallate reduces the progression of posttraumatic osteoarthritis thanks to its potent anti-inflammatory and antioxidant properties [136]. For osteoarthritis, epigallocatechin 3-gallate inhibits IL-1β-induced phosphorylation as well as proteolysis of nuclear factor-kappa B inhibitor alpha (IκBα) to prevent NF-κB nuclear translocation [138]. Epigallocatechin 3-gallate shows the potential for anti-inflammatory treatment from medicinal plants.

9.4.4 Guggulsterone

Guggulsterone (4,17(20)-pregnadiene-3,16-dione) is a phytosteroid found within the resin of *Commiphora mukul*. It can exist as either of two stereoisomers, E-guggulsterone and Z-guggulsterone [139]. Guggulsterone has important pharmacologic actions like being antineoplastic, antioxidant, antidiabetic, and cardioprotective. It is a compound that is biologically active in inhibiting the inflammatory response. Guggulsterone has been shown to potently inhibit the activation of NF-kB, which is a critical regulator of inflammatory responses. It also downregulates the expression of inflammatory gene products like COX-2, which are major players in the development of arthritis [140, 141]. Guggulsterone is a potent anti-inflammatory mediator with fewer side effects than classic steroids [142]. It inhibits NF-kB activation while suppressing the expression of antiapoptotic gene products and enhancing apoptosis [143]. Guggulsterone significantly suppressed elevated production of proinflammatory mediators in LPS-stimulated RAW264.7 cells [144]. It attenuated colitis in mice through inhibition of NF-kB activation [145]. Guggulsterone has potential for anti-inflammatory treatment from medicinal plants, but further preclinical and clinical studies are needed in the future.

9.4.5 Resveratrol

Resveratrol (3,5,4′-trihydroxy-trans-stilbene) belongs to the stilbenoids group of polyphenols, with two phenol rings connected by an ethylene bridge [146]. More than 70 plant species have been found to contain this natural polyphenol, including grapes, mulberries, peanuts, rhubarb, and several others. Diabetes, obesity, cardiovascular disease, neurodegeneration, and cancer, among other chronic diseases associated with inflammation, may benefit from resveratrol [147]. This compound has many properties, including activity against glycation, oxidative stress, inflammation, neurodegeneration, several types of cancer, and aging [148]. The mechanism of action of resveratrol involves multiple immune responses and signaling pathways; it is quickly absorbed and metabolized into various derivatives [149]. Resveratrol inhibits immune cell regulation, the synthesis of proinflammatory cytokines, and gene expression, all of which affect immunity. Sirtuin, adenosine monophosphate kinase, NF-κB, inflammatory cytokines, antioxidant enzymes, and cellular processes like gluconeogenesis, lipid metabolism, mitochondrial biogenesis, angiogenesis, and apoptosis are all targets at the molecular level.

Resveratrol can suppress TLR and proinflammatory gene expression [147]. Anti-inflammatory drugs and monoclonal antibodies targeting pro-inflammatory cytokines have been shown to prevent inflammatory bowel disease symptoms, as has resveratrol [149]. The immunomodulatory effect in immunosuppressive mice induced by cyclophosphamide and anti-inflammatory activities were assessed using a resveratrol dry suspension (RDS) [150]. In immunosuppressive mice, RDS inhibited resveratrol degradation and increased the

CD3+ CD4+/CD3+ CD8+ ratio, spleen index, IL-2 level, and splenic lymphocytes. RDS (0.1 g/kg/d) inhibited acetic acid-induced capillary permeability and suppressed ear swelling and granuloma formation in immunocompromised mice at doses of 0.33 and 1 g/kg/d [150]. Resveratrol increased collagen I levels and decreased matrix metalloproteinase (MMP)-2, transforming growth factor (TGF)-β, and NF-κB expression levels in monocular form deprivation (MFD)-induced myopia. It also suppressed the levels of TNF-α, IL-66, and IL-1β [151]. Resveratrol increases endogenous cellular antioxidant defenses in the CNS, modulating multiple synergistic pathways that are responsible for its antioxidant, anti-inflammatory, and antiapoptotic properties [152].

9.4.6 Thymoquinone

Thymoquinone (TQ, 2-isopropyl-5-methyl-1, 4-benzoquinone), a monoterpene molecule mostly found in the essential oil of *Nigella sativila* L. [153], is a potent herb with numerous promising pharmacologic properties against several diseases such as asthma, hypertension, diabetes, inflammation, cough, bronchitis, headache, eczema, influenza, and fever, and has anticancer and anticonvulsant properties [154]. TQ exerts a spread of pharmacologic effects, including being anti-inflammatory, antioxidant, antimicrobial, analgesic, histamine release inhibitory, hypoglycemic, anticarcinogenic, antihypertensive, immune-activating, and hepatoprotective effects [155]. Previous studies have shown that TQ was able to reduce the expression of proinflammatory and proliferative mediators such as TNF, iNOS, COX-2, 5-LOX, and cyclin D1 [156].

A recent study concluded that the anti-inflammatory and antioxidative effects of TQ were exhibited by inhibiting 12-O-tetradecanoylphorbol-13-acetate-stimulated COX-2 expression and NF-κB activation, as well as enhancing cytoprotective protein expression in nude mouse skin [157]. Moreover, TQ reduced the release of IL-10, IL-12, and TNF-α from LPS-induced dendritic cells (DCs), inhibited globin transcription factor (GATA) binding at the IL-5 LPS-stimulated promoter, and suppressed phosphorylation of prosurvival factors Akt and ERK1/2, but induced caspase-3 and caspase-8 activity in DCs [158]. In addition, it has been demonstrated that TQ prevented LT formation in human body cells through a decrease in synthase 5-lipoxygenase (5-LO) and leukotriene B4 (LTB4) in a time- and dose-dependent manner [154]. Furthermore, TQ plays a critical role in the inhibition of the NF-kB activation pathway via IkBα kinase inactivation, phosphorylation, and degradation of IkBα, inhibiting p65 phosphorylation that protects p65 from binding to DNA and the reporter gene [156]. Parla et al. showed that TQ inhibits both ileum tissue and plasma TNF-α and IL-1β expression, together with downregulating malondialdehyde quantity, restoring glutathione level, and suppressing MPO activity in intestinal ischemia/reperfusion injury [159]. In osteoarthritis treatment, TQ is claimed to become a potentially useful agent by preventing the production of inflammatory mediators (NO, iNOS, COX-2, PGE1, PGE2) induced by IL-1β via suppression of MAPK and NF-κB signaling routes in osteoarthritis chondrocytes [154].

9.5 Conclusion

Inflammation is the body's natural defense mechanism, and it is a healthy response to a disturbance or disease. However, some inflammatory disorders, such as rheumatoid arthritis, osteoarthritis, inflammatory bowel disease, retinitis, multiple sclerosis, psoriasis, and atherosclerosis, produce negative effects from the inflammatory process. Anti-inflammatory drugs can help to reduce tissue damage and improve patient comfort by interfering with the pathophysiology of inflammation. Plants with anti-inflammatory properties are extremely useful in treatment. The application of natural methods is a cutting-edge strategy, so we have included some herbal medicinal plants in this chapter that can help with inflammation.

References

1 Vishal, V., Ganesh, S., Mukesh, G., and Ranjan, B. (2014). A review on some plants having anti-inflammatory activity. *J. Phytopharmacol.* 3 (3): 214–221.

2 Kumar, S., Bajwa, B., Kuldeep, S., and Kalia, A. (2013). Anti-inflammatory activity of herbal plants: a review. *Int. J. Adv. Pharm. Biol. Chem.* 2 (2): 272–281.

3 Tasneem, S., Liu, B., Li, B. et al. (2019). Molecular pharmacology of inflammation: medicinal plants as anti-inflammatory agents. *Pharmacol. Res.* 139: 126–140.

4 Bindu, S., Mazumder, S., and Bandyopadhyay, U. (2020). Non-steroidal anti-inflammatory drugs (NSAIDs) and organ damage: a current perspective. *Biochem. Pharmacol.* 180: 114147.

5 Filiou, M.D., Arefin, A.S., Moscato, P., and Graeber, M.B. (2014). Neuroinflammation'differs categorically from inflammation: transcriptomes of Alzheimer's disease, Parkinson's disease, schizophrenia and inflammatory diseases compared. *Neurogenetics* 15 (3): 201–212.

6 Ronchetti, S., Migliorati, G., and Delfino, D. (2017). Association of inflammatory mediators with pain perception. *Biomed. Pharmacother.* 96: 1445–1452.

7 Ansar, W. and Ghosh, S. (2016). Inflammation and inflammatory diseases, markers, and mediators: role of CRP in some inflammatory diseases. In: *Biology of C Reactive Protein in Health and Disease* (ed. W. Ansar and S. Ghosh), 67–107. Berlin: Springer.

8 DiDonato, J.A., Mercurio, F., and Karin, M. (2012). NF-κB and the link between inflammation and cancer. *Immunol. Rev.* 246 (1): 379–400.

9 Scott, A., Khan, K., Roberts, C. et al. (2004). What do we mean by the term "inflammation"? A contemporary basic science update for sports medicine. *Br. J. Sports Med.* 38 (3): 372–380.

10 Glass, C.K., Saijo, K., Winner, B. et al. (2010). Mechanisms underlying inflammation in neurodegeneration. *Cell* 140 (6): 918–934.

11 Fernandes, L.M., Lima, R.R., Monteiro, M.C. et al. (2014). Alcohol in pregnancy and its repercussion on central nervous system. In: *Advances in Medicine and Biology* (ed. L.V. Berhardt), 25–42. Hauppauge, NY: Nova Science.

12 González, H. and Pacheco, R. (2014). T-cell-mediated regulation of neuroinflammation involved in neurodegenerative diseases. *J. Neuroinflammation* 11 (1): 1–11.

13 Jones, M.K., Nair, A., and Gupta, M. (2019). Mast cells in neurodegenerative disease. *Front. Cell. Neurosci.* 13: 171.

14 Thomsen, A.R. (2009). Lymphocytic choriomeningitis virus-induced central nervous system disease: a model for studying the role of chemokines in regulating the acute antiviral CD8+ T-cell response in an immune-privileged organ. *J. Virol.* 83 (1): 20–28.

15 Huber, S., Schramm, C., Lehr, H.A. et al. (2004). Cutting edge: TGF-β signaling is required for the in vivo expansion and immunosuppressive capacity of regulatory CD4+ CD25+ T cells. *J. Immunol.* 173 (11): 6526–6531.

16 Morris, G., Berk, M., Galecki, P. et al. (2016). The neuro-immune pathophysiology of central and peripheral fatigue in systemic immune-inflammatory and neuro-immune diseases. *Mol. Neurobiol.* 53 (2): 1195–1219.

17 Hou, Y., Dan, X., Babbar, M. et al. (2019). Ageing as a risk factor for neurodegenerative disease. *Nat. Rev. Neurol.* 15 (10): 565–581.

18 Vitner, E.B., Farfel-Becker, T., Eilam, R. et al. (2012). Contribution of brain inflammation to neuronal cell death in neuronopathic forms of Gaucher's disease. *Brain* 135 (6): 1724–1735.

19 Persidsky, Y. and Gendelman, H.E. (2003). Mononuclear phagocyte immunity and the neuropathogenesis of HIV-1 infection. *J. Leukoc. Biol.* 74 (5): 691–701.

20 Carrano, A., Hoozemans, J.J., Van Der Vies, S.M. et al. (2012). Neuroinflammation and blood-brain barrier changes in capillary amyloid angiopathy. *Neurodegen. Dis.* 10 (1–4): 329–331.

21 Graeber, M.B., Li, W., and Rodriguez, M.L. (2011). Role of microglia in CNS inflammation. *FEBS Lett.* 585 (23): 3798–3805.

22 Simpson, D.S. and Oliver, P.L. (2020). ROS generation in microglia: understanding oxidative stress and inflammation in neurodegenerative disease. *Antioxidants* 9 (8): 743.

23 Hanna, L., Poluyi, E., Ikwuegbuenyi, C. et al. (2022). Peripheral inflammation and neurodegeneration; a potential for therapeutic intervention in Alzheimer's disease (AD), Parkinson's disease (PD) and amyotrophic lateral sclerosis (ALS). *Egyptian J. Neurosurg.* 37 (1): 1–9.

24 Montesinos, J., Alfonso-Loeches, S., and Guerri, C. (2016). Impact of the innate immune response in the actions of ethanol on the central nervous system. *Alcohol. Clin. Exp. Res.* 40 (11): 2260–2270.

25 Tansey, M.G., McCoy, M.K., and Frank-Cannon, T.C. (2007). Neuroinflammatory mechanisms in Parkinson's disease: potential environmental triggers, pathways, and targets for early therapeutic intervention. *Exp. Neurol.* 208 (1): 1–25.

26 Parada, E., Egea, J., Buendia, I. et al. (2013). The microglial α7-acetylcholine nicotinic receptor is a key element in promoting neuroprotection by inducing heme oxygenase-1 via nuclear factor erythroid-2-related factor 2. *Antioxid. Redox Signal.* 19 (11): 1135–1148.

27 Ikeda, K. and Iwasaki, Y. (2015). Edaravone, a free radical scavenger, delayed symptomatic and pathological progression of motor neuron disease in the wobbler mouse. *PLoS One* 10 (10): e0140316.

28 Goldenberg, M.M. (2011). Pharmaceutical approval update. *Pharmacy Ther.* 36 (12): 798–800.

29 Johnson, J.W. and Kotermanski, S.E. (2006). Mechanism of action of memantine. *Curr. Opin. Pharmacol.* 6 (1): 61–67.

30 Betchel, N.T., Fariba, K.A., and Saadabadi, A. (2022). Lamotrigine. In: *StatPearls*. Treasure Island, FL: StatPearls Publishing.

31 Tolou-Ghamari, Z., Zare, M., Habibabadi, J.M., and Najafi, M.R. (2013). A quick review of carbamazepine pharmacokinetics in epilepsy from 1953 to 2012. *J. Res. Med. Sci.* 18 (Suppl 1): S81–S85.

32 Gupta, M. and Tripp, J. (2022). Phenytoin. In: *StatPearls*. Treasure Island, FL: StatPearls Publishing.

33 Berkovich, R. and Agius, M.A. (2014). Mechanisms of action of ACTH in the management of relapsing forms of multiple sclerosis. *Ther. Adv. Neurol. Disord.* 7 (2): 83–96.

34 Mulero, P., Midaglia, L., and Montalban, X. (2018). Ocrelizumab: a new milestone in multiple sclerosis therapy. *Ther. Adv. Neurol. Disord.* 11: 1756286418773025.

35 Chisari, C.G., Sgarlata, E., Arena, S. et al. (2022). Rituximab for the treatment of multiple sclerosis: a review. *J. Neurol.* 269 (1): 159–183.

36 Gerlach, M., Youdim, M.B., and Riederer, P. (1996). Pharmacology of selegiline. *Neurology* 47 (6 Suppl 3): S137–S145.

37 Finberg, J.P. (2010). Pharmacology of Rasagiline, a new MAO-B inhibitor drug for the treatment of Parkinson's disease with neuroprotective potential. *Rambam Maimonides Med. J.* 1 (1): e0003.

38 Mounsey, R.B. and Teismann, P. (2012). Chelators in the treatment of iron accumulation in Parkinson's disease. *Int. J. Cell Biol.* 2012: 983245.

39 Tenenbaum, L. and Humbert-Claude, M. (2017). Glial cell line-derived neurotrophic factor gene delivery in Parkinson's disease: a delicate balance between neuroprotection, trophic effects, and unwanted compensatory mechanisms. *Front. Neuroanat.* 11: 29.

40 Schenk, D.B., Koller, M., Ness, D.K. et al. (2017). First-in-human assessment of PRX002, an anti-alpha-synuclein monoclonal antibody, in healthy volunteers. *Mov. Disord.* 32 (2): 211–218.

41 Nunes, C.R., Barreto Arantes, M., de Faria, M. et al. (2020). Plants as sources of anti-inflammatory agents. *Molecules* 25 (16): 3726.

42 Recio, M.C., Andujar, I., and Rios, J.L. (2012). Anti-inflammatory agents from plants: progress and potential. *Curr. Med. Chem.* 19 (14): 2088–2103.

43 O'Brien, P. (2007). Garlic in traditional Chinese medicine. *Spezzatino* 2010: 26–29.

44 Huang, Y., Yao, P., Leung, K.W. et al. (2018). The Yin-Yang property of Chinese medicinal herbs relates to chemical composition but not anti-oxidative activity: an illustration using spleen-Meridian herbs. *Front. Pharmacol.* 9: 1304.

45 Batiha, G.E.S., Beshbishy, A.M., Wasef, L.G. et al. (2020). Chemical constituents and pharmacological activities of garlic (*Allium sativum* L.): a review. *Nutrients* 12 (3): 872.

46 Ayaz, E. and Alpsoy, H.C. (2007). Garlic (*Allium sativum*) and traditional medicine. *Turkiye Parazitol. Derg.* 31 (2): 145–149.

47 Shang, A., Cao, S.Y., Xu, X.Y. et al. (2019). Bioactive compounds and biological functions of garlic (*Allium sativum* L.). *Foods* 8 (7): 246.

48 Arreola, R., Quintero-Fabián, S., López-Roa, R.I. et al. (2015). Immunomodulation and anti-inflammatory effects of garlic compounds. *J. Immunol. Res.* 2015: 401630.

49 Shao, X., Sun, C., Tang, X. et al. (2020). Anti-inflammatory and intestinal microbiota modulation properties of Jinxiang garlic (*Allium sativum* L.) polysaccharides toward dextran sodium sulfate-induced colitis. *J. Agric. Food Chem.* 68 (44): 12295–12309.

50 Kalkal, A., Allawadhi, P., Pradhan, R. et al. (2021). *Allium sativum* derived carbon dots as a potential theranostic agent to combat the COVID-19 crisis. *Sens. Int.* 2: 100102.

51 Dehghani, S., Alipoor, E., Salimzadeh, A. et al. (2018). The effect of a garlic supplement on the pro-inflammatory adipocytokines, resistin and tumor necrosis factor-alpha, and on pain severity, in overweight or obese women with knee osteoarthritis. *Phytomedicine* 48: 70–75.

52 Braga, T.M., Rocha, L., Chung, T.Y. et al. (2021). *Azadirachta indica* A. Juss. In vivo toxicity – an updated review. *Molecules* 26 (2): 252.

53 Saleem, S., Muhammad, G., Hussain, M.A., and Bukhari, S.N.A. (2018). A comprehensive review of phytochemical profile, bioactives for pharmaceuticals, and pharmacological attributes of *Azadirachta indica*. *Phytother. Res.* 32 (7): 1241–1272.

54 Nigussie, D., Makonnen, E., Tufa, T.B. et al. (2021). Systematic review of Ethiopian medicinal plants used for their anti-inflammatory and wound healing activities. *J. Ethnopharmacol.* 276: 114179.

55 Shewale, S.P., Panadare, D., and Rathod, V.K. (2021). Extraction of total antioxidants from *Azadirachta indica* (neem) using three phase partitioning and its process intensification using ultrasound. *Prep. Biochem. Biotechnol.* 52 (5): 534–539.

56 Nagaraj, A. and Samiappan, S. (2019). Presentation of antibacterial and therapeutic anti-inflammatory potentials to hydroxyapatite via biomimetic with *Azadirachta indica*: an in vitro anti-inflammatory assessment in contradiction of LPS-induced stress in RAW 264.7 cells. *Front. Microbiol.* 10: 1757.

57 Eze, M.O., Ejike, C.E., Ifeonu, P. et al. (2022). Anti-COVID-19 potential of *Azadirachta indica* (neem) leaf extract. *Sci. Afr.* e01184.

58 Anyebe, D.A., Tajudeen, Y.O., Shemishere, U.B. et al. (2021). Methanol leaf extract of *Cassia tora* ameliorates dextran sulfate sodium-induced ulcerative colitis in BALB/c mice. *Sci. Afr.* 13: e00865.

59 Lee, K.H., Jang, J.H., Woo, K.W. et al. (2019). Anti-inflammatory effects of Aurantio-obtusin isolated from *Cassia tora* L. in RAW264. 7 cells. *Korean J. Pharmacognosy* 50 (1): 11–17.

60 Lee, M.-J., Nho, J.-H., Yang, B.-D. et al. (2019). Subchronic toxicity evaluation of ethanol extract of *Cassia tora* L. seeds in rats. *Regul. Toxicol. Pharmacol.* 109: 104487.

61 Shadab, M., Shamsi, S., and Ahmad, I. (2019). *Cassia tora* Linn: a medicinal herb for skin diseases. *J. Emerg. Technol. Innovat. Res.* 6 (3): 97–104.

62 Lee, E.K., Song, J., Seo, Y. et al. (2020). Inhibitory effects of AF-343, a mixture of *Cassia tora* L., *Ulmus pumila* L., and *Taraxacum officinale*, on compound 48/80-mediated allergic responses in RBL-2H3 cells. *Molecules* 25 (10): 2434.

63 Antonisamy, P., Dhanasekaran, M., Kim, H.-R. et al. (2017). Anti-inflammatory and analgesic activity of ononitol monohydrate isolated from *Cassia tora* L. in animal models. *Saudi J. Biol. Sci.* 24 (8): 1933–1938.

64 Kumar, S., Malhotra, R., and Kumar, D. (2010). *Euphorbia hirta*: its chemistry, traditional and medicinal uses, and pharmacological activities. *Pharmacogn. Rev.* 4 (7): 58.

65 Sood, S.K., Bhardwaj, R., and Lakhanpal, T. (2005). *Ethnic Indian Plants in Cure of Diabetes*. Jodhpur: Scientific Publishers.

66 Williamson, E. (2002). *Major Herbs of Ayurveda*. Edinburgh: Churchill Livingstone.

67 Liu, Y., Murakami, N., Ji, H. et al. (2007). Antimalarial Flavonol glycosides from *Euphorbia hirta*. *Pharm. Biol.* 45 (4): 278–281.

68 Sharma, N., Samarakoon, K.W., Gyawali, R. et al. (2014). Evaluation of the antioxidant, anti-inflammatory, and anticancer activities of *Euphorbia hirta* ethanolic extract. *Molecules* 19 (9): 14567–14581.

69 Shih, M.-F., Cheng, Y.-D., Shen, C.-R., and Cherng, J.-Y. (2010). A molecular pharmacology study into the anti-inflammatory actions of *Euphorbia hirta* L. on the LPS-induced RAW 264.7 cells through selective iNOS protein inhibition. *J. Nat. Med.* 64 (3): 330–335.

70 Martínez-Vázquez, M., Apan, T.O.R., Lazcano, M.E., and Bye, R. (1999). Anti-inflammatory active compounds from the n-hexane extract of *Euphorbia hirta*. *J. Mex. Chem. Soc.* 43: 103–105.

71 Ahmad, S.F., Attia, S.M., Bakheet, S.A. et al. (2014). Anti-inflammatory effect of *Euphorbia hirta* in an adjuvant-induced arthritic murine model. *Immunol. Invest.* 43 (3): 197–211.

72 Chen, J., Er, H., Mohamed, S., and Chen, Y. (2015). In vitro anti-inflammatory activity of fractionated *Euphorbia hirta* aqueous extract on rabbit synovial fibroblasts. *Biom. J.* 38 (4): 301–306.

73 Xia, M., Liu, L., Qiu, R. et al. (2018). Anti-inflammatory and anxiolytic activities of *Euphorbia hirta* extract in neonatal asthmatic rats. *AMB Express* 8 (1): 1–11.

74 Chaverri, J.P., Rodriguez, N.C., Ibarra, M.O., and Rojas, J.M.P. (2019). Medicinal properties of mangosteen (Garcinia mangostana). *Food Chem. Toxicol.* 46 (10): 3227–3239.

75 Ansori, A.N.M., Fadholly, A., Hayaza, S. et al. (2020). A review on medicinal properties of mangosteen (*Garcinia mangostana* L.). *Res. J. Pharmacy Technol.* 13 (2): 974–982.

76 Ovalle-Magallanes, B., Eugenio-Perez, D., and Pedraza-Chaverri, J. (2017). Medicinal properties of mangosteen (*Garcinia mangostana* L.): a comprehensive update. *Food Chem. Toxicol.* 109 (Pt 1): 102–122.

77 Rizaldy, D., Hartati, R., Nadifa, T., and Fidrianny, I. (2021). Chemical compounds and pharmacological activities of mangosteen (*Garcinia mangostana* L.) – updated review. *Biointerface Res. Appl. Chem.* 12 (2): 2503–2516.

78 Chomnawang, M.T., Surassmo, S., Nukoolkarn, V.S., and Gritsanapan, W. (2007). Effect of *Garcinia mangostana* on inflammation caused by Propionibacterium acnes. *Fitoterapia* 78 (6): 401–408.

79 Gutierrez-Orozco, F., Chitchumroonchokchai, C., Lesinski, G.B. et al. (2013). Alpha-Mangostin: anti-inflammatory activity and metabolism by human cells. *J. Agric. Food Chem.* 61 (16): 3891–3900.

80 Karunakaran, T., Ee, G.C.L., Ismail, I.S. et al. (2018). Acetyl- and O-alkyl-derivatives of beta-mangostin from *Garcinia mangostana* and their anti-inflammatory activities. *Nat. Prod. Res.* 32 (12): 1390–1394.

81 Pedraza-Chaverri, J., Cárdenas-Rodríguez, N., Orozco-Ibarra, M., and Pérez-Rojas, J.M. (2008). Medicinal properties of mangosteen (*Garcinia mangostana*). *Food Chem. Toxicol.* 46 (10): 3227–3239.

82 Lim, Y.K., Yoo, S.Y., Jang, Y.Y. et al. (2020). Anti-inflammatory and in vitro bone formation effects of *Garcinia mangostana* L. and propolis extracts. *Food Sci. Biotechnol.* 29 (4): 539–548.

83 Tatiya-Aphiradee, N., Chatuphonprasert, W., and Jarukamjorn, K. (2019). Anti-inflammatory effect of *Garcinia mangostana* Linn. pericarp extract in methicillin-resistant *Staphylococcus aureus*-induced superficial skin infection in mice. *Biomed. Pharmacother.* 111: 705–713.

84 Eghbali, S., Askari, S.F., Avan, R., and Sahebkar, A. (2021). Therapeutic effects of *Punica granatum* (pomegranate): an updated review of clinical trials. *J. Nutr. Metab.* 2021: 5297162.

85 Akhtar, N. and Haqqi, T.M. (2012). Current nutraceuticals in the management of osteoarthritis: a review. *Ther. Adv. Musculoskelet. Dis.* 4 (3): 181–207.

86 Saeed, M., Naveed, M., BiBi, J. et al. (2018). The promising pharmacological effects and therapeutic/medicinal applications of *Punica granatum* L. (pomegranate) as a functional food in humans and animals. *Recent Pat. Inflamm. Allergy Drug Discov.* 12 (1): 24–38.

87 Vucic, V., Grabez, M., Trchounian, A., and Arsic, A. (2019). Composition and potential health benefits of pomegranate: a review. *Curr. Pharm. Des.* 25 (16): 1817–1827.

88 Mastrogiovanni, F., Bernini, R., Basirico, L. et al. (2020). Antioxidant and anti-inflammatory effects of pomegranate peel extracts on bovine mammary epithelial cells BME-UV1. *Nat. Prod. Res.* 34 (10): 1465–1469.

89 Colombo, E., Sangiovanni, E., and Dell'agli, M. (2013). A review on the anti-inflammatory activity of pomegranate in the gastrointestinal tract. *Evid. Based Complement. Alternat. Med.* 2013: 247145.

90 Cao, Y., Chen, J., Ren, G. et al. (2019). Punicalagin prevents inflammation in LPS-induced RAW264.7 macrophages by inhibiting $FoxO_3a$/autophagy signaling pathway. *Nutrients* 11 (11): 2794.

91 BenSaad, L.A., Kim, K.H., Quah, C.C. et al. (2017). Anti-inflammatory potential of ellagic acid, gallic acid and punicalagin A&B isolated from *Punica granatum*. *BMC Complement. Altern. Med.* 17 (1): 47.

92 Polito, L., Bortolotti, M., Battelli, M.G. et al. (2019). Ricin: an ancient story for a timeless plant toxin. *Toxins* 11 (6): 324.

93 Abdul, W.M., Hajrah, N.H., Sabir, J.S. et al. (2018). Therapeutic role of *Ricinus communis* L. and its bioactive compounds in disease prevention and treatment. *Asian Pac. J. Trop. Med.* 11 (3): 177.

94 Yeboah, A., Ying, S., Lu, J. et al. (2020). Castor oil (*Ricinus communis*): a review on the chemical composition and physicochemical properties. *Food Sci. Technol.* https://doi.org/10.1590/fst.19620.

95 Elkousy, R.H., Said, Z.N., and Abd El-Baseer, M.A. (2021). Antiviral activity of castor oil plant (*Ricinus communis*) leaf extracts. *J. Ethnopharmacol.* 271: 113878.

96 Hussain, A., Aslam, B., Muhammad, F. et al. (2021). Anti-arthritic activity of *Ricinus communis* L. and *Withania somnifera* L. extracts in adjuvant-induced arthritic rats via modulating inflammatory mediators and subsiding oxidative stress. *Iran. J. Basic Med. Sci.* 24 (7): 951.

97 Zhao, T., Tang, H., Xie, L. et al. (2019). *Scutellaria baicalensis* Georgi (Lamiaceae): a review of its traditional uses, botany, phytochemistry, pharmacology and toxicology. *J. Pharm. Pharmacol.* 71 (9): 1353–1369.

98 Wang, Z.-L., Wang, S., Kuang, Y. et al. (2018). A comprehensive review on phytochemistry, pharmacology, and flavonoid biosynthesis of *Scutellaria baicalensis*. *Pharm. Biol.* 56 (1): 465–484.

99 Liao, H., Ye, J., Gao, L., and Liu, Y. (2021). The main bioactive compounds of *Scutellaria baicalensis* Georgi. For alleviation of inflammatory cytokines: a comprehensive review. *Biomed. Pharmacother.* 133: 110917.

100 Maleki, S.J., Crespo, J.F., and Cabanillas, B. (2019). Anti-inflammatory effects of flavonoids. *Food Chem.* 299: 125124.

101 Li, Y., Liu, T., Li, Y. et al. (2020). Baicalin ameliorates cognitive impairment and protects microglia from LPS-induced neuroinflammation via the SIRT1/HMGB1 pathway. *Oxid. Med. Cell. Longev.* 2020: 4751349.

102 Li, D., Shi, G., Wang, J. et al. (2019). Baicalein ameliorates pristane-induced lupus nephritis via activating Nrf2/HO-1 in myeloid-derived suppressor cells. *Arthritis Res. Ther.* 21 (1): 1–14.

103 Khan, N.M., Haseeb, A., Ansari, M.Y. et al. (2017). Wogonin, a plant derived small molecule, exerts potent anti-inflammatory and chondroprotective effects through the activation of ROS/ERK/Nrf2 signaling pathways in human osteoarthritis chondrocytes. *Free Radic. Biol. Med.* 106: 288–301.

104 Xu, X., Xia, J., Zhao, S. et al. (2021). Qing-Fei-Pai-Du decoction and wogonoside exert anti-inflammatory action through down-regulating USP14 to promote the degradation of activating transcription factor 2. *FASEB J.* 35 (9): e21870.

105 Ye, M., Wang, Q., Zhang, W. et al. (2014). Oroxylin A exerts anti-inflammatory activity on lipopolysaccharide-induced mouse macrophage via Nrf2/ARE activation. *Biochem. Cell Biol.* 92 (5): 337–348.

106 Eck, J.V. and Snyder, A. (2006). Eggplant (*Solanum melongena* L.). *Agrobacterium Protocols* 343: 439–448.

107 Igwe, S., Akunyili, D., and Ogbogu, C. (2003). Effects of *Solanum melongena* (garden egg) on some visual functions of visually active Igbos of Nigeria. *J. Ethnopharmacol.* 86 (2–3): 135–138.

108 Das, M. and Barua, N. (2013). Pharmacological activities of *Solanum melongena* Linn. (brinjal plant). *Int. J. Green Pharmacy (IJGP)* 7 (4): 274.

109 Song, Y., Mei, T., Liu, Y. et al. (2021). Metabolites identification of chemical constituents from the eggplant (*Solanum melongena* L.) calyx in rats by UPLC/ESI/qTOF-MS analysis and their cytotoxic activities. *Front. Pharmacol.* 12: 655008.

110 Umamageswari, M. and Maniyar, Y.A. (2015). Evaluation of anti-inflammatory activity of aqueous extract of leaves of *Solanum melongena* Linn. in experimental animals. *J. Clin. Diagn. Res.* 9 (1): FF01.

111 Bai, R., Yao, C., Zhong, Z. et al. (2021). Discovery of natural anti-inflammatory alkaloids: potential leads for the drug discovery for the treatment of inflammation. *Eur. J. Med. Chem.* 213: 113165.

112 Liu, Y., Yin, X., Sun, Y.-p. et al. (2022). Chemical constituent from the roots of *Solanum melongena* L. and their potential anti-inflammatory activity. *Nat. Prod. Res.* 36 (7): 1757–1764.

113 Sun, J., Gu, Y.-F., Su, X.-Q. et al. (2014). Anti-inflammatory lignanamides from the roots of *Solanum melongena* L. *Fitoterapia* 98: 110–116.

114 Han, S.-W., Tae, J., Kim, J.-A. et al. (2003). The aqueous extract of *Solanum melongena* inhibits PAR2 agonist-induced inflammation. *Clin. Chim. Acta* 328 (1–2): 39–44.

115 Patel, S.S., Acharya, A., Ray, R. et al. (2020). Cellular and molecular mechanisms of curcumin in prevention and treatment of disease. *Crit. Rev. Food Sci. Nutr.* 60 (6): 887–939.

116 Kocaadam, B. and Şanlier, N. (2017). Curcumin, an active component of turmeric (*Curcuma longa*), and its effects on health. *Crit. Rev. Food Sci. Nutr.* 57 (13): 2889–2895.

117 Fallahi, F., Borran, S., Ashrafizadeh, M. et al. (2021). Curcumin and inflammatory bowel diseases: from in vitro studies to clinical trials. *Mol. Immunol.* 130: 20–30.

118 Peng, Y., Ao, M., Dong, B. et al. (2021). Anti-inflammatory effects of curcumin in the inflammatory diseases: status, limitations and countermeasures. *Drug Des. Devel. Ther.* 15: 4503.

119 Vallée, A. and Lecarpentier, Y. (2020). Curcumin and endometriosis. *Int. J. Mol. Sci.* 21 (7): 2440.

120 Campbell, M.S., Carlini, N.A., and Fleenor, B.S. (2021). Influence of curcumin on performance and post-exercise recovery. *Crit. Rev. Food Sci. Nutr.* 61 (7): 1152–1162.

121 Rinkunaite, I., Simoliunas, E., Alksne, M. et al. (2021). Anti-inflammatory effect of different curcumin preparations on adjuvant-induced arthritis in rats. *BMC Complement. Med. Therapies* 21 (1): 1–12.

122 Sadeghi, N., Mansoori, A., Shayesteh, A., and Hashemi, S.J. (2020). The effect of curcumin supplementation on clinical outcomes and inflammatory markers in patients with ulcerative colitis. *Phytother. Res.* 34 (5): 1123–1133.

123 Barboza, J.N., da Silva Maia Bezerra Filho, C., Silva, R.O. et al. (2018). An overview on the anti-inflammatory potential and antioxidant profile of eugenol. *Oxid. Med. Cell. Longev.* 2018: 3957262.

124 Mohammadi Nejad, S., Ozgunes, H., and Basaran, N. (2017). Pharmacological and toxicological properties of eugenol. *Turk. J. Pharm. Sci.* 14 (2): 201–206.

125 Ulanowska, M. and Olas, B. (2021). Biological properties and prospects for the application of eugenol – a review. *Int. J. Mol. Sci.* 22 (7): 3671.

126 Han, X. and Parker, T.L. (2017). Anti-inflammatory activity of clove (*Eugenia caryophyllata*) essential oil in human dermal fibroblasts. *Pharm. Biol.* 55 (1): 1619–1622.

127 Estevao-Silva, C.F., Kummer, R., Fachini-Queiroz, F.C. et al. (2014). Anethole and eugenol reduce in vitro and in vivo leukocyte migration induced by fMLP, LTB4, and carrageenan. *J. Nat. Med.* 68 (3): 567–575.

128 Huang, X., Liu, Y., Lu, Y., and Ma, C. (2015). Anti-inflammatory effects of eugenol on lipopolysaccharide-induced inflammatory reaction in acute lung injury via regulating inflammation and redox status. *Int. Immunopharmacol.* 26 (1): 265–271.

129 Magalhaes, C.B., Casquilho, N.V., Machado, M.N. et al. (2019). The anti-inflammatory and anti-oxidative actions of eugenol improve lipopolysaccharide-induced lung injury. *Respir. Physiol. Neurobiol.* 259: 30–36.

130 Arana-Argáez, V.E., Domínguez, F., Moreno, D.A. et al. (2020). Anti-inflammatory and antinociceptive effects of an ethanol extract from *Senna septemtrionalis*. *Inflammopharmacology* 28 (2): 541–549.

131 Nguyen, T.H., Nachtergael, A., Nguyen, T.M. et al. (2020). Anti-inflammatory properties of the ethanol extract from *Clerodendrum cyrtophyllum* Turcz based on in vitro and in vivo studies. *J. Ethnopharmacol.* 254: 112739.

132 Dos Santos, E., Silva-Filho, S.E., Radai, J.A.S. et al. (2021). Anti-inflammatory properties of ethanolic extract from *Vatairea macrocarpa* leaves. *J. Ethnopharmacol.* 278: 114308.

133 Thomas, P., Essien, E., Udoh, A. et al. (2021). Isolation and characterization of anti-inflammatory and analgesic compounds from *Uapaca staudtii* Pax (Phyllanthaceae) stem bark. *J. Ethnopharmacol.* 269: 113737.

134 Alonso-Castro, A.J., Arana-Argáez, V., Yáñez-Barrientos, E. et al. (2021). Antinociceptive and anti-inflammatory effects of *Cuphea aequipetala* Cav (Lythraceae). *Inflammopharmacology* 29 (1): 295–306.

135 Chakrawarti, L., Agrawal, R., Dang, S. et al. (2016). Therapeutic effects of EGCG: a patent review. *Expert Opin. Ther. Pat.* 26 (8): 907–916.

136 Huang, H.-T., Cheng, T.-L., Yang, C.-D. et al. (2021). Intra-articular injection of (−)-epigallocatechin 3-gallate (EGCG) ameliorates cartilage degeneration in Guinea pigs with spontaneous osteoarthritis. *Antioxidants* 10 (2): 178.

137 Ohishi, T., Goto, S., Monira, P. et al. (2016). Anti-inflammatory action of green tea. *Anti-Inflammatory Anti-Allergy Agents Med. Chem. (Formerly Current Medicinal Chemistry-Anti-Inflammatory and Anti-Allergy Agents)* 15 (2): 74–90.

138 Ahmed, S. (2010). Green tea polyphenol epigallocatechin 3-gallate in arthritis: progress and promise. *Arthritis Res. Ther.* 12 (2): 1–9.

139 Kunnumakkara, A.B., Banik, K., Bordoloi, D. et al. (2018). Googling the Guggul (Commiphora and Boswellia) for prevention of chronic diseases. *Front. Pharmacol.* 9: 686.

140 Arora, R., Malhotra, P., Sharma, A. et al. (2013). Medicinal efficacy of Indian herbal remedies for the treatment of arthritis. In: *Bioactive Food as Dietary Interventions for Arthritis and Related Inflammatory Diseases* (ed. R.R. Watson and V.R. Preedy), 601–617. St. Louis, MO: Elsevier.

141 Deng, R. (2007). Therapeutic effects of guggul and its constituent guggulsterone: cardiovascular benefits. *Cardiovasc. Drug Rev.* 25 (4): 375–390.

142 Gebhard, C., Stämpfli, S.F., Gebhard, C.E. et al. (2009). Guggulsterone, an anti-inflammatory phytosterol, inhibits tissue factor and arterial thrombosis. *Basic Res. Cardiol.* 104 (3): 285–294.

143 Shishodia, S., Sethi, G., Ahn, K.S., and Aggarwal, B.B. (2007). Guggulsterone inhibits tumor cell proliferation, induces S-phase arrest, and promotes apoptosis through activation of c-Jun N-terminal kinase, suppression of Akt pathway, and downregulation of antiapoptotic gene products. *Biochem. Pharmacol.* 74 (1): 118–130.

144 Zhang, J.-H., Shangguan, Z.-S., Chen, C. et al. (2016). Anti-inflammatory effects of guggulsterone on murine macrophage by inhibiting LPS-induced inflammatory cytokines in NF-κB signaling pathway. *Drug Des. Devel. Ther.* 10: 1829.

145 Kim, J.M., Kim, S.H., Ko, S.H. et al. (2013). The guggulsterone derivative GG-52 inhibits NF-κB signaling in gastric epithelial cells and ameliorates ethanol-induced gastric mucosal lesions in mice. *Am. J. Physiol. Gastrointest. Liver Phys. Ther.* 304 (2): G193–G202.

146 Salehi, B., Mishra, A.P., Nigam, M. et al. (2018). Resveratrol: a double-edged sword in health benefits. *Biomedicines* 6 (3): 91.

147 Malaguarnera, L. (2019). Influence of resveratrol on the immune response. *Nutrients* 11 (5): 946.

148 Galiniak, S., Aebisher, D., and Bartusik-Aebisher, D. (2019). Health benefits of resveratrol administration. *Acta Biochim. Pol.* 66 (1): 13–21.

149 Nunes, S., Danesi, F., Del Rio, D., and Silva, P. (2018). Resveratrol and inflammatory bowel disease: the evidence so far. *Nutr. Res. Rev.* 31 (1): 85–97.

150 Chen, M., Fu, Q., Song, X. et al. (2020). Preparation of resveratrol dry suspension and its immunomodulatory and anti-inflammatory activity in mice. *Pharm. Biol.* 58 (1): 8–15.

151 Hsu, Y.-A., Chen, C.-S., Wang, Y.-C. et al. (2021). Anti-inflammatory effects of resveratrol on human retinal pigment cells and a myopia animal model. *Curr. Issues Mol. Biol.* 43 (2): 716–727.

152 Miguel, C., Noya-Riobó, M., Mazzone, G. et al. (2021). Antioxidant, anti-inflammatory and neuroprotective actions of resveratrol after experimental nervous system insults. Special focus on the molecular mechanisms involved. *Neurochem. Int.* 150: 105188.

153 Ahmad, A., Mishra, R.K., Vyawahare, A. et al. (2019). Thymoquinone (2-Isopropyl-5-methyl-1, 4-benzoquinone) as a chemopreventive/anticancer agent: chemistry and biological effects. *Saudi Pharm. J.* 27 (8): 1113–1126.

154 Shaterzadeh-Yazdi, H., Noorbakhsh, M.F., Hayati, F. et al. (2018). Immunomodulatory and anti-inflammatory effects of thymoquinone. *Cardiovasc. Hematol. Disord. Drug Targets* 18 (1): 52–60.

155 Mekhemar, M., Hassan, Y., and Dorfer, C. (2020). *Nigella sativa* and thymoquinone: a natural blessing for periodontal therapy. *Antioxidants (Basel)* 9 (12): 1260.

156 Khan, A. (2018). Antioxidant and anti-inflammatory action of thymoquinone. In: *Molecular and Therapeutic Actions of Thymoquinone: Actions of Thymoquinone* (ed. H. Younus), 41–56. Singapore: Springer.

157 Kohandel, Z., Farkhondeh, T., Aschner, M., and Samarghandian, S. (2021). Anti-inflammatory effects of thymoquinone and its protective effects against several diseases. *Biomed. Pharmacother.* 138: 111492.

158 Khazdair, M.R., Ghafari, S., and Sadeghi, M. (2021). Possible therapeutic effects of Nigella sativa and its thymoquinone on COVID-19. *Pharm. Biol.* 59 (1): 696–703.

159 Parlar, A. and Arslan, S.O. (2020). Thymoquinone reduces ischemia and reperfusion-induced intestinal injury in rats, through anti-oxidative and anti-inflammatory effects. *Turk. J. Surg.* 36 (1): 96–104.

10

Plant-Based Products and Phytochemicals against Viral Infections of the Central Nervous System

Santwana Palai[1], Shyam S. Kesh[2], and Mithun Rudrapal[3]

[1] *Department of Veterinary Pharmacology & Toxicology, College of Veterinary Science and Animal Husbandry, Odisha University of Agriculture and Technology, Bhubaneswar, Odisha, India*
[2] *Department of Veterinary Clinical Complex (Veterinary Biochemistry), Faculty of Veterinary and Animal Sciences, West Bengal University of Animal and Fishery Sciences, Kolkata, West Bengal, India*
[3] *Department of Pharmaceutical Sciences, School of Biotechnology and Pharmaceutical Sciences, Vignan's Foundation for Science, Technology & Research (Deemed to be University), Guntur, Andhra Pradesh, India*

According to a World Health Organization (WHO) assessment, traditional herbs are used by 80% of the population in poorer nations for health purposes. One-quarter of regularly used medications contain plant-derived chemicals. Phytoconstituents derived from extracts of specific plant parts like stems, bark, seeds, flowers, fruits, roots, and rhizomes can create plant-based products to be used as dietary supplements and nutraceuticals. These have a wide range of applications in treating disorders of infectious and non-infectious origin. Emetine derived from *Cephaelis ipecacuanha* (amoebicidal and antibacterial drug), quinine derived from the bark of the cinchona tree (malaria), and vincristine also derived from the bark of the cinchona tree (cancer) are three classic examples of herbal leads that have been translated into potential drug candidates. Plant-derived chemicals have also been used to make aspirin, morphine, and paclitaxel [1].

According to the WHO, "infectious illnesses constitute the sixth greatest cause of early mortality worldwide." Infectious illnesses that are developing or reemerging continue to represent a persistent hazard to the human race. Among infectious illnesses, viral infections produced by various novel and ancient infectious viruses pose a threat to human survival. Antiviral effects of several plant extracts and phytochemicals have been researched and proved to be effective against deadly viral infections. Several herbal plants with broad-spectrum antiviral action have been documented in the ethnomedicinal literature. With the advent and reemergence of highly infectious viruses, the exploration of medicinal plants for their antiviral activity has increased tremendously [2]. In an era of antiviral medicine resistance, herbal antiviral compounds are the emerging alternatives.

Herbal treatments fortify the immunity of the body in fighting against invasive infective viruses due to their comprehensive approach. There are several potential herbal remedies for viral disorders, with evidence of effectiveness and safety. This chapter summarizes information on plants with antiviral characteristics from traditional knowledge [3]. With the goal of prolonging drug therapy efficacy, researchers are trying to develop potential antiviral drugs

Phytochemical Drug Discovery for Central Nervous System Disorders: Biochemistry and Therapeutic Effects,
First Edition. Edited by Chukwuebuka Egbuna and Mithun Rudrapal.

from natural substances that were used as antiviral agents in the past. In the absence of efficient antiviral medications in the domain of escalating resistance of many new and reemergent viral disorders, these herbal sources provide researchers with a tremendous chance to investigate viable alternatives against viral infections of the central nervous system (CNS). These plant-based products and phytochemicals that are effective against viral infections of the CNS (Table 10.1) can be used by drug developers and researchers in novel drug development for CNS viral diseases.

10.1 Viral Infections of the Central Nervous System

Infectious diseases are the top cause of death around the world. Viral infections are still a major public health concern that can lead to pandemics. The most common viruses implicated in CNS illnesses are SARS-Cov-2 virus, Japanese encephalitis virus (JEV), West Nile virus (WNV), tick-borne encephalitis (TBE), herpes simplex virus (HSV), rabies virus, varicella zoster virus (VZV), poliomyelitis virus, and human immunodeficiency virus (HIV). Viruses infecting the mammalian CNS exhibit acute lethal consequences to chronic or few or no symptoms. Vaccines and antiviral drugs are both important strategies for preventing and treating viral infections [28].

10.2 Plant and Phytochemicals as Antiviral Agents for Central Nervous System Viral Infections

Natural compounds derived from plant resources have been utilized to treat viral CNS illnesses since ancient times. These chemically diverse, naturally occurring compounds have antiviral characteristics that encourage their use in therapies for viral infections. The active principles with antiviral activity are the phytochemicals found in diverse parts of plants [29]. Phytochemicals like phenolic compounds, alkaloids, flavonoids, saponins, quinines, tannins, terpenes, proanthocyanidins, proanthocyanins, lignins, glycosides, steroids, organic acids, coumarins, and thiosulfonates have potent antiviral action [30]. Plant-based natural substances such as caffeic acid, griffithsin, isobavachalcone, myricetin, psoralidin, quercetin, saikosaponin B2, scutellarein, silvestrol, and tryptanthrin have shown encouraging results against fatal CNS viruses in humans in the ongoing hunt for antiviral medicines (Table 10.1). Bioactives generated from plants have antiviral efficacy against viral targets such as viral entrance and replication in host cells. The suppression of viral attachment to cell membranes, the blocking of viral genome replication, the prevention of viral protein production, and the scavenging of reactive oxygen species are the primary mechanisms by which medicinal plant-based products inhibit CNS viruses (Figure 10.1). Phytoconstituents also have the ability to increase bodily immunity [30].

10.2.1 SARS Cov-2 Virus

Coronaviruses cause a wide spectrum of viral infections in humans and animals. Among these are severe acute respiratory syndrome (SARS), avian infectious bronchitis (IB) in poultry, Middle East respiratory syndrome (MERS), and coronavirus disease 2019 (COVID-19) in

Table 10.1 Plant-based products and phytochemicals against viral infections of the central nervous system.

Phytochemical	Source	Structure	Mechanism	References
Psoralidin	*Psoralea corylifolia*	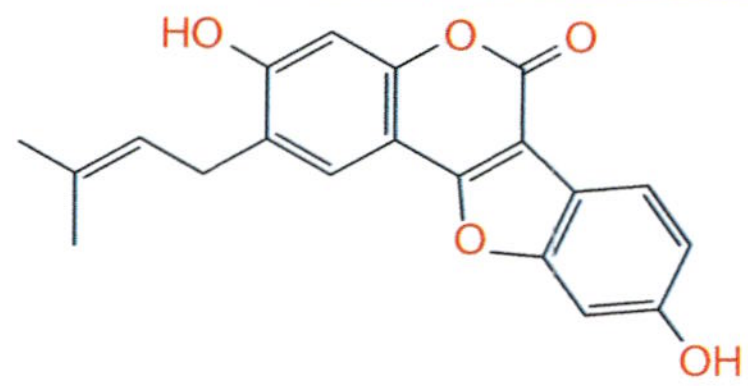	Demonstrated inhibition of SARS-CoV-2 papain-like protease	[4]
Scutellarein	*Scutellaria baicalensis*	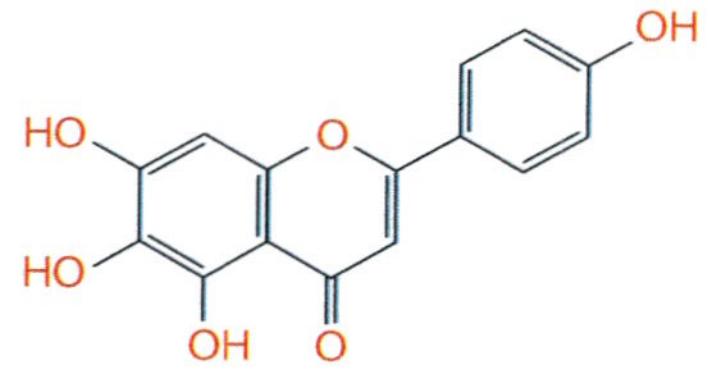	Reduces SARS-CoV-2 helicase ATPase activity	[5]
Silvestrol	*Aglaia* sp.	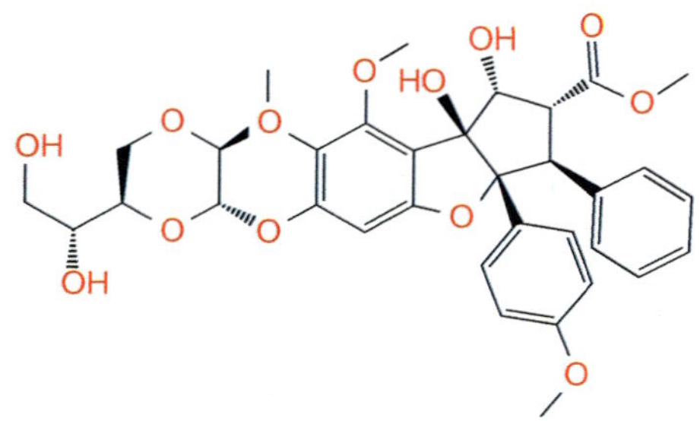	Acts as specific inhibitor of RNA helicase eIF4A	[6]

(*Continued*)

Table 10.1 (Continued)

Phytochemical	Source	Structure	Mechanism	References
Tryptanthrin	*Strobilanthes cusia*	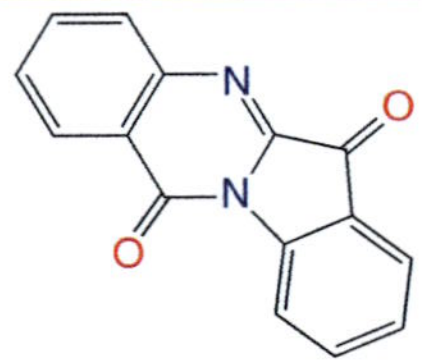	Exhibits blocking activity of papain-like protease 2 and viral RNA genome synthesis	[7]
Caffeic acid	*Ocimum basilicum*	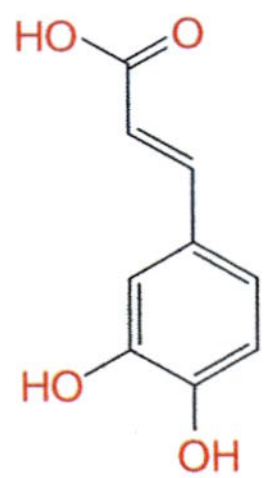	Acts through increased intracellular Zn^{2+} level	[8]
Quercetin	*Allium cepa*	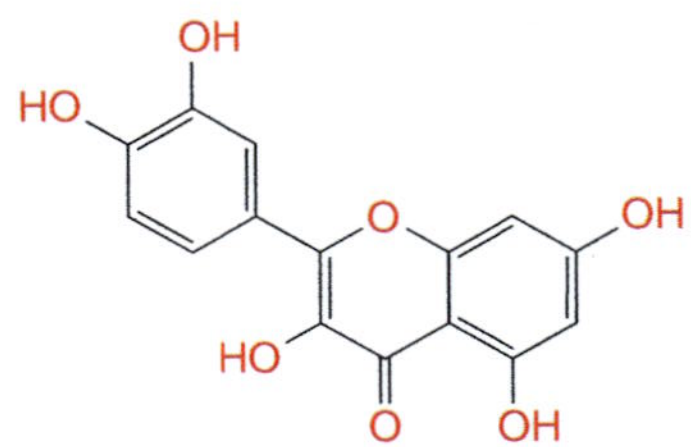	Acts as coronavirus protease inhibitor	[9]
Myricetin	*Camellia sinensis*	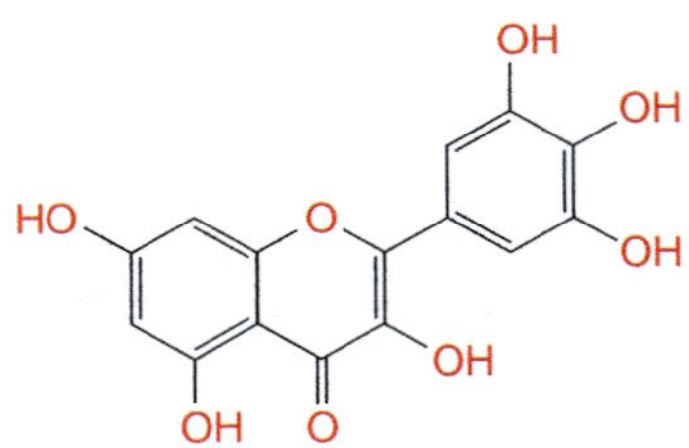	SARS-CoV-2 helicase ATPase activity reduced	[5]

Saikosaponin b2	*Scrophularia scorodonia*, *Bupleurum* spp., *Heteromorpha* spp.	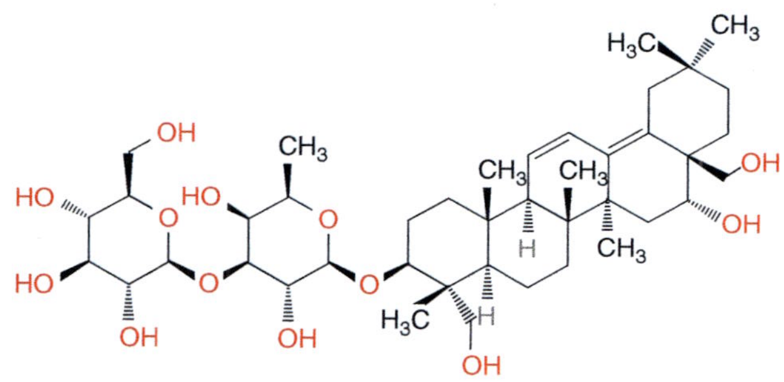	Has moderate antiviral potential against HCoV-229E, as inhibits viral attachment and penetration	[10]
Griffithsin	*Griffithsia* sp.	—	Stops viruses from attaching to host cells	[11]
Isobavachalcone	*Angelica keiskei*	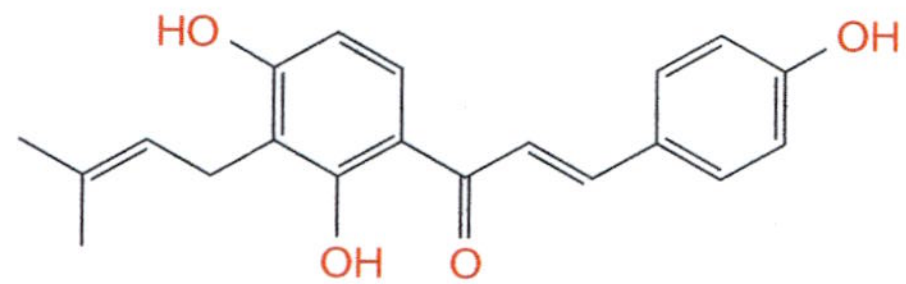	Acts by inhibiting the SARS-CoV-2 3CLpro activity and cysteine proteases of SARS-CoV-2	[4, 9]
Luteolin	*Capsicum annuum*, *Apium graveolens*, *Perilla frutescens*	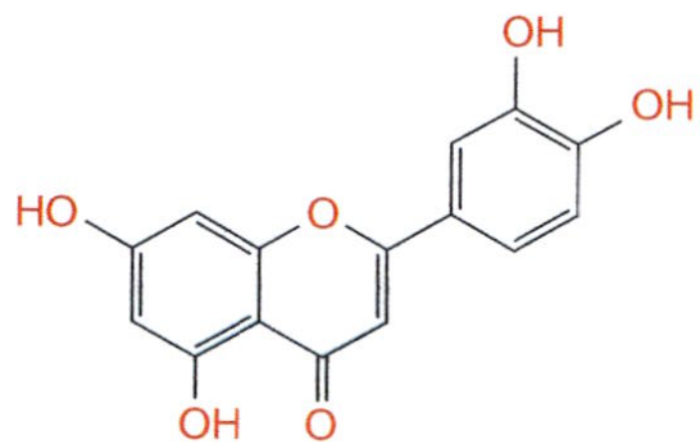	Slows viral replication after virus has entered cell. At time of viral entrance, luteolin had no effect on viral replication	[12]

(*Continued*)

Table 10.1 (Continued)

Phytochemical	Source	Structure	Mechanism	References
Myricitrin	*Euphorbia hirta*	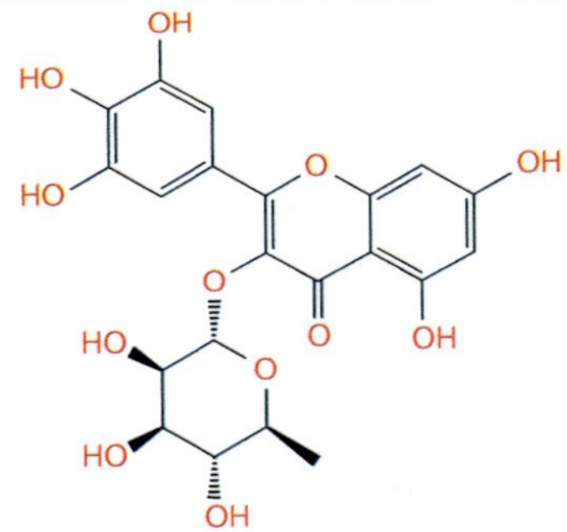	Inhibited generation of infectious virus particles effectively at 100 M	[13]
Baicalein	—	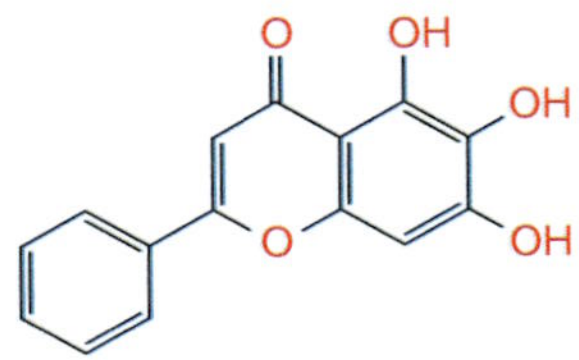	Has virucidal and antiadsorption activity	[14]
Palmatine	*Coptis chinensis* Franch.	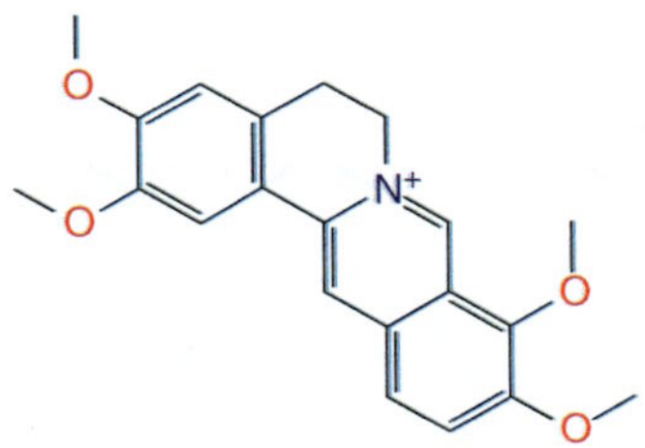	Inhibits WNVNS2B-NS3 protease activity in uncompetitive manner, with 50% inhibitory concentration	[15]

Hebulinic acid	*Terminalia chebula*	—	Possesses inhibitory activity against tick-borne encephalitis virus	[16]
Panicalagin	*Terminalia chebula*	—	Possesses inhibitory activity against tick-borne encephalitis virus	[16]
Baikalein	*Terminalia chebula*	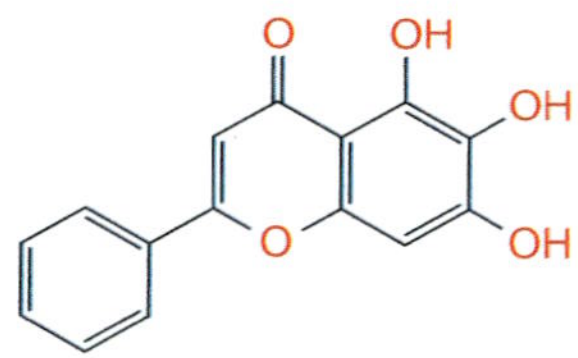	Possesses inhibitory activity against tick-borne encephalitis virus	[16]
Baikalskullcap	*Scutellaria baicalensis*	—	Extremely potent antiviral medicine with direct viricidal activity as causes direct inhibition of tick-borne encephalitis virus adsorption and intracellular replication	[17]
Genistein	—	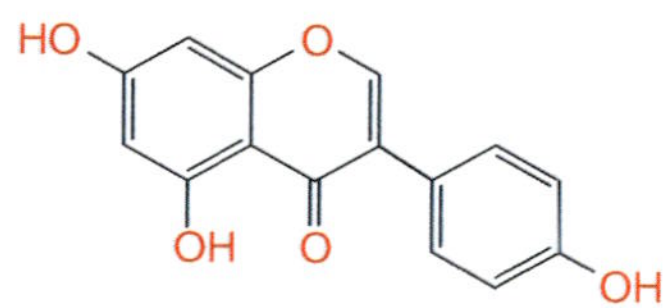	Operates in concert with antiviral medications to limit effective dose, preventing B virus plaque formation and lowering virus production	[18]
Isorhamnetin	*Ginkgo biloba*	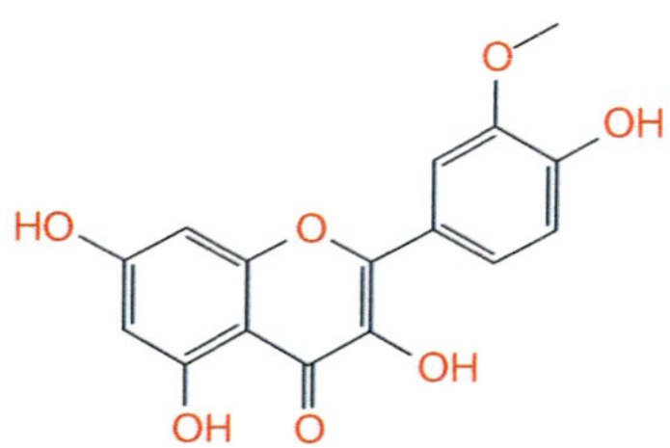	At non-cytotoxic concentrations, harmed viruses before they adsorb to cell surface	[19]

(*Continued*)

Table 10.1 (Continued)

Phytochemical	Source	Structure	Mechanism	References
7-Galloyl catechin	*Euphorbia coopire, Morus alba*	—	Acts by interfering with viral replication or genome synthesis	[20]
Gallic acid	*Euphorbia coopire, Morus alba*	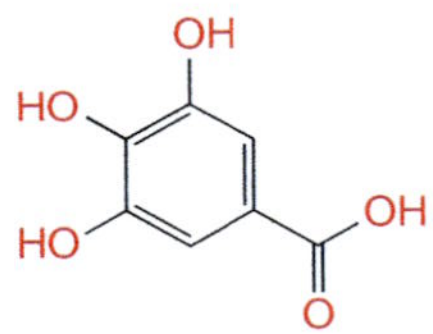	Acts by interfering with viral replication or genome synthesis	[20]
Kaempferol 3-O-(6″-galloyl)-β-D-glucopyranoside	*Euphorbia coopire, Morus alba*	—	Acts by interfering with viral replication or genome synthesis	[20]
Quercetin 3-O-(6″-galloyl)-β-D-glucopyranoside	*Euphorbia coopire, Morus alba*	—	Acts by interfering with viral replication or genome synthesis	[20]
Curcumin	*Euphorbia coopire, Morus alba*	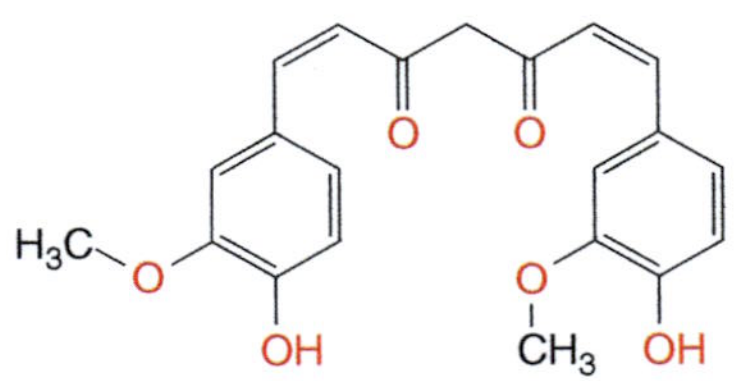	Acts by interfering with viral replication or genome synthesis	[20]
Quercetin	*Euphorbia coopire, Morus alba*	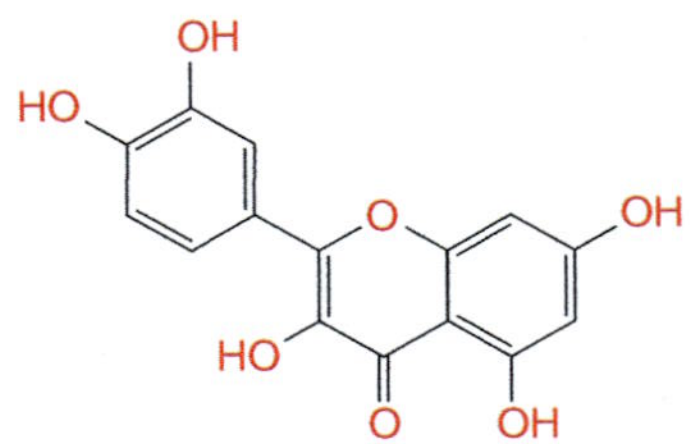	Acts by interfering with viral replication or genome synthesis	[20]

Kaempferol	*Euphorbia coopire, Morus alba*	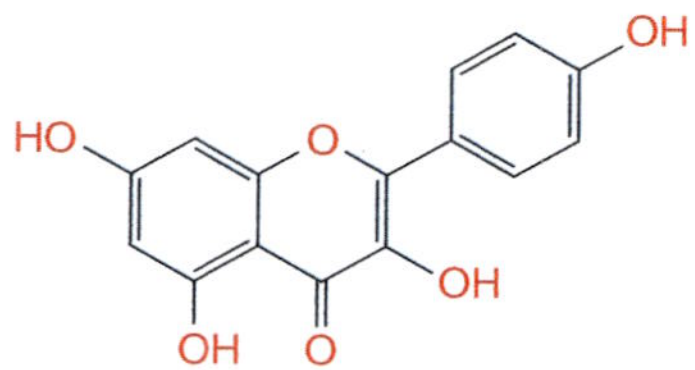	Acts by interfering with viral replication or genome synthesis	[20]
Glycyrrhizin	*Glycyrrhiza glabra*	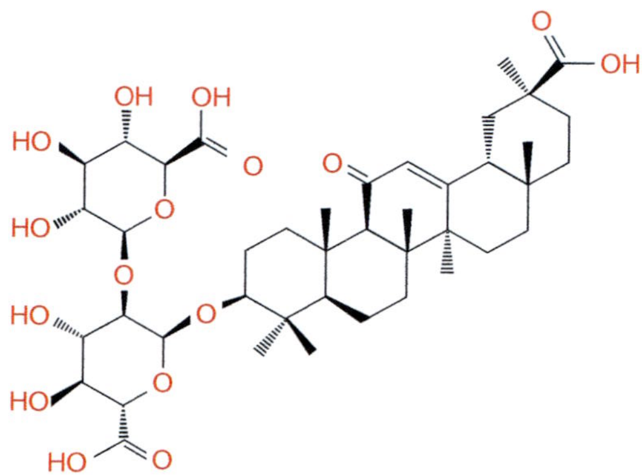	Inhibits virus gene expression and replication, lowering adhesion force and stress, and decreasing HMGB1 DNA binding	[21]
18β-glycyrrhetinic acid	*Glycyrrhiza glabra*	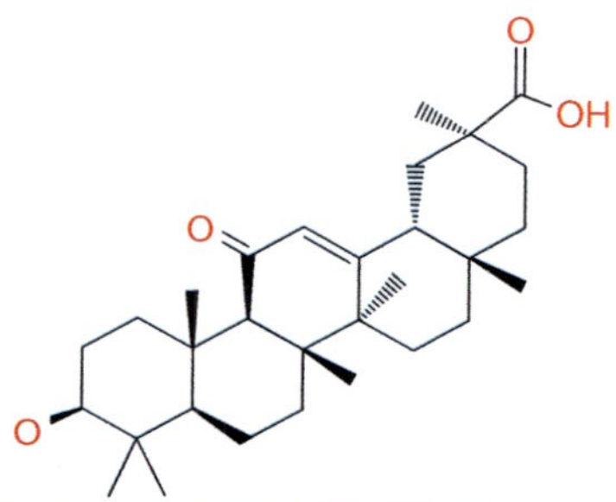	Acts through blocking breakdown of IB enzyme, involved in propagation of cellular response to inflammation, promoting T-lymphocyte proliferation, reducing host cell death, and enhancing host cell activity	[22]

(*Continued*)

Table 10.1 (Continued)

Phytochemical	Source	Structure	Mechanism	References
Galanthamine	*Ungernia victoris, Leucojum aestivum, Lycoris radiata, Narcissus* spp.	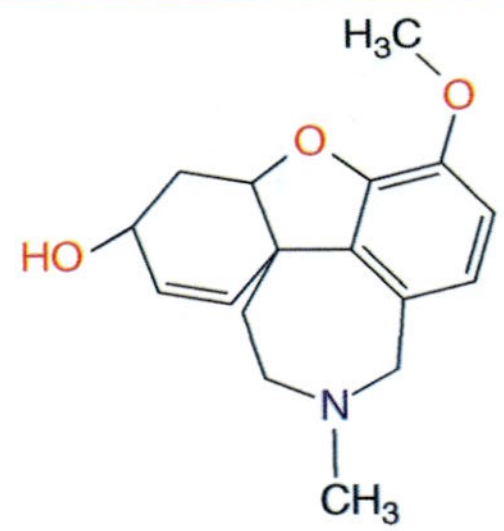	Acts as acetylcholinesterase inhibitor	[23]
Michellamines (A, B, and C)	*Ancistrocladus korupensis*	—	HIV-induced cell death and viral replication are inhibited by these alkaloids in human cell lines	[24]
Cepharanthine	*Stephania cepharantha*	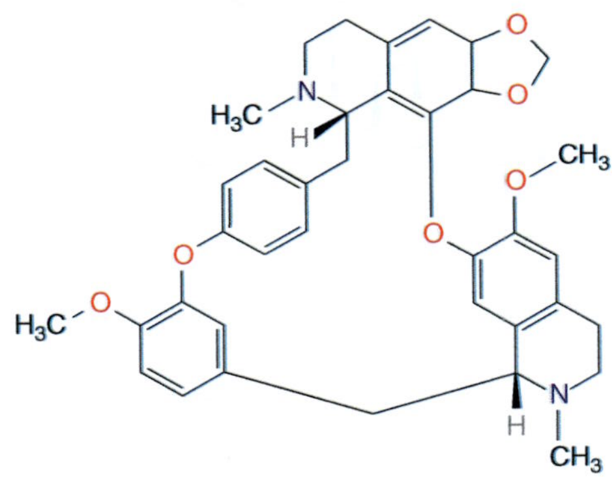	Inhibits nuclear factor kappa B (NF-kB) activation, which decreases HIV-1 long terminal repeat (LTR)-driven gene expression	[25]

Calanolide	*Calophyllum lanigerum*	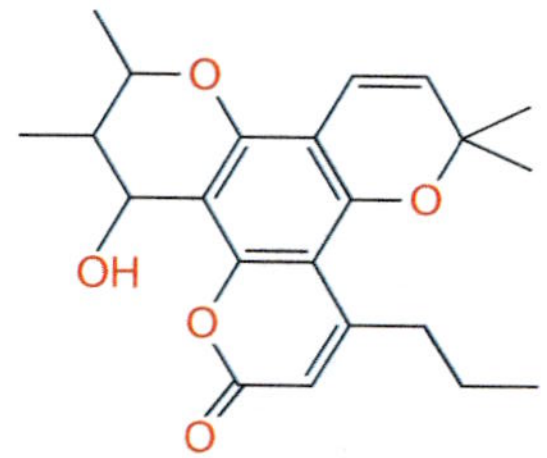	Inhibits HIV-1 reverse transcriptase	[26]
Bevirimat	*Syzygium claviflorum*	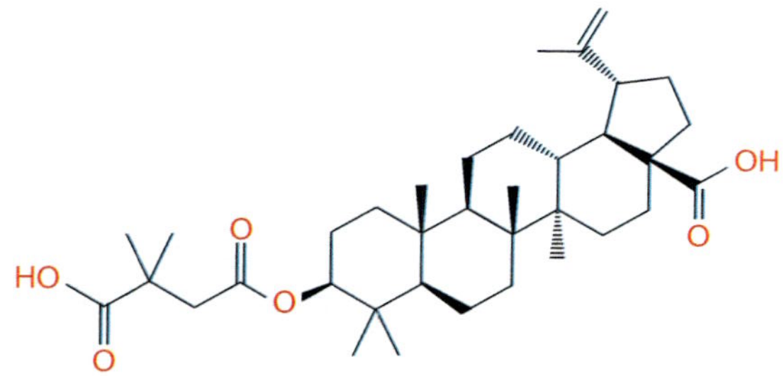	Targets maturation of freshly produced HIV particles and has strong potential for targeting more than one step in HIV cycle	[27]

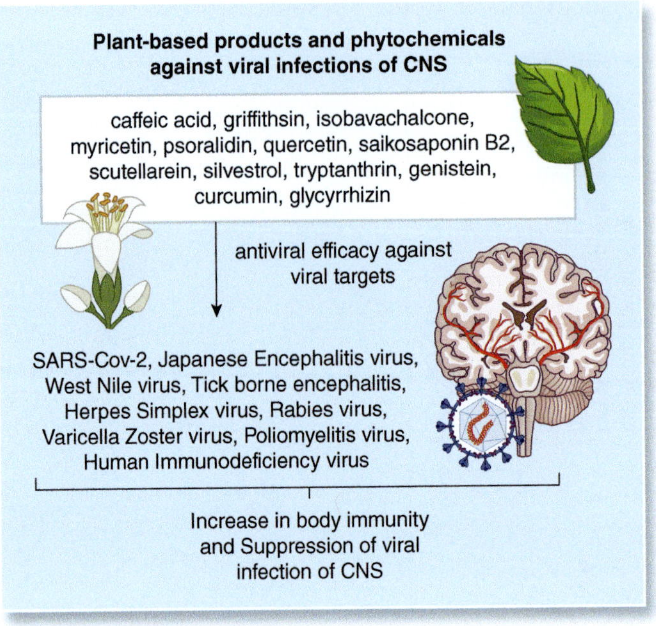

Figure 10.1 Plant-based products and phytochemicals against viral infections of the central nervous system (CNS).

humans. The most common symptoms of COVID-19 infection are cough, fever, and sore throat. Effective antiviral medicines for treating various coronavirus types are still scarce. These results in high mortality and serious socioeconomic consequences.

The importance of the innate immune response and diverse host effector pathways involved in virus replication in various CNS cell types is seen through coronavirus infection in the mouse. The cytolytic effector pathways found during acute infection and non-lytic humoral immunity predominate in remission [31]. Researchers have been working non-stop since the onset of the Covid-19 pandemic to uncover active chemicals, mostly from plants, that could slow or stop the spread of the causative agent (SARS-CoV-2). Herbal medicinal natural treatment aims to reduce the use of preservatives, excipients, and their adverse effects. The most promising phytochemicals are polyphenols, which are tiny molecules with conjugated fused ring structures. These polar phytochemicals are effective against the majority of coronavirus, MERS, and SARS [6].

10.2.2 Japanese Encephalitis Virus

JEV is a Flaviviridae flavivirus causing Japanese encephalitis. JEV is a major cause of human viral encephalitis with a high death rate. JEV is a virus that is encapsulated and has 11 Kb single-stranded positive-sense RNA. The viral illness is spread by mosquito. Japanese encephalitis is present in many regions of the world, including East and Southeast Asia. It is presently considered a public health threat globally, with a 30% mortality rate and long-term neurologic or psychiatric effects [32]. Because there is no effective therapy for Japanese encephalitis, an effective antiviral medication to inhibit JEV replication is required. Live attenuated vaccines, inactivated baby hamster kidney (BHK)-21 cell-derived vaccines, chimeric-live attenuated vaccines, inactivated mouse brain-derived vaccinations, Vero cell-derived inactivated vaccines, and other vaccines are presently available to prevent JEV infection [33].

Despite the availability of effective vaccines, effective antiviral treatment for JE has not been developed. It is thus important to develop effective antiviral medications to fight JEV infection. Luteolin, belonging to the flavone group of flavonoids, is found in celery (*Apium graveolens*), perilla leaf (*Perilla frutescens*), chamomile tea (*Matricaria chamomilla*), and green pepper (*Capsicum annuum*). Lutein, a benign dietary flavonoid, has antiviral activity against JEV. The virucidal effect of luteolin was discovered in JEV-infected A549 cell lines with an IC50 of 4.56 g/ml. Luteolin has a virucidal effect against extracellular JEV particles and inhibits viral replication at the postentry stage, but has no effect on viral replication during viral entrance and binding [12]. Thus, luteolin could be regarded as a possible JEV infection inhibitor and may be exploited to produce a JEV antiviral medication [29].

The antiviral activity of baicalein and quercetin was examined in the reproduction of JEV in vitro. The anti-JEV activities were investigated in Vero cells using the foci forming unit reduction assay (FFURA) and quantitative reverse transcription polymerase chain reaction (RT-PCR) at many phases of the JEV replication sequence. Following JEV adsorption in Vero cells, baicalein demonstrated considerable antiviral activity with an IC50 of 14.28 g/ml, while quercetin showed modest antiviral activity with an IC50 of 212.1 g/ml. Baicalein inhibited JEV adsorption significantly, whereas quercetin had little antiadsorption action. Baicalein had direct extracellular virucidal action against JEV, so it could be a good option for additional research into cellular and molecular mechanisms of action and in vivo testing to produce an efficient antiviral drug against JEV [14].

10.2.3 West Nile Virus

WNV belongs to the Flavivirus genus and family Flaviviridae. The principal hosts for viral infection are humans and monkeys. It is transmitted by arthropods. This family includes yellow fever virus, JEV, St. Louis encephalitis virus, dengue virus, and TBE. The first instance of WNV was discovered in the Western Hemisphere in 1999, and it has since spread around the world, particularly over Northern and Central America [15]. Three methodologies are used to find potential anti-WNV inhibitors: enzyme assay, virtual screening, and viral replication-based screening. Despite intensive efforts to create WNV inhibitors, no effective treatment for WNV yet exists for humans. As a result, developing an effective therapeutic treatment as well as antiviral medications for WNV infection is critical.

The two-component non-structural protein NS2B-NS3 protease, required for virus replication and assembly, is highly conserved across all flaviviruses. With the support of the host proteases, it plays a vital role in digesting the polyprotein precursor into individual functional proteins like furin and signalase. So NS2B-NS3 protease could be used as a therapeutic target. Also, the non-structural protein (NS5RNA)–dependent RNA polymerase (RdRp) activity is crucial for WNV replication. NS5 methyltransferase could be an essential target for antiflaviviral medication development. Palmatine, an isoquinoline alkaloid from *Coptis chinensis* Franch. of the protoberberine class, demonstrated antiviral activity in enzyme assays and viral titer reduction studies. Palmatine inhibited the activity of NS2B-NS3 protease in a dose-dependent manner, with an IC50 of 96 lM. Because flaviviruses (WNV, dengue virus, and yellow fever virus) all encode the same functional NS3 protease, palmatine can suppress them by preventing its activity. Palmatine suppresses WNV with virus-encoded proteases, the viral reverse genetics system, and characterization of palmatine-resistant viruses. Thus, palmatine could be a viable therapy option for flavivirus infections [15].

10.2.4 Tick-Borne Encephalitis Virus

TBE is a well-known viral contaminant carried through ixodid tick bites. It may result in the patient's disability or death in severe cases. The TBE virus belongs to the Flaviviridae family. Despite the relevance of TBE in human illness, the only approach to prevent and cure it is to use donor immunoglobulin, which has a number of drawbacks.

Herbal remedies have been studied for their antiviral properties with the possibility of their use in new TBE treatments. The antiviral activity was discovered by studying whole plant extracts and the biologic activity of purified extract components. Determining the target substance concentration and preclinical testing in cell cultures and laboratory animals will enable the safe use of phytochemicals as therapeutic agents. Anti-dengue flavivirus herbal agents like hebulinic acid, punicalagin, and baicalein can be effective against TBE virus [16]. The flavonoids in dry Baikal skullcap extract, which contain up to 90% baicalin, have a complex inhibitory effect on TBE virus in vitro, have direct viricidal activity, and affect the virus's adsorption and intracellular replication, all of which contribute to their value as a highly effective antiviral drug. In addition, in vivo testing of the extract's properties will allow it to be recommended as therapy against TBE virus and other flaviviruses for the creation of extremely efficient antiviral medications [17].

10.2.5 Herpes Simplex Virus

Infection with HSV is caused by either HSV type 1 or type 2. HSV-1 is a highly infectious virus that can be found all over the world. HSV-1 is spread via infected oral secretions, leading to more serious problems like gingivostomatitis, encephalitis, and keratitis. Human alpha herpesvirus 1 (HHV-1) and human alpha herpesvirus 2 (HHV-2) are two of the most frequent herpesviruses that cause lifelong sickness in humans [34]. Following initial infection through the bruised skin or mucous membrane, viruses become dormant in the trigeminal ganglia (HHV-1) or lumbar-sacral ganglia (HHV-2). Ultraviolet (UV) light exposure, stress, injury, and immunologic deterioration cause viral reactivation in the original infection site, leading to discomfort and skin problems [35].

Antiviral medications are to be tried against viral infections for which no vaccination exists, such as HHV-1. Antivirals including acyclovir, valacyclovir, cidofovir, and penciclovir are available to treat HHV-1 and HHV-2 infections, but they quickly acquire resistance and have hazardous drug-related adverse effects. On the other side, antiherpes drugs have been linked to the formation of drug-resistant strains, particularly in immunocompromised persons. Pure plant-derived chemicals could be used as novel antiviral medications, decreasing viral replication or viral genome synthesis in HSV strains. During host cell infection, HSV-1 and other viruses use glycosaminoglycan (GAG) as an initial attachment receptor. Polyphenols aim for HSV-1 glycoproteins that relate to GAGs, preventing them from interacting with cell surface glycosaminoglycans and succeeding in binding receptors. This inhibitory effect is observed against cell-free viruses, during the attachment and fusion phases of the virus, and during the proliferation of HSV-1 glycoprotein-mediated intercellular links.

Flavonoids, alkaloids, saponins, terpenes, quinones, lignans, polysaccharides, tannins, thiosulfinates, steroidal glycoside, and proanthocyanidin are phytochemicals that have antiviral activity against HSV types. Viruses like HSV, HIV, and influenza virus were all found to be suppressed during viral RNA synthesis [35]. HSV-1 and HSV-2 were strongly reduced by *L. candidum* leaf extracts [36]. Antiherpetic action is shown by *Morus alba* (Moraceae) and *Euphorbia coopire* (Euphorbiaceae). Pure chemicals including 7-galloyl catechin, gallic acid, curcumin, quercetin, and kaempferol are responsible for the antiviral activity [20]. *Ginkgo biloba* is a commonly used phytotherapeutic that contains flavonoids, terpenes, and organic acids in the form of dried green leaf extract. Flavonoids such as kaempferol, quercetin, and isorhamnetin, as well as terpene trilactones such as ginkgolides A, B, C, bilobalide, and ginkgolic acids, have cytotoxic properties that diminish HHV-1 and HHV-2 infectivity [19]. Alkaloids, steroids, flavones, aglycones, tannins, saponins, and terpenoids found in aqueous extract of *Phaleria macrocarpa* fruit enhance cytotoxicity and antiviral effects against HSV-1, as measured by plaque reduction assay [37]. Antiviral activity of *Mentha pulegium* L. extract on HSV-1 following infection shows the best antiviral activity, where the HSV-1 titer is dramatically lowered after one hour of direct treatment with the extract [38].

10.2.6 Rabies Virus

Rabies, a neglected viral zoonosis, is still a major public health concern around the world, killing over 50 000 people a year, primarily in developing Asian and African countries. It is caused by a neurotropic, non-segmented, single-stranded RNA virus of the Lyssavirus genus of the Rhabdoviridae family and Mononegavirale order that infects neurons in the CNS and causes physiologic issues. It is kept alive in nature by a variety of animal reservoirs and primarily infects the CNS, causing progressive encephalopathy and eventually death in afflicted humans. Focused brainstem signals, myoclonus, hemichorea, and Horner's syndrome signs are all symptoms of rabies caused by bat rabies virus strains. It is most commonly spread via a bite from an infected animal, although non-bite exposures can happen in rare cases. Exposure can occur by direct contact with saliva, nervous or other sick tissues from a rabid animal, or scratching. The illness is nearly always lethal once symptoms arise. In unimmunized persons, death is almost always unavoidable; only supportive treatments are indicated after the development of neurologic signs and symptoms. Even after high-risk exposure, early vaccination with rabies immunoglobulins and proper wound care following exposure prevent rabies. Rabies is widespread in impoverished countries due to a lack of risk communication, insufficient monitoring of rabies-related viruses and diagnostic skills, low vaccination coverage, and failure to immunize domestic dogs, which transmit the disease to people.

Because of the disease's continuous impact, low-cost protective biologics and rabies treatment regimens are needed. Many traditional medicinal herbs have antiviral properties and are used to treat viral infections in both animals and humans. *Phytolacca dodecandra*, *Justicia schimperiana*, *Ricinus communis*, *Brucea antidysenterica*, *Croton macrostachyus*, and *Cucumis ficifolius* were the most regularly listed medicinal plant species used to treat rabies in Ethiopia [39]. Extracts from the leaves of *Justicia schimperiana* and *R. communis*, as well as *C. macrostachyus* stem bark, have antirabies activity due to a range of phytochemical components, acute toxicity, and antirabies potential. Methanol plant extracts outperformed ethanol and water extracts in a mouse model in terms of antirabies effectiveness [40]. Datura has traditionally been used to treat asthma symptoms. Atropine is an antiasthmatic drug that paralyzes the pulmonary branches of the lungs, preventing the spasms that trigger asthma attacks. The potential of Datura plant extract has significant antirabies efficacy in vitro. In a mouse model, the Datura extract increased survival time by four days in the postexposure treatment [41]. *Carissa* L., a member of the Apocynaceae family, is a tropical and subtropical evergreen shrub or small tree found in Africa, Asia, and Oceania, which is used to treat rabies and other disorders such as headaches, gonorrhea, oedema, syphilis, chest complaints, and rheumatism. The 123 chemicals found in nine Carissa species include simple phenolic compounds, lignans, esters, sterols, terpenes, flavonoids, and fatty acids [42].

10.2.7 Varicella Zoster Virus

VZV is a DNA virus belonging to the Herpesviridae family of alpha herpesviruses. VZV is a highly infectious virus that is found all over the world. "Chickenpox" or acute varicella is caused by direct contact through skin lesions or airborne transmission through respiratory droplets. After initial infection, VZV develops lifelong latency in the cranial nerve and dorsal root ganglia, and can resurface later as herpes zoster or shingles. Fever and a broadly diffused vesicular, pruritic rash involving the face and trunk are characteristic signs of primary varicella. The symptoms normally go away in 7–10 days, but primary varicella can cause additional serious disease and visceral invasion in rare situations. Hepatitis, encephalitis, pancreatitis, and pneumonitis are all rare complications, although they can be fatal.

The drug acyclovir and its derivatives like valacyclovir are nucleoside analogues used in the treatment of VZV that exhibit their antiviral activity through the viral-encoded thymidine kinase enzyme, which phosphorylates acyclovir (acycloguanosine) into its monophosphate form, which is subsequently phosphorylated by cellular kinases to target viral DNA polymerase, thereby preventing further viral DNA replication. Numerous side effects are seen like renal failure, neurotoxicity, thrombocytopenia, renal dysfunction, nephrotoxicity, and hypokalemia. Foscarnet, a pyrophosphate inhibitor, is commonly employed for treatment in resistant cases, although acute renal failure and nephrotoxicity are the adverse effects associated with it. Therefore, there is a necessity to develop new potential antiviral agents for VZV infections [43].

Biflavonoid groups like robustaflavone, agathisflavone, rhusflavanone, amentoflavone, and succedanea flavanone isolated from medicinal plants like *Rhus succedanea* and *Garcinia multiflora* showed antiviral activity against various viruses, including respiratory viruses and VZV. Rhusflavanone and succedanea flavanone exhibited inhibitory effects against measles and VZV, respectively [44]. Purified extracts of *Rhamnus frangula*, *Cassia angustifolia*, *Rhamnus purshianus*, *Rheum officinale*, and *Aloe barbadensis* produce aloe-like emodin (aglycones or anthraquinones) to inactivate VZV. The polyphenols rosmarinic acid, caffeic acid, and chlorogenic acid and their derived products are responsible for this antiviral activity [45]. *Glycyrrhiza glabra* extracts show antiviral activity against VZV due to two triterpenoids, glycyrrhizin and 18 glycyrrhetinic acid, which suppress virus gene expression and reproduction. They reduce the virus's adhesion force and stress, as well as its adsorption and penetration, in the early stages of the replicative cycle [22]. *Ficus binjamina* leaf and fruit extracts suppressed VZV, with the plant extracts having high interactions with the virus and weak contacts with the cell surface [36].

10.2.8 Poliomyelitis Virus

The poliovirus with an RNA genome belongs to the Enterovirus subgroup of the Picornaviridae family. Enteroviruses are temporary gastrointestinal tract residents that thrive in an acidic environment. There are three polio virus serotypes (P1, P2, and P3), each of which has low heterotypic immunity (i.e. immunization to one serotype does not result in considerable immunity to the other). The three poliovirus serotypes are distinguishable from other enteroviruses by their ability to be neutralized by serotype-specific antisera and their proclivity for causing paralysis. The polioviruses, the genus Enterovirus, and the family Picornaviridae all started with the Mahoney strain of type 1 poliovirus. It is one of the most well-studied and well-characterized human disease agents. Fever, headaches, and a sore throat are common symptoms. Abortive poliomyelitis is a type of polio infection in which the CNS is invaded, yet the patient recovers completely in less than a week. Flaccid paralysis occurs in less than 1% of all polio infections. Any youngster under the age of 15 or any age with sudden onset of flaccid/floppy paralysis is considered as having an acute flaccid paralysis (AFP) condition (muscular weakness). This infection's neurologic phase is assumed to be an unintentional diversion from regular gastrointestinal sickness. Individuals who are infected with the virus or who have been immunized with the polio vaccine develop immunity. Immunoglobulin (Ig)A antibodies against polio virus are found in the tonsils and gastrointestinal tracts of immune people and can stop the virus from replicating; IgG and IgM antibodies against polio virus can stop the virus from spreading to the CNS motor neurons.

Poliomyelitis, often known as polio or infantile paralysis, is a severe viral infection that is a major public health concern. Oral polio vaccine (OPV) has long been the vaccination of choice for poliomyelitis control in many countries; however, the attenuated virus in OPV can revert to a form that can paralyze on rare occasions. As a result, there is a pressing need to find a safe and effective poliomyelitis drug, and medicinal plants appear to be a promising alternative source of antiviral medicines [46]. The poliomyelitis virus is treated with methanol, water, and petroleum ether extracts of *Moringa oleifera* Lam. (Moringaceae) leaves. In L20B cell lines, the endpoint cytopathic effect assay revealed that it had antiviral activity against the three poliovirus serotypes (SL1, SL2, and SL3). Saponins, flavonoids, tannins, reducing sugar, carbohydrates, alkaloids, acidic chemicals, glycosides, resins, lipids, and proteins were found in phytochemical examination of the extracts. The extracts selectively inhibited the virus due to their powerful antiviral potential. With little toxicity to the host cells, moringa could be developed as a drug for poliomyelitis [47]. Galanthamine is an alkaloid of the Amaryllidaceae family that is used to treat poliomyelitis, Alzheimer's disease, and other neurologic illnesses. It is a selective, long-acting, reversible, and competitive acetylcholinesterase (AChE) inhibitor. Other galanthamine-like alkaloids like narwedine or N-desmethyl galanthamine can be employed as galanthamine precursors [23].

10.2.9 Human Immunodeficiency Virus

HIV is a double-stranded RNA virus belonging to the Retrovirus family. It is a sexually transmitted infection that spreads by contact with infected blood, sperm, and vaginal fluid. People all across the world are still being affected by the HIV epidemic. In 2020, nearly 37.6 million people were living with HIV, and roughly 36.3 million people have died from acquired immune deficiency syndrome (AIDS)-related illnesses since the epidemic began [48]. Combination antiretroviral therapy (cART) is now used as an antiviral treatment. Two nucleoside or nucleotide reverse transcriptase inhibitors plus a third medication from a different group make up cART. This combination has been shown to considerably reduce virus replication, resulting in considerable improvements in HIV infection clinical care through delay of disease progression and increase in overall quality of survival and life [49]. Phytochemicals' potential as anti-HIV agents is still being investigated. Some antiviral assays are performed to analyze these phytochemicals for HIV enzyme inhibition. Medicinal botanicals with efficacy for HIV management are linked to treatment. In Southern Africa, two plants, *Hypoxis hemerocallidea* (African potato) and *Sutherlandia frutescens*, have dominated traditional HIV management [50]. *Plectranthus barbatus*, *Siphonochilus natalensis*, *Abrus precatorius*, *Erythrina abyssinica*, *Cassia fistula*, *Bulbine alooides*, and *Dodonaea angustifolia* are some of the other plants utilized in the prophylaxis of HIV in southern Africa [27]. Alkaloids, quinones, coumarins, flavonoids, saponins, phenolics, tannins, and lignans are some of the promising plant-derived anti-HIV phytochemicals. HIV patients in India have employed herbs like *Azadirachta indica*, *Amukkara chooranum*, *Andrographis paniculata*, *Boxus sempervirens*, *Rasagandhi mezhuga*, and *Nellikkai lehyam* to treat their infection [51]. The gaps in the development of plant-based medications to be used in HIV therapy and management necessitate concentrated research on traditional therapy of the disease. in vitro assays are a critical tool for bettering our understanding of phytomedicines' modes of action and levels of safety. The present push to sell phytochemicals comes from the widespread use of medicinal plants in old-style HIV/AIDS treatment.

10.3 Controlling Vectors of Viral Diseases of the Central Nervous System

Mosquitoes are the primary vectors of CNS viral illnesses. Arthropod-borne viruses (Arboviruses) like JEV, eastern equine encephalomyelitis virus, WNV, St. Louis encephalitis virus, and La Crosse encephalitis virus are transmitted by them. Aedes and Haemogogus are mosquito species that are the vectors that transmit yellow fever virus and serve as a virus reservoir. In the absence of particular antivirals or vaccines, mosquito management is the mainstay of mosquito-borne virus prevention. Insect/mosquito control using modern synthetic pesticides could yield fast results. These pesticides pose an irreversible environmental risk, however, as well as serious side effects such as severe neurotoxicity and pernicious poisoning in humans and beneficial creatures. They are also costly. The advent of widespread pesticide resistance and likely environmental risks connected with commonly used insecticides (malathion, dichlorodiphenyltrichloroethane [DDT]) highlights the urgent need for alternative mosquito control methods.

Plant products are inexpensive, biodegradable, and eco-friendly. Mosquito repellents made from natural ingredients are target specific, biodegradable, environmentally friendly, and botanically derived [52]. Rotenone was discovered as an active ingredient in the Japanese plant *Rhododendron hortense* in 1902. The pyrethrum extract containing pyrethrins is derived from ancient pesticidal plants of the genus *Chrysanthemum* spp. The herb *A. indica* is important for pest control in Ayurveda because it contains azadirachtin, which is an active ingredient. The plant *Eucalyptus citriodora* is grown all over the world to harvest the mosquito-repellent properties of its essential oil due to the presence of a modest amount of p-menthane-3,8-diol. *Tanacetum cinerariifolium* and *Tanacetum coccineum* pyrethrins are particularly efficient pesticides. Derris, hellebore, camphor, quassia, anabasine, turpentine, and D-limonene are also important phytochemical insecticides [53].

The neurologic system of insects is immediately attacked and damaged by the majority of synthetic insecticides. Mosquitoes' lactic acid receptors are blocked by the botanical repellents, preventing upwind flying and interaction with the host. Female mosquitoes are attracted to warm-blooded animals because their sweat contains lactic acid-like excretory secretions. Botanical repellants affect mainly the mosquito larvae midgut epithelium, along with some effects on gastric caeca and Malpighian tubules [52]. The transmission in arbovirus is disrupted at two points in the virus's lifecycle by lowering or stopping virus replication in vector mosquitoes exposed to the virus, and by reducing or blocking the number of vectors. This technique could evolve into effective control for arbovirus with more research on antiviral plant-derived chemicals [54].

Computational choices of pharmacologic inhibitors for the sterol carrier protein-2 (AeSCP-2) are a new and extremely reliable way of finding mosquito-controlling drugs. The phenolic compounds triterpenoid, tricin, stigmasterol, and pyrethrum were evaluated against WNV replication and indicated that these ligand molecules had high binding affinity with the NS5 methyltransferase protein. The chemicals generated from mangroves bind to NS5 methyltransferase and prevent WNV replication. As a result, chemicals generated from mangroves should be effective against WNV. For the foreseeable future, in vitro experimentation approaches will be required [55]. Screening of the mosquitocidal ability of isolated natural products could lead to the discovery of novel lead compounds, which will be of tremendous economic benefit to developing nations, allowing impoverished farmers to gain economic empowerment and, in turn, improving their economies.

10.4 Future Perspectives

Although herbal plant treatments are extensively utilized in many regions of the globe, either alone or in combination, there is no evidence on the interactions of these therapeutic plant-based drugs in living systems. As a result, traditional medicine can be combined with scientific research to facilitate contemporary medication development from phytochemicals. Researchers and policymakers should have access to scientific data on the full pharmacokinetics and pharmacodynamics of plant-based drugs and their medicinal preparations in order to conduct randomized multicenter clinical studies. Optimal treatments, dosages, and formulas for plant-based drugs and phytochemicals are needed for use in CNS diseases caused by viruses.

10.5 Conclusion

Scientists from all around the globe are working to develop the most powerful antiviral medications to battle viruses that attack the CNS. Plant-based products and phytochemicals are secondary plant metabolites that show antiviral action against viral infections of the CNS by limiting viral pathogenesis and reproduction cycle. Plants are a vastly underappreciated source of bioactive chemicals with antiviral activity. These plant secondary metabolites may be a promising therapeutic candidate for CNS viral infections. Advanced extraction, purification, and characterization procedures for secondary plant metabolites, along with interdisciplinary knowledge, would be advantageous in novel drug discovery against viral infections of the CNS.

References

1 Iwu, M.W., Duncan, A.R., and Okunji, C.O. (1999). New antimicrobials of plant origin. In: *Perspectives on New Crops and New Uses* (ed. J. Janick), 457–462. Alexandria, GA: ASHS Press.

2 Malekmohammad, K., Rafieian-Kopaei, M., Sardari, S., and Sewell, R.D. (2020). Effective antiviral medicinal plants and biological compounds against central nervous system infections: a mechanistic review. *Curr. Drug Discov. Technol.* 17 (4): 469–483.

3 Ganjhu, R.K., Mudgal, P.P., Maity, H. et al. (2015). Herbal plants and plant preparations as remedial approach for viral diseases. *Virus Dis.* 26 (4): 225–236.

4 Kim, D.W., Seo, K.H., Curtis-Long, M.J. et al. (2014). Phenolic phytochemical displaying SARS-CoV papain-like protease inhibition from the seeds of *Psoralea corylifolia*. *J. Enzyme Inhib. Med. Chem.* 29 (1): 59–63.

5 Yu, M.S., Lee, J., Lee, J.M. et al. (2012). Identification of myricetin and scutellarein as novel chemical inhibitors of the SARS coronavirus helicase, nsP13. *Bioorg. Med. Chem. Lett.* 22 (12): 4049–4054.

6 Attia, Y.A., Alagawany, M.M., Farag, M.R. et al. (2020). Phytogenic products and phytochemicals as a candidate strategy to improve tolerance to coronavirus. *Front. Vet. Sci.* 7: 573159.

7 Tsai, Y.C., Lee, C.L., Yen, H.R. et al. (2020). Antiviral action of tryptanthrin isolated from *Strobilanthes cusia* leaf against human coronavirus NL63. *Biomolecules* 10 (3): 366.

8 Clergeaud, G., Dabbagh-Bazarbachi, H., Ortiz, M. et al. (2016). A simple liposome assay for the screening of zinc ionophore activity of polyphenols. *Food Chem.* 197: 916–923.

9 Park, J.Y., Ko, J.A., Kim, D.W. et al. (2016). Chalcones isolated from *Angelica keiskei* inhibit cysteine proteases of SARS-CoV. *J. Enzyme Inhib. Med. Chem.* 31 (1): 23–30.

10 Müller, C., Schulte, F.W., Lange-Grünweller, K. et al. (2018). Broad-spectrum antiviral activity of the eIF4A inhibitor silvestrol against corona-and picornaviruses. *Antiviral Res.* 150: 123–129.

11 Millet, J.K., Séron, K., Labitt, R.N. et al. (2016). Middle East respiratory syndrome coronavirus infection is inhibited by griffithsin. *Antiviral Res.* 133: 1–8.

12 Fan, W., Qian, S., Qian, P., and Li, X. (2016). Antiviral activity of luteolin against Japanese encephalitis virus. *Virus Res.* 220: 112–116.

13 Tayone, W.C., Ishida, K., Goto, S. et al. (2020). Anti-Japanese encephalitis virus (JEV) activity of triterpenes and flavonoids from Euphorbia hirta. *Philipp J. Sci.* 149 (3): 603–613.

14 Johari, J., Kianmehr, A., Mustafa, M.R. et al. (2012). Antiviral activity of baicalein and quercetin against the Japanese encephalitis virus. *Int. J. Mol. Sci.* 13 (12): 16785–16795.

15 Jia, F., Zou, G., Fan, J., and Yuan, Z. (2010). Identification of palmatine as an inhibitor of West Nile virus. *Arch. Virol.* 155 (8): 1325–1329.

16 Solovarov, I.S., Khasnatinov, M.A., Danchinova, G.A. et al. (2017). Assessment of neutralizing properties of DNA-aptamers and extracts of medicinal herbs against the tick-borne encephalitis virus. *Biomed. Biochim. Acta* 2 (1): 84–88.

17 Leonova, G.N., Shutikova, A.L., Lubova, V.A., and Maistrovskaya, O.S. (2020). Inhibitory activity of *Scutellaria baicalensis* flavonoids against tick-borne encephalitis virus. *Bull. Exp. Biol. Med.* 168 (5): 665–668.

18 LeCher, J.C., Diep, N., Krug, P.W., and Hilliard, J.K. (2019). Genistein has antiviral activity against herpes B virus and acts synergistically with antiviral treatments to reduce effective dose. *Viruses* 11 (6): 499.

19 Sochocka, M., Sobczyński, M., Ochnik, M. et al. (2019). Hampering herpesviruses HHV-1 and HHV-2 infection by extract of Ginkgo biloba (EGb) and its phytochemical constituents. *Front. Microbiol.* 10: 2367.

20 El-Toumy, S.A., Salib, J.Y., El-Kashak, W.A. et al. (2018). Antiviral effect of polyphenol rich plant extracts on herpes simplex virus type 1. *Food Sci. Hum. Wellness* 7 (1): 91–101.

21 Wang, L., Yang, R., Yuan, B. et al. (2015). The antiviral and antimicrobial activities of licorice, a widely-used Chinese herb. *Acta Pharm. Sin. B* 5 (4): 310–315.

22 Pastorino, G., Cornara, L., Soares, S. et al. (2018). Liquorice (*Glycyrrhiza glabra*): a phytochemical and pharmacological review. *Phytother. Res.* 32 (12): 2323–2339.

23 Berkov, S., Georgieva, L., Kondakova, V. et al. (2009). Plant sources of galanthamine: phytochemical and biotechnological aspects. *Biotechnol. Biotechnol. Equip.* 23 (2): 1170–1176.

24 Daniyal, M., Akram, M., Hamid, A. et al. (2016). Comprehensive review on treatment of HIV. *Pak. J. Pharm. Sci.* 29 (4): 1331–1338.

25 Cos, P., Maes, L., Vlietinck, A., and Pieters, L. (2008). Plant-derived leading compounds for chemotherapy of human immunodefiency virus (HIV) infection – an update (1998–2007). *Planta Med.* 74 (11): 1323–1337.

26 César, G.Z., Alfonso, M.G., Marius, M.M. et al. (2011). Inhibition of HIV-1 reverse transcriptase, toxicological and chemical profile of *Calophyllum brasiliense* extracts from Chiapas, Mexico. *Fitoterapia* 82 (7): 1027–1034.

27 Fonteh, P., Chingwaru, W., Meyer, D., and Kapewangolo, P. (2019). In-vitro assays for determining anti-HIV potential of phytochemicals. In: *Phytochemistry: An In-Silico and In-Vitro Update* (ed. S. Kumar and C. Egbuna), 299–313. Singapore: Springer.

28 Pan, Y., Gao, F., Zhao, S. et al. (2021). Role of the SphK-S1P-S1PRs pathway in invasion of the nervous system by SARS-CoV-2 infection. *Clin. Exp. Pharmacol. Physiol.* 48 (5): 637–650.

29 Ghildiyal, R., Prakash, V., Chaudhary, V.K. et al. (2020). Phytochemicals as antiviral agents: recent updates. In: *Plant-Derived Bioactives* (ed. M.K. Swamy), 279–295. Singapore: Springer.

30 Santhi, V.P., Masilamani, P., Sriramavaratharajan, V. et al. (2021). Therapeutic potential of phytoconstituents of edible fruits in combating emerging viral infections. *J. Food Biochem.* 45 (8): e13851.

31 Bergmann, C.C., Lane, T.E., and Stohlman, S.A. (2006). Coronavirus infection of the central nervous system: host–virus stand-off. *Nat. Rev. Microbiol.* 4 (2): 121–132.

32 Solomon, T., Dung, N.M., Kneen, R. et al. (2000). Japanese encephalitis. *J. Neurol. Neurosurg. Psychiatry.* 68 (4): 405–415.

33 Yun, S.I. and Lee, Y.M. (2014). Japanese encephalitis: the virus and vaccines. *Hum. Hum Vaccin. Immunother.* 10 (2): 263–279.

34 Rechenchoski, D.Z., Faccin-Galhardi, L.C., Linhares, R.E., and Nozawa, C. (2017). Herpesvirus: an underestimated virus. *Folia Microbiol.* 62 (2): 151–156.

35 Kumar, S. and Pandey, A.K. (2013). Chemistry and biological activities of flavonoids: an overview. *Sci. World J.* 2013: 162750.

36 Yarmolinsky, L., Zaccai, M., Ben-Shabat, S. et al. (2009). Antiviral activity of ethanol extracts of *Ficus binjamina* and *Lilium candidum* in vitro. *Nat. Biotechnol.* 26 (6): 307–313.

37 Ismaeel, M.Y., Dyari, H.R., Yaacob, W.A., and Ibrahim, N. (2018). in vitro antiviral activity of aqueous extract of Phaleria macrocarpa fruit against herpes simplex virus type 1. In: *The 2017 UKM FST Postgraduate Colloquium*, American Institute of Physics Conference Series, vol. 1940, 020076. Boston, MA: AIP Publishing.

38 Parsania, M., Rezaee, M.B., Hamidreza Monavari, S. et al. (2017). Antiviral screening of four plant extracts against acyclovir resistant herpes simplex virus type-1. *Pak. J. Pharm. Sci.* 30 (4): 1407–1411.

39 Meresa, A., Degu, S., Tadele, A. et al. (2017). Medicinal plants used for the management of rabies in Ethiopia – a review. *J. Med. Chem.* 7: 795–806.

40 Tesera, Y., Desalegn, A., Tadele, A. et al. (2021). Phytochemical screening, acute toxicity and anti-rabies activities of extracts of selected Ethiopian traditional medicinal plants. *Research Square* https://doi.org/10.21203/rs.3.rs-512523/v1.

41 Roy, S., Samant, L., Ganjhu, R. et al. (2018). Assessment of in vivo antiviral potential of *Datura metel* Linn. extracts against rabies virus. *Pharmacogn. Res.* 10 (1): 109.

42 Kaunda, J.S. and Zhang, Y.J. (2017). The genus Carissa: an ethnopharmacological, phytochemical and pharmacological review. *Nat. Prod. Bioprospect.* 7 (2): 181–199.

43 Divyadarshini, A., Indu, P., Rajarajan, S., and Sridhar, R. (2016). Screening of potential phytochemicals against protease of Varicella-Zoster virus-an in silico approach. *World J. Pharm. Res.* 5: 679–693.

44 Lin, Y.M., Flavin, M.T., Schure, R. et al. (1999). Antiviral activities of biflavonoids. *Planta Med.* 65 (02): 120–125.

45 Jassim, S.A. and Naji, M.A. (2003). Novel antiviral agents: a medicinal plant perspective. *J. Appl. Microbiol.* 95 (3): 412–427.

46 Adamu, H.M., Iliyasu, M.Y., Yakubu, M.N. et al. (2020). In-vitro evaluation of antiviral activity of *Moringa oleifera* extracts against polio virus. *J. Pharm. Res. Int.* 2020: 101–109.

47 Okoye, E.L. (2014). Inhibition of the poliomyelitis viral-induced cytopathic effect by extracts of *Moringa oleifera* Lam. *Aust. J. Herb. Med.* 26 (3): 95–99.

48 Nyaga, R.K., Kimani, D.N., Mwabu, G., and Kimenyi, M.S. (2004). *HIV/AIDS in Kenya: A Review of Research and Policy Issues*. Nairobi: Kenya Institute for Public Policy Research and Analysis (KIPPRA) https://pdf.usaid.gov/pdf_docs/pnads077.pdf.

49 Cihlar, T. and Fordyce, M. (2016). Current status and prospects of HIV treatment. *Curr. Opin. Virol.* 18: 50–56.

50 Maroyi, A. (2014). Alternative medicines for HIV/AIDS in resource-poor settings: insight from traditional medicines use in sub-Saharan Africa. *Trop. J. Pharm. Res.* 13: 1527–1536.

51 Fritts, M., Crawford, C.C., Quibell, D. et al. (2008). Traditional Indian medicine and homeopathy for HIV/AIDS: a review of the literature. *AIDS Res. Ther.* 5: 1–9.

52 Kishore, N., Mishra, B.B., Tiwari, V.K. et al. (2014). Natural products as leads to potential mosquitocides. *Phytochem. Rev.* 13 (3): 587–627.

53 Shaalan, E.A., Canyon, D., Younes, M.W. et al. (2005). A review of botanical phytochemicals with mosquitocidal potential. *Environ. Int.* 31 (8): 1149–1166.

54 Eastep, N.E., Albert, R.E., and Anderson, J.R. (2012). Modulation of La Crosse virus infection in *Aedes albopictus* mosquitoes following larval exposure to coffee extracts. *Front. Physiol.* 3: 66.

55 Senthilraja, P., Suganya, K., Manikandaprabhu, S. et al. (2014). Phytochemicals inhibit against West Nile virus replication in-silico analysis. *Int. J. Plant Anim. Environ. Sci.* 2: 5–11.

11

Fruits and Nutraceuticals for the Prevention and Treatment of Central Nervous System Disorders

Abeer M. A. El Sayed[1] *and Chukwuebuka Egbuna*[2,3]

[1] *Department of Pharmacognosy, Faculty of Pharmacy, Cairo University, Cairo, Egypt*
[2] *Department of Biochemistry, Chukwuemeka Odumegwu Ojukwu University, Uli, Anambra State, Nigeria*
[3] *Africa Centre of Excellence for Public Health and Toxicological Research (ACE-PUTOR), University of Port Harcourt, Choba, Rivers State, Nigeria*

There is clear evidence that a diet rich in specific nutritional food groups (here, fruit) can reduce the frequency and occurrence of some of the main clinical consequences, such as neurodegenerative disorders [1]. The fruits of many medicinal plant species, including *Aegle marmelos* [2], *Citrullus lanatus*, *Citrus grandis*, *Phoenix sylvestris*, *Phyllanthus emblica*, *Solanum torvum*, and *Terminalia chebula*, are predominantly used as foods and sometimes included in the management of nervous system disorders. Certain fruits such as oranges, bell peppers, guava, kiwi, tomatoes, and strawberries contain high amounts of vitamin C. Vitamin C helps prevent brain cells from becoming damaged and supports overall brain health.

Many plants in traditional medicine have been used to cure neurodegenerative diseases such as Alzheimer's disease and other memory-related disorders. The major components of these plants are carotenoids, monoterpenes, and polyphenol compounds that enhance neural functions. These medicinal plants decrease oxidant levels and inhibit acetylcholinesterase activity in the neural system. Therefore, the effects of these medicinal plants and their active constituents have been shown to improve neurodegenerative diseases, which indicates their therapeutic potential in disorders associated with neuro-inflammation and neurotransmitter deficiency.

The word "nutraceuticals" refers to foods or food products that rational clinical evidence proposes may provide health and medical benefits, including for prevention and treatment of disease [3]. Such products may be categorized as dietary supplements (extracts or concentrates), herbal products, or processed foods. Nutraceuticals can provide neuroprotection via a wide range of proposed mechanisms, such as scavenging of free radicals and reactive oxygen species (ROS), chelation of iron, modulation of cell-signaling pathways, and inhibition of inflammation [4]. The focus of the ongoing research in the field of nutraceuticals is the investigation of molecules isolated from traditional medicines and how they can be helpful in debilitating and degenerative pathologies [5]. This chapter therefore highlights the potential role of nutraceuticals in brain health and neurologic disorders.

Phytochemical Drug Discovery for Central Nervous System Disorders: Biochemistry and Therapeutic Effects, First Edition. Edited by Chukwuebuka Egbuna and Mithun Rudrapal.

11.1 Fruits for Cognition and Brain Health

Fruits and fruit juice have been used for cognition and brain health. Keservani et al. [6] reported the use of various fruits for brain health. Keservani and Sharma [7] highlighted the role of vitamin C and polyphenols found in citrus fruits and blueberries in mental performance.

11.1.1 *Aegle marmelos*

A. marmelos fruit extract produced a reduction in acetylcholinesterase (AChE) and glutamate levels, but significantly increased the dopamine level. Increasing AChE activity indicates that it ameliorates cholinergic function by inhibiting sodium nitrite, which induces an increase in AChE activity [2].

11.1.2 *Citrullus lanatus*

The neuroprotective potential of *C. lanatus* seed extract (watermelon seed) and vitamin E on mercury chloride intoxication on the frontal cerebral cortex of male rats was investigated. *C. lanatus* seed extract and vitamin E mitigated $HgCl_2$-induced degeneration of frontal cerebral cortical neurons, thus demonstrating their neuropotential capacity to protect cerebral cortex neurons from mercury toxicity. However, the neuroprotective effect of *C. lanatus* seed extract against bilateral common carotid artery occlusion induced cognitive impairment and associated oxidative damage was previously reported [8].

11.1.3 *Citrus grandis*

A series of chromatographic separations performed on the ethanol extracts of the peel of *C. grandis* led to the characterization of 40 compounds, including 17 coumarins, 8 flavonoids, 2 triterpenoids, 4 benzenoids, 2 steroids, 1 lignan, 1 amide, and 5 other compounds. Most of the isolated compounds were examined for their inhibition of superoxide anion generation and elastase release by human neutrophils [9].

11.1.4 *Phoenix sylvestris*

Al-Alawi et al. [10] highlight the benefits of date fruit extracts in individual health-promoting conditions and applications and their usefulness to the pharmaceutical and nutraceutical industries in the development of natural compound-based industrial products. Cerebral ischemia was significantly attenuated by 15 days' pretreatment with methanolic extract of *Phoenix dactylifera* fruits (100 and 300 mg/kg). These results suggest the possible use of *P. dactylifera* against bilateral common carotid artery occlusion-induced oxidative stress and neuronal damage [11].

11.1.5 *Phyllanthus emblica*

The edible fruit of *P. emblica* (known as amla and Indian gooseberry) is widely used in Eastern traditional medicinal systems for a variety of ailments. An amla juice powder was previously reported to show neuroprotective effects in several in vitro and in vivo assay models, but its chemical constituents and their neuroprotective activity remain unknown. A phytochemical investigation of amla juice powder was conducted and evaluated the antioxidant and neuroprotective effects of the isolates.

Ellagic acid reduced β-amyloid-induced neurotoxicity in a transgenic *Caenorhabditis elegans* model by increasing their survival rate by 28.3% compared to the control group [12]. Pretreatment with *Phyllanthus amarus* extract for 14 and 28 days was comparable with pretreatment with ibuprofen in prevention of memory impairment and alleviation of neuroinflammatory responses. Further studies are essential to identify the bioactive phytochemicals and the precise underlying mechanisms [13]. In neuroprotection bioactivity, *P. emblica* extracts had higher percentages of pheochromocytoma cell protection and PC12 cell protection than commercial extracts, Whereas *P. emblica* hydroalcoholic extracts had a neuroprotective effect against oxidative damage, which could be due to their antioxidant and anti-inflammatory activity [14].

11.1.6 *Emblica officinalis*

E. officinalis fruit extract can be considered a novel herbal medicine candidate for managing neuroinflammatory diseases [15]. In lipopolysaccharide (LPS)-treated BV2 cells, *E. officinalis* fruit extract reduced nitric oxide, interleukin-6, and tumor necrotic factor-α production. It also enhanced the neurite length of Neuro2a cells, which was linked to the upregulation of TuJ1 and MAP2 expression. *E. officinalis* fruits also have promising neuroprotective potential to exhibit anti-neuroinflammation activity and an accelerative effect on neurite outgrowth in vitro.

11.1.7 *Solanum torvum*

A comparative study of the phytoconstituents of the fruits/berries of three different species of *Solanum* (*torvum*, *erianthum*, and *macrocarpon*) at different maturity stages was previously reported and revealed that the levels of antioxidants, total phenolics, and flavonoids are significantly affected by both the physiologic maturity stage and extraction solvents [16]. The mechanism of neuroprotection mediated by methyl caffeate was reported to be activated after H_2O_2 exposure in previous studies, which verified the involvement of various signaling pathways by using inhibitors of calpain, cathepsin D, or the PI3–K/Akt pathway [17].

11.1.8 *Terminalia chebula*

The methanolic and water extracts of *T. chebula* and ellagic acid exhibited the strongest neuroprotective activity against amyloid-beta (Aβ)25–35-induced PC12 cell damage and enhanced cell viability thorough two key mechanisms, by inhibiting ROS production and reducing calcium ion influx [18]. Ellagic acid also exhibited partial neuroprotective activity against H_2O_2-induced PC12 cell damage.

11.1.9 Blackberries

Berry anthocyanins improve neuronal and cognitive brain function and ocular health, as well as protecting genomic DNA integrity [19]. Neurogenesis acting on the hippocampus may represent one mechanism by which blueberry flavonoids improve memory. There is strong evidence suggesting that blueberry can improve memory and learning through modulation of important structural and synaptic plasticity markers [20]. One of the roles of anthocyanins in neuroprotection could be mediated through phospholipase A2 inhibition [1], which is negatively involved in a complex network of signaling pathways linking receptor agonists, oxidants, and proinflammatory cytokines to the release of arachidonic acid and eicosanoid synthesis [21].

11.1.10 Blueberries

The blueberry anthocyanins and cocoa flavanols promote positive effects on cognitive outcomes, especially in aged populations [22]. Blueberry juice in older adults with slight cognitive destruction yielded working memory improvements and further reduced depressive symptoms. An improved short-term cognitive action with high-flavonoid fruit juices, including blueberry juice, was observed in a few trials in a geriatric population [23]. Confirming these results, berries are rich in a subclass of flavonoids known as anthocyanidins, which can cross the blood–brain barrier and concentrate in regions of learning and memory [24].

11.1.11 Strawberries

The strawberry, *Fragaria ananassa*, is a broadly grown crossbred species of the genus Fragaria [25]. Diets high in strawberries can possess the ability to gain benefits to the aging brain [26]. A few studies have evaluated the influences of antioxidant- and polyphenol-rich materials on the cognitive process. The evidence suggest that these components may certainly affect cognition acutely (when given in high doses) as well as on long-term intake [27].

11.1.12 Raspberries

Hybrids between *Rubus idaeus* and *R. strigosus* are commercially employed to obtain red raspberries. Raspberries possess anthocyanin pigments, ellagic acid (from ellagitannins, e.g. the polyphenol ellagitannin), quercetin, gallic acid, cyanidins, pelargonidins, catechins, kaempferol, and salicylic acid [28]. Yellow raspberries and other yellow-colored fruits are poor in anthocyanins [29]. Yellow as well as red raspberries possess carotenoids, often lutein esters, but these are covered by anthocyanins in red raspberries [30]. The berry ingredients are suggested to protect against damage caused by ROS, which are known to be involved in the progress of neurologic diseases like Alzheimer's disease [31]. The hypotheses from in vitro investigations have also been exploited to reinforce the probability that polyphenols present in berries may positively remodel Aβ aggregation [32] in vitro, a process that eventually leads to brain damage in Alzheimer's disease. Blueberry extracts defended against damage caused by inflammation of microglial cells via a decrease in inflammatory mediators [33].

11.1.13 Cherries

The majority of edible cherries are obtained either from *Prunus avium*, the sweet cherry (also known as wild cherry), or from *Prunus cerasus*, the bitter cherry. It is thought that several pathways work together to improve cognition and prevent degeneration of the brain.

First, the high antioxidant content of these fruits may scavenge free radicals and reduce inflammation in the brain. Additionally, flavonoids in the fruit have the potential to inhibit cell death of nerve cells, and improve connections between the neurons, especially in the areas of the brain associated with learning and memory (hippocampus). Flavonoids may also disrupt the aggregation of Aβ in the brain and thereby prevent the formation of amyloid plaques [34].

11.1.14 Oranges

Oranges are the most famous citrus harvest and make up 75% of all cultivated citrus fruits. Hesperetin, a citrus flavanone, exhibited a promising protective role in neuroinflammation and neurodegeneration [35]. The neuroprotective role and potential mechanisms of hesperetin against LPS-induced Toll-like receptor (TLR)-4-mediated neuroinflammation and related neurodegeneration was reported [35]. A variety of phytochemicals, such as carotenoids (β-carotene, lutein, and β-cryptoxanthin), flavonoids (e.g. naringenin) [36], and several volatile organic substances are present in oranges [37].

11.1.15 Plums

Plums are high in flavonoids and anthocyanins, and could offer a good supply of antioxidants to protect the brain. The plum comes in the subgenus Prunus. Plums, which contain anti-inflammatory and antioxidant substances, have been shown to protect the brain, helping to prevent neurodegenerative diseases and improving memory and cognitive function [38]. In fact, the intake of plums has helped to control the formation of accumulations of β-amyloid proteins, which are a characteristic of Alzheimer's disease [39]. Not only that, plum juice and fresh plums have been shown to improve cognitive function even in elderly people [40].

11.1.16 Prunes

A prune is a dried plum of any cultivar, mostly *Prunus domestica* or European plum. The medicinal significance of prunes and their products is also reported [41]. Prunes provide major nutrients, embracing carbohydrates, many amino acids, vitamins, and dietary fiber. The pathogenesis of anxiety conditions has recently been found to implicate the role of oxidative stress in the brain. The presence of chlorogenic acid and the ability to enhance antioxidant protection make prunes promising in anxiety disorders [42].

11.1.17 Red Grapes

Investigations have revealed that manifestations of degenerative nerve disease, viral infections, and Alzheimer's disease are inhibited by phytochemicals present in grapes like resveratrol [43]. A common activity of resveratrol is responsible for defense of the genome via antioxidant activity [44]. Grapes possess several biologic activities, owing to their high polyphenol components, the majority of which are present in the seeds and pericarp. Guo et al. [45] investigated the probable shielding outcomes of grape seed oligomer and polymer procyanidin fractions against neurodegeneration, where the consumption of grape seed procyanidins or other natural antioxidants could safeguard the brain.

11.1.18 Pomegranates

The edible seeds of pomegranate (*Punica granatum*) are a brilliant source of dietary fiber. The most common phytochemicals in pomegranate juice are polyphenols, encompassing the hydrolyzable tannins known as ellagitannins, also called punicalagins [46]. The neuroprotective features of pomegranate polyphenols were assessed in an animal model of Alzheimer's disease, which showed 50% lower deposition of soluble Aβ and fewer hippocampal amyloid accumulations than mice consuming sugar water, proposing that pomegranate juice can be neuroprotective [47]. These findings advise that pomegranate juice possesses antioxidant-governed neuroprotective activity transferred from mother to neonate. Naveen et al. [48] studied the antidepression influence of polyphenols and omega-3 fatty acid from pomegranate peel and flax seed, which were able to decrease all the outcomes of chronic mild stress. Pomegranates from Oman decreased brain oxidative loss in a transgenic mouse model of Alzheimer's disease, and it was proposed that the therapeutic caliber of 4% pomegranate in the treatment of Alzheimer's disease might be related to combating oxidative stress by the active phytochemicals therein [49].

11.2 Nutraceuticals in Ameliorating Neurodegeneration

Nutraceuticals can tend to modify the cellular and molecular cascade and can lead to the prevention of neurodegeneration by targeting proteins that are misfolded practically at all levels and act as supplementation therapy. Nutraceuticals, when used for their therapeutic potential, can easily replace synthetic drug ingredients. Hence, nutraceuticals offer an all-in-one effective alternative in the management of neurologic disorders due to their affordable prices and availability and decreased side effects [50].

11.2.1 Quercetin and Kaempferol

Generation of free radicals in the brain leads to inhibition of β-amyloid 1–42 (Aβ1–42) proteins and their aggregation that leads to fibril destabilization. Quercetin and kaempferol have been proven to decrease the levels of free radicals remarkably [51]. They also prevent the activation of nuclear factor kappa B (NF-κB), which further prevents the activation of proinflammatory cytokines, mainly interleukins. They are highly effective in improving the circulation of blood in the brain and prevent the progression of Alzheimer's disease [52]. In vivo studies have demonstrated that quercetin could have a role in vascular dementia by decreasing the size of ischemic lesions [53]. Quercetin improves memory and hippocampal synaptic plasticity in models of impairment induced by chronic lead exposure, and it could have a role in neuronal repair, as shown in spinal cord injury models [54]. Kaempferol improves cognitive learning and memory capability in mice [55].

11.2.2 Withanine

Withanine is the chief steroidal alkaloid obtained from ashwagandha, also known as Indian ginseng [56], which has been used for its memory-boosting and neurocognitive-enhancing properties. It possesses high antioxidant potential and can be used to improve oxidative stress-mediated neurodegeneration. Withanine inhibits the activation of nitric oxide, which further reverses oxidative stress, and presents remarkable neuroprotective effects. Somniferine, also obtained from ashwagandha, is widely used for its neuroprotective and memory-enhancing effects.

11.2.3 Asiatic Acid

Centella asiatica, commonly known as gotu kola, has been used for its memory-enhancing properties in Ayurveda and aids in improving learning. Its principal phytoconstituent, asiatic acid, is chiefly responsible for the neuroprotective actions. It acts by decreasing the levels of malondialdehyde while simultaneous increasing glutathione. Malondialdehyde is a byproduct formed post peroxidation of lipids that acts as a very important marker for the detection of free radicals of oxidative stress-mediated neurodegeneration. Asiatic acid increases the levels of free radical scavenging enzymes, such as glutathione, and augments the antioxidant medication protection of neurodegeneration [57].

11.2.4 Bhilavanol A and B

Bhilavanol A and B are chiefly obtained from *Semecarpus anacardium*. They belong to the Anacardaceae family. Since bhilavanol is very high in potency, it is used only after purification procedures. It inhibits the activation of acetylcholinesterase and is highly effective against stress-mediated neurodegeneration. Ingredients of the Mediterranean diet, such as coffee, extra virgin olive oil, and walnuts, also improve memory and are highly beneficial. The phenolic compounds extracted from plants are highly emphasized as they possess maximum therapeutic benefits [57].

11.3 Nutraceuticals in Alzheimer's Disease

Pronounced nutraceuticals that are helpful in the management of Alzheimer's disease include absolutely essential antioxidants, which can be employed in the treatment of all chronic diseases due to oxidative stress, which plays a crucial part in neurologic disorders, including Alzheimer's disease [1]. Various studies have reported an association between the intake of higher amounts of dietary antioxidants and diminished risk in patients with Alzheimer's disease [58]. The consumption of food products that are rich in polyunsaturated fatty acids and low in saturated and trans fatty acids tends to suppress neurodegeneration [59].

11.3.1 Flavonoids

The main flavonoids in neurogenerative disorders, mainly Alzheimer's, include catechin, epicatechin, epigallocatechin, and epigallocatechin gallate. These are a group of commonly found polyphenolic compounds mainly extracted from the human diet. Flavonoids and their metabolic products possess neuromodulating actions and have been shown to interact with the neuronal–glial signaling pathway, which is mainly involved in the survival and functioning of neurons [60]. Flavonoids induce changes in cerebral blood flow [61], upregulate antioxidant enzymes and proteins involved in synaptic plasticity and neuronal repair [62], and inhibit neuropathologic processes in brain regions typically involved in Alzheimer's pathogenesis [63].

11.3.2 Apigenin

Apigenin was shown to protect neurons against Aβ-mediated toxicity induced by copper, to increase neuronal viability, as well as to relieve mitochondrial membrane dissipation and neuronal nuclear condensation [64]. Apigenin also modulates GABAergic and glutamatergic transmission in cultured cortical neurons [65].

11.3.3 Genistein, Daidzein, Glycitin

Isoflavones appear to improve cognitive function by mimicking the effects of estrogen, through the estrogen receptor β in the brain, to reduce age-related neuronal loss and cognition decline in male rats [66]. Previous studies revealed that soy isoflavones improve visual spatial memory and learning ability as well as the memory of humans [67].

11.3.4 Resveratrol

There is evidence that resveratrol shows beneficial effects in cell culture and in vivo models of neurotoxicity and neurodegeneration. Recent evidence has shown that resveratrol can increase serotonin (5-HT) activity, which could explain its antidepressant properties [68]. The protection of organotypic hippocampal slices from hydroperoxide insults has also been observed [53]. Other possible mechanisms for the neuroprotective action displayed by resveratrol are related to its antioxidant properties and its capacity to modulate Aβ processing and upregulate the longevity-linked gene sirtuin 1 [69].

11.3.5 Curcumin

Curcumin seems to act with different mechanisms, including antioxidant, anti-inflammatory and antiamylodogenic. For example, curcumin enhances cell viability by decreasing ROS and inhibiting proapoptotic signals in mouse models of encephalitis [53]. It also reduces Aβ-related cerebral burden and inflammation in transgenic AD mice [70]. It has been proposed that curcumin exerts most of its effects by inhibiting monoamine oxidase (MAO) levels, thereby reducing depression also because it modulates serotonergic, dopaminergic, and noradrenergic transmission [68].

11.3.6 Carotenoids

Major carotenoids in the human organism are lycopene, lutein, zeaxanthin, β-cryptoxanthin, α-carotene, and the most prominent carotenoid, β-carotene. They are fat-soluble pigments and can mainly be extracted from fruits and vegetables that are orange, deep yellow, and red in color [51]. Astaxanthin, a seafood-derived carotenoid, has been extensively studied for its anti-inflammatory and antioxidant potential in in vivo and in vitro animal models, and its microcirculatory-protective and mitochondrial-protective functions have been identified, suggesting it is a potent neuroprotective compound. Patients with severe or moderate Alzheimer's disease lack major carotenoids, such as lutein and β-carotene, compared with patients with mild Alzheimer's disease [71]. Among six carotenoids tested, lycopene was the only carotenoid inversely associated with quality of cognitive performance, as assessed by both the mini mental state examination, dementia detection test, and clock drawing test in healthy subjects from 45 to 102 years of age [72].

11.3.7 Crocin

Crocin is the main chemical compound identified in saffron, whose scientific name is *Crocus sativus*. Crocin is the actual active component involved in both the improvement of learning and memory and preventing the effect of long-term potentiation blocked by ethanol and its potential in the treatment of neurodegenerative diseases like Alzheimer's disease [73]. In animal models, the effectiveness of crocin has been shown in antagonizing the cognitive deficits caused by neurotoxic agents like streptozocin [74]. Crocin improved cognition as evaluated by means of ADAS-Cog and CDR-SB in subjects with mild to moderate Alzheimer's disease [75]. In a recent in vivo study, it has been demonstrated that crocin significantly modulates the levels of oxidative markers in the hippocampus, abolishing the deleterious effects of chronic stress on learning and memory [76]. Through various studies it has been concluded that crocin can significantly alter the levels of oxidative markers in the region of the hippocampus and abolish the deleterious effects on learning and memory due to chronic stress [77].

11.3.8 Carnosic Acid and Rosmarinic Acid

Carnosic and rosmarinic acids are two of the most important antioxidant compounds in rosemary. Both carnosic and rosmarinic acids showed neuroprotective action in both in vitro models of neuronal death and in vivo models of neurodegeneration. They scavenge reactive nitrogen species and protect neuroblastoma cells from hydrogen peroxide-induced oxidative stress. Diterpenes significantly alleviate memory impairment associated with Aβ neurotoxicity in Alzheimer's disease and significantly delays the onset of the disease [53].

11.3.9 Alkaloids

Among the alkaloid compounds used in neurodegenerative diseases are indole, oxindole, pyrroloindole, isoquinoline, piperidine, β-carboline, and methylxanthine derivatives. These compounds are muscarinic and adenosine receptor agonists, anti-amyloid and MAO inhibitors, acetylcholinesterase and butyrylcholinesterase inhibitors, an inhibitor of α-synuclein aggregation, dopaminergic and nicotine agonists, and N-methyl-D-aspartate antagonists. Therefore, they play positive roles in improving the pathophysiology of this kind of diseases, including Alzheimer's disease, Parkinson's disease, Huntington's disease, schizophrenia, epilepsy, and stroke [78] (Table 11.1).

11.4 Nutraceuticals in Parkinson's Disease

Parkinson's disease is a neurologic disease with impaired dopaminergic neurons in the substantia nigra par compacta region of the brain, leading to drastic depletion of dopamine. Factors such as oxidative stress, depletion of antioxidants, and damage to mitochondria contribute to the neurodegeneration leading to Parkinson's disease [79].

11.4.1 Vitamins: Folate, Cobalamin, Pyridoxin

Folate (vitamin B9), cobalamin (vitamin B12), and pyridoxin (vitamin B6) are the most studied B vitamins in the field of cognitive decline and dementia [80]. They are essential for maintaining the integrity of the nervous and hematopoietic systems [81].

Table 11.1 Nutraceuticals and their mode of action and specific disease activity.

Disease	Mechanism of action	Nutraceuticals
Neurodegenerative disorders	Neurodegenerative disorders are mainly developed by protein misfolding. Nutraceuticals mainly prevent misfolding of proteins by inhibiting the activation and synthesis of proinflammatory cytokines and associated pathways	Bacoside A, bacoside B, brahmine, quercetin, kaempferol, withanine, somniferine, asiatic acid, bhilavanol A and B.
Alzheimer's disease	Alzheimer's disease is mainly associated with increase in oxidative stress and free radicals. The nutraceuticals typically antioxidant in nature are mostly employed in the management of this disease	Flavonoids (fruits, vegetables, tea, wine, coffee); carotenoids (lutein, zeaxanthin, lycopene, β-cryptoxanthin, including α- and β-carotenes); anthocyanidins (cyanidin); flavones (luteolin, apigenin)
Parkinson's disease	Uninhibited oxidative stress and free radicals in association with abnormally misfolded proteins, neuroinflammation, and dysfunctional mitochondria lead to compromised cellular metabolism and energy, thereby impacting the functioning of the brain and leading to neurodegenerative disorders, including Parkinson's disease	Vitamin A, omega-3 fatty acids, lycopene, vincamine, gallic acid, curcumin
Depression	Nutraceuticals that act by inhibiting reuptake of monoamines possess anti-inflammatory and antioxidant properties that are well suited for management of depression	Omega-3 fatty acids, folic acid, S-adenosyl methionine, zinc, N-acetyl cysteine, L-tryptophan
Psychosis	Nutraceuticals that can improve neurotransmission in dopaminergic serotoninergic neurons can be employed in management of psychosis	Vitamins and omega-3 fatty acids

Vitamin B6 – comprising three chemically distinct compounds, pyridoxal, pyridoxamine, and pyridoxine – is involved in the regulation of mental function and mood. It is also an essential homocysteine remethylation cofactor, and its deficiency is associated with an increase in blood homocysteine levels [81]. Vitamin B12 is involved in the methylation of homocysteine to methionine for the synthesis of methyl acceptors such as membrane phospholipids, myelin, and neurotransmitters.

Several vitamins, including vitamin B3, vitamin B9 or folate, vitamin B12, vitamin B6, vitamin D, vitamin E, and vitamin C, can be used in Parkinson's disease [82]. The anti-Parkinson drugs currently employed prevent disease progression by providing symptomatic relief only. A wide number of herbal and natural products have been studied clinically for use in Parkinson's disease [83] to evaluate and clarify if such herbal molecules can be implemented as an independent or adjunctive therapy in disease management [84]. During clinical trials, synthesized herbal products restrict the occurrence of identified lead molecules in the market [85]. Natural products exert favorable effects in Parkinson's disease by reducing the different pathologic pathways inducing it, such as oxidative stress, dysfunction of mitochondria, neuroinflammation, and apoptosis.

11.5 Nutraceuticals in Depression

Omega-3 fatty acids and folic acids have generally been effective for unipolar depression, particularly as an adjunctive therapy, with increasing evidence for their efficacy as monotherapy. The nutrients obtained from dietary products are critical for proper brain functioning, as a relationship between the quality of food and brain health and mood has been identified and studied [86], leading to the application of nutraceuticals as supplements. A whole-grain diet rich in nutrients, such as zinc, folic acid, omega-3 fatty acids, and several other essential macro- and micronutrients, can trigger the functioning of the brain and has evidently shown results in the management of depression [87].

11.6 Nutraceuticals in Psychotic Disorders

Nutraceuticals, besides the functional roles studied, also play a key role in the management of mood disorders and psychotic disorders, such as schizophrenia and bipolar disorder [88]. Nutraceuticals strongly amplify the therapeutic efficacy of the medications employed by strengthening neuroprotection, enhancing the inhibited reuptake of monoamines and showing neurobiological effects [89], thereby improving the efficacy of psychiatric medicines and ameliorating their side effects.

The commonly used nutraceuticals in psychosis include omega-3 fatty acids and vitamins. There are two main types of polyunsaturated fatty acids in the human body: those of the omega-6 series, such as arachidonic acid, obtained from linoleic acid; and those of the omega-3 series, obtained as α-linolenic acid. The latter includes eicosapentaenoic acid and docosahexaenoic acid [90]. Omega-3 eicosapentaenoic and docosahexaenoic acid have properties that are of interest in psychotic disorders. They improve dopaminergic and serotoninergic neurotransmission. They decrease microinflammatory and oxidative stress. They modulate the functioning of mitochondria, which are the main source of oxidative stress [91]. Additionally, they protect against toxicity due to apoptosis and regulate gene expression of brain-derived neurotrophic factor. Figure 11.1 represents nutraceuticals that can be utilized as adjunctive therapy in the management of neurological degeneration, Parkinson's disease, depression, Alzheimer's disease, and psychosis simultaneously.

11.7 Conclusion

Nature has provided us with valuable herbal molecules with high potential in the cure and prevention of life-threatening diseases and lifestyle-related disorders, including neurodegeneration. Fruit and fruit juice would impart a beneficial effect on brain functioning. The chemical contents have properties that alleviate the harmful effect of disease by reducing oxidative stress or by other mechanisms. The influences of dietary factors on neuronal processes and synaptic flexibility have exhibited a few of the essential underlying mechanisms

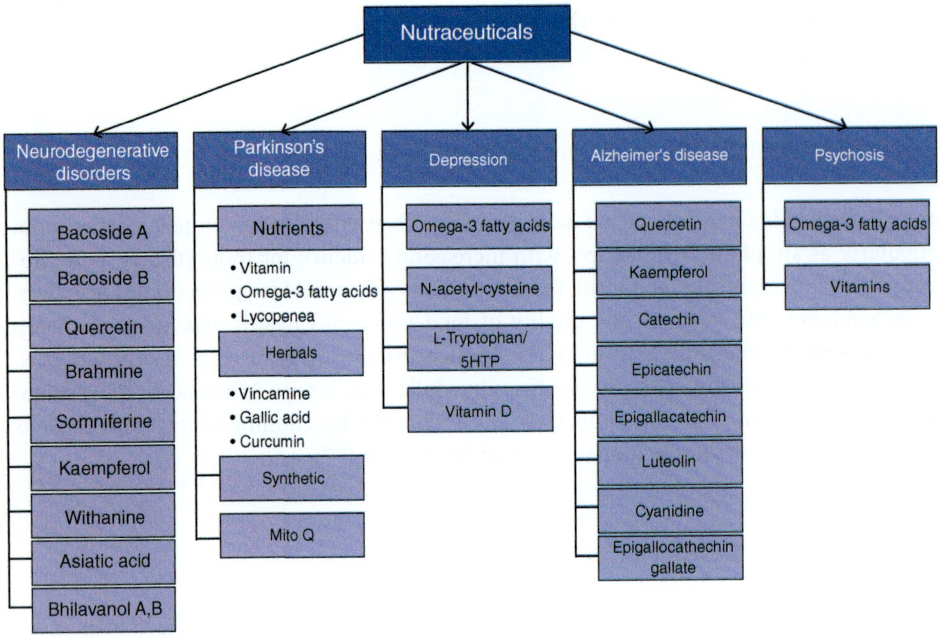

Figure 11.1 Nutraceuticals that can be utilized as adjunctive therapy in the management of neurological degeneration, Parkinson's disease, depression, Alzheimer's disease, and psychosis simultaneously.

for the role of diet in maintaining brain health and mental activity. More investigations are warranted to elucidate the role of phytoconstituents in therapeutics of brain disorders. The curative effects of nutraceuticals can be attributed to their neuroprotective, anti-inflammatory, antioxidant, hypolipidemic, and healing properties, which target different ligands and receptors to enhance protein synthesis, which ultimately leads to neuroprotection. This chapter has highlighted the merits and demerits of nutraceutical therapy and its susceptibility to preventing disease progression in neurologic disorders.

References

1 Frisardi, V., Panza, F., Solfrizzi, V. et al. (2010). Plasma lipid disturbances and cognitive decline. *J. Am. Geriatr. Soc.* 58: 2429–2430.

2 Raheja, S., Girdhar, A., Kamboj, A. et al. (2019). *Aegle marmelos* leaf extract ameliorates the cognitive impairment and oxidative stress induced by intracerebroventricular streptozotocin in male rats. *Life Sci.* 221: 196–203.

3 Calabrese, V., Cornelius, C., Mancuso, C. et al. (2009). Vitagenes, dietary antioxidants and neuroprotection in neurodegenerative diseases. *Front. Biosci.* 14: 376–397.

4 Chao, J., Leung, Y., Wang, M., and Chang, R.C. (2012). Nutraceuticals and their preventive or potential therapeutic value in Parkinson's disease. *Nutr. Rev.* 70 (7): 373–386.

5 Abdel-Daim, M.M., El-Tawil, O.S., Bungau, S.G., and Atanasov, A.G. (2019). Applications of antioxidants in metabolic disorders and degenerative diseases: mechanistic approach. *Oxid. Med. Cell. Longev.* 2019: 4179676.

6 Keservani, R.K., Kesharwani, R.K., Sharma, A.K., and Ahmad, M.F. (2015). Stress, Parkinson's, Alzemer's disease: role of dietary supplements. In: *Nutraceuticals and Functional Foods in Human Health and Disease Prevention* (ed. D. Bagchi, H.G. Preuss, and A. Swaroop), 241–254. Boca Raton, FL: CRC Press.

7 Keservani, R.K. and Sharma, A.K. (2014). Flavonoids: emerging trends and potential health benefits. *J. Chin. Pharm. Sci.* 23 (12): 815–822.

8 Pashikanti, G., Ajitha, M., and Puchchakayala, G. (2019). Neuroprotective effect of *Citrullus lanatus* seed extracts on cerebral ischemic reperfusion injury induced cognitive impairment and oxidative stress. *Int. J. Pharm. Pharm. Sci.* 11 (6): 38–44.

9 Kuo, P.C., Liao, Y.R., Hung, H.Y. et al. (2017). Anti-inflammatory and neuroprotective constituents from the peels of *Citrus grandis*. *Molecules* 22 (6): 967.

10 Al-Alawi, R.A., Al-Mashiqri, J.H., Al-Nadabi, J.S.M. et al. (2017). Date palm tree (*Phoenix dactylifera* L.): natural products and therapeutic options. *Front. Plant Sci.* 8: 845.

11 Pujari, R.R., Vyawahare, N.S., and Kagathara, V.G. (2011). Evaluation of antioxidant and neuroprotective effect of date palm (*Phoenix dactylifera* L.) against bilateral common carotid artery occlusion in rats. *Indian J. Exp. Biol.* 49 (8): 627–633.

12 Rose, K., Wan, C., Thomas, A. et al. (2018). Phenolic compounds isolated and identified from Amla (*Phyllanthus emblica*) juice powder and their antioxidant and neuroprotective activities. *Nat. Prod. Commun.* 13 (10): 1309–1311.

13 Alagan, A., Jantan, I., Kumolosasi, E. et al. (2019). Protective effects of *Phyllanthus amarus* against lipopolysaccharide-induced neuroinflammation and cognitive impairment in rats. *Front. Pharmacol.* 10: 632.

14 Li, P.-H., Wang, C.-W., Lu, W.-C. et al. (2022). Antioxidant, anti-inflammatory activities, and neuroprotective behaviors of *Phyllanthus emblica* L. fruit extracts. *Agriculture* 12 (5): 588.

15 Phochantachinda, S., Chatchaisak, D., Temviriyanukul, P. et al. (2021). Ethanolic fruit extract of *Emblica officinalis* suppresses neuroinflammation in microglia and promotes neurite outgrowth in Neuro2a cells. *J. Evid. Based Complement. Altern. Med.* 2021: 6405987.

16 Kortei, N.K., Suetor, J.M., Aboagye, G. et al. (2020). Comparative study of the bioactive and chemical properties of three different *Solanum* spp. from Ghana. *J. Food Sci.* 4: 1773–1784.

17 Jantas, D., Chwastek, J., Malarz, J. et al. (2020). Neuroprotective effects of methyl Caffeate against hydrogen peroxide-induced cell damage: involvement of caspase 3 and Cathepsin D inhibition. *Biomolecules* 10 (11): 1530.

18 Shen, Y.C., Juan, C.W., Lin, C.S. et al. (2017). Neuroprotective effect of *Terminalia chebula* extracts and Ellagic acid in PC12 cells. *Afr. J. Tradit. Complement. Altern. Med.* 14 (4): 22–30.

19 Zafra-Stone, S., Yasmin, T., Bagchi, M. et al. (2007). Berry anthocyanins as novel antioxidants in human health and disease prevention. *Mol. Nutr. Food Res.* 51: 675–683.

20 Rendeiro, C., Guerreiro, J.D., Williams, C.M., and Spencer, J.P. (2012). Flavonoids as modulators of memory and learning: molecular interactions resulting in behavioural effects. *Proc. Nutr. Soc.* 71: 246–262.

21 Sun, G.Y., Xu, J., Jensen, M.D., and Simonyi, A. (2004). Phospholipase A2 in the central nervous system: implications for neurodegenerative diseases. *J. Lipid Res.* 45: 205–213.

22 Field, D.T., Williams, C.M., and Butler, L.T. (2011). Consumption of cocoa flavanols results in an acute improvement in visual and cognitive functions. *Physiol. Behav.* 103: 255–260.

23 Krikorian, R., Nash, T.A., Shidler, M.D. et al. (2010). Concord grape juice supplementation improves memory function in older adults with mild cognitive impairment. *Br. J. Nutr.* 103: 730–734.

24 Andres-Lacueva, C., Shukitt-Hale, B., Galli, R.L. et al. (2005). Anthocyanins in aged blueberry-fed rats are found centrally and may enhance memory. *Nutr. Neurosci.* 8 (2): 111–120.

25 Manganaris, G.A., Goulas, V., Vicente, A.R., and Terry, L.A. (2014). Berry antioxidants: small fruits providing large benefits. *J. Sci. Food Agric.* 94 (5): 825–833.

26 Hannum, S.M. (2004). Potential impact of strawberries on human health: a review of the science. *Crit. Rev. Food Sci. Nutr.* 44 (1): 1–17.

27 Collie, A. and Morley, G. (2007). Do polyphenols affect human cognitive function? *Curr. Top. Nutraceutical Res.* 5 (4): 145–148.

28 Mazur, S.P., Nes, A., Wold, A.-B. et al. (2014). Quality and chemical composition of ten red raspberry (*Rubus idaeus* L.) genotypes during three harvest seasons. *Food Chem.* 160: 233–240.

29 Carvalho, E., Fraser, P.D., and Martens, S. (2013). Carotenoids and tocopherols in yellow and red raspberries. *Food Chem.* 139 (1–4): 744–752.

30 Carvalho, E., Franceschi, P., Feller, A. et al. (2013). A targeted metabolomics approach to understand differences in flavonoid biosynthesis in red and yellow raspberries. *Plant Physiol. Biochem.* 72: 79–86.

31 Miller, M.G. and Shukitt-Hale, B. (2012). Berry fruit enhances beneficial signaling in the brain. *J. Agric. Food Chem.* 60 (23): 5709–5715.

32 Ladiwala, A.R., Mora-Pale, M., Lin, J.C. et al. (2011). Polyphenolic glycosides and aglycones utilize opposing pathways to selectively remodel and inactivate toxic oligomers of amyloid β. *Chembiochem.* 12 (11): 1749–1758.

33 Lau, F.C., Bielinski, D.F., and Joseph, J.A. (2007). Inhibitory effects of blueberry extract on the production of inflammatory mediators in lipopolysaccharide-activated BV2 microglia. *J. Neurosci. Res.* 85 (5): 1010–1017.

34 Gonçalves, B., Correia, C.M., Silva, A.P. et al. (2008). Leaf structure and function of sweet cherry tree (*Prunus avium* L.) cultivars with open and dense canopies. *Sci. Hortic.* 116 (4): 381–387.

35 Scoditti, E. (2020). Neuroinflammation and neurodegeneration: the promising protective role of the citrus flavanone Hesperetin. *Nutrients* 2 (8): 2336.

36 Aschoff, J.K., Kaufmann, S., Kalkan, O. et al. (2015). In vitro bioaccessibility of carotenoids, flavonoids, and vitamin C from differently processed oranges and orange juices [*Citrus sinensis* (L.) Osbeck]. *J. Agric. Food Chem.* 63 (2): 578–587.

37 Perez-Cacho, P.R. and Rouseff, R.L. (2008). Fresh squeezed orange juice odor: a review. *Crit. Rev. Food Sci. Nutr.* 48 (7): 681–695.

38 Igwe, E.O. and Charlton, K.E. (2016). A systematic review on the health effects of plums (*Prunus domestica* and *Prunus salicina*). *Phytother. Res.* 30 (5): 701–731.

39 Murphy, M.P. and LeVine, H. 3rd. (2010). Alzheimer's disease and the amyloid-beta peptide. *J. Alzheimers Dis.* 19 (1): 311–323.

40 Shukitt-Hale, B., Lau, F.C., Carey, A.N. et al. (2008). Blueberry polyphenols attenuate kainic acid-induced decrements in cognition and alter inflammatory gene expression in rat hippocampus. *Nutr. Neurosci.* 11: 172–182.

41 Li, T.S.C. (2008). *Vegetables and Fruits: Nutritional and Therapeutic Values*, vols. 169, 195, 198. Boca Raton, FL: Taylor and Francis.

42 Bouayed, J., Rammal, H., and Soulimani, R. (2009). Oxidative stress and anxiety, relationship and cellular pathways. *Oxid. Med. Cell. Longev.* 2: 63–67.

43 Shankar, S., Singh, G., and Srivastava, R.K. (2007). Chemoprevention by resveratrol: molecular mechanisms and therapeutic potential. *Front. Biosci.* 12 (12): 4839–4854.

44 Gatz, S.A. and Wiesmüller, L. (2008). Take a break—resveratrol in action on DNA. *Carcinogenesis* 29 (2): 321–332.

45 Guo, L., Li, H.W., Sun, B. et al. (2007). Direct in vivo evidence of protective effects of grape seed procyanidin fractions and other antioxidants against ethanol-induced oxidative DNA damage in mouse brain cells. *J. Agric. Food Chem.* 55 (14): 5881–5891.

46 Singh, R.P., Murthy, K.N.C., and Jayaprakasha, G.K. (2002). Studies on the antioxidant activity of pomegranate (*Punica granatum*) peel and seed extracts using in vitro models. *J. Agric. Food Chem.* 50 (1): 81–86.

47 Hartman, R.E., Shah, A., Fagan, A.M. et al. (2006). Pomegranate juice decreases amyloid load and improves behavior in a mouse model of Alzheimer's disease. *Neurobiol. Dis.* 24 (3): 506–515.

48 Naveen, S., Siddalingaswamy, M., Singsit, D., and Khanum, F. (2013). Anti-depressive effect of polyphenols and omega-3 fatty acid from pomegranate peel and flax seed in mice exposed to chronic mild stress. *Psychiatry Clin. Neurosci.* 67 (7): 501–508.

49 Subash, S., Essa, M., Al-Asmi, A. et al. (2014). Pomegranate from Oman alleviates the brain oxidative damage in transgenic mouse model of Alzheimer's disease. *J. Tradit. Complement. Med.* 4 (4): 232–238.

50 Grassi, D., Ferri, C., and Desideri, G. (2016). Brain protection and cognitive function: cocoa flavonoids as nutraceuticals. *Curr. Pharm. Des.* 22: 145–151.

51 Bungau, S., Abdel-Daim, M.M., Tit, D.M. et al. (2019). Health benefits of polyphenols and carotenoids in age-related eye diseases. *Oxid. Med. Cell. Longev.* 2019: 9783429.

52 Purza, L., Abdel-Daim, M., Belba, A. et al. (2019). Monitoring the effects of various combination of specific drug therapies at different stages of Alzheimer's dementia. *Farmacia* 67: 477–481.

53 Kelsey, N.A., Wilkins, H.M., and Linseman, D.A. (2010). Nutraceutical antioxidants as novel neuroprotective agents. *Molecules* 15: 7792–7814.

54 Schültke, E., Kendall, E., Kamencic, H. et al. (2003). Quercetin promotes functional recovery following acute spinal cord injury. *J. Neurotrauma* 20: 583–591.

55 Lei, Y., Chen, J., Zhang, W. et al. (2012). In vivo investigation on the potential of galangin, kaempferol and myricetin for protection of D-galactose-induced cognitive impairment. *Food Chem.* 135: 2702–2707.

56 Sivasankarapillai, V.S., Madhu Kumar Nair, R., Rahdar, A. et al. (2020). Overview of the anticancer activity of Withaferin A, an active constituent of the Indian ginseng Withania somnifera. *Environ. Sci. Pollut. Res.* 27 (21): 26025–26035.

57 Preethi Pallavi, M.C. and Sampath Kumar, H.M. (2018). Nutraceuticals in prophylaxis and therapy of neurodegenerative diseases. In: *Discovery and Development of Neuroprotective Agents from Natural Products* (ed. G. Brahmachari), 359–376. Amsterdam: Elsevier.

58 Nunomura, A., Perry, G., Aliev, G. et al. (2001). Oxidative damage is the earliest event in Alzheimer disease. *J. Neuropathol. Exp. Neurol.* 60: 759–767.

59 Dohrmann, D.D., Putnik, P., Bursać Kovačević, D. et al. (2019). Japanese, Mediterranean and Argentinean diets and their potential roles in neurodegenerative diseases. *Food Res. Int.* 120: 464–477.

60 Glevitzky, I., Dumitrel, G.A., Glevitzky, M. et al. (2019). Statistical analysis of the relationship between antioxidant activity and the structure of flavonoid compounds. *Rev. Chim.* 70: 3103–3107.

61 Williams, R.J. and Spencer, J.P. (2012). Flavonoids, cognition, and dementia: actions, mechanisms, and potential therapeutic utility for Alzheimer disease. *Free Radic. Biol. Med.* 52: 35–45.

62 Spencer, J.P. (2009). The impact of flavonoids on memory: physiological and molecular considerations. *Chem. Soc. Rev.* 38: 1152–1161.

63 Wang, H., Wang, H., Cheng, H., and Che, Z. (2016). Ameliorating effect of Luteolin on memory impairment in an Alzheimer's disease model. *Mol. Med. Rep.* 13: 4215–4220.

64 Zhao, L., Wang, J.L., Wang, Y.R., and Fa, X.Z. (2013). Apigenin attenuates copper-mediated β-amyloid neurotoxicity through antioxidation, mitochondrion protection and MAPK signal inactivation in an AD cell model. *Brain Res.* 1492: 33–45.

65 Losi, G., Puia, G., Garzon, G. et al. (2004). Apigenin modulates GABAergic and glutamatergic transmission in cultured cortical neurons. *Eur. J. Pharmacol.* 502: 41–46.

66 Lee, Y.B., Lee, H.J., Won, M.H. et al. (2004). Soy isoflavones improve spatial delayed matching-to-place performance and reduce cholinergic neuron loss in elderly male rats. *J. Nutr.* 134: 1827–1831.

67 Bansal, N. and Parle, M. (2010). Soybean supplementation helps reverse age- and scopolamine-induced memory deficits in mice. *J. Med. Food* 13: 1293–1300.

68 Ogle, W.O., Speisman, R.B., and Ormerod, B.K. (2013). Potential of treating age-related depression and cognitive decline with nutraceutical approaches: a mini-review. *Gerontology* 59: 23–31.

69 Pocernich, C.B., Lange, M.L., Sultana, R., and Butterfield, D.A. (2011). Nutritional approaches to modulate oxidative stress in Alzheimer's disease. *Curr. Alzheimer Res.* 8: 452–469.

70 Lim, G.P., Chu, T., Yang, F. et al. (2001). The curry spice curcumin reduces oxidative damage and amyloid pathology in an Alzheimer transgenic mouse. *J. Neurosci.* 21: 8370–8377.

71 Castellani, R.J., Harris, P.L., Sayre, L.M. et al. (2001). Active glycation in neurofibrillary pathology of Alzheimer disease: Nε-(carboxymethyl) lysine and hexitol-lysine. *Free Radic. Biol. Med.* 31: 175–180.

72 Polidori, M.C., Praticó, D., Mangialasche, F. et al. (2009). High fruit and vegetable intake is positively correlated with antioxidant status and cognitive performance in healthy subjects. *J. Alzheimers Dis.* 17: 921–927.

73 Khalili, M. and Hamzeh, F. (2010). Effects of active constituents of *Crocus sativus* L., crocin on streptozocin-induced model of sporadic Alzheimer's disease in male rats. *Iran. Biomed. J.* 14: 59–65.

74 Naghizadeh, B., Mansouri, M.T., Ghorbanzadeh, B. et al. (2013). Protective effects of oral crocin against intracerebroventricular streptozotocin-induced spatial memory deficit and oxidative stress in rats. *Phytomedicine* 20: 537–542.

75 Akhondzadeh, S., Sabet, M.S., Harirchian, M.H. et al. (2010). Saffron in the treatment of patients with mild to moderate Alzheimer's disease: a 16-week, randomized and placebo-controlled trial. *J. Clin. Pharm. Ther.* 35: 581–588.

76 Ghadrdoost, B., Vafaei, A.A., Rashidy-Pour, A. et al. (2011). Protective effects of saffron extract and its active constituent crocin against oxidative stress and spatial learning and memory deficits induced by chronic stress in rats. *Eur. J. Pharmacol.* 667: 222–229.

77 Boccardi, V., Tinarelli, C., and Mecocci, P. (2017). Nutraceuticals and cognitive dysfunction. In: *Neuroprotective Effects of Phytochemicals in Neurological Disorders* (ed. D.A.A. Farooqui and T. Farooqui), 561–579. Chichester: Wiley.

78 Hussain, G., Rasul, A., Anwar, H. et al. (2018). Role of plant derived alkaloids and their mechanism in neurodegenerative disorders. *Int. J. Biol. Sci.* 14 (3): 341–357.

79 Mythri, R.B., Joshi, A.K., and Bharath, M.M.S. (2015). Nutraceuticals and other natural products in Parkinson's disease therapy: focus on clinical applications. In: *Bioactive Nutraceuticals and Dietary Supplements in Neurological and Brain Disease* (ed. R.R. Watson and V.R. Preedy), 421–431. New York: Academic Press.

80 Brown, L.A., Riby, L.M., and Reay, J.L. (2010). Supplementing cognitive aging: a selective review of the effects of ginkgo biloba and a number of everyday nutritional substances. *Exp. Aging Res.* 36: 105–122.

81 Malouf, R. and Areosa, S.A. (2003). Vitamin B12 for cognition. *Cochrane Database Syst. Rev.* 3: CD004326.

82 Tatton, W., Chalmers-Redman, R., Brown, D., and Tatton, N. (2003). Apoptosis in Parkinson's disease: signals for neuronal degradation. *Ann. Neurol.* 53 (Suppl. 3): S61–S70.

83 Jenner, P. and Olanow, C.W. (1996). Oxidative stress and the pathogenesis of Parkinson's disease. *Neurology* 47: 161S–170S.

84 Henchcliffe, C. and Beal, M.F. (2008). Mitochondrial biology and oxidative stress in Parkinson disease pathogenesis. *Nat. Clin. Pract. Neurol.* 4: 600–609.

85 Lama, A., Pirozzi, C., Avagliano, C. et al. (2020). Nutraceuticals: an integrative approach to starve Parkinson's disease. *Brain Behav. Immun. Health* 2: 100037.

86 Ceskova, E. and Silhan, P. (2018). Novel treatment options in depression and psychosis. *Neuropsychiatr. Dis. Treat.* 14: 741.

87 van der Burg, K.P., Cribb, L., Firth, J. et al. (2021). Nutrient, and genetic biomarkers of nutraceutical treatment response in mood and psychotic disorders: a systematic review. *Nutr. Neurosci.* 24 (4): 279–295.

88 Cloutier, M., Aigbogun, M.S., Guerin, A. et al. (2016). The economic burden of schizophrenia in the United States in 2013. *J. Clin. Psychiatry* 77: 764–771.

89 Savitz, A.J., Xu, H., Gopal, S. et al. (2016). Efficacy and safety of Paliperidone palmitate 3-month formulation for patients with schizophrenia: a randomized, multicenter, double-blind, noninferiority study. *Int. J. Neuropsychopharmacol.* 19 (7): pyw018.

90 Brown, H.E. and Roffman, J.L. (2016). Emerging treatments in schizophrenia: highlights from recent supplementation and prevention trials. *Harv. Rev. Psychiatry* 24: e1–e7.

91 Sarris, J., Logan, A.C., Akbaraly, T.N. et al. (2015). Nutritional medicine as mainstream in psychiatry. *Lancet Psychiatry* 2: 271–274.

12

Neurorestorative Potential of Medicinal Plants and Their Phytochemicals

Babatunde O. Adetuyi[1], *Kehinde A. Odelade*[1], *Grace O. Odine*[1], *Oluwatosin A. Adetuyi*[2], *Semiloore O. Omowumi*[1], *Olubanke O. Ogunlana*[3], *and Chukwuebuka Egbuna*[4,5]

[1] *Department of Natural Sciences, Faculty of Pure and Applied Sciences, Precious Cornerstone University, Ibadan, Oyo State, Nigeria*
[2] *Department of Biochemistry, Osun State University, Osogbo, Osun State, Nigeria*
[3] *Department of Biochemistry, Covenant University, Ota, Ogun State, Nigeria*
[4] *Africa Centre of Excellence for Public Health and Toxicological Research (ACE-PUTOR), University of Port Harcourt, Choba, Rivers State, Nigeria*
[5] *Department of Biochemistry, Chukwuemeka Odumegwu Ojukwu University, Uli, Anambra State, Nigeria*

The term "medicinal plant' or "therapeutic plant" encapsulates a wide variety of different kinds of plant life utilized in what is called herbalism ("natural medication" or "herbology"). Herbalism is the utilization of vegetation (plants) for therapeutic purposes. It can also be said to be the practice or study of therapeutic plants [1]. Different parts of plants such as fruit, branch, grain, petal, leaf, seed, root, stem, or semi-plant can all be utilized as herbs. Historically, the word "herb" was reserved for semi-plants, including those obtained from shrubs and trees. Plants have been utilized for restorative purposes extending back to antiquity. Egyptian papyruses, Chinese works, and Unani texts in ancient times have been found to contain descriptions of phytonutrients [1, 2]. Different researchers have shown that plants have been utilized as medication by Europeans, those around the Mediterranean, Indians, and Unani Hakims for the past four millennia [2]. Plants were frequently utilized for restorative practices by indigenous peoples, including Africans, Romans, Egyptians, and Native Americans. However, others created customary medical service organizations like Ayurvedic medicine, Chinese traditional medicine, and Unani, in which traditional supplements were utilized in a deliberate fashion [1].

Conventional healing strategies are still universally recognized on a multitude of occasions. Rapid population developments, insufficient drug accessibility, outrageously high cost of treatment, detrimental effects from several synthetic drugs, and also the advancing resilience to drugs that are administered presently to treat communicable sicknesses influence the utilization of traditional prescriptions as a natural medicine for an assortment of neurologic illnesses and conditions. Treatments with plant extracts or fractions are said to be extremely healthy because there are few or no implications or adverse effects attached to them. The main advantage of these medications is the fact that they are in accordance with nature; another brilliant illustration is the way therapeutic plants are utilized by individuals of all ages and sexes [3].

Phytochemical Drug Discovery for Central Nervous System Disorders: Biochemistry and Therapeutic Effects, First Edition. Edited by Chukwuebuka Egbuna and Mithun Rudrapal.

Therapeutic plants such as ginger, turmeric, *Aloe vera*, and neem contain different phytochemical constituents and are utilized in the treatment of various infections. These are viewed as natural treatments in numerous regions of the world. Research has shown that basil/tulsi is utilized to make drugs, for pooja, as black tea, and for different purposes in their everyday way of life [1]. It is believed that therapeutic plants are a rich store of chemicals that have the ability to be utilized in the formulation of artificial drugs, pharmaceutical drugs, and non-pharmacopeial drugs. Besides that, such vegetation assumes a significant part in the advancement of human societies everywhere. Moreover, a few medicinal plants are seen as essential origins of supplements or nutrients, and as a result they are proposed for their therapeutic potential. Some of these medicinal plants include walnuts, ginger, turmeric, green tea, and pepper. Certain medicinal plants and their metabolites are regarded as important origins of phytochemicals (active compounds) utilized in toothpaste and headache pain relief (aspirin), in addition to other things. Several ways in which medicinal plants can be used include in beverages, meals, scents/fragrance, pesticides, natural dyes, and so on. Different therapeutic plants are utilized in different countries to eliminate and keep mice, flies, and insects away from working environments and homes. Nowadays, restorative plants are key suppliers of unrefined substances for drug manufacturers [4].

12.1 Therapeutic Value of Some Medicinal Plants and their Importance

Different medicinal plants with therapeutic values and their important uses are discussed in this section. Medicinal plants such as *Aloe vera*, red clover, myrrh, black pepper, cinnamon, bayberry, and sandalwood are utilized in the treatment of wounds, sores, injuries, and boils. A few significant therapeutic plants that can be cultivated in the domestic garden include thyme, cilantro, apple mint, rosemary, variegated lemon balm, and basil. They are easy to cultivate, look beautiful, have nice scents, and taste delicious. A considerable number of them attract butterflies and honey bees [1]. Numerous therapeutic plants are utilized as blood cleansing agents, or "blood purifiers," to treat or cure a prolonged illness by removing biochemical poisons or toxins. Many plants boost the body's immune system, for instance lowering the temperature.

Certain botanicals also have antibacterial qualities. An example is turmeric, which can help to prevent the proliferation of pathogens, bacteria, and pathogenic microorganisms. It is frequently utilized for wound treatment and cut healing at home [2, 3]. Medicinal plants that have antipyretic properties like black pepper, safflower, chirayta, and sandalwood have been offered by traditional Indian medical specialists to decrease the level of fever and heat produced by the ailment. Aside from being scented, sandalwood and cinnamon are superb astringents. Sandalwood is exceptionally valuable in preventing the release of blood, bodily fluid, mucus, and other natural liquids. Marshmallow root and leaf are examples of therapeutic plants that can be utilized to neutralize the stomach's acid production. These plants act as antacids and help to maintain the healthy stomach acid required for optimal digestion [1].

Indian sages were reputed to have plant-based cures that were effective against snake bites and animal poisons. Coriander and cardamom are well known for their therapeutic properties. Specific plants such as turmeric, khare khasak, sandalwood, and aloe are widely used as disinfectants and have great therapeutic qualities. Certain cough syrups contain ginger and cloves, well known for their antiemetic properties, which stimulate the thinning of mucus and its release from the lungs, bronchi, and windpipe. Expectorants incorporate cloves, cardamom, eucalyptus, wild cherry, basil, calamus, chrysanthemum, chamomile, and other plants. Medicinal plants that are beneficial to blood dissemination and circulation include cinnamon, turmeric, peppermint, fennel, spearmint, and ginger. Therefore, they are utilized to stimulate the heart.

Certain therapeutic plants have disinfective qualities that eliminate infection-causing microscopic organisms. They also prevent pathogenic microorganisms that cause communicable diseases from developing. Alternative medicine experts propose calming plants that affect the body. They are often utilized as tranquilizers or sedatives. Mild tonics are made from aromatic medicinal plants like barberry, golden seal, chirayta, and aloe. Toxins in the blood are reduced by the pungency of such plants. They are also beneficial in the elimination of infection. Certain plants, like cayenne (Lal), are utilized as energizers to support the movement of a framework or organ. Tonics are produced using a range of medicinal plants, which include aloe, giloy, barberry, and brilliant seal. They can likewise be nutritious and restorative to both healthy and ailing people. Fresh cuts and wounds can be appropriately treated with honey, turmeric, marshmallow, and liquorice. These are known as defensive plants [3].

12.2 Types of Medicinal Plants and Their Uses

Herbal medicine refers to the capacity to generate a large range of synthetic mixtures that may be employed to perform basic organic tasks as well as to protect against hunters/predators like herbivorous warm-blooded animals, bugs, parasites, insects, and fungi. When ingested by humans, a large number of these phytochemicals offer long-term health benefits and can be used to treat human ailments efficiently [5]. Active chemical substances in vegetation/plants regulate their impacts on each individual body by actions similar to those known for their chemical mixtures/chemical substances in regular pharmaceutical drugs; thus, natural prescriptions do not vary much in terms of how they work from customary medications. This permits natural medications to be similarly efficient and successful to ordinary drugs, while likewise having a similar capacity to create negative side effects [6].

Present-day medication will in general utilize the dynamic elements (active compounds) of plants as opposed to the entire plants. These active compounds are also called "phytochemicals." They can be formulated, compounded, or generally changed to make drugs. A few illustrations of such subordinate compounds include aspirin, which is artificially connected to the salicylic acid seen in white willow; digitoxin, which is obtained from digitalis; and capsaicin, which is obtained from chili. Table 12.1 further describes the uses of medicinal plants.

12.3 Phytochemicals

Phytochemicals are plant-derived chemicals [17]. "Phytochemical" is derived from the Greek *phyto*, meaning "plant," and chemical. Phytochemicals are chemicals produced by plants and they are the outcome of either primary or secondary metabolism. They generally have metabolic cycles in the plant host and also help the plant develop or safeguard itself

Table 12.1 List of medicinal plants and their uses.

Scientific name	Common name	Uses
Allium sativum	Garlic	Usually utilized as an antitoxin and mainly in the treatment of coronary illness [7]. Could also be utilized as a natural stimulant or sedative because is monoamine oxidase inhibitor [8]
Aloe vera	Aloe vera	Aloe vera leaves are broadly utilized to treat wounds, injuries, burns, and other skin diseases [9]
Azadirachta indica	Neem	Utilized for treatment of malaria, worms, skin diseases, and rheumatism in India [10]
Carica papaya	Papaya	Utilized for treatment of stomach aches and wounds [11]
Cannabis	Hemp, cannabis, marijuana, Indian hemp	Utilized overall since ancient times as therapy for different illnesses, such as inflammation, pain, irritation, gastrointestinal illnesses like irritable bowel syndrome, muscle relaxation, stress, dementia, attention-deficit hyperactivity disorder, chemical imbalance, malignant growth, cerebral paralysis, repeating migraines, epilepsy, glaucoma, sleep deprivation
Citrus limon	Lemon	Used alongside honey to treat sore throats and coughs [12]
Curcuma longa	Turmeric	Helps in easy absorption of food, regulates functions of liver, controls menstrual cycle, and assuages joint inflammation pain
Lavandula angustifolia	Lavender	Used as medication for mental health purposes and also as antiseptic. Was utilized in ancient Egypt for embalming dead bodies
Ginkgo biloba	Ginkgo	Plant fractions utilized for the of bronchitis, tinnitus, Alzheimer's disease, and asthma
Momordica charantia	Bitter melon	Utilized as agent to decrease level of glucose in blood
Moringa oleifera	Drumstick plant	Utilized for food and traditional medication. Is in primary trials to explore likely properties of its supplements and phytochemicals
Panax sp.	Ginseng	Utilized therapeutically, especially in Asia, for more than two millennia [13]
Rosmarinus officinalis	Rosemary	Utilized restoratively from ancient times
Rosa majalis	Cinnamon rose	Produces edible hips plentiful in L-ascorbic acid, utilized in medication and to produce rose hip syrup
Silybum marianum	Milk thorn	Utilized for millennia for assortment of therapeutic purposes, specifically liver issues
Syzygium aromaticum	Clove	Plant utilized for stomach ache and as expectorant, among different purposes. Plant's oil utilized for treatment of toothache
Vernonia amygdalina	Bitter leaf	Utilized by indigenous groups in Africa to treat digestive diseases like looseness of the bowels (dysentery) [14]
Thymus vulgaris	Thyme	Utilized to treat cough and bronchitis, acting as antispasmodic and expectorant. Has additionally been utilized in numerous other therapeutic functions in Asian and Ayurvedic medicine, despite not having been demonstrated to be powerful in non-respiratory restorative functions [15]
Zingiber officinale	Ginger	Efficient for treatment of nausea [16]

Table 12.2 Different phytochemicals and their potential benefits.

Phytochemicals	Food varieties	Benefits
β-Carotene (carotenoids)	Carrots, green plants, sweet potatoes, orange squash, boiled tomatoes	Can restrain malignant growth cell development, lessen hazard of cardiovascular infection, and enhance immune systems [21]
Flavonoids	Apples, coffee, walnuts, whole grains, berries, other fruits	Can reduce irritation, lessen harm to DNA, and retard cancer development [22]
Anthocyanins	Berries	Aid in reduction of high blood pressure [23]
Sulforaphane/ isothiocyanates	Cruciferous plants like kale, broccoli, and cauliflower	May safeguard against malignant growth and cardiovascular illness [24]
Zeaxanthin and lutein	Chard and spinach	Enhances wellbeing of eyes

against contenders, microorganisms, or hunters [18]. Since confirmation of their conceivable wellbeing impacts presently cannot seem to be obtained, phytochemicals are by and large viewed as scientific constituents instead of key supplements [19, 20].

The two main categories of phytochemicals that are being scrutinized are carotenoids and polyphenols, which include lignans, flavonoids, stilbenes, and phenolic acids. Flavonoids are classified in groups on the basis of their chemical structural formula, for example flavanones, anthocyanins, flavones, flavanols, and isoflavanones. Flavanols are partitioned into three kinds: catechins, epicatechins, and proanthocyanidins [19, 20]. Table 12.2 shows a few phytochemicals and their potential benefits.

12.4 Phytochemical Constituents in Some Medicinal Plants

12.4.1 Onions

Since ancient times, onions (*Allium cepa* L.) have been esteemed as both a restorative plant and also as food. Onions are widely grown vegetable bulb crops, which are next to tomatoes in popularity. They are known to most societies and consumed worldwide. It is a low-latitude horticultural crop with a short growing season [25]. Onion is universally referred to as the "queen of the kitchen" due to its exceptionally esteemed flavor, scent, and distinct taste, as well as the therapeutic qualities of its flavor compounds [26]. Over time, it has been utilized in curries, as flavoring, in plates of mixed greens, as a condiment, or garnished with different vegetables, baked or boiled. It is additionally utilized in an assortment of forms like flakes, pickles, powders, and pastes, and it is likewise known for its therapeutic properties.

12.4.1.1 Phytochemical Constituents in Onions

Saponin, quercetin, and anthocyanin are the most well-known phytochemicals found in onions. The entire onion bulb contains glutamyl peptide and (+)-S-alk(en)yl-L-cysteine sulfoxide, which represent over 70% of the absolute sulfur in onions [27]. It contains three significant non-unstable and unscented alk(en)yl cysteine sulfoxides: S-methyl cysteine sulfoxide (methiin), S-propyl cysteine sulfoxide (propiin), and S-trans-prop-1-enyl cysteine sulfoxide (isoalliin). The most well-known flavor antecedent, which accounts for over 80% of the aggregate sum of alk(en)yl cysteine sulfoxides, is isoalliin. Methiin is available in reduced concentrations, while alliin and propiin are just in minute quantities [28]. Allinase enzymes degrades alk(en) yl-L-cysteine sulfoxides during the destruction of cells and tissue. This results in the development of new chemical mixtures, for example alkyl alkane-thiosulfinates, which influence the scent and kind of alliums. All these compositions can disintegrate into other organosulfur chemical mixtures, for example ajoene, 1-propanethial-S-oxide, allicin, dipropyl disulfide, propyl-1-propanyl thiosulfinate, methyl propanyl disulfide, and diallyl sulfide. At the same time, γ-glutamyl cysteine is changed over time into different organosulfur chemical constituents, which include S-allyl mercaptocysteine and S-allyl cysteine [29].

Fructans, organo-sulfur chemical constituents, and flavonoids are extremely valuable phytochemicals in perfect proportion. These compounds are thought to have beneficial properties. Onions are commonly identified as a source of nutrient flavonoids because they contain three different and hugely beneficial phytochemicals in the right proportions: organo-sulfur chemical constituents, flavonoids, and fructans. All these three chemical substances have important benefits to human health. Onions contain a diverse range of organosulfur compounds, basically four kinds of diallyl sulfides: diallylmonosulfide (DMS), diallyldisulfide (DDS), diallyltrisulfide (DTS), and diallyltetrasulfide (DTTS). Another phytochemical compound found in onions is flavanol, which is one of the subclasses of flavonoids that incorporate quercetin. Other flavanols, for example kaempferol and isorhamnetin, have been estimated to be in onions in addition to quercetin [30]. Other sulfoxides found in onions incorporate S-propyl-L-cysteine sulfoxide, (+)-S-(1-propenyl)-L-cysteine sulfoxide (PRENCSO), S-methyl-L-cysteine sulfoxide, S-propenyl-L-cysteines, and (+)-S-methyl-L-cysteine sulfoxide (MCSO) [31].

12.4.2 Turmeric

Turmeric is a restorative plant produced using the ribosomal part of the tropical plant *Curcuma longa* Linn., which has a place in the ginger family (Zingiberaceae). Turmeric's bright yellow-orange color is primarily because of the polyphenolic pigment (fat soluble) curcuminoid. Curcumin, the main curcuminoid that turmeric contains, is widely regarded as its most active component [32]. Demethoxycurcumin and bisdemethoxycurcumin are two other curcuminoids found in turmeric. Turmeric has been utilized for centuries in India for therapeutic reasons, notwithstanding its utilization as a color pigment and a spice [33]. Curcumin has of late ignited restored interest in its capability to forestall and treat illness, on account of indications that it might have anticancer and anti-inflammatory properties. Curcumin additionally has promising therapeutic and diagnostic potential for various significant human conditions, including cardiovascular diseases, inflammation, antimicrobial resistance, obesity, tumorigenesis, chronic tiredness, depression and neurophysiologic conditions, anxiety, muscle and bone loss, and neuropathic pain.

12.4.2.1 Phytochemical Constituents in Turmeric

Turmeric consists of the phytochemicals saponin (0.4%), alkaloid (0.76%), sterol (0.03%), tannin (1.08%), flavonoid (0.40%), phytic acid (0.82%), and phenol (0.08%). The presence of alkaloids (0.76%) in turmeric plants suggests that turmeric can be used to treat hypertension-related headaches, colds, and chronic migraine. Because of the antioxidant properties of saponins, flavonoids, and tannins, turmeric plants can improve sex hormones, lower cholesterol, prevent harmful cytotoxins, and reduce inflammation. Turmeric plants have antioxidant properties and can be utilized in the treatment of gastrointestinal illnesses like loose diarrhea and dysentery [34].

12.4.3 Ginger

Ginger (*Zingiber officinale* Roscoe), is a member of the family Zingiberaceae and the genus Zuniber. It has long been utilized as a spice and in Ayurvedic medicine. The ribosomal part (root) of ginger is utilized to avert and treat a variety of normal infections, including nausea, migraines (headache), emesis, and colds. Numerous bioactive chemical constituents have been identified in ginger, like terpene constituents and phenolic content [35]. The phenolic contents that account for ginger's different bioactivities are primarily gingerols, shogaols, and paradols. Ginger has as of late been found to have natural functions like antimicrobial, anticancer, anti-inflammatory, and antioxidant properties. Moreover, research has shown that ginger has the ability to maintain and inhibit an assortment of ailments, including neurodegenerative illnesses, obesity, cardiovascular infections, diabetes mellitus, chemotherapy-incited nausea and emesis, and respiratory issues [36]. Other gastrointestinal diseases that ginger is utilized to cure are colic, morning ailment, stomach ache, loss of appetite, indigestion, flatulence, dyspepsia (discomfort after eating), heartburn, and bloating. Ginger is suggested by the Indian Ayurvedic medical system to assist with absorption. Fresh ginger juice has been shown to help in the treatment of skin burns [37]. Ginger's phytochemicals are utilized as a diuretic and an acid neutralizer. Ginger is likewise utilized to regulate the body's temperature (keep the body warm) in order to improve circulation and reduce high blood pressure. It acts as an anti-inflammatory due to its warming effect. Ginger is likewise utilized as an enhancing specialist (flavoring agent) in food sources and drinks and as a scent in cleansers (soaps) and beauty care products (cosmetics).

12.4.3.1 Phytochemical Constituents in Ginger

Ginger contains a high number of phytochemicals (active compounds), including phenols and terpenoids. The phenolic constituents in ginger are primarily paradols, gingerols, and shogaols. Examples of gingerols, which are the most bountiful polyphenols in ginger, are 6-gingerol, 8-gingerol, and 10-gingerol. Gingerols can be changed into related shogaols through heat treatment or long-term stockpiling. Shogaols can be changed into paradols after hydrogenation. Other phenolic constituents include gingerenone-A, zingerone, 6-dehydrogingerdione, and quercetin. In addition, ginger contains many terpenoids, including curcumene, farnesene, bisabolene, sesquiphellandrene, and zingiberene, which are believed to be the primary constituents of ginger rejuvenating balms (essential oils). Moreover, ginger contains lipids, natural acids, polysaccharides, and fibers [37].

12.4.4 Garlic

Garlic (*Allium sativum* L.) is a plant that has been studied widely for a long time and has been utilized to battle infectious diseases. It is a member of the Amaryllidaceae family [38]. It is a food that is widely available from one side of the planet to the other. Garlic began in Asia and was among the first developed plants [39]. For millennia, garlic has been perceived as having likely therapeutic value for different microorganisms. For instance, it has all-round antibacterial, antifungal, anthelmintic, antiviral, antiseptic, and anti-inflammatory properties. Moreover, garlic extracts inhibited the growth of both Gram-negative and Gram-positive microscopic organisms. It is utilized to cure stomach discomfort, looseness of the bowels, otitis media, and respiratory problems in Africa, particularly in agricultural nations like Nigeria. It has been utilized to treat normal colds, hay fever, and asthma in Europe and India [40].

Garlic has antiviral, antibacterial, and antifungal properties, notwithstanding its prevalence as a quality food. Moreover, antiatherosclerotic and anticancer properties have been illustrated. Many studies have shown that it is effective and has an expansive range of antimicrobial potential against a wide scope of viruses, parasites, fungi, and bacteria. Garlic fractions prevent the improvement of Gram-positive and Gram-negative microbes, for example *Staphylococcus aureus, Pseudomonas, Enterobacter, Klebsiella, Streptococcus aureus, Escherichia, Micrococcus, Salmonella, Proteus, Helicobacter pylori, Lactobacillus*, and *Shigella* [22].

12.4.4.1 Phytochemical Constituents in Garlic

A. sativum's color, flavor, and smell are due to phytochemicals, which are essential for the plant's normal defense mechanism and shield it from herbivorous bugs and animals with a backbone (vertebrates), parasites, and pathogenic microbes [41]. According to this examination, phytochemical compounds like flavonoids, steroids, saponins, tannins, terpenoids, and reducing sugar were identified in *A. sativum* fractions. The phytochemical content of *A. sativum* extract showed that alkaloids (7.2%) were the most plentiful phytochemical constituents, followed by tannin compound (4.8%), saponin compound (4.3%), and flavonoid compound (2.18%).

Terpenoids can be found in both plant and leaf extracts. Terpenoids have indeed been proven to be useful in the treatment and prevention of a variety of infections, including malignant growths (cancer). Terpenoids have antiviral, antimicrobial, antispasmodic, antihyperglycemic, antifungal, antiparasitic, antiallergenic, anti-inflammatory, and immunomodulatory properties [42]. Flavonoids are likewise detectable in the leaf extracts and act as a strong water-solvent cancer prevention agent and free extreme scrounger, preventing oxidative cell harm and having solid anticancer properties [43]. They likewise help with the administration of diabetes-instigated oxidative pressure. Steroids are significant in drugs since they contain substances like sex hormones and can be utilized in drug manufacturing [44]. *A. sativum* extract also contained tannins and saponins. Tannins prevent the development of numerous fungi, yeasts, bacteria, and viruses. Saponins have anticholesterolemic and antibiotic properties. In addition, saponins have antimutagenic, antitumor, and antioxidant properties and may lessen the risk of human cancers by impairing cancer cell development [45].

12.5 The Brain

The organ that acts as the core of the sensory/nervous system in every vertebrate (with a backbone) and invertebrate (without a backbone) animal is the brain. The brain is located in the head, close to the sensory organs used for various senses such as vision. The brain is the most complicated part of the body in vertebrates. The cerebral cortex in each individual consists of 14–16 billion neurons and the cerebellum is assessed to contain 55–70 billion neurons [46]. Synapses connect every neuron to a huge number of different neurons. These neurons typically interact with each other through lengthy fibers called axons, which convey tons of signal heartbeats called action potentials to far-off parts of the body or brain that target explicit areas.

Neurologically, other organs of the body are influenced by the functions of the brain. The brain has control over the remaining parts of the body by invigorating the secretion of synthetic substances called hormones and also causing muscle–mind signals. This centralized influence empowers rapid and facilitated reactions to environmental changes. A few essential kinds of responsive, such as reflexes, can be intercepted by the peripheral ganglia and spinal cord, yet refined intentional conduct control on the basis of complex sensory input necessitates the data-incorporating abilities of a centralized brain. The functions of each brain cell are currently being interpreted, but how they collaborate in groups of millions remains a mystery [47]. Ongoing models in present-day neuroscience treat the mind/brain as an organic personal computer: the process is distinct from that of an electronic computer, but comparative because it procures data from the environment, saves those data, and also processes them in various patterns.

12.5.1 Brain Physiology

The capability of nerve cells to communicate signals to different cells, and their readiness to answer appropriately the electrochemical signals obtained from different cells, is fundamental for the essential functions of the brain. The electrical characteristics of nerve cells are controlled by a wide range of different forms of metabolism, almost all synaptic interactions between neurotransmitters and receptors [47].

12.5.1.1 Neurotransmitters and Receptors

Neurotransmitters are synthetic substances delivered at neural connections (synapses) when the topical film (membrane) demagnetizes and Ca^{2+} moves into the tissue or cell, commonly when an activity potential shows up at the synapse. Neurotransmitters are joined to receptor molecules on the layer of the neural connection's objective cell (or cells), changing the electrical or synthetic substance properties of the receptor particles. In a couple of special cases, every nerve cell in the brain delivers a similar chemical compound synapse (neurotransmitter), or a mix of synapses, at all synaptic associations it structures with different nerve cells; this is called Dale's principle [47]. Hence, the synapses delivered by a nerve cell can be utilized to characterize drugs, including nicotine, cannabinoids, alcohol, heroin, chlorpromazine, fluoxetine, cocaine, and a variety of different drugs. Almost all psychoactive medications work by altering explicit receptors [48].

The two most abundant neurotransmitters in a vertebrate animal's brain are glutamate, which quite often has stimulatory results on target nerve cells, and γ-aminobutyric acid (GABA), which quite often has inhibitory impacts. Nerve cells that utilize these transmitters may be identified in all areas of the brain. Many other neurotransmitters are utilized in many other confined regions in the brain, frequently regions committed to a specific activity. Serotonin, for instance, which is an essential objective of numerous medications and nutritional supplements, is derived completely from a little brainstem region known as the raphe nuclei [49]. Norepinephrine, which plays a part in arousal, is derived completely from a close-by little region known as the locus coeruleus. Other neurotransmitters with various sources in the mind, like acetylcholine and dopamine, are not quite so widely disseminated as GABA and glutamate [50].

12.5.1.2 Electrical Activity

At the point when the brain tissue is active, it generates electrical fields as a result of the electrochemical processes utilized by nerve cells for flagging/signaling. When a huge number of nerve cells fire at the same time, the electrons they create can be large enough to recognize in the external skull utilizing magnetoencephalography (MEG) or electroencephalography (EEG) [51]. Every region of the cerebrum exhibits a blend of rhythmic and nonrhythmic interaction, which varies depending on cognitive and behavioral state. The cerebral cortex of mammals exhibits enormous sluggish δ-waves during rest, quicker α-waves when the creature is alert yet not concentrating, and turbulent sporadic action when the creature is effectively occupied with an errand, alluded to as β- and γ-waves. In an epileptic convulsion, the mind's inhibitory control systems fizzle and electrical action ascends to neurotic levels, shown on EEG to be followed by an enormous wave and spike design not found in a solid cerebrum [51].

12.5.1.3 Metabolism

Each vertebrate has a blood–brain barrier, which permits metabolic action in the cerebrum to contrast with metabolic exercises in different parts of the individual body. Glial cells assume a significant part in cerebrum digestion/metabolism by managing the synthetic substance component of the liquid that encompasses nerve cells, which includes ion and nutrient levels [52]. Huge brains put serious metabolic demands on creatures due to the fact that the cerebrum tissue consumes a lot of energy in relation to its volume. Smaller brains in certain species like bats result in restricted body weight in order to fly [53]. The brain typically obtains the majority of its energy from glucose (i.e. blood sugar) metabolism (which is oxygen dependent); however ketones, alongside commitments from medium-chain unsaturated fats (caprylic acid), provide a significant alternative source [54].

12.5.2 Functions of the Brain

12.5.2.1 Perception

Proprioception (which is the position of the body in space), sound, elements in the atmosphere, light, chemical components of the blood, temperature, and other elements are taken care of in the human brain. Different creatures have extrasensory organs, for example the infrared hotness feeling of snakes, the attractive field feeling of certain birds, or the electrical field sensing found fundamentally in aquatic animals. All sense organs start with specific sensory receptors, like the light receiver cells (photoreceptor cells) in the retina of the eyes, as well as the vibration-sensitive hair cells in the cochlea of the ear [55]. Sensory receptor cell axons move to the brain or spinal cord, where messages are sent to the first-order sensory nucleus devoted to a single sensory methodology. The sensory information core communicates with higher-order sensory regions devoted to a similar methodology. In the end, the signals are sent from the cerebral cortex via a thalamic way-station, where they are produced to recover specific highlights and incorporated with signals from other sensory frameworks [56].

12.5.3 Motor Control

The motor frameworks are parts of the cerebrum responsible for the initiation of body movements, or firing muscles. Aside from the muscles that influence the eyes, which are controlled by midbrain cores, all muscles under willful control in the body are straightforwardly innervated by motor nerve cells in the hindbrain and spinal cord [57]. Spinal motor nerve cells are influenced by neural circuits within the spinal cord as well as contributions from the brain. Many reflex responses are implemented by the intrinsic spinal circuits, which additionally consist of design generators for rhythmic motions like swimming or strolling [58].

A few motor regions in the cerebrum have specific activity in relation to the spinal cord. Different motor regions in the medulla and pons control generalized motions such as walking, gulping, and breathing at the most minimal level. Different regions in the midbrain, such as the red nucleus, are answerable for organizing arm and leg motions at a more elevated level. The essential motor cortex, a piece of tissue situated at the back edge of the front-facing projection (lobe), is situated at a more significant level. The main motor cortex projects toward the subcortical motor regions, yet it additionally projects massively toward the spinal cord via the pyramidal tract. This immediate corticospinal projection empowers exact intentional control of the development of fine subtleties. A few motor-related synapses optionally affect the essential motor regions by projecting. The premotor cortex, advantageous motor region, cerebellum, and basal ganglia are among the most important alternative areas [48].

12.5.4 Homeostasis

The little area at the center of the anterior part of the brain (hypothalamus), whose size does not mirror its intricacy or the significance of its capacity, is the most essential region of the brain in vertebrates [59]. The hypothalamus is made up of many small nuclei, large numbers of which are associated with fundamental biochemical processes. A portion of these capacities are associated with social collaboration or excitement, like hostility, sexuality, or maternal ways of behaving, yet many are associated with homeostasis. Several hypothalamic nuclei receive information such as glucose level, blood oxygen level, or sodium level, and different factors from various sensors in the lining of blood vessels. These hypothalamic nuclei transmit feedback signals to motor areas, which can cause measures to be taken to correct inadequacies. A portion of the results are likewise coordinated to the pituitary organ, a little organ found straight underneath the hypothalamus in the brain. Hormones are released by the pituitary organ into the circulatory system, where they disseminate all through the body and cause changes in cell behavior [60].

12.5.5 Motivation

The motivational system is primarily based on a reward/punishment mechanism. At the point when a specific way of behaving is followed by ideal results, the brain's reward mechanism is initiated, causing primary changes in the brain that lead a similar way of behaving to be repeated later when that happens again. Whenever a way of behaving is followed by troublesome outcomes, the cerebrum's punishment mechanism is initiated, making primary changes that cause the conduct to be suppressed later on when comparable circumstances emerge [61].

The reward/punishment framework in vertebrates is executed by a particular arrangement of brain structures, at the heart of which are the basal ganglia, a collection of interconnected regions at the foundation of the anterior part of the brain [62]. The basal ganglia form the focal dynamic site in the brain: the basal ganglia impose an inhibitory influence over the greater part of the motor frameworks in the cerebrum; when this restraint is delivered, a motor framework is permitted to complete the activity it is modified to do.

12.5.6 Learning and Memory

Neuroscientists have identified a few kinds of memory and learning that are carried out by the cerebrum in specific ways:

- *Working memory* – the brain's ability to keep an impermanent description of the data about the assignment that a creature is presently performing. The synthesis of cell assemblies – a group of actuated nerve cells that keep up with their movement by continually stimulating each other – is remembered to mediate this kind of dynamic memory [63].
- *Episodic memory* – the process of retrieving specific details about events. This type of memory has the potential to last a lifetime. Many studies have found that the hippocampus plays an important role: people with severe hippocampus damage may experience amnesia, or lack the ability to develop new enduring episodic memories.
- *Semantic memory* – the capability of learning relationships and realities. This kind of memory is presumably located to a great extent in the cerebral cortex, intervened in by changes in associations between cells that address distinct sorts of data [64].
- *Instrumental learning* – the capacity for rewards and punishments to alter conduct. It is carried out by the organization of brain regions focused on the basal ganglia [65].
- *Motor learning* – the capability to modify processes of bodily motion through practice or, all the more extensively, recurrence. Various brain regions are involved, which include the basal ganglia, premotor cortex, and, in particular, the cerebellum, which functions as a huge memory bank for microadjustments to movement parameters. Motor learning and the capability to finetune patterns of the body's motion arise through practice, or more broadly redundancy [66].

12.6 Brain Conditions

Some common brain conditions are as follows:

- *Headache* – there are many kinds of headaches; a few may be important, yet most are not and are by and large treated with analgesics/pain relievers.
- *Stroke (brain infarction)* – when the bloodstream and oxygen cease to reach the region of brain tissue, this brain tissue is unexpectedly cut off, and the tissue dies. Most strokes are caused by blood coagulation or bleeding in the brain. An aneurysm is a weak spot in a brain artery that grows in a balloon-like shape. A damaged brain aneurysm can result in a stroke.
- *Subdural hematoma* – when there is internal bleeding or bleeding beneath the dura, the lining of the skull. A subdural hematoma can cause neurologic problems by putting pressure on the brain. Bleeding occurs within the dura and in the skull itself, particularly soon after head damage. If left untreated, gentle side effects can rapidly advance to unconsciousness and death.
- *Intracerebral hemorrhage* – bleeding that occurs in the brain.
- *Concussion/blackout* – brain damage that leads to an impermanent aggravation in the cerebrum. A traumatic head injury causes most concussions. Horrible head injury leads to most blackouts/concussions.
- *Cerebral edema* – expansion of the brain tissue due to electrolyte awkwardness or injury.

- *Parkinson's disease* – nerves in a focal region of the brain decline gradually and cause some issues with coordination and movement. Shaking hands are a typical early sign.
- *Alzheimer's disease* – for mixed reasons, nerves in specific brain regions degenerate, causing moderate dementia. Alzheimer's disease is the most well-known type of dementia [67].

12.7 Protective Effects of Medicinal Plants on the Brain

Neurodegenerative diseases like Alzheimer's disease, multiple sclerosis, and Parkinson's disease cause steady neuronal demise, which is joined by mental and tactile decline. Recently these diseases have been linked to a variety of multifactorial etiologies, as well as social and financial issues [68]. Anti-inflammatory drugs have been proposed to slow the progression of neurodegenerative diseases like Alzheimer's [69]. Obsessive cycles including irritation, oxidative pressure, apoptosis, mitochondrial breakdown, and hereditary variables lead to neuronal degeneration in Parkinson's disease. It has been proposed that inordinate lipid peroxidation might obliterate cholinergic neurons in AD7 and dopaminergic neurons in Parkinson's disease [70]. Different enzymatic cancer prevention agents, for example superoxide dismutase (grass), and non-enzymatic cell reinforcement, for example complete thiol gatherings [71], exist in the cerebrum. Low cell reinforcement action of the cerebrum concerning different tissues has made the brain tissue helpless to oxidative harm.

In conventional medication, the organs of plants, for example leaves, stems, roots, blossoms, and fruits, were utilized as an option and correlative treatment. A few compounds from spices, for example resveratrol, curcumin, ginsenoside, polyphenols, triptolide, and so forth, have neuroprotective impacts. Natural products have complex dynamic constituents like flavonoids, alkaloids, and isoprenoids. Thus, it is often hard to figure out what part of the spice has more organic action [72].

12.7.1 *Crocus sativus*

C. sativus L., regularly known as saffron, has a place in the Iridaceae family. The Crocoideae superfamily is grown in numerous nations, including Turkey, Iran, Spain, and Afghanistan. It is utilized for the most part as a natural medication in different locales on the planet [73].

Saffron has 150 unique chemical constituents, including lipids, carbohydrates, water, polypeptides, nutrients, and minerals. Crocins are the primary organically dynamic elements of saffron, a group of red-hued and water-dissolvable carotenoids, which are on the whole glycosides of crocetin. Additionally, saffron has four primary bioactive parts: crocin, crocetin, picrocrocin, and safranal. One more constituent of saffron is picrocrocin, which has an unpleasant taste [74].

12.7.1.1 Medicinal Properties of *Crocus sativus*

C. sativus is utilized to treat mental problems. Recently *C. sativus* constituents have been utilized to treat some brain issues and to loosen up the smooth muscle. *C. sativus* and its primary constituent, crocin, have powerful cell reinforcement impacts by decreasing the malondialdehyde (MDA) level [75].

12.7.2 *Nigella sativa*

Nigella sativa L. is a yearly herbaceous in the Ranunculaceae family, which is widely grown in Mediterranean countries, Western Asia, and Central and Eastern Europe. *N. sativa* seeds have been added as a flavoring to a range of Persian food sources, for example bread, pickles, sauces, and mixed greens [76]. Synthetic parts of *N. sativa* seeds incorporate oil, protein, sugar, and fiber. The compounds in *N. sativa* oil are linoleic acid, myristic acid, oleic acid, stearic acid, palmitic acid, arachidic acid, and eicosadienoic acid. The significant phenolic contents of *N. sativa* seeds are thymoquinone (13.7%), thymol (0.33%), p-cymene (37.3%), and carvacrol (11.77%) [77].

12.7.2.1 Medicinal Properties of *Nigella sativa*

N. sativa as a restorative plant is notable for its strong antioxidative impacts. It has been shown that *N. sativa* is defensive against renal harm. *N. sativa* further promotes scopolamine-prompted learning and memory disability as well as decreasing throbbing movement and oxidative pressure of the brain [78]. The cancer-preventive impacts of *N. sativa* oil on patients with rheumatoid arthritis (RA) joint pain showed that *N. sativa* decreased the serum level of interleukin (IL)-10 and MDA. *N. sativa* additionally worked on provocative reactions and decreased oxidative pressure in patients with RA [79].

12.7.3 *Coriandrum sativum*

Coriander (*Coriandrum sativum* L.) is an annual spice from the parsley family (Apiaceae). It is for called *geshniz* in Persian. *C. sativum* is local to the Mediterranean area and is widely grown everywhere [80]. Coriander is additionally a possible source of lipids, for example petroselinic acid, and has a high content of medicinal essential oils that are vital for development and cerebrum capacities. The principal coriander essential oils are linalool, linoleic, and linolenic acids. Coriander seed oil contains linalool (60–70%) and 20% hydrocarbons, yet the organization of the spice oil is totally different from the seed oil [81].

12.7.3.1 Medicinal Properties of *Coriandrum sativum*

C. sativum has for some time been utilized as a traditional cure. *C. sativum* seed has been utilized in creams and shampoos and has antimicrobial and antirheumatoid properties. *C. sativum* has been proposed as a therapy for constant weakness [82].

12.7.4 *Ferula assafoetida*

Asafoetida (*Ferula assafoetida* L.) is a spice in the Apiaceae family that develops from the exudates of the plant's living tap roots or the ribosomal part of the plant. In India and Nepal it has been utilized in domestic medicine and as an enhancer in different food sources. The hydrodistilled oil has 25 constituents, including E-1-propyl sec-butyl disulfide, which is a key component. *F. assafoetida*'s primary constituents are E-1-propenyl sec-butyl disulfide (40.0%) and germacrene B (7.8%) [83].

12.7.4.1 Medicinal Properties of *Ferula assafoetida*

Researchers are keen on *F. assafoetida* as a result of its nutritive and restorative properties. The plant's roots, young shoots, and leaves are consumed as vegetables. *F. assafoetida* leaves have anthelmintic, carminative, and diaphoretic properties, and the plant's root is utilized as an antipyretic. Also, *F. assafoetida* is utilized in natural medication to treat an assortment of illnesses like asthma, epilepsy, stomach ache, flatulence, gastrointestinal parasites, malabsorption, and influenza [84].

12.7.5 *Thymus vulgaris*

T. vulgaris is a spice in the Lamiaceae family that is profoundly delightful. This leaf has around 38 species and is found in subtropical nations [85]. *T. vulgaris* is essentially composed of phenols, thymol (40%) and carvacrol (15%). During the virus season, it has lower levels of phenol. The rejuvenating balm likewise contains thymol methyl ether (2%) as well as cineol, cymen, pinene, borneol, and esters [85].

12.7.5.1 Medicinal Properties of *Thymus vulgaris*

Bioactive thyme components, such as thyme essential oil constituents, flavonoids and phenolic acids, natural terpenoid thymol, and phenol isomer carvacrol, have been shown to have antioxidant, antimicrobial, antitussive, antispasmodic, as well as expectorant characteristics. It has also been reported that thymol treatment improves antioxidant potential in the brain [86].

12.7.6 *Curcuma longa*

C. longa is a plant in the Zingiberaceae family that is grown in Southeast Asian nations [87]. Turmeric's powerful fixings are the flavonoid curcumin (diferuloylmethane) and an assortment of natural oil, for example, tumerone, atlantone, and zingiberone. Sugars, proteins, and saps are among different parts. Curcumin, which represents 0.3–5.4% of crude turmeric, is the most completely investigated strong ingredients [87].

12.7.6.1 Medicinal Properties of *Curcuma longa*

Curcumin is an unadulterated polyphenol and non-flavonoid compound found in plants, for example *C. longa*. Curcumin is notable for its likely therapeutic properties, for example calming and cell reinforcement impacts. Curcumin's beneficial property in neurologic problems has aroused the curiosity of numerous researchers recently. Curcumin water solvent concentrate can expand dopamine, norepinephrine, and serotonin (5-HT) levels in the cerebrum [88]. *C. longa* extract(1000 mg/kg body weight per ingestion) has been shown to safeguard against oxidative and renal harm. Curcumin has likewise been demonstrated to have helpful potential in neurologic circumstances and cerebral ischemia. Curcumin has been shown to safeguard the mind from oxyhemoglobin-incited neuronal harm and oxidative pressure. Curcumin shielded neurons from reactive oxygen species by fixing glutathione levels that had been reduced. Curcumin's neuroprotective impacts might worked by overexpression of B-cell lymphoma protein 2 (Bcl-2), an inducible nitric oxide synthase antagonist. Subsequently, curcumin can ease nitric oxide-induced degeneration [89].

12.8 Conclusion

Various medicinal herbs have demonstrated anti-inflammatory and antioxidant effects, suggesting that they may have medicinal benefits in a variety of neurologic diseases. Several researchers also believe that the plants' potential benefits in neurologic diseases like Alzheimer's and Parkinson's diseases are primarily due to communications with the cholinergic, dopaminergic, and glutamatergic systems. The plants' communication with the GABA and opioid systems may have anticonvulsant and analgesic consequences.

References

1 Khan, M.A. (2016). *Introduction and Importance of Medicinal Plants and Herbs.* Zahid: Unani.

2 Mohanty, B., Puri, S., and Kesavan, V. (2018). A review on therapeutic potential of *Artemisia nilagirica*. *J. Plant Biochem. Physiol.* 6 (205): 2.

3 Puri, D., Kumar, A., and Virmani, J. (2019). Classification of leaves of medicinal plants using laws' texture features. *Int. J. Inf. Technol.* 14 (2): 1–12.

4 Adetuyi, B.O., Adebisi, O.A., Awoyelu, E.H. et al. (2021). Phytochemical and toxicological effect of ethanol extract of *Heliotropium indicum* on liver of male Albino rats. *Lett. Appl. Biosci.* 10 (2): 2085–2095.

5 Bhat, A., Mahalakshmi, A.M., Ray, B. et al. (2019). Benefits of curcumin in brain disorders. *Biofactors* 45 (5): 666–689.

6 Jiang, T.A. (2019). Health benefits of culinary herbs and spices. *J. AOAC Int.* 102 (2): 395–411.

7 Lawson, L.D. and Hunsaker, S.M. (2018). Allicin bioavailability and bioequivalence from garlic supplements and garlic foods. *Nutrients* 10 (7): 812.

8 Maleki, S.J., Crespo, J.F., and Cabanillas, B. (2019). Anti-inflammatory effects of flavonoids. *Food Chem.* 299: 125124.

9 Chandran, S., Sunny, J.C., Chandran, S., and Bellan, C. (2018). Enhanced antimicrobial activity of Aloe vera blended zinc oxide nanoparticles in PVA matrix. *Mater. Today: Proc.* 5 (8): 16190–16198.

10 Aremu, M.O. (2018). The potential of neem seed oil (*Azadirachta indica*) as natural household insecticide. *LAUTECH J. Eng. Technol.* 12 (1): 30–35.

11 Khaw, K.Y., Chear, N.J.Y., Maran, S. et al. (2020). Butyrylcholinesterase inhibitory activity and GC-MS analysis of Carica papaya leaves. *Nat. Prod. Sci.* 26 (2): 165–170.

12 Musara, C., Aladejana, E.B., and Mudyiwa, S.M. (2020). Review of the nutritional composition, medicinal, phytochemical and pharmacological properties of *Citrus reticulata* Blanco (Rutaceae). *F1000 Research* 9 (1387): 1387.

13 Guo, X. and Mei, N. (2016). Aloe vera: a review of toxicity and adverse clinical effects. *J. Environ. Sci. Health C* 34 (2): 77–96.

14 Oyeyemi, I.T., Akinlabi, A.A., Adewumi, A. et al. (2018). Vernonia amygdalina: a folkloric herb with anthelminthic properties. *Beni-suef Univ. J. Basic Appl. Sci.* 7 (1): 43–49.

15 Nawaz, A., Bano, S., Sheikh, Z.A. et al. (2014). Evaluation of acute and repeated dose toxicity of the polyherbal formulation Linkus syrup in experimental animals. *Chin. Med.* 5 (04): 179.

16 Nikkhah-Bodaghi, M., Maleki, I., Agah, S., and Hekmatdoost, A. (2019). Zingiber officinale and oxidative stress in patients with ulcerative colitis: A randomized, placebo-controlled, clinical trial. *Complement. Ther. Med.* 43: 1–6.

17 Breslin, A. (2017). *The chemical composition of green plants. Sciencing, Jan.* Santa Monica, CA: Leaf Group.

18 Hossen, I., Hua, W., Ting, L. et al. (2022). Phytochemicals and inflammatory bowel disease: a review. *Crit. Rev. Food Sci. Nutr.* 60 (8): 1321–1345.

19 Sharma, P. (2018). Preliminary phytochemical screening of Ankola seed. *Pharma. Innov. J.* 7 (6): 657–660.

20 Taher, A. and Evans, C. (2018). The effects of type of school lunch consumed on British adolescent's diet quality P18-051. *Curr. Dev. Nutr.* 2 (11): 31–32.

21 Bogacz-Radomska, L. and Harasym, J. (2018). β-Carotene-properties and production methods. *Food Qual. Saf.* 2 (2): 69–74.

22 Benmeziane, F., Adamou Hassan, K., Zeghad, H., and Djermoune-Arkoub, L. (2018). Evaluation of antibacterial activity of aqueous extract and essential oil from garlic against some pathogenic bacteria. *Int. Food Res. J.* 25 (2): 561–564.

23 Tarone, A.G., Cazarin, C.B.B., and Junior, M.R.M. (2020). Anthocyanins: new techniques and challenges in microencapsulation. *Food Res. Int.* 133: 109092.

24 Uddin, M.S., Al Mamun, A., Jakaria, M. et al. (2020). Emerging promise of sulforaphane-mediated Nrf2 signaling cascade against neurological disorders. *Sci. Total Environ.* 707: 135624.

25 Brewster, J.L. and Rabinowitch, H.D. (2020). *Onions and Allied Crops: Volume III: Biochemistry, Food Science, and Minor Crops*. Boca Raton, FL: CRC Press.

26 Sudha, G.S., Ramesh, P., Sekhar, A.C. et al. (2019). Genetic diversity analysis of selected onion (*Allium cepa* L.) germplasm using specific RAPD and ISSR polymorphism markers. *Biocatal. Agric. Biotechnol.* 17: 110–118.

27 Efiong, E.E., Akumba, L.P., Chukwu, E.C. et al. (2020). Comparative qualitative phytochemical analysis of oil, juice and dry forms of garlic (*Allium sativum*) and different varieties of onions (*Allium cepa*) consumed in Makurdi metropolis. *Int. J. Plant Physiol. Biochem.* 12 (1): 9–16.

28 Ko, E.Y. and Nile, S.H. (2022). Effect of different cooking methods on S-Alk (en) yl-L-cysteine sulfoxides profile and antiplatelet activity of onion. *S. Afr. J. Bot.* 151 (Part B): 19–24.

29 Dennis, K.K., Liu, K.H., Uppal, K. et al. (2021). Distribution of phytochelatins, metal-binding compounds, in plant foods: a survey of commonly consumed fruits, vegetables, grains and legumes. *Food Chem.* 339: 128051.

30 Kim, J., Woo, S., Uyeh, D.D. et al. (2019). Analyses of garlic stem strength for development of cutting machine. In: *2019 ASABE Annual International Meeting. St.* Joseph, MI: American Society of Agricultural and Biological Engineers.

31 Mateljan, G. (2015). *The World's Healthiest Foods: The Force for Change to Health-Promoting Foods and New Nutrient-Rich Cooking*, 2e. GMF Publishing: Chicago, IL.

32 Hoseini, S.M., Gupta, S.K., Yousefi, M. et al. (2022). Mitigation of transportation stress in common carp, *Cyprinus carpio*, by dietary administration of turmeric. *Aquaculture* 546: 737380.

33 Kar, S., Tudu, B., Bag, A.K., and Bandyopadhyay, R. (2018). Application of near-infrared spectroscopy for the detection of metanil yellow in turmeric powder. *Food Anal. Methods* 11 (5): 1291–1302.

34 Hewlings, S.J. and Kalman, D.S. (2017). Curcumin: A review of its effects on human health. *Foods* 6 (1): 92.

35 Mao, Q.Q., Xu, X.Y., Cao, S.Y. et al. (2019). Bioactive compounds and bioactivities of ginger (*Zingiber officinale Roscoe*). *Foods* 8 (6): 185.

36 Shahrajabian, M.H., Sun, W., and Cheng, Q. (2019). Clinical aspects and health benefits of ginger (*Zingiber officinale*) in both traditional Chinese medicine and modern industry. Acta agriculturae scandinavica, section b. *Soil Plant Sci.* 69 (6): 546–556.

37 Adetuyi, B.O. and Farombi, E.O. (2021). 6-Gingerol, an active constituent of ginger, attenuates lipopolysaccharide-induced oxidation, inflammation, cognitive deficits, neuroplasticity, and amyloidogenesis in rat. *J. Food Biochem.* 45 (4): e13660.

38 Chinedu, A.F. (2019). Comparative assessment of the antibacterial profile of onion (*Allium cepa*) and garlic (*Allium sativum*) against *Pseudomonas aeruginosa*, *Staphylococcus aureus* and *Candida albicans*. *Int. J. Biotechnol. Food Sci.* 7 (5): 65–71.

39 Armentia, A., Martín-Armentia, S., Pineda, F. et al. (2020). Allergic hypersensitivity to garlic and onion in children and adults. *Allergol. Immunopathol.* 48 (3): 232–236.

40 El-Beltagy, A.E.F., Mohamed, G., Akeel, M.A. et al. (2022). Modulatory role of garlic (Allium sativum) extract against cisplatin-induced nephrotoxicity in female albino rats and their offspring. *Research Square* https://doi.org/10.21203/rs.3.rs-1295740/v2.

41 Ibrahim, T.A., Dada, I.B.O., and Adejare, R.A. (2010). Comparative phytochemical properties of crude ethanolic extracts and physicochemical characteristics of essential oils of *Myristical fragrans* (nutmeg) seeds and *Zingiber officinate* (ginger) roots. *Electronic J. Environ. Agric. Food Chem.* 9 (6): 1110–1116.

42 Naik, B.N.R., Jyothi, D., and Vishnuvardhan, Z. (2021). Quantification of major bioactive compounds from Diospyros chloroxylon Roxb. *J. Plant Sci. Res.* 37 (1): 203–207.

43 Ushie, O.A., Longbap, B.D., Iyen, S.I. et al. (2022). Phytochemical screening and antioxidants investigations of ethyl acetate and acetone leaf extracts of *Thaumatococcus danielli*. *Res. J. Pharm. Biol. Chem. Sci.* 3 (1): 1–6.

44 Salah, M., Abdelsamie, A.S., and Frotscher, M. (2019). Inhibitors of 17β-hydroxysteroid dehydrogenase type 1, 2 and 14: structures, biological activities and future challenges. *Mol. Cell. Endocrinol.* 489: 66–81.

45 Ma, X., Bi, S., Wang, Y. et al. (2019). Combined adjuvant effect of ginseng stem-leaf saponins and selenium on immune responses to a live bivalent vaccine of Newcastle disease virus and infectious bronchitis virus in chickens. *Poult. Sci.* 98 (9): 3548–3556.

46 von Bartheld, C.S., Bahney, J., and Herculano-Houzel, S. (2016). The search for true numbers of neurons and glial cells in the human brain: a review of 150 years of cell counting. *J. Comp. Neurol.* 524 (18): 3865–3895.

47 Farahany, N.A., Greely, H.T., Hyman, S. et al. (2018). The ethics of experimenting with human brain tissue. *Nature* 556 (7702): 429–432.

48 Borchard, S., Bork, F., Rieder, T. et al. (2018). The exceptional sensitivity of brain mitochondria to copper. *Toxicol. in vitro* 51: 11–22.

49 Soslau, G. (2022). Cardiovascular serotonergic system: evolution, receptors, transporter, and function. *J. Exp. Zool. Part A: Ecol. Integr. Physiol.* 337 (2): 115–127.

50 Begum, Z. and Younus, I. (2018). *Hibiscus rosa sinensis* mediate anxiolytic effect via modulation of ionotropic GABA-A receptors: possible mechanism of action. *Metab. Brain Dis.* 33 (3): 823–827.

51 Askari, S., Bastany, Z., Holsti, L. et al. (2020). Combined low-frequency EEG and NIRS during hypoxia. In: *Biophotonics in Exercise Science, Sports Medicine, Health Monitoring Technologies, and Wearables*, vol. 11237, 1123706. Bellingham, WA: International Society for Optics and Photonics.

52 Egle, T., Cavanaugh, T.J., and Friend, C.M. (2022). Comment on "STM study of the (111) and (100) surfaces of PdAg, *Surf. Sci.* 417 (1998) 292–300" and references therein). *Surf. Sci.* 720: 122048.

53 Safi, K., Seid, M.A., and Dechmann, D.K.N. (2005). Bigger is not always better: when brains get smaller. *Biol. Lett.* 1 (3): 283–286.

54 Abrahao, V.P., Ballen, G.A., Pastana, M.N., and Shibatta, O.A. (2021). Ontogeny of the brain of *Microglanis garavelloi* Shibatta and Benine 2005 (Teleostei: Siluriformes: Pseudopimelodidae). *J. Morphol.* 282 (4): 489–499.

55 McDonald, S., Hassell, K., and Cresswell, T. (2021). Effect of short-term dietary exposure on metal assimilation and metallothionein induction in the estuarine fish *Pseudogobius* sp. *Sci. Total Environ.* 772: 145042.

56 Harshman, L.A. and Hooper, S.R. (2020). The brain in pediatric chronic kidney disease – the intersection of cognition, neuroimaging, and clinical biomarkers. *Pediatr. Nephrol.* 35 (12): 2221–2229.

57 Grooms, D.R., Diekfuss, J.A., Ellis, J.D. et al. (2019). A novel approach to evaluate brain activation for lower extremity motor control. *J. Neuroimaging* 29 (5): 580–588.

58 Kim, C.S. (2021). Bayesian mechanics of perceptual inference and motor control in the brain. *Biol. Cybern.* 115 (1): 87–102.

59 Dougherty, P. (2020). Hypothalamus: structural organization. *Neuroscience Online*. https://nba.uth.tmc.edu/neuroscience/m/s4/chapter01.html (accessed 9 March 2023).

60 Dougherty, P. (2020). Hypothalamic control of pituitary hormone. *Neuroscience Online*. https://nba.uth.tmc.edu/neuroscience/m/s4/chapter02.html (accessed 9 March 2023).

61 Berridge, K.C. (2019). Affective valence in the brain: modules or modes? *Nat. Rev. Neurosci.* 20 (4): 225–234.

62 Riga, S., Nishimaru, H., Takamura, Y. et al. (2021). Neuronal representation of locomotion during motivated behavior in the mouse anterior cingulate cortex. *Front. Syst. Neurosci.* 15: 23.

63 Rahmati, M., Sprague, T.C., Sreenivasan, K.K., and Curtis, C.E. (2019). The nature of top-down signals during non-spatial working memory. *J. Vis.* 19 (10): 245a–245a.

64 Loprinzi, P.D. and Edwards, M.K. (2018). Exercise and cognitive-related semantic memory function. *J. Cognit. Behav. Psychotherap. Res.* 7 (2): 51–52.

65 Peak, J., Hart, G., and Balleine, B.W. (2019). From learning to action: the integration of dorsal striatal input and output pathways in instrumental conditioning. *Eur. J. Neurosci.* 49 (5): 658–671.

66 de Brouwer, A.J., Areshenkoff, C.N., Rashid, M.R. et al. (2022). Human variation in error-based and reinforcement motor learning is associated with entorhinal volume. *Cereb. Cortex* 32 (16): 3423–3440.

67 Britannica (2023). Dementia. *Encyclopedia Britannica*. https://www.britannica.com/science/dementia (accessed 9 March 2023).

68 Adetuyi, B.O., Toloyai, P.E.Y., Ojugbeli, E.T. et al. (2021). Neurorestorative roles of microgliosis and astrogliosis in neuroinflammation and neurodegeneration. *Scicom J. Med. Appl. Med. Sci.* 1 (1): 1–5.

69 Firdaus, Z. and Singh, T.D. (2021). An insight in pathophysiological mechanism of Alzheimer's disease and its management using plant natural products. *Mini Rev. Med. Chem.* 21 (1): 35–57.

70 Wang, D., Zhang, L., Huang, J. et al. (2021). Cardiovascular protective effect of black pepper (*Piper nigrum* L.) and its major bioactive constituent piperine. *Trends Food Sci. Technol.* 117: 34–45.

71 Choi, S.H., Lee, A.Y., Park, C.H. et al. (2018). Protective effect of *Carthamus tinctorius* L. seed on oxidative stress and cognitive impairment induced by chronic alcohol consumption in mice. *Food Sci. Biotechnol.* 27 (5): 1475–1484.

72 Gupta, P., Harky, A., Jahangeer, S. et al. (2018). Varying evidence on deep hypothermic circulatory arrest in thoracic aortic aneurysm surgery. *Tex. Heart Inst. J.* 45 (2): 70–75.

73 Guclu, G., Kelebek, H., and Selli, S. (2020). Saffron (*Crocus sativus* L.): its aroma and key odorants. In: *Saffron: The Age-Old Panaces in a New Light* (ed. M. Sarwat and S. Sumaiya), 69–82. Cambridge, MA: Academic Press.

74 Esmaeelian, M., Jahani, M., Feizy, J., and Einafshar, S. (2021). Effects of ultrasound-assisted and direct solvent extraction methods on the antioxidant and antibacterial properties of Saffron (*Crocus sativus* L.) corm extract. *Food Anal. Methods* 14 (1): 74–87.

75 Tamaddonfard, E., Farshid, A.A., Eghdami, K., and Erfanparast, A. (2013). Comparison of the effects of crocin, safranal and diclofenac on local inflammation and inflammatory pain responses induced by carrageenan in rats. *Pharmacol. Rep.* 65 (5): 1272–1280.

76 Yimer, E.M., Tuem, K.B., Karim, A. et al. (2019). *Nigella sativa* L. (black cumin): a promising natural remedy for wide range of illnesses. *Evid. Based Complement. Altern. Med.* 2019:1528635.

77 Rahman, M.T. (2020). Potential benefits of combination of *Nigella sativa* and Zn supplements to treat COVID-19. *J. Herb. Med.* 23: 100382.

78 Hosseini, M., Mohammadpour, T., Karami, R. et al. (2015). Effects of the hydro-alcoholic extract of *Nigella sativa* on scopolamine-induced spatial memory impairment in rats and its possible mechanism. *Chin. J. Integr. Med.* 21: 438–444.

79 Hadi, S., Mirmiran, P., Hosseinpour-Niazi, S. et al. (2015). Effect of nigella sativa oil extract on lipid profiles in type 2 diabetic patients: a randomized, double blind, placebo-controlled clinical trial. *Iran. J. Endocrinol. Metab.* 16 (6): Pe411–Pe418.

80 Wei, J.N., Liu, Z.H., Zhao, Y.P. et al. (2019). Phytochemical and bioactive profile of *Coriandrum sativum* L. *Food Chem.* 286: 260–267.

81 Prachayasittikul, V., Prachayasittikul, S., Ruchirawat, S., and Prachayasittikul, V. (2018). Coriander (*Coriandrum sativum*): a promising functional food toward the well-being. *Food Res. Int.* 105: 305–323.

82 Wu, T., Feng, S.-Y., Yang, Q.-H. et al. Integration of the metabolome and transcriptome reveals the metabolites and genes related to nutritional and medicinal value in *Coriandrum sativum*. *J. Integ. Agric.* 20 (7): 1807–1818.

83 Khajeh, M., Yamini, Y., Bahranifar, N. et al. (2005). Comparison of essential oils compositions of *Ferula assa-foetida* obtained by supercritical carbon dioxide extraction and hydrodistillation methods. *Food Chem.* 91: 639–644.

84 Lee, C.L., Chiang, L.C., Cheng, L.H. et al. (2009). Influenza A (H(1)N(1)) antiviral and cytotoxic agents from *Ferula assa-foetida*. *J. Nat. Prod.* 72 (9): 1568–1572.

85 Azaz, A.D., Irtem, H.A., Kurkcuoglu, M., and Baser, K.H.C. (2004). Composition and the *in vitro* antimicrobial activities of the essential oils of some Thymus species. *Z. Naturforsch. C* 59: 75–80.

86 Youdim, K.A. and Deans, S.G. (2000). Effect of thyme oil and thymol dietary supplementation on the antioxidant status and fatty acid composition of the ageing rat brain. *Br. J. Nutr.* 83 (1): 87–93.

87 Soleimani, V., Sahebkar, A., and Hosseinzadeh, H. (2018). Turmeric (*Curcuma longa*) and its major constituent (curcumin) as nontoxic and safe substances. *Phytother. Res.* 32 (6): 985–995.

88 Kulkarni, S., Akula, K., and Deshpande, J. (2012). Evaluation of antidepressant-like activity of novel water-soluble curcumin formulations and St. John's wort in behavioral paradigms of despair. *Pharmacology* 89: 83–90.

89 Chen, C., Long, L., Zhang, F. et al. (2018). Antifungal activity, main active components and mechanism of *Curcuma longa* extract against *Fusarium graminearum*. *PLoS One* 13 (3): e0194284.

13

Neurotransmitter Modulation by Phytochemicals

Prachee Dubey[1] and Kanti Bhooshan Pandey[2]

[1] *Government G.I. College, Malwan, Fatehpur, India*
[2] *CSIR – Central Salt & Marine Chemicals Research Institute, Bhavnagar, Gujarat, India*

Phytochemicals are naturally occurring bioactive compounds that have effects on living organisms. They have the ability to modulate biologic mechanisms either positively or negatively. Antiaging, anticancer, hepatoprotective, and cardioprotective are some very well studied and established biologic actions of plant-derived bioactive chemicals [1–4]. Studies have provided experimental evidence that the health-promotive effects of bioactive phytochemicals are elicited by their antioxidant, anti-inflammatory, ascorbate recycling, regulating gene expression, and hormetic effects [1, 5–7]. In the last decades, biologic effects of phytochemicals on the central nervous system (CNS) and thus mediating neuro-functions have attracted researchers across the world. The present chapter describes the role of phytochemicals in modulating neurotransmitters and their receptors.

13.1 Sources, Structures, and Classifications of Phytochemicals

Phytochemicals are found naturally in almost all parts of plants, including seeds, fruits, flowers, leaves, stems, and roots. Most of the phytochemicals are responsible for the specific color, texture, and taste of the fruits, leaves, and flowers [8]. Polyphenols are the most abundant phytochemicals and are most studied for their health-promoting biologic activities. They are synthesized in plants during adverse conditions as defense molecules against various pathogens. Plant polyphenols are further divided into four groups: phenolic acids, flavonoids, stilbenes, and lignans [1]. Flavonoids comprise the largest group of polyphenols. Table 13.1 summarizes different classes of phytochemicals, their sources, and their structures.

Phytochemical Drug Discovery for Central Nervous System Disorders: Biochemistry and Therapeutic Effects, First Edition. Edited by Chukwuebuka Egbuna and Mithun Rudrapal.

Table 13.1 Structure, sources, and biologic actions of major phytochemicals.

Phyto-chemical	Structural formula	Sources	Biologic actions	References
Curcumin	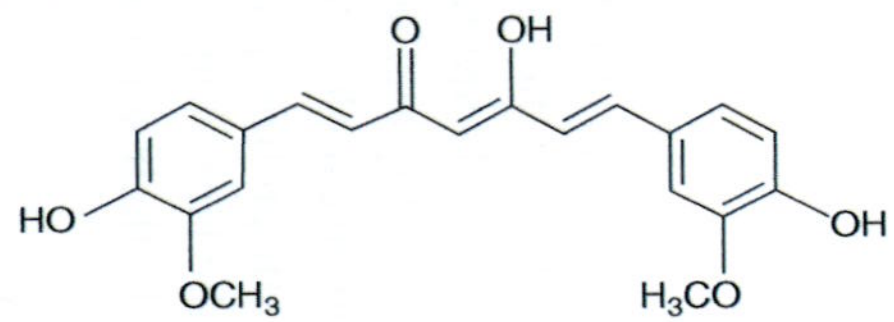	Turmeric, mango, ginger, and curry powder	Affects glutathione signaling, inhibits Ca ion voltage channels, influences NMDAR expression, increases serotonin level; inhibits glutamate release in nerve terminals, induces expression of serotonergic receptors, influences dopaminergic signaling	[9–14]
Quercetin	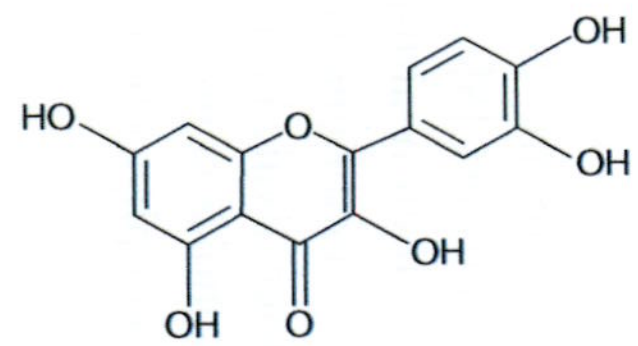	Grapes, oranges, apples, berries, cherries, green tea, red wine, and onions	Inhibits AChE, protects against AChR degradation, affects GABA receptors and glutathione signaling, inhibits glutamate release in nerve terminals	[15–17]
Myricetin	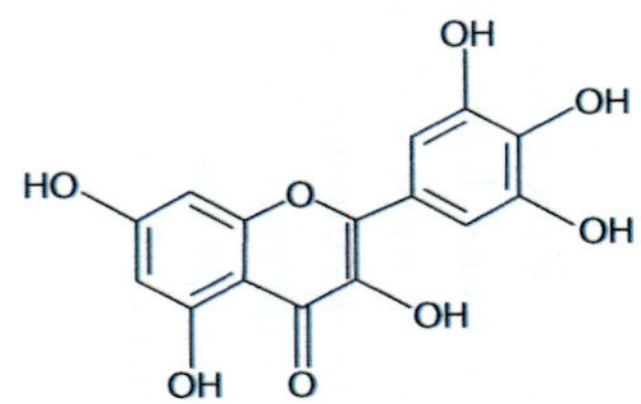	Blueberry, turnip, peas, garlic, cauliflower, spinach, grapes, and tea	Prevents deactivation of AChR	[18, 19]
Resveratrol	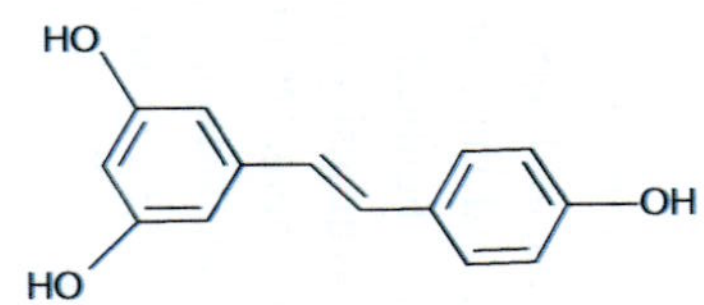	Grapes, blueberries, cranberries, peanuts, red wine, and dark chocolate	Upregulates AChE, mediates GABAergic signaling, influences glutamate release, modulates hippocampal KAR expression, promotes serotonin release	[20–24]
Luteolin	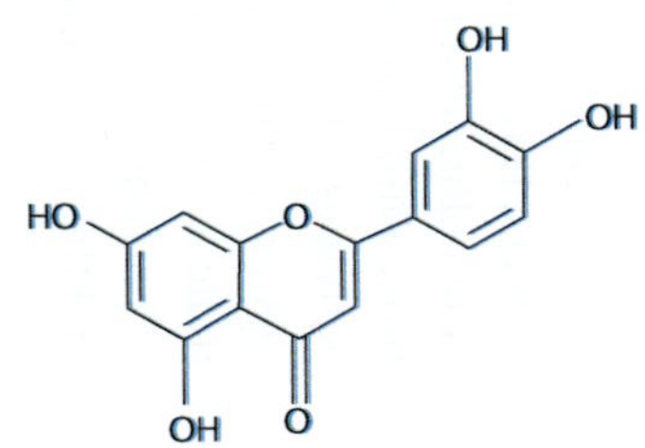	Apples, onions, carrots, celery, cabbage, parsley, broccoli, and chrysanthemum flowers	Upregulates AChE, mediates GABAergic signaling, inhibits glutamate release	[25–27]

Apigenin	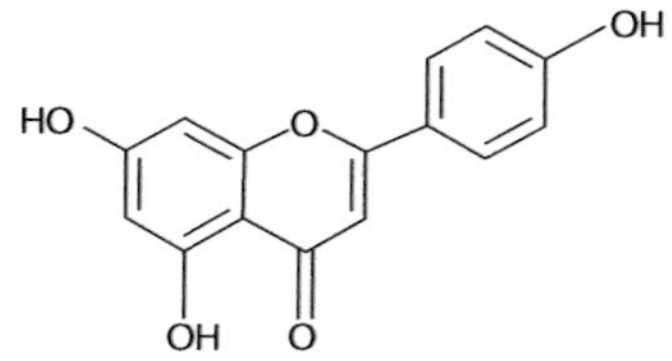	Celery, artichoke, parsley, chamomile, vine spinach, and oregano	Agonistic action on $GABA_AR$, inhibits glutamate release in hippocampus, promotes dopamine level	[28–30]
Baicalein and baicalin	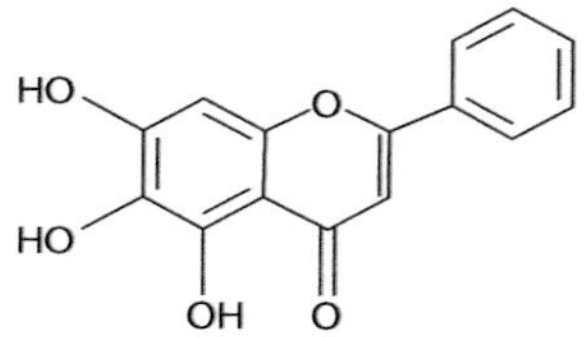 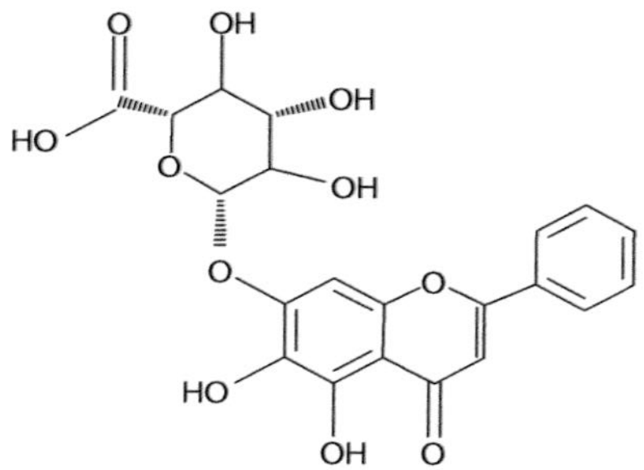	Roots of *Scutellaria baicalensis* and *Scutellaria lateriflora*	Modulates GABAergic signaling	[31–33]
Spinosin	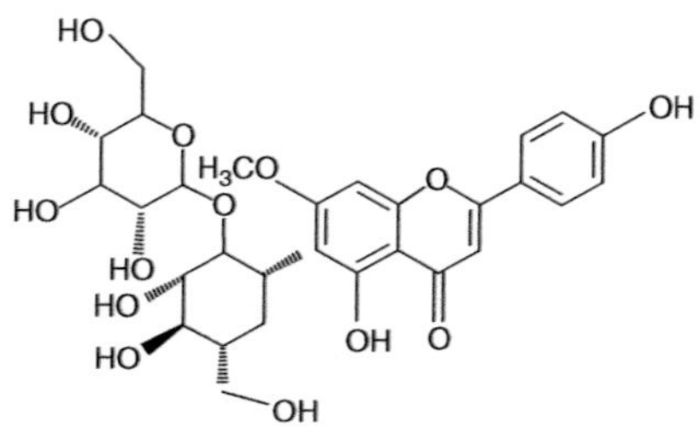	*Zizyphus jujube*	Upregulates AChE, serotoin receptor agonist	[34, 35]

(*Continued*)

Table 13.1 (Continued)

Phyto-chemical	Structural formula	Sources	Biologic actions	References
Naringenin	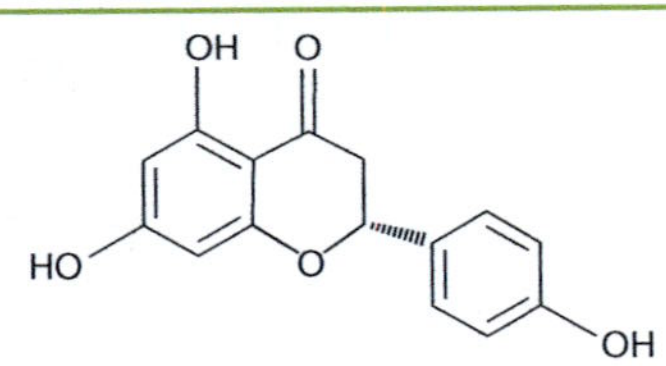	Grapefruit, orange, tart cherries, tomatoes, Greek oregano	Suppresses AChE activity, stimulates serotonergic neurotransmission, influences dopaminergic signaling	[11, 36–38]
Tea catechins (−)-Epicatechin (EC)	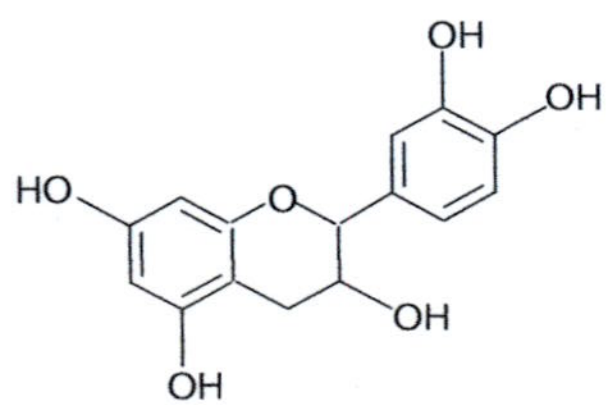	Tea leaves, grapes, berries, apricots, apples, pears, broad beans, red wine, and green tea	Suppresses β-amyloid-induced apoptotic cortical neuronal death, activates α7nAChR signaling cascade, prevents oxidative injury of neuro receptors, influences AChE expression, inhibits Aβ peptide aggregation, modulates GABAergic signaling, inhibits glutamate dehydrogenase, activates NMDAR1 expression, influences dopaminergic signaling	[39–46]
(−)-Epigallocatechin (EGC)	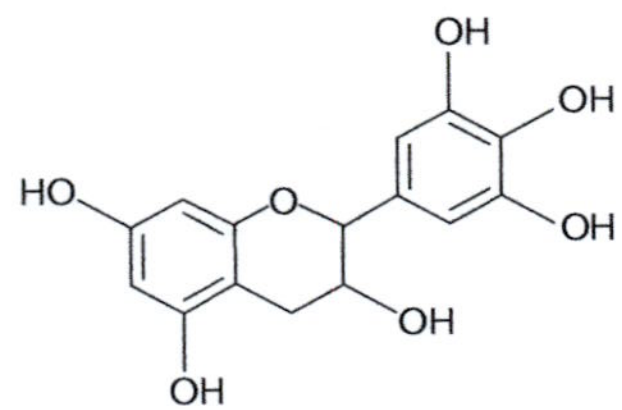			
(−)-Epicatechin gallate (ECG)	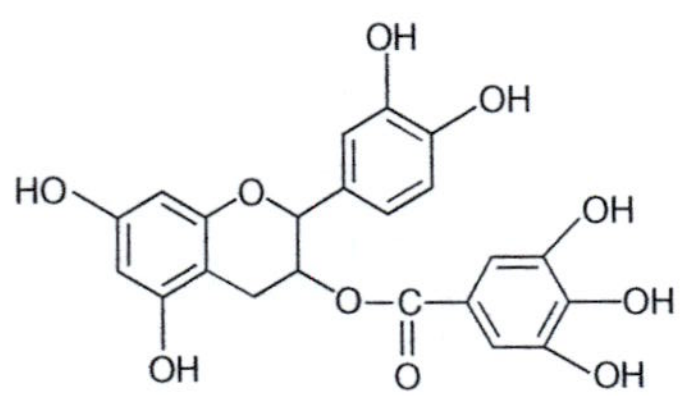			

(−)-Epigallocatechin gallate (EGCG)	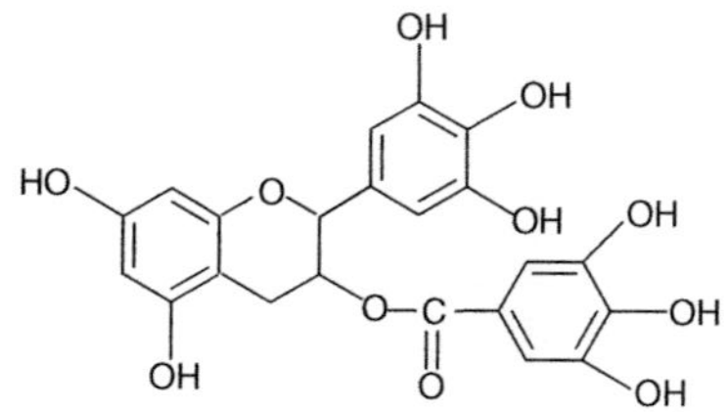			
Caffeine	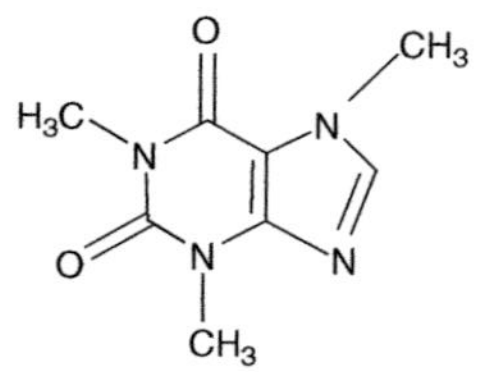	Coffee plant (*Caffia arabica*) seeds	Inhibits AChE	[47]

α7nAChR, α7 subtype of nicotinic ionotropic acetylcholine receptor; Aβ, amyloid-β; AChE, acetylcholine esterase; AChR, acetylcholine receptor; GABA, γ-aminobutyric acid; $GABA_AR$, $GABA_A$ receptor; KAR, kainate receptor; NMDAR, N-methyl-D-aspartate receptor.

13.2 Neurotransmitters and Their Functions

Neurotransmitters are signaling molecules secreted by nerve cells/neurons to affect the other cells, glands, or parts of the body. In the axon terminal of neurons, these neurotransmitters are enclosed in vesicles; at the time of signaling these molecules are released into the synaptic cleft and once in the synaptic cleft, these transmitters interact with the receptors of the target cell or, if that target cell is a neuron, then the receptors of dendrite. The effect of the message remitted by neurotransmitters is determined by the receptors to which they bind (Figure 13.1).

Neurotransmitters are essential for the proper functioning of the complex nervous system of the body. Glutamate, acetylcholine (ACh), γ-aminobutyric acid (GABA), serotonin, and dopamine are some vital neurotransmitters in humans. It has been proposed that more than 100 neurotransmitters are present in humans; however, the exact number of neurotransmitters involved in the human nervous system are still under investigation [48].

ACh was the first neurotransmitter to be discovered. It occurs in both the CNS and the peripheral nervous system (PNS). ACh is involved in the stimulation of motor neurons that are connected with movements of the body, thus in general it is known as a connector neurotransmitter that connects motor nerves to muscles. ACh in the brain affects neuronal excitability and synaptic plasticity and influences synaptic transmission, thereby modulating neuronal networks' state in the brain and also their response to internal and external inputs [49].

GABA is another important neurotransmitter that delivers fast responses in every part of the brain. Seductive drugs in general act by enhancing the effect of GABA [50]. In the brain and spinal cord, glutamate or glutamic acid is utilized at the fast excitatory synapses. Over-release of glutamate may overstimulate the brain and a condition of excitotoxicity arises,

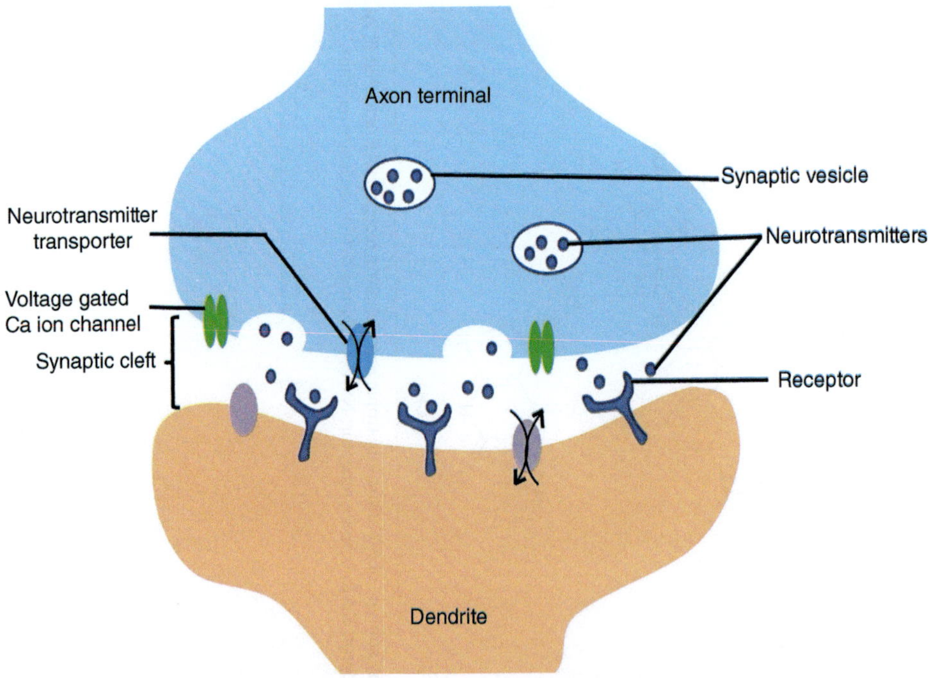

Figure 13.1 Diagrammatic representation of a typical chemical synapse.

which may result in ischemic stroke, Alzheimer's disease, epilepsy, Huntington's disease, and Parkinson's disease [51]. Serotonin (5HT) and dopamine are monoamine neurotransmitters; 5-HT is mostly produced in the intestine, although it also occurs in blood platelets and the CNS. The biologic roles of 5-HT and dopamine are complex and diverse, including mediation in learning, memory, appetite, sleep, and vasoconstriction; however, they are well reported in modulation of mood, where they act as an antidepressant so are sometimes called happy molecules [52, 53].

13.3 Modulation of Cholinergic Signaling by Phytochemicals

One of the most abundant neurotransmitters found in the human body, ACh plays an important role in the brain, where this neurotransmitter is associated with learning, memory, and cognitive functions. Due to direct mediation with the β-amyloid protein (Aβ), ACh is also referred as an anti-Alzheimer's agent [54]. The signaling mediated by ACh is known as cholinergic signaling. Studies have reported that receptors of ACh, muscarinic metabotropic receptor (mAChR) and nicotinic ionotropic receptor (nAChR), are both involved in Alzheimer's disease development; however, involvement of the α7 subtype of nAChR is critical. Researchers engaged in understanding cognitive impairments have documented that upregulation/overexpression of α7nAChR is vital since α7nAChR binds with Aβ and inhibits the deposition of amyloid plaques [55, 56]. A significant decrease in the M1 and M4 subtypes of mAChR has been reported during the development of Alzheimer's disease [57], which suggests that candidates with the ability to upregulate the expression of these receptors may be of therapeutic interest against disorders associated with their diminished activities.

Many bioactive phytochemicals are reported to possess the ability to stimulate the expression of ACh receptors as well as their density. A study performed by Nebrisi et al. reported that curcumin (1,7-bis(4-hydroxy-3-methoxyphenyl)-1,6-heptadiene-3,5-dione), a principal yellow-colored bioactive phytochemical that occurs in the turmeric plant (*Curcuma longa*), and its metabolites such as demethylcurcumin, tetrahydrocurcumin, and didemethylcurcumin, caused potentiation of choline-induced Ca ion transients in human neuroblastoma cells through activation of α7nAChR [9].

Bioactive phytochemicals of green tea (*Camellia sinesis*) known as catechins, are established antioxidants [58]. The four main catechins, (−)-epicatechin (EC), (−)-epigallocatechin (EGC), (−)-epicatechin gallate (ECG), and (−)-epigallocatechin gallate (EGCG), are documented to modulate cholinergic signaling [59]. An experiment on rat primary cortical neurons documents suppression of β-amyloid-induced apoptotic cortical neuronal insults by EGCG. The study explored how this effect of EGCG was mediated by activation of the α7nAChR signaling cascade. Treatment with EGCG also increased cell viability and reduced oxidative stress load [39].

Quercetin (2-(3,4-dihydroxyphenyl)-3,5,7-trihydroxychromen-4-one), a bioactive phytochemical found largely in onions, grapes, and berries, is a secondary metabolite that elicits potent antioxidant and anti-inflammatory properties [60]. Quercetin is also reported to affect cholinergic signaling. The protective effect of quercetin in cadmium-mediated cholinergic dysfunction has been documented in the rat brain [61]. There are studies that provide experimental evidence that quercetin possesses great potential to be developed as a neuroprotective drug by exerting potent antioxidant effects and by preventing deactivation of nAChR and mAChR in the Alzheimer's brain [15]. In addition, upregulation of the M1 subtype of mAChR in C1 hippocampal and medial prefrontal cortex cells of the aging brain was reported in rats after administration of grape seed proanthocyanidin extract (GSPE) orally [62]. The role of oxidative stress cannot be denied in the pathology of cognitive impairments. It has been reported that many bioactive phytochemicals such as carnosol, a phenolic diterpene found in the herbs rosemary (*Rosmarinus officinalis*) and sage (*Salvia pachyphylla*), myricetin (3,5,7,3′,4′,5′-hexahydroxyflavone) from red wine, black plum (*Syzygium cumini*), and quercetin can protect brain cells as well as preventing deactivation of ACh receptors by low molecular inhibitor–mediated generation of free radicals in the human brain [18].

13.3.1 Effect of Phytochemicals on Acetylcholinesterase

Acetylcholinesterase (AChE) works to terminate cholinergic synaptic transmission according to need, by degrading ACh. Malfunctioned AChE activity has been reported in many neuronal ailments, including Alzheimer's disease, Parkinson's disease, and other dementias [63]. Some phytochemicals are reported to influence AChE by inhibiting its activity directly or indirectly by acting as competitive/non-competitive inhibitors or changing their expression at a gene level [64]. Caffeine (1,3,7-trimethylpurine-2,6-dione), the natural chemical present in the seed of the coffee plant (*Coffea arabica*), is well known as a CNS stimulant. Caffeine acts as a neurotransmission modulator by inhibiting AChE and thus increasing the ACh level [47].

The direct inhibition of AChE activity was elicited by many bioactive chemicals extracted from various plant species, such as quercetin (black chokeberry), rosmarinic acid (*Paulownia barbatus* leaves), galangin (*Alpiniae officinarum* rhizome), and linarin (*Linaria* sp. aerial parts) [65, 66]. Phytochemicals from *Prunella vulgaris* and *Oenothera biennis* have also been reported to improve memory and overall cognitive health by the mechanism that involves inhibition of AChE [67].

Different tea catechins including EGCG are reported to diminish AChE expression and to inhibit Aβ peptide aggregation, thereby preventing neurologic loss, as a result providing prevention against worsening Alzheimer's disease symptoms [40, 41]. Naringenin ((2S)-4′,5,7-trihydroxyflavan-4-one), an important phytochemical found in citrus fruits such as grapes, oranges, and tomatoes, and liquiritigenin ((2S)-4′,7-dihydroxyflavan-4-one), present in different plants including *Glycyrrhiza glabra*, are also documented to suppress AChE activity [36, 37, 68].

In contrast, some phytochemicals are reported to enhance the activity of AChE, such as resveratrol (3,5,4′-trihydroxy-trans-stilbene), a natural phytoalexin present abundantly in grapes, peanuts, blueberries, and raspberries; luteolin (3′,4′,5,7-tetrahydroxyflavone), present in olive oils, carrots, thyme, and peppers; and spinosin, a flavonoid isolated from *Zizyphus jujube*. Their neuroprotective effect was justified by the activation of extracellular signal-regulated kinase signaling pathways; however, no confirmatory and document-based mechanisms have yet been proposed to explain their cognitive improving effects [20, 25, 34].

13.4 Effect of Phytochemicals on GABAergic Signaling

GABA is another primary neurotransmitter in the human CNS. In most neuro-related ailments such as anxiety, schizophrenia, and Alzheimer's disease, disturbance in the GABAergic system has been observed. Bioactive chemicals from various plants have been reported as modulators of GABAergic signaling with different mechanisms of $GABA_A$ receptor ($GABA_AR$) inhibition. Some phytochemicals use a flumazenil-sensitive high-affinity benzodiazepine binding site to act, while others use a flumazenil-insensitive low-affinity benzodiazepine site [42, 69, 70].

Green tea EGCG is reported to elicit its neuroprotective, anxiolytic, sedative, hypnotic properties by interaction with the benzodiazepine binding site [42, 43]. Baicalein and baicalin, the two other bioactives found largely in *Scutellaria baicalensis*, showed neuroprotective as well as anticonvulsant effects by acting on the benzodiazepine site on the α2 and α3 subunits of $GABA_AR$ [31, 71]. Interestingly, the sedative effects of baicalein may also be exerted by its interaction with GABAergic non-benzodiazepine sites [32].

Plant flavone, apigenin (4′,5,7-trihydroxyflavone), is naturally present in many plants and has been documented to elicit neuroprotective properties. Studies have documented that the antianxiety activity of apigenin is mediated through agonistic action on $GABA_AR$ [72]. Amentoflavone, a bioactive phenolic compound found in *Selaginella tamariscina* and *Ginkgo biloba* leaves, is also documented as a very significant modulator of $GABA_AR$. Its affinity to bind the benzodiazepine site of $GABA_AR$ is similar to diazepam, a drug frequently used as an anxiolytic [70]; however, its activity may be flumazenil dependent or independent [42]. Many other phytochemicals are reported to act as agonists of $GABA_AR$ such as hispidulin, skrofulein, and daidzein [26, 73].

13.5 Effect of Phytochemicals on Glutamatergic Signaling

Glutamic acid or glutamate is an excitatory neurotransmitter that plays a vital role in maintaining proper functioning of the CNS via influencing memory, the learning process, and synaptic plasticity. Unregulated secretion of glutamate invited several cognitive ailments. Excess release of glutamate followed by overactivated glutamate receptors, N-methyl-D-aspartate receptors (NMDAR), kainate receptors (KAR), and α-amino-3-hydroxy-5-methyl-4-isoxazole propionic acid receptors (AMPAR) is associated with Alzheimer's disease, Parkinson's disease, epilepsy, and many other neuro-diseases due to more intensive calcium influx [74].

On the other hand, very low concentrations of glutamate or diminished activities of glutamate receptors are reported to disturb long-term potentiation in the hippocampus, impairing learning and memory functions. So modulation of glutamate amount and/or glutamate receptors may be a strategy to combat neuro-impairments related to this phenomenon [75].

It has been investigated that presynaptic glutamate secretion is highly dependent on Ca^{2+} influx through voltage-gated Ca^{2+} channels, both Cav2.1 and Cav2.2. Studies on the effect of luteolin on synaptosomes in rats revealed that luteolin significantly decreased glutamate release and the action was mediated by the effect of luteolin on Cav channels [27]. A similar inhibitory effect was elicited by curcumin, apigenin, and acacetin on the release of glutamate hippocampal synaptosomes [28, 76].

Interesting dose-dependent effects were exerted by resveratrol on glutamate uptake in cultured cortical astrocytes. Resveratrol at low doses increased the uptake of glutamate, while at higher doses it diminished the uptake [21]. As a control mechanism, excess glutamate is removed by glutamate dehydrogenase (GDH) by a reversible oxidation deamination mechanism in which glutamate is converted into α-ketoglutarate. EGCG and ECG are found to be inhibitory on GDH activity, thus preventing the removal of glutamate [44].

Zhang and colleagues have reported that phytochemicals from green tea, especially EGCG, ECG, EC, and EGC, restored the spatial learning and memory impairments induced by ethanol in rats by a mechanism that involved enchantment in the expression of NMDAR1 and the cyclic adenosine monophosphate response element binding protein (CREB) [45].

Curcumin and tannic acid have been reported to ameliorate the neurodestructive action of lead and aluminum metals in the rat hippocampus by preventing the inhibition of NR2A and NR2B subunits' expression of NMDAR. This mechanism of both phytochemicals is assumed to enhance learning and memory in the brain [77]. Resveratrol modulates the hippocampal KAR expression in rats, which may be threptic in epilepsy [22].

13.6 Modulation of Serotonergic and Dopaminergic Signaling by Phytochemicals

Serotonin and dopamine play a very critical role in movement, behavior control, and mood and are mostly associated with depression [78]. Dysfunction of the serotonergic and dopaminergic systems may lead to various mental and motor disorders. The role of dopamine is critical in movement and behavior control. An inhibited dopaminergic system due to diminished receptors or loss of dopamine mediates many neuronal disorders, including Parkinson's disease [79]. Many phytochemicals are reported to modulate the secretion and expression of serotonin and dopamine and their acceptors thus may be used in therapeutic purposes. Bioactive phytochemicals may induce the synthesis of monoamines and/or expression of serotonergic receptors (5-HTR). Curcumin has been reported to affect A, B, and C subtypes of serotonergic receptors [80].

Chang and co-workers in the study exploring the antidepressant potential of curcumin have reported that high-dose administration of curcumin significantly induced the expression of 5-HTR density in the hippocampus and frontal cortex of expression-induced male albino Wistar rats [10].

Nobiletin (an active phytochemical from citrus peels) has been reported to elicit potent antidepressant activity that involves the participation of 5-HT1AR and 5-HT2R and the α1-adrenoceptor, D1, and D2 subtypes of the dopamine receptors [81]. Spinosin is found to be effective in treatment of Alzheimer's disease and improving cognitive functions, acting as a 5-HT1A receptor agonist [35]. The grapefruit phytochemical naringenin is also reported to modulate mood by stimulating serotonergic neurotransmission in the brain [38]. Stilbene, resveratrol obtained from *Polygonum cuspidatum*, has been reported to elicit the antidepressant-like effect of trans-resveratrol on mouse models and significantly increased serotonin and noradrenaline levels in the brain [23].

The CNS contains neurons involved in dopaminergic signaling in enormous amounts. Five subtypes of dopamine receptors are reported, which vary from D1 to D5; all use G proteins. Malfunctioning dopaminergic signaling due to imbalanced dopamine concentration and/or dopamine transporter/receptor dysregulation has been reported in many neurogenic impairments, including multiple sclerosis [82].

Studies have reported that many plant bioactives possess the ability to modulate the dopamine level and thus affect dopaminergic signaling. Yi and colleagues have reported the antidepressant-like behavioral and neurochemical effects of apigenin, a citrus phytochemical in chronic mild stress (CMS)–exposed rats that showed an elevated dopamine level in the frontal cortex and nucleus accumbens. Administration of apigenin significantly protected from CMS-induced alterations in serotonin and dopamine levels in distinct rat brain regions [29].

An interesting study by Zbarsky and co-workers on a 6-hydroxydopamine-induced Parkinson's disease model claims that curcumin and naringenin pretreated rat models provided significant protection against loss of tyrosine hydroxylase–positive cells in the substantia nigra and dopamine content in the striata [11]. The dose-dependent inhibitor effect of EGCG on dopamine uptake happens by induction of dopamine transporter uptake [46].

13.7 Conclusion

Many plant bioactive molecules possess the ability to modulate different neurotransmitters, either directly affecting their levels or influencing the expression of their receptors, and thus influence their effects and overall neurophysiology. In this chapter only evidence-based actions of phytochemicals are discussed. Studies on different model systems have revealed that phytochemicals elicit multifaceted action, and may up- or down-regulate neurotransmitters. In addition, their dose is also a determining factor of their actions. Although most of the studies/results related to the action of phytochemicals on neurotransmitters are still ongoing, the documented results strongly claim potent neuroprotective effects of these natural molecules that may be utilized in the development of neuro-drugs with natural origins.

Acknowledgments

PD acknowledges the support of her college and DSE, UP, and KBP is thankful to the director of CSIR-CSMCRI for his support and encouragement.

References

1 Pandey, K.B. and Rizvi, S.I. (2009). Plant polyphenols as dietary antioxidants in human health and disease. *Oxid. Med. Cell Longev.* 2: 270–278.

2 Mishra, J., Srivastava, S.K., and Pandey, K.B. (2021). Dietary polyphenols in the intervention of gestational diabetes. *Curr. Tradit. Med.* 7: 8–15.

3 Cosme, F., Pinto, T., Aires, A. et al. (2022). Red fruits composition and their health benefits – a review. *Foods* 11: 644.

4 Emire, Z. and Yabalak, E. (2022). Can Origanum be a hope for cancer treatment? A review on the potential of Origanum species in preventing and treating cancers. *Int. J. Environ. Health Res.* 13: 1–17.

5 Pandey, K.B. and Rizvi, S.I. (2013). Resveratrol up-regulates the erythrocyte plasma membrane redox system and mitigates oxidation-induced alterations in erythrocytes during aging in humans. *Rejuvenation Res.* 16: 232–240.

6 Alì, S., Davinelli, S., Accardi, G. et al. (2021). Healthy ageing and Mediterranean diet: a focus on hormetic phytochemicals. *Mech. Ageing Dev.* 200: 111592.

7 Guo, P., Zhang, B., Zhao, J. et al. (2022). Medicine-food herbs against Alzheimer's disease: a review of their traditional functional features, substance basis, clinical practices and mechanisms of action. *Molecules* 27: 901.

8 Pandey, K.B. and Rizvi, S.I. (2012). Recent advances in health promoting effect of dietary polyphenols. *Curr. Nutr. Food Sci.* 8: 254–264.

9 Nebrisi, E.E., Al Kury, L.T., Yang, K.S. et al. (2018). Curcumin potentiates the function of human α7-nicotinic acetylcholine receptors expressed in SH-EP1 cells. *Neurochem. Int.* 114: 80–84.

10 Chang, X.R., Wang, L., Li, J., and Wu, D.S. (2016). Analysis of antidepressant potential of curcumin against depression induced male albino Wistar rats. *Brain Res.* 1642: 219–225.

11 Zbarsky, V., Datla, K.P., Parkar, S. et al. (2005). Neuroprotective properties of the natural phenolic antioxidants curcumin and naringenin but not quercetin and fisetin in a 6-OHDA model of Parkinson's disease. *Free Radic. Res.* 39: 1119–1125.

12 Singh, P., Pandey, K.B., and Rizvi, S.I. (2016). Curcumin: the yellow molecule with pleiotropic biological effects. *Lett. Drug Des. Discov.* 13: 170–177.

13 Stanić, Z. (2017). Curcumin, a compound from natural sources, a true scientific challenge – a review. *Plant Foods Hum. Nutr.* 72: 1–12.

14 Lin, T.Y., Lu, C.W., Wang, C.C. et al. (2011). Curcumin inhibits glutamate release in nerve terminals from rat prefrontal cortex: possible relevance to its antidepressant mechanism. *Prog. Neuropsychopharmacol. Biol. Psychiatry.* 35: 1785–1793.

15 Bakoyiannis, I., Daskalopoulou, A., Pergialiotis, V., and Perrea, D. (2019). Phytochemicals and cognitive health: are flavonoids doing the trick? *Biomed. Pharmacother.* 109: 1488–1497.

16 Dabeek, W.M. and Marra, M.V. (2019). Dietary quercetin and Kaempferol: bioavailability and potential cardiovascular-related bioactivity in humans. *Nutrients* 11: 2288.

17 Lu, C.W., Lin, T.Y., and Wang, S.J. (2013). Quercetin inhibits depolarization evoked glutamate release in nerve terminals from rat cerebral cortex. *Neurotoxicology* 39: 1–9.

18 Fawcett, J.R., Bordayo, E.Z., Jackson, K. et al. (2002). Inactivation of the human brain muscarinic acetylcholine receptor by oxidative damage catalyzed by a low molecular weight endogenous inhibitor from Alzheimer's brain is prevented by pyrophosphate analogs, bioflavonoids and other antioxidants. *Brain Res.* 950: 10–20.

19 Taheri, Y., Suleria, H.A.R., Martins, N. et al. (2020). Myricetin bioactive effects: moving from preclinical evidence to potential clinical applications. *BMC Complement. Med. Ther.* 20: 241.

20 Liu, B., Xu, L., Guo, M. et al. (2017). Resveratrol improves cognition of rats impaired by carotid artery stenosis through the cholinergic system. *Biomed. Res.* 1 (Spec Iss): 286–293.

21 de Almeida, L.M.V., Piñeiro, C.C., Leite, M.C. et al. (2007). Resveratrol increases glutamate uptake, glutathione content, and S100B secretion in cortical astrocyte cultures. *Cell. Mol. Neurobiol.* 27: 661–668.

22 Li, Z., You, Z., Li, M. et al. (2017). Protective effect of resveratrol on the brain in a rat model of epilepsy. *Neurosci. Bull.* 33: 273–280.

23 Xu, Y., Wang, Z., You, W. et al. (2010). Anti depressant like effect of trans-resveratrol: involvement of serotonin and noradrenaline system. *Eur. Neuropsychopharmacol.* 20: 405–413.

24 Pandey, K.B. and Rizvi, S.I. (2009). Pleiotropic biological effects of resveratrol: implications for human health. *Natl. Acad. Sci. Lett.* 32: 321–326.

25 El Omri, A., Han, J., Kawada, K. et al. (2012). Luteolin enhances cholinergic activities in PC12 cells through ERK1/2 and PI3K/Akt pathways. *Brain Res.* 1437: 16–25.

26 Coleta, M., Campos, M.G., Cotrim, M.D. et al. (2008). Assessment of luteolin (3′,4′,5,7-tetrahydroxyflavone) neuropharmacological activity. *Behav. Brain Res.* 189: 75–82.

27 Lin, T.Y., Lu, C.W., Chang, C.C. et al. (2011). Luteolin inhibits the release of glutamate in rat cerebrocortical nerve terminals. *J. Agric. Food Chem.* 59: 8458–8466.

28 Chang, C.Y., Lin, T.Y., Lu, C.W. et al. (2015). Apigenin, a natural flavonoid, inhibits glutamate release in the rat hippocampus. *Eur. J. Pharmacol.* 762: 72–81.

29 Yi, L.T., Li, J.M., Li, Y.C. et al. (2008). Antidepressant-like behavioral and neurochemical effects of the citrusassociated chemical apigenin. *Life Sci.* 82: 741–751.

30 Shankar, E., Goel, A., Gupta, K., and Gupta, S. (2017). Plant flavone apigenin: an emerging anticancer agent. *Curr. Pharmacol. Rep.* 3: 423–446.

31 Wang, F., Xu, Z., Ren, L. et al. (2008). GABA A receptor subtype selectivity underlying selective anxiolytic effect of baicalin. *Neuropharmacology* 55: 1231–1237.

32 de Carvalho, R.S.M., Duarte, F.S., and de Lima, T.C.M. (2011). Involvement of GABAergic non-benzodiazepine sites in the anxiolytic-like and sedative effects of the flavonoid baicalein in mice. *Behav. Brain Res.* 221: 75–82.

33 Tao, Y., Zhan, S., Wang, Y. et al. (2018). Baicalin, the major component of traditional Chinese medicine *Scutellaria baicalensis* induces colon cancer cell apoptosis through inhibition of oncomi RNAs. *Sci. Rep.* 8: 14477.

34 Ko, S.Y., Lee, H.E., Park, S.J. et al. (2015). Spinosin, a C-glucosylflavone, from *Zizyphus jujube* var. spinosa ameliorates Aβ142 oligomer-induced memory impairment in mice. *Biomol. Ther.* 23 (2): 156–164.

35 Jung, I.H., Lee, H.E., Park, S.J. et al. (2014). Ameliorating effect of spinosin, a C-glycoside flavonoid, on scopolamine- induced memory impairment in mice. *Pharmacol. Biochem. Behav.* 120: 88–94.

36 Liaquat, L., Batool, Z., Sadir, S. et al. (2018). Naringenin-induced enhanced antioxidant defence system meliorates cholinergic neurotransmission and consolidates memory in male rats. *Life Sci.* 194: 213–223.

37 Rahigude, A., Bhutada, P., Kaulaskar, S. et al. (2012). Participation of antioxidant and cholinergic system in protective effect of naringenin against type-2 diabetes-induced memory dysfunction in rats. *Neuroscience* 226: 62–72.

38 Nouri, Z., Fakhri, S., El-Senduny, F.F. et al. (2019). On the neuroprotective effects of naringenin: pharmacological targets, signaling pathways, molecular mechanisms, and clinical perspective. *Biomolecules* 9: E690.

39 Zhang, X., Wu, M., Lu, F. et al. (2014). Involvement of α7 nAChR signaling cascade in epigallocatechin gallate suppression of β-amyloid-induced apoptotic cortical neuronal insults. *Mol. Neurobiol.* 49: 66–77.

40 Xiao, J., Chen, X., Zhang, L. et al. (2008). Investigation of the mechanism of enhanced effect of EGCG on huperzine A's inhibition of acetylcholinesterase activity in rats by a multispectroscopic method. *J. Agric. Food Chem.* 56: 910–915.

41 Wang, S.H., Dong, X.Y., and Sun, Y. (2012). Thermodynamic analysis of the molecular interactions between amyloid β-protein fragments and (−)-epigallocatechin-3-gallate. *J. Phys. Chem. B.* 116: 5803–5809.

42 Hanrahan, J.R., Chebib, M., and Johnston, G.A. (2011). Flavonoid modulation of GABA(A) receptors. *Br. J. Pharmacol.* 163: 234–245.

43 Adachi, N., Tomonaga, S., Tachibana, T. et al. (2006). (−)-Epigallocatechin gallate attenuates acute stress responses through GABAergic system in the brain. *Eur. J. Pharmacol.* 531: 171–175.

44 Li, C., Li, M., Chen, P. et al. (2011). Green tea polyphenols control dysregulated glutamate dehydrogenase in transgenic mice by hijacking the ADP activation site. *J. Biol. Chem.* 286: 34164–34174.

45 Zhang, Y., He, F., Hua, T., and Sun, Q. (2018). Green tea polyphenols ameliorate ethanol-induced spatial learning and memory impairments by enhancing hippocampus NMDAR1 expression and CREB activity in rats. *Neuroreport* 29: 1564–1570.

46 Li, R., Peng, N., Li, X.P., and Le, W.D. (2006). (−)-Epigallocatechin gallate regulates dopamine transporter internalization via protein kinase C dependent pathway. *Brain Res.* 1097: 85–89.

47 Pohanka, M. and Dobes, P. (2013). Caffeine inhibits acetylcholinesterase, but not butyrylcholinesterase. *Int. J. Mol. Sci.* 14: 9873–9882.

48 Feher, J. (2012). Cells, synapses, and neurotransmitters. In: *Quantitative Human Physiology* (ed. J. Feher), 307–320. New York: Academic Press.

49 Picciotto, M.R., Higley, M.J., and Mineur, Y.S. (2012). Acetylcholine as a neuromodulator: cholinergic signaling shapes nervous system function and behaviour. *Neuron* 76: 116–129.

50 Stephens, D.N., King, S.L., Lambert, J.J. et al. (2017). GABAA receptor subtype involvement in addictive behaviour. *Genes Brain Behav.* 16: 149–184.

51 Yang, J.L., Sykora, P., Wilson, D.M. 3rd et al. (2011). The excitatory neurotransmitter glutamate stimulates DNA repair to increase neuronal resiliency. *Mech. Ageing Dev.* 132: 405–411.

52 Fischer, A.G. and Ullsperger, M. (2017). An update on the role of serotonin and its interplay with dopamine for reward. *Front. Hum. Neurosci.* 11: 484.

53 Seo, D., Patrick, C.J., and Kennealy, P.J. (2008). Role of serotonin and dopamine system interactions in the neurobiology of impulsive aggression and its comorbidity with other clinical disorders. *Aggress. Violent Behav.* 13: 383–395.

54 Grimaldi, M., Marino, S.D., Florenzano, F. et al. (2016). β-Amyloid-acetylcholine molecular interaction: new role of cholinergic mediators in anti-Alzheimer therapy? *Future Med. Chem.* 8: 1179–1189.

55 Ma, K.G. and Qian, Y.H. (2019). Alpha 7 nicotinic acetylcholine receptor and its effects on Alzheimer's disease. *Neuropeptides* 73: 96–106.

56 Parri, H.R., Hernandez, C.M., and Dineley, K.T. (2011). Research update: Alpha7 nicotinic acetylcholine receptor mechanisms in Alzheimer's disease. *Biochem. Pharmacol.* 82: 931–942.

57 Lebois, E.P., Thorn, C., Edgerton, J.R. et al. (2018). Muscarinic receptor subtype distribution in the central nervous system and relevance to aging and Alzheimer's disease. *Neuropharmacology* 136 (Pt C): 362–373.

58 Pandey, K.B. and Rizvi, S.I. (2012). Upregulation of erythrocyte ascorbate free radical reductase by tea catechins: correlation with their antioxidant properties. *Food Res. Int.* 46: 46–49.

59 Kim, J.M. and Heo, H.J. (2022). The roles of catechins in regulation of systemic inflammation. *Food Sci. Biotechnol.* 31: 957–970.

60 Salehi, B., Machin, L., Monzote, L. et al. (2020). Therapeutic potential of quercetin: new insights and perspectives for human health. *ACS Omega* 5: 11849–11872.

61 Gupta, R., Shukla, R.K., Chandravanshi, L.P. et al. (2017). Protective role of quercetin in cadmium-induced cholinergic dysfunctions in rat brain by modulating mitochondrial integrity and MAP kinase signaling. *Mol. Neurobiol.* 54: 4560–4583.

62 Abhijit, S., Subramanyam, M.V.V., and Devi, S.A. (2017). Grape seed proanthocyanidin and swimming exercise protects against cognitive decline: a study on M1 acetylcholine receptors in aging male rat brain. *Neurochem. Res.* 42: 3573–3586.

63 Marucci, G., Buccioni, M., Ben, D.D. et al. (2021). Efficacy of acetylcholinesterase inhibitors in Alzheimer's disease. *Neuropharmacology.* 190: 108352.

64 Kim, H.R., Kim, W.K., and Ha, A.W. (2019). Effects of phytochemicals on blood pressure and neuroprotection mediated via brain renin–angiotensin system. *Nutrients* 11: 2761.

65 Asif, M., Qazi, M.H., Qazi, A.M. et al. (2014). Recent updates in the treatment of neurodegenerative disorders using natural compounds. *Evid. Based Complement. Altern. Med.* 2014: 979730.

66 Rebas, E., Rzajew, J., Radzik, T., and Zylinska, L. (2020). Neuroprotective polyphenols: a modulatory action on neurotransmitter pathways. *Curr. Neuropharmacol.* 18: 431–445.

67 Lee, S., Lee, D., Baek, J. et al. (2017). in vitro assessment of selected Korean plants for antioxidant and antiacetylcholinesterase activities. *Pharm. Biol.* 55: 2205–2210.

68 Ko, Y.H., Kwon, S.H., Lee, S.Y., and Jang, C.G. (2017). Liquiritigenin ameliorates memory and cognitive impairment through cholinergic and BDNF pathways in the mouse hippocampus. *Arch. Pharm. Res.* 40: 1209–1217.

69 Hood, S.D., Norman, A., Hince, D.A. et al. (2014). Benzodiazepine dependence and its treatment with low dose flumazenil. *Br. J. Clin. Pharmacol.* 77 (2): 285–294.

70 Wasowski, C. and Marder, M. (2012). Flavonoids as GABAA receptor ligands: the whole story? *J. Exp. Pharmacol.* 4: 9–24.

71 Hansen, R.S., Paulsen, I., and Davies, M. (2005). Determinants of amentoflavone interaction at the GABA(A) receptor. *Eur. J. Pharmacol.* 519 (3): 199–207.

72 Campbell, E.L., Chebib, M., and Johnston, G.A.R. (2004). The dietary flavonoids apigenin and (−)-epigallocatechin gallate enhance the positive modulation by diazepam of the activation by GABA of recombinant GABA(A) receptors. *Biochem. Pharmacol.* 68: 1631–1638.

73 Kavvadias, D., Sand, P., Youdim, K.A. et al. (2004). The flavone hispidulin, a benzodiazepine receptor ligand with positive allosteric properties, traverses the blood–brain barrier and exhibits anticonvulsive effects. *Br. J. Pharmacol.* 142: 811–820.

74 Lewerenz, J. and Maher, P. (2015). Chronic glutamate toxicity in neurodegenerative diseases – what is the evidence? *Front Neurosci.* 9: 469.

75 Katagiri, H., Tanaka, K., and Manabe, T. (2001). Requirement of appropriate glutamate concentrations in the synaptic cleft for hippocampal LTP induction. *Eur. J. Neurosci.* 14: 547–553.

76 Lin, T.Y., Huang, W.J., Wu, C.C. et al. (2014). Acacetin inhibits glutamate release and prevents kainic acid-induced neurotoxicity in rats. *PLoS One* 9: e88644.

77 Tüzmen, M.N., Yücel, N.C., Kalburcu, T., and Demiryas, N. (2015). Effects of curcumin and tannic acid on the aluminum- and lead-induced oxidative neurotoxicity and alterations in NMDA receptors. *Toxicol. Mech. Methods* 25: 120–127.

78 Khan, H., Perviz, S., Sureda, A. et al. (2018). Current standing of plant derived flavonoids as an antidepressant. *Food Chem. Toxicol.* 119: 176–188.

79 Juárez Olguín, H., Calderón Guzmán, D., Hernández García, E., and Barragán, M.G. (2016). The role of dopamine and its dysfunction as a consequence of oxidative stress. *Oxid. Med. Cell Longev.* 2016: 9730467.

80 Wang, R., Xu, Y., Wu, H.L. et al. (2008). The antidepressant effects of curcumin in the forced swimming test involve 5-HT1 and 5-HT2 receptors. *Eur. J. Pharmacol.* 578: 43–50.

81 Yi, L.T., Xu, H.L., Feng, J. et al. (2011). Involvement of monoaminergic systems in the antidepressant-like effect of nobiletin. *Physiol. Behav.* 102: 1–6.

82 Klein, M.O., Battagello, D.S., Cardoso, A.R. et al. (2019). Dopamine: functions, signaling, and association with neurological diseases. *Cell. Mol. Neurobiol.* 39: 31–59.

14

Antipyretic Agents from Plant Origins

Kamoru A. Adedokun[1#], Sikiru O. Imodoye[2#], Akeem O. Busari[3], Malik A. Sanusi[4], Abdullah Olawuyi[5], and Maroof G. Oyeniyi[6]*

[1] *Department of Immunology, Roswell Park Comprehensive Cancer Center, Buffalo, NY, USA*
[2] *Department of Oncological Sciences, Huntsman Cancer Institute, University of Utah, Salt Lake City, UT, USA*
[3] *Department of Medical Laboratory Science, Ladoke Akintola University of Technology, Ogbomosho, Oyo State, Nigeria*
[4] *Department of Biochemistry and Molecular Biology, Obafemi Awolowo University, Ile-Ife, Osun State, Nigeria*
[5] *Department of Medical Laboratory Science, Lead City University, Ibadan, Oyo State, Nigeria*
[6] *Department of Medical Laboratory Science, General Hospital, Offa, Kwara State, Nigeria*
[#] *First author position*
**Corresponding author position*

Fever is a multistep adaptive response to immunologic perturbations that creates an environment in which infectious pathogens or injured tissue are unable to survive due to the elevated body temperature. Infectious microorganisms and non-infectious agents such as neoplasia, trauma, and inflammatory conditions promote fever [1]. It is commonly defined as an elevation in body temperature beyond daily physiologic fluctuations due to a rise in thermoregulatory set point [2–4]. Fever is considered a surrogate marker for underlying disease processes rather than a disease in itself. To highlight the adaptive nature of febrile responses, the International Union of Physiological Sciences Commission for Thermal Physiology defined fever in 2001 as an elevated core temperature that is frequently, but not always, part of multicellular organisms' (host) defensive responses to invasive agents, whether live microorganisms or inanimate objects [5].

Owing to their potency to abate pains and inflammations, non-steroidal anti-inflammatory drugs (NSAIDs) are among the most favored therapeutics for treating fever, being on the World Health Organization's (WHO) Model List of Essential Medicines [6]. However, NSAIDs are quickly becoming infamous, despite their widespread therapeutic utility, as alarming reports from multiple placebo-controlled trials and meta-analytic studies demonstrate their adverse effects on multisystemic levels such as gastrointestinal, cardiovascular, hepatic, renal, cerebral, and pulmonary disorders [7–9]. These complications have been associated with activation of mitochondrial oxidative stress (MOS), a cytopathologic condition characterized by severe mitochondrial damage due to activation of redox-active chain reactions following inhibition of the mitochondrial electron transport chain (ETC) complex-I by NSAIDs [6]. Ultimately, mitochondrial stress results in oxidative damage to cellular biomolecules, including DNA, protein, lipids, and carbohydrates, triggering an apoptotic pathway.

Phytochemical Drug Discovery for Central Nervous System Disorders: Biochemistry and Therapeutic Effects, First Edition. Edited by Chukwuebuka Egbuna and Mithun Rudrapal.

As a result, clinicians are open to new agents devoid of side effects, which has increased the search for alternative therapies worldwide. Research into the efficacy of plant-based drugs used in the treatment of pain, fever, and inflammatory ailments has thus gained renewed attention because these drugs are cheap and have minimal or no side effects. Several publications have highlighted many plants that may offer antipyretic roles in animal models, but the exact mechanism for their actions and the isolation of individual compounds from some of those plants have not been elucidated. In this chapter, we discuss the mechanisms of fever development and the roles of secondary plant metabolites as antipyretics under their botanical origins and provide some important insights.

14.1 Pyrexia Development, Its Mechanisms, and the Roles of Plant Metabolites as Antipyretics

14.1.1 Fever Development

Pyrexia is the medical term for fever. Fever is defined as a core body temperature of ≥38.3 °C, according to the American College of Critical Care Medicine, the International Statistical Classification of Diseases, and the Infectious Diseases Society of America [10]. Fever arises as a symptom of various medical problems, including infection, cancer, and autoimmune diseases such as systemic lupus erythematosus, rheumatoid arthritis, and inflammatory bowel disease. Sometimes, fever is provoked by the side effects of medication, hormone imbalances, and seizures, among other factors. Pathophysiologically, fever can also be triggered by an immune reaction to foreign agents such as bacteria, fungi, viruses, poisons, and even medications in an attempt to eliminate the pyrogen. In other words, fever may be regarded as a desirable physiologic reaction rather than an illness, especially at specific temperatures.

Fever develops when the thermoregulatory set point (in the hypothalamus) changes as a result of a systemic inflammatory response to stimuli. Even though the molecular events leading to increased body temperature are not fully understood, current knowledge indicates that infection activates the innate immune system by implicating Toll-like receptor 4 (TLR-4), which promotes the release of pyrogenic cytokines such as interleukins (UL)-1 and -6 and tumor necrosis factor (TNF). These cytokines impact the supraoptic crest (organum vasculosum laminae terminalis, OVLT) to produce prostaglandin E2 (PGE2) by activating the cyclooxygenase (COX)-2 enzyme (Figure 14.1). By acting on the hypothalamus, PGE2 release is linked to a rise in the new temperature set point. Meanwhile, the mechanisms of fever are complex, sometimes involving both humoral and neurologic pathways in some cases.

14.1.2 Mechanisms of Fever Development

Fever signals are induced by exogenous and endogenous pyrogens, which ultimately reset the thermoregulatory networks via two fundamental pathways: humoral and neural.

14.1.2.1 Humoral Pathway

According to the classic model of fever pathogenesis, circulating endogenous pyrogenic cytokines, such as IL-1, IL-6, TNF-α, and interferon (IFN)-γ are secreted by macrophages in response to exogenous pyrogens such as viruses and bacteria toxins. These cytokines influence the thermoregulatory center through direct and indirect pathways. In the indirect

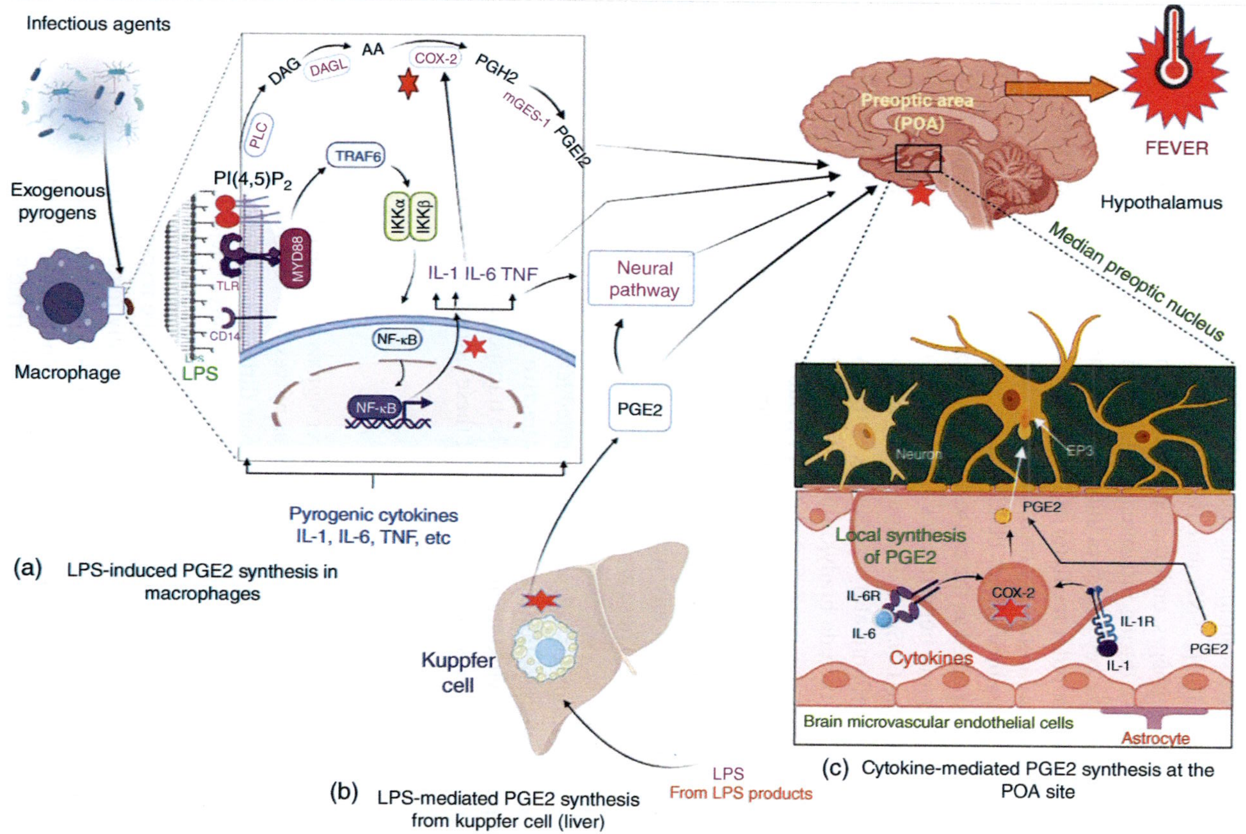

Figure 14.1 Potential pathogenetic mechanisms in the development of pyrexia. (a) Toll-like receptor 4 (TLR-4) recognizes lipopolysaccharides (LPS) produced by bacteria. TNF receptor-associated factor 6 (TRAF6) and IκB kinase (IKK) family kinases transmit signals to increase nuclear factor B (NF-B)-dependent gene expression, resulting in the production of pyrogenic tumor necrosis factor (TNF), interleukin (IL)1-, and IL-6. Phospholipase C (PLC) hydrolyzes phosphatidylinositol 4,5-bisphosphate (PI[4,5]P2) to diacylglycerol (DAG) and inositol triphosphate (IP3) as a result of TLR4 activation. Diacylglycerol lipase (DAGL) hydrolyzes DAG in the membrane, yielding free arachidonic acid (AA). Following that, AA is converted to prostaglandin H2 (PGH2) by cyclooxygenase (COX-2). PGH2 is then converted to prostaglandin E2 (PGE2), a key modulator of fever, by microsomal PGE2 synthase (mPGES-1). (b) Exogenous pyrogens induce the release of endogenous pyrogens in sentinel cells, which stimulate fever production in the preoptic region (POA) of the hypothalamus via the second messenger, PGE2. In response to stimulation from LPS, which also stimulates the POA via the vagus nerve, Kupffer cells in the liver produce PGE2. (c) A body stressor (such as infection, damage, event trauma) causes lymphocytes, macrophages, and other immune cells to release cytokines, which stimulates the hypothalamus. The vascular organ of lamina terminalis, or supraoptic crest (OVLT), within the hypothalamus stimulates cyclooxygenases, which catalyze the synthesis of prostaglandins (particularly PGE2). Furthermore, prostanoids (such as PGE2 and other arachidonate metabolites) may reach the POA via the bloodstream and influence the temperature set point. The red stars illustrate some potential targets for plant-derived secondary metabolites. Other possible pathways or molecular targets have received less attention.

pathway, pyrogenic cytokines act outside the brain by binding and activating cytokine receptors located on the fenestrated capillaries of the circumventricular organ, leading to the release of PGE2 [3, 4, 11]. In the direct pathway, circulating cytokines circumvent the blood–brain barrier and find their way to the brain, where they bind to the cytokine receptors located in the vascular, glial, and neuronal structures [11]. Activation of these receptors results in the central synthesis of PGE2 (Figure 14.1).

In both cases, the product of these pathways is PGE2, which diffuses into the adjacent preoptic area (POA) to turn up the temperature set point and promote fever development. It is worth mentioning that prostaglandin secretion from the arachidonic acid pathway requires COX-2, which is the main target of most antipyretic agents. Furthermore, COX-2 is inducible by inflammatory signals such as IL-1β, TNF, and IL-6, and bacterial lipopolysaccharide (LPS). Many cells, including macrophages, endothelial cells, and chondrocytes, can rapidly upregulate the expression of COX-2 during inflammation [12], affirming that suppressing inflammatory responses is one of the key targets of antipyretic therapy.

Even though PGE2 is the most popular mediator of febrile response, it is now clear that some cytokines and many other inflammatory mediators may activate the fever response independent of PGE2 [13]. For instance, direct PGE2-independent activation of thermal neurons by cytokines has been suspected as the cause of hyperpyrexia seen in infections of the central nervous system (CNS) and hemorrhages [3]. In these conditions, the physiologic (antipyretic) properties of the CNS are disrupted, leading to an unregulated rise in body temperature. Examples of inflammatory mediators that may reset the thermal balance point independent of PGE2 include bradykinin, corticotrophin-releasing hormone, nitric oxide, IL-6, IL-8, preformed pyrogenic factors (PFPF), substance P, and endothelin 1. It has also been proposed that scorpion venom may induce fever via a direct effect on the hypothalamus.

This understanding presents a unique opportunity to conceive effective antipyretic agents capable of interrupting pyrexogenesis at any step that connects peripheral inflammation with the central production of PGE2. Those agents might blunt peripheral inflammation, suppress central pyrogenic signals, or even both. Inhibiting the central production of PGE2 is the focus of the well-known antipyretic agents. Still, activated leukocytes and endothelial cells in peripheral areas of inflammation may also be attractive potential drug targets. Furthermore, the fever-promoting cytokines are believed to be transported from the blood to the brain via specific carriers [14], hence characterization of these carriers may be necessary to expand the scope of fever treatment.

14.1.2.2 Neural Pathway

This pathway relies on the communication of peripheral fever signals with the CNS through peripheral nerves, including the cutaneous sensory and vagus nerves. It is believed that activation of the neural pathway leads to the rapid onset of fever through local production of PGE2 at inflammation sites by activating cold-sensitive cutaneous nerves, which in turn transmit fever signals to the hypothalamus [15] (Figure 14.1). On the other hand, the role of vagal afferents in fever generation is based on the attenuation of febrile responses to pyrogenic signals in experimental studies in rats whose vagus is surgically removed. However, the current view is that lack of febrile response to pyrogenic signals is due to the side effects of vagotomy, such as malnutrition [16, 17]. This was evidenced when such side effects of vagotomy were removed in a rat model, and the vagotomy failed to abort the febrile response to pyrogenic signals such as intravenous PGE2 [16]. This implies that vagus nerves are not essential in the transmission of pyrogenic signals through the neural pathway.

14.1.3 Roles of Plant Metabolites as Antipyretics

During fever development, cytokine induction and the inflammatory response are beneficial in attacking foreign agents in the body. While the fever of sepsis is generally beneficial, there is growing evidence that the CNS is especially prone to injury from the associated hyperthermia. High temperature has a notably detrimental effect on the cerebellum and the tendency to cause cognitive and neurologic effects, which could ultimately lead to brain damage [18]. High temperatures have the potential to cause protein denaturing in brain cells. The most temperature-sensitive cellular features of brain cells are mitochondrial and plasma membranes, with irreversible changes in protein structure occurring at temperatures above 40 °C [19]. Although this process may be reversible, denatured proteins eventually form aggregates that interfere with normal cellular function, impede replication, and eventually cause cell death [19].

Meanwhile, non-pharmacologic therapies such as cooling by bathing or sponging with lukewarm or cool water can decrease body temperature in people suffering from heat exhaustion, but never in those suffering from certain fevers. According to the literature, the standard measure for thermal dose evaluation is cumulative equivalent minutes at 43 °C, which corresponds well with thermal injury in a variety of tissues [20]. While a fever cannot itself impact the brain negatively, the associated altered temperature set points above 42 °C may cause brain damage if maintained for a long time, while temperatures above 43 °C could result in death. As a result, it is rational to take immediate action to suppress any febrile rise in body temperature. Nonetheless, according to a study published by the Royal Society, fever suppression causes at least 1% more influenza mortality in the United States each year, as a consequence of 700 extra deaths [21]. Since fever is part of the immune response to infection, there is some controversy about when antipyretics should be used, most notably for clinical purposes [22, 23].

Moreover, the majority of antipyretic drugs are synthetic, and they cause stomach ulcers, liver necrosis, and kidney damage among other side effects. Herbal medications have a wide range of biologic and pharmacologic properties, as well as greater safety and lower costs. In addition, most antipyretics have anti-inflammatory and analgesic properties. The traditional use of plant bioactive compounds as febrifuges in fevers has a long history. A febrifuge is a term used in ethnobotany to designate a plant with antipyretic properties. Extensive phytochemical analyses of various plant bioactive compounds reveal the presence of alkaloids, glycosides, anthracene derivatives, flavonoids and proanthocyanidins, saponins, tannins, phenolics, triterpenoids, diterpenoids, mono- and sesquiterpenoids, phytosterols, and volatile oils (see Table 14.1). Many of these metabolites have been used in animal models to establish their antipyretic effects.

Antipyretic drugs have diverse modes of action, such as (i) inhibiting the enzyme COX; (ii) lowering PGE2 levels in the hypothalamus; (iii) reducing proinflammatory cytokines; (iv) augmenting anti-inflammatory responses at injury sites; and (v) elevating antipyretic signals in the brain [22]. However, plant metabolites are also associated with a similar variety of mechanisms, including (i) preventing LPS-induced PGE2 production by flavonoid compounds such as quercetin and kaempferol; (ii) inhibiting LPS-induced PGE2 production in macrophages by compounds such as isoquercetin, rutin, α-tocopherol, and myricetin; (iii) suppressing PGE2 synthesis by alkaloid, steroid, and flavonoid compounds; (iv) suppressing various pyrogens such as proinflammatory cytokines (IL-1β, IL-6, IFN-α, and TNF-α) by flavonoids; and (vi) reducing the levels of COX by constituents of black pepper (Figure 14.1). Furthermore, in non-classic thermoregulatory pathways, some metabolites act as antagonists against receptors that aid in the development of pyrexia, such as the transient receptor potential vanilloid-1 (TRPV1) receptor and cannabinoid receptor 1 (CB1). These metabolites through various botanical origins are discussed in greater detail later in this chapter.

Table 14.1 List of various plant secondary metabolites as antipyretics.

Plant name	Bioactive agents	Solvent extraction	Experimental model	Inducer of pyrexia	Routes of administration	Proposed mechanism of action	References
Acacia leucophloea	Flavonoids, glycosides, alkaloids	Distilled water	Swiss albino mice	15% saline suspension brewer's yeast	Subcutaneous	Blocks prostaglandin biosynthesis	[24]
Acacia mellifera (stem bark)	Alkaloids, cardiac glycosides, flavonoids, saponins, steroids, terpenoids, tannins, and phenolic compounds	1 : 1 Dichloromethane : methanol	Swiss albino mice	20% turpentine	Intraperitoneal	Inhibits prostaglandin synthesis	[25]
Albizia zygia	Triterpenoids, saponins	Hydroethanol	Sprague-Dawley rats	135 mg/kg brewer's yeast	Oral	Inhibits prostaglandin synthesis	[26]
Alstonia boonei (stem bark)	Alkaloids, steroids, cardiac glycosides, flavonoids, terpenoids, and saponins	Methanol	Mice	20 ml/kg of 20% aqueous yeast solution	Oral	Inhibits prostaglandin synthesis	[27]
Amaranthus spinosus (leaves)	Alkaloids, steroids, glycosides, flavonoids, phenolic compounds, and terpenoids	Methanol	Albino Wistar Rats	20 ml/kg of 20% aqueous yeast solution	Oral	Inhibition of prostaglandin synthesis by blocking cyclooxygenase (COX) enzyme activity	[28]
Anogeissus latifolia	Phytoconstituents	Ethanol	Rat	10 ml/kg of 20% aqueous yeast solution	Subcutaneous	Inhibition of prostaglandin synthesis	[29]
Anonna senegalensis (leaves)	Alkaloids, terpenoids, saponins, tannins, flavonoids	Methanol	Wistar rats	2,4-dinitrophenol (DNP)	Oral	Inhibits prostaglandin synthesis	[30]

Arbutus andrachne (leaves)	Linoleic acid, arbutin, hydroquinone, β-sitosterol, ursolic acid, quercetin, gallic acid, catechin gallate, and α-tocopherol	Methanol	Male Wistar rats	20 ml/kg 20% baker's yeast solution	Intraperitoneal	Inhibits prostaglandin synthesis by interacting with transient receptor potential vanilloid-1 (TRPV1) and cannabinoid receptor 1 (CR1)	[31]
Argemone Mexicana	Alkaloids	Methanol	Mice	20 ml/kg 20% baker's yeast solution	Subcutaneous	Inhibits COX-1 and -2 enzymes	[32]
Azima tetracantha (leaves)	Phytoconstituents	Ethanol	Albino rats	10 ml/kg 10% aqueous brewer's yeast solution	Oral	Inhibits prostaglandin synthesis	[33]
Bacopa monniera	Phytoconstituents	Ethanol	Rat	10 ml/kg yeast solution	Subcutaneous	Inhibits prostaglandin synthesis	[34]
Bauhinia racemosa	Flavanols, coumarins, triterpenoids, steroids, and tannins	Methanol	Wistar albino rats	10 ml/kg 15% aqueous brewer's yeast solution	Oral	Inhibits prostaglandin synthesis	[35]
Bauhinia purpurea (leaves)	Flavonoids, saponins, steroids, and triterpenes	Water	BALB/C mice	10 ml/kg 10% aqueous brewer's yeast solution	Oral	Inhibits prostaglandin synthesis	[36]
Cadaba trifoliata (leaves)	Alkaloids, tannins, lactones, steroids, and flavonoids	Ethanol	Albino Wistar rats	10 ml/kg of 15% w/v brewer's yeast	Oral	Inhibits prostaglandin synthesis	[37]
Calotropis gigantea (aerial part)	Isorhamnetin-3-O-rutinoside, taraxasterol acetate, and flavanol trisaccharide	Water : ethanol (1 : 1)	Albino Swiss rats	Brewer's yeast and TAB (typhoid) vaccine	Intraperitoneal	Inhibits prostaglandin synthesis	[38]

(Continued)

Table 14.1 (Continued)

Plant name	Bioactive agents	Solvent extraction	Experimental model	Inducer of pyrexia	Routes of administration	Proposed mechanism of action	References
Capparis zeylanica	Flavonoids	Methanol	Wistar albino rats	15% suspension of brewer's yeast in 0.5% w/v methylcellulose solution	Oral	Blocks prostaglandin synthesis	[39]
Capparis zeylanica (leaves)	Flavonoids	Ethanol	Wistar rats	20% aqueous suspension of brewer's yeast (10 ml/kg subcutaneously)	Oral	Inhibits prostaglandin synthesis	[40]
Carissa edulis (leaf and root bark)	Steroids, saponins, phenolics, alkaloids, flavonoids, and terpenoids	1 : 1 Dichloromethane : methanol	Wistar albino rats	20 ml/kg 20% turpentine solution	Intraperitoneal	Inhibits COX-2, thereby reducing prostaglandin synthesis	[41]
Castanopsis tribuloides	Quercetins, flavonoids	Methanol	Swiss albino mice	10 ml/kg 30% brewer's yeast solution	Oral	Inhibits COX-1/ COX-2, leading to reduction in prostaglandin synthesis	[42]
Cissum assamica	Steroids, saponins, alkaloids, flavonoids, and triterpenes	Methanol	Swiss albino mice	10 ml/kg 30% brewer's yeast solution	Oral	Inhibits prostaglandin synthesis	[43]
Cleome rutidosperma (aerial part)	Terpenoids, steroids, flavonoids, and tannins	Ethanol	Rabbits	3 ml/kg of 10% yeast suspension	Subcutaneous	Inhibits prostaglandin synthesis	[44]
Clerodendron serratum (root)	Saponins, flavonoids, and phenolics	Ethanol	Rabbits	0.1 ml TAB vaccine	Oral	Inhibits prostaglandin synthesis	[45]

Clitoria ternatea Linn (root)	Flavonoids, tannins, steroids, and saponins	Methanol	Wistar rats	10 ml/kg body weight of 15% w/v yeast solution	Oral	Inhibits prostaglandin synthesis	[46]
Codiaeum variegatum (leaves)	Alkaloids flavonoids, saponins, phenolics, tannins, and terpenoids	Ethanol	Mice	DPT-HB-Hib vaccine (0.05 ml/20 g)	Oral	Inhibits prostaglandin synthesis	[47]
Colocasia esculenta (leaves)	Alkaloids, saponins, and tannins	Hydro-alcohol	Wistar albino rats	15% w/v of brewer's yeast suspension (10 mg/kg)	Oral	Inhibits prostaglandin synthesis	[48]
Corchuros capsuaris (leaves)	Flavonoids, triterpenes, saponins, and steroids	Water	Sprague-Dawley rats	10 ml/kg 10% aqueous brewer's yeast solution	Oral	Inhibits prostaglandin synthesis	[49]
Costus speciosus (rhizome)	Alkaloids, saponins, carbohydrates, and phytosterols	Ethanol	Wistar albino rats	10 ml/kg 12% yeast solution	Oral	Inhibits prostaglandin synthesis	[50]
Crataeva magna (bark)	Terpenoids, flavonoids, and quercetin	Ethanol	Rabbit	0.5 ml TAB vaccine	Oral	Inhibits prostaglandin synthesis	[51]
Curcuma zedoaria (rhizome)	Tannins, flavonoids, saponins, alkaloids, and terpenoids	Ethanol	Mice	20 ml/kg of 20% solution	Oral	Inhibits prostaglandin synthesis	[52]
Dicranopteris linearis (leaves)	Flavonoids	Water	Sprague-Dawley rats	10 ml/kg 10% aqueous brewer's yeast solution	Subcutaneous	Inhibits prostaglandin synthesis	[53]
Diospyros lotus L. (roots)	Naphthoquinones, naphthalene	Ethyl acetate	BALB/C male and female mice	20% aqueous suspension of brewer's yeast (10 ml/kg)	Intraperitoneal	Inhibits prostaglandin synthesis	[54]
Emblica officinalis	Alkaloids	Methanol	Wistar Albino rats	5 ml/kg yeast	Subcutaneous	Inhibits prostaglandin synthesis	[55]

(*Continued*)

Table 14.1 (Continued)

Plant name	Bioactive agents	Solvent extraction	Experimental model	Inducer of pyrexia	Routes of administration	Proposed mechanism of action	References
Feronia limonia	Alkaloids	Methanol	Wistar Albino rats	1 ml 20% brewer's yeast	Intraperitoneal	Inhibits prostaglandin	[56]
Fumaria Parviflora	Flavonoids and phenolic compounds	Ethyl acetate fraction (at dose of 162 mg/kg)	Wistar albino rats	1 ml/100 g brewer's yeast	Subcutaneous	Not investigated	[57]
Gardenia jasminoides Ellis (fruit)	Gardoside, geniposide, p-coumaroylgenipin gentiobioside, crocins I and II	Ethanol	Male Sprague-Dawley rats	15% saline suspension of yeast (10 ml/kg)	Dorsal	Suppression of pyrogenic cytokines (interleukin 1 and tumor necrosis factor 6)	[58]
Gmelina arborea	Alkaloids	Ethanol	Wistar rats	10 ml/kg 20% brewer's yeast solution	Subcutaneous	Inhibits prostaglandin synthesis by blocking COX enzyme activity	[59]
Hedydium Spicatum	Tannins, triterpenoids, flavonoids	Methanol	Wistar rat	Brewer's yeast	Subcutaneous	Inhibits synthesis of inflammatory mediators	[60]
Lens culinaris	Triterpenoids	Ethanol	Mice	Saponins	Subcutaneous	Inhibits biosynthesis of inflammatory mediators	[61]
Litchi chinensis (leaves)	Saponins	Petroleum ether	Sprague-Dawley rats	1 ml 20% brewer's yeast solution	Intraperitoneal	Inhibits prostaglandin synthesis	[62]
Lygodium microphyllum (leaves)	Flavonoids, phenolic glycosides, quercetin, stigma-5(6)-en-3β-ol or β-sitosterol, stigmast-4-en-3	Methanol	Mice	10 ml/kg 20% aqueous brewer's yeast solution	Oral	Suppresses COX-2 and enhances body's antipyretic substances such as arginine and vasopressin	[63]

Madhuca indica	Flavonoids, steroids, and triterpenes	Methanol	Albino male Wistar rats	Yeast	Oral	Inhibition of production of prostaglandin	[64]
Mangifera indica (stem bark)	Saponins, flavonoids, alkaloids, cardiac glycosides, terpenoids, phenols, coumarins, and steroids	Methanol	Rats	20% distilled turpentine	Intraperitoneal	Inhibits both prostaglandin synthase and/or COX-2 enzyme	[65]
Marsilea trifolia	Flavonoids	Aqueous extract	Male Sprague-Dawley rats	10 ml/kg 20% aqueous suspension of brewer's yeast	Oral	Inhibits prostaglandin synthesis	[66]
Melastoma malabathricum (leaves)	Flavonoids, triterpenes, tannins, saponins, and steroids	Water	BALB/C Mice	10 ml/kg 10% aqueous brewer's yeast solution	Oral	Inhibits prostaglandin synthesis	[67]
Mimosa Pudica	Alkanoid	Methanol	Mice	10 µl/g brewer's yeast extract	Subcutaneous	Inhibits synthesis cf inflammatory mediators	[68]
Mollugo pentaphylla (aerial parts)	Alkanoid, flavonoids, saponins, terpenoids, glycosides, and tannins	Ethanol	Swiss albino rats	2 ml/kg 20% aqueous brewer's yeast solution	Oral	Inhibits prostaglandin synthesis	[69]
Morinda lucida (leaves)	Steroids, phenols, flavonoids, glycosides, and tannins	Water	Rabbits	0.1 ml TAB vaccine	Oral	Inhibits prostaglandin synthesis	[70]
Muntingia calabura (leaves)	Flavonoids, saponins, tannins, and triterpenes	Water	Sprague-Dawley rats	10 ml/kg 10% aqueous brewer's yeast solution	Oral	Inhibits central prostaglandin synthesis	[71]
Nelumbo nucifera (seeds)	Possibly steroids, alkaloids, saponins, flavonoids, and terpenoids	Alcohol	Rats	20% brewer's yeast solution	Oral	Inhibits prostaglandin synthesis	[72]

(Continued)

Table 14.1 (Continued)

Plant name	Bioactive agents	Solvent extraction	Experimental model	Inducer of pyrexia	Routes of administration	Proposed mechanism of action	References
Ocimum suave and *Ocimum lamiifolium* (leaves)	Phenolic acids, flavonoids, glycosides, and sugars	Water and ethanol	Mice	30% (w/v) suspension of yeast in 0.9% NaCl in dose of 10 ml/kg subcutaneously	Oral	Inhibition of prostaglandin synthesis	[73]
Oryza barthii	Alkaloids	Ethanol	Rats	Brewer's yeast	Subcutaneous	Inhibits the synthesis of inflammatory mediators	[74]
Phrygilanthus acutifolius (flowers)	Flavonoids	Aqueous and ethanol extraction	Wistar rats	10 ml/kg 20% aqueous suspension of brewer's yeast	Oral	Inhibits prostaglandin synthesis	[75]
Piper interruptum Opiz (stem) and *Piper chaba* Linn. (fruit)	Flavonoids, phenols	Ethanol	Sprague-Dawley rats	10 ml/kg 20% brewer's yeast solution	Oral	Inhibits synthesis and release of local prostaglandins into preoptic area of anterior hypothalamus	[76]
Piper nigrum	Cepharadione A, paprazine, piperine, sylvamide, paracetamol, pellitorine, piperolactam D, and tricosanone	Methanol	Swiss albino mice	10 ml/kg brewer's yeast extract	Oral	Cepharadione binding affinity with prostaglandin E synthase-1 (4YK5) proteins	[77]
Plumeria rubra (leaves)	Tannins, saponins, flavonoids, and alkaloids	Ethanol	Rabbits	0.5 ml/kg boiled cow's milk	Intraperitoneal	Inhibits prostaglandin synthesis	[78]
Prosopis cineraria (leaves and fruits)	Saponins, tannins, flavonoids, alkaloids, steroids, and cardiac glycosides	Ethanol	Albino rats	Intraperitoneal 20% w/v brewer's yeast at dose of 10 ml/kg	Oral	Inhibits prostaglandin synthesis	[79]

Pseudocedrela kotschyi (leave)	Flavonoids, tannins, saponins, alkaloids, glycosides, anthraquinones, and terpenoids	Ethanol	Wistar albino rats	20 ml/kg 15% yeast solution	Oral	Inhibits prostaglandin synthesis	[80]
Pterocarpus santalinus L.	Steroids, anthocyanins, triterpenoids, flavonoids, saponins, glycosides, and glycerides	Methanol	White albino rats	10 ml/kg 30% brewer's yeast solution	Oral	Inhibits prostaglandin synthesis	[81]
Putranjiva roxburghii (leaves)	Triterpenes and flavonoids	Ethanol	Wistar rats	10 ml/kg 20% aqueous brewer's yeast solution	Oral	Inhibits prostaglandin synthesis	[82]
Schoenoplectus grossus (root)	Alkaloids, flavonoids, and tannins	Methanol	Albino Wistar rats	10 ml/kg 15% brewer's yeast solution	Oral	Inhibits prostaglandin E2 via eicosanoid pathway	[83]
Solanum incanum (root)	Flavonoids, chlorogenics, alkaloids	1 : 1 Dichloromethane : methanol	Sprague-Dawley rats	50 μg lipopolysaccharide (LPS)	Intraperitoneal	Inhibits prostaglandin biosynthesis	[84]
Solanum nigrum (leaves)	Flavonoids, saponins, and triterpenes	Water	Sprague-Dawley rats	10 ml/kg 10% aqueous brewer's yeast solution	Oral	Inhibits central COX	[85]
Sonchus oleraceus (leaves)	Alkaloids, flavonoids, steroids, tannins, anthraquinones, and terpenes	Hydroethanol	Wistar rats	50 μg/kg LPS	Oral	Inhibits prostaglandin synthesis	[86]
Taxus wallichiana (leaves)	Alkaloids, phenols, polyphenols, saponins, tannins, anthraquinones, and steroids	Methanol	Wistar rats	10 ml/kg 15% brewer's yeast solution	Intraperitoneal	Inhibits prostaglandin synthesis	[87]
Teclea nobilis (leaves)	Furoquinoline alkaloids: tecleabine, tecleoxine, isotecleoxine, and chlorodesnkolbisine	Ethanol	Male albino Wistar-Nossan rats	10 mg/kg 30% w/v yeast	Oral	Inhibits prostaglandin synthesis	[88, 89]

(Continued)

Table 14.1 (Continued)

Plant name	Bioactive agents	Solvent extraction	Experimental model	Inducer of pyrexia	Routes of administration	Proposed mechanism of action	References
Tecomaria capensis	Alkaloids	Methanol	Rats	10 ml/kg brewer's yeast solution	Subcutaneous	Inhibits histamine release and prostaglandin E2	[90]
Tectona grandis (leaves)	Phytoconstituents	Methanol	Wistar albino rats	Yeast suspension (500 mg/kg)	Oral	Inhibits prostaglandin synthesis	[91]
Vernonia cinerea	Flavonoids and terpenoids	Methanol, chloroform	Adult albino rats, Wistar albino rats	10 ml/kg 15% yeast solution	Oral, intraperitoneal	Inhibits prostaglandin synthesis	[92, 93]
Viola betonicifolia b (whole plant)	Alkaloids, saponins, flavonoids, tannins, and phenols	Methanolic extract of whole plant	BALB/C mice	10 ml/kg subcutaneous 20% aqueous suspension of brewer's yeast	Intraperitoneal	Inhibits prostaglandin synthesis	[94]
Echinops kebericho (root)	Flavonoids	Methanol	Swiss albino mice	20% suspension of brewer's yeast	Oral	Inhibits COX-2	[95]
Ximenia americana (leaves and stem bark)	Alkaloids, flavonoids, saponins, terpenoids, and steroids	1:1 Dichloromethane : methanol	Wistar albino rats	Turpentine	Intraperitoneal	Inhibits prostaglandin synthesis	[96]
Ziziphus jujube	Alkaloids, tannins, flavonoids, and phenolic compounds	Methanol	Wistar albino rats	20% suspension of brewer's yeast	Oral	Inhibits prostaglandin synthesis	[97]
Ziziphus nummularia (root)	Alkanoid	Methanol	Mice	1 ml 20% brewer's yeast	Intraperitoneal	Inhibits prostaglandin	[98]
Zanthoxylum armatum	Alkaloids	Methanol	Mouse	10 ml/kg 15% yeast suspension	Oral	Inhibits COX-2	[99]

14.2 Antipyretic Agents of Plant Origin

Plants are responsible for over 25% of all medications prescribed worldwide, with 121 active chemicals already in use. Of the 252 medications designated basic and essential by the WHO, 11% are 100% plant based, while a large percentage are synthetic drugs developed from natural precursors [100]. Salicin, an analgesic and antipyretic molecule isolated from willow (*Salix* sp.) and utilized as a model for the synthesis of acetylsalicylic acid, is a well-known example of secondary plant metabolites with pharmacologic activity (in aspirin). Also, artemisinin from *Artemisia annua* L., atropine from *Atropa belladonna*, digoxin from *Digitalis* sp., quinine and quinidine from *Cinchona* sp., morphine and codeine from *Papaver somniferum*, and vincristine and vinblastine from *Catharanthus roseus* are just a few other examples of important drugs obtained from plants [101–103].

Concerning the antipyretic effect exhibited by some plant products, many bioactive compounds have been reported in the literature. For instance, the reversal of yeast-evoked pyrexia was demonstrated with virgin coconut oil containing lauric, myristic, and linoleic acids as active ingredients [104]. Additionally, Fawad et al. reported that two derivatives of hydroquinone significantly reduced febrile response in rats [105]. Again, linalool, rutin, ursolic acid, β-sitosterol, isoquercetin, quercetin, gallocatechin, kaempferol, and gallic acid have also been identified in many plant extracts that demonstrated potent antipyretic effects in previous studies [106–110]. In the same vein, alkaloids reduce body temperature, but they cause gastric ulcers, which suggests that they have the same mechanism as NSAIDs. Alkaloids in this case are one of the active constituents that have been reported to lower temperature [111].

Furthermore, Hämäläinen and colleagues [112] reported that several flavonoids (e.g. quercetin and kaempferol) prevented LPS-induced PGE2 production. Similar findings were reported for the effect of other compounds, including isoquercetin, rutin, and myricetin, on LPS-induced PGE2 production in macrophages [113]. Similarly, treating macrophages with α-tocopherol has been demonstrated to reduce PGE2 levels for a long time [114], and also the administration of epigallocatechin-3-gallate has been shown to attenuate the levels of PGE2 induced by photofrin-mediated photodynamic therapy [115]. While the extraction methods are a determining factor in the isolation of a specific compound, components of plant metabolites differ from plant species to species. As a result, these agents are methodically described under selected botanical origins (with their main phytoconstituents listed in Figure 14.2) as follows.

14.2.1 *Arbutus andrachne* (Family Ericaceae)

Arbutus andrachne, otherwise known as the Greek strawberry tree, is an evergreen shrub or small tree native to the Mediterranean region and the Middle East. One report claims it is found in Jordan [116]. The plant can grow up to 12 m tall in its native area, but grows much smaller in cultivation. The smooth red bark on the trunk peels off once in a while. The plant is harvested in the wild for use as a source of food and a material resource in the local community. It is frequently grown as an ornamental, regarded for its attractive leaves, as well as its flower and fruit displays. *A. andrachne* thrives in long, hot, dry summers and mild, damp winters. It can withstand temperatures as low as −15 °C. The plant is edible and has been used in traditional medicine and the stain industry [117].

Generally, the leaves and fruits of *A. andrachne* are consumed as laxatives, blood tonics, diuretics, and antiseptics [117]. Most importantly, the extract showed strong antinociceptive [31]

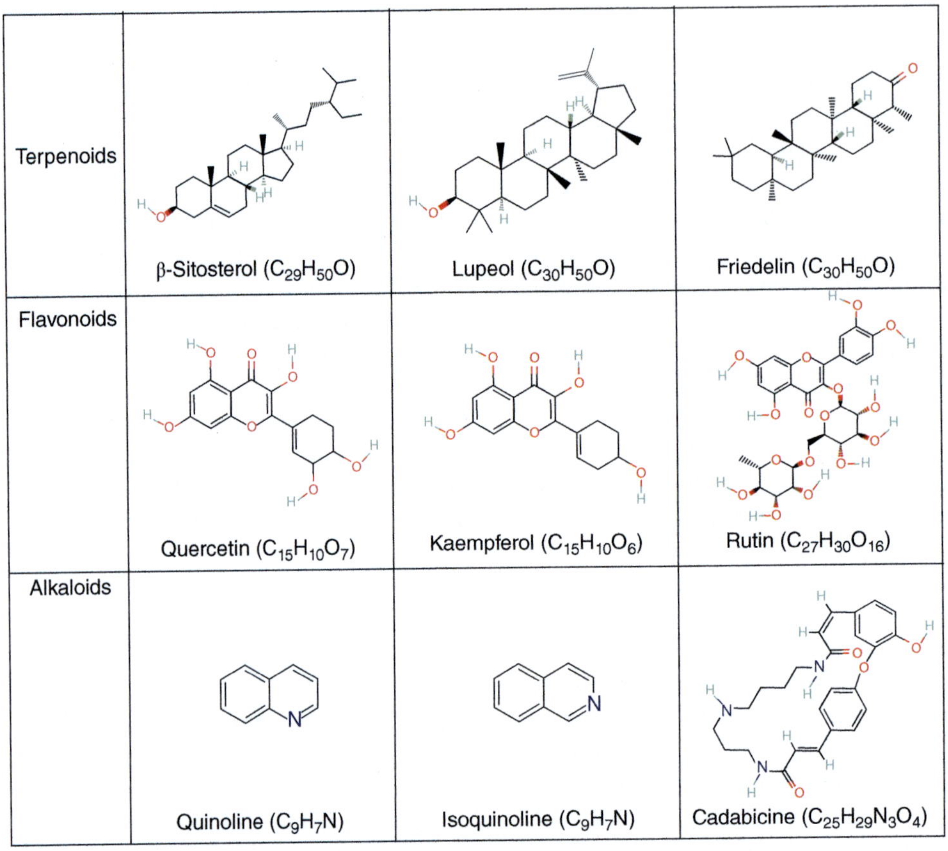

Figure 14.2 Some common plant compounds with potential antipyretic activities.

and antipyretic activities [118]. Several biologic studies were conducted to explore the effects of *A. andrachne* in different models. Jaffal et al. [31] used liquid chromatography–mass spectrometry (LC–MS) to identify the bioactive compounds responsible for these biologic effects. Several phytoconstituents such as linalool, lauric acid, myristic acid, linoleic acid, rutin, arbutin, hydroquinone, β-sitosterol, ursolic acid, isoquercitrin, myricetin-3-O-α-L-rhamnosides, quercetin, gallic acid, (+)-gallocatechin, kaempferol, catechin gallate, and α-tocopherol were isolated. Tannins can also be found in the leaves [119].

Tannic acid (25 mg/kg) has been demonstrated to be an effective antipyretic in an animal model [120]. In a separate report, Jaffal et al. showed that *A. andrachne* extract decreased yeast-induced pyrexia after 1–4 hours of injections, with the maximum antipyretic effect being achieved by 300 mg/kg body weight (BW) of the extract [118]. This was demonstrated by a dramatic decrease in PGE2 levels in the animal's serum. The study went a step further by assessing the involvement of certain receptors in the animals' response to the plant extract. The TRPV1 and CB1 antagonists were administered before 300 mg/kg BW *A. andrachne* and were able to overturn the antipyretic effect of the plant extract. This suggests that both TRPV1 and CB1 receptors are essential for the plant to mediate its antipyretic roles. Surprisingly, much is unknown about the roles of these receptors in the classic fever pathways. Although TRPV1 was earlier associated with many processes, including thermoregulation in inflammatory conditions [121], it is often not mentioned in the classic thermoregulatory pathways.

Furthermore, CB1 has been implicated in inflammation and pyrexia [122]. Notably, it was reported that the endocannabinoid arachidonoyl ethanolamine enhances LPS-induced fever by activation of the CB1 receptor [122]. Collectively, this study provides clear evidence that TRPV1A and CB1 are required for optimal antipyretic action of *A. andrachne* extract. Further work is needed to investigate the exact component that mediates this interaction.

14.2.2 *Berberis* spp. (Family Berberidaceae)

Berberis, also known as barberry, is a large genus of evergreen and deciduous trees found in many parts of temperate and subtropical regions globally. This genus includes spiny deciduous evergreen shrubs featuring yellow wood and blooms [123]. Various species are more prevalent in Asia and South America; other continents such as Africa, Europe, and North America also have a large number of native species. Berberis has been employed in traditional and folk medicine for its numerous pharmacologic and biologic activities, which make it an effective therapy for a variety of ailments due to the presence of quinoline and isoquinoline alkaloids, as well as sterols and triterpenes.

In 2014, a comprehensive list of phytochemical and pharmacologic compounds was compiled and published, and suggested that the major natural phytochemicals are responsible for a wide range of therapeutic qualities [123]. Fourier transform infrared spectroscopy (FITR)-spectral analysis established that berberine is the major constituent of berberis in addition to a few other chemical compounds [124]. Chemical analysis revealed the presence of other alkaloids, amino acids, tannins, terpenes, resins, phenols, and reducing sugars as major compounds in the plant genus. Meanwhile, berberine is an isoquinoline alkaloid found in the root, rhizome, and stem bark of a variety of plants, including *Beta vulgaris*. In *B. vulgaris*, the literature shows that berberine and berbamine are the major constituents responsible for its antipyretic activity [123]. Many other Berberis species, such as *B. lyceum*, *B. koreana* Palib., *B. calliobotrys*, and *B. crataegina*, have been studied for antipyretic properties. Alamgeer et al. [125] show that aqueous methanolic extract of *B. calliobotrys* also has antipyretic efficacy against yeast-induced pyrexia, particularly at a 500 mg/kg dose. Berberine, berbamine, palmatine, oxyacanthine, magnoflorine, and columbamine characterized from *B. crataegina* were earlier found to have significant antipyretic effects in a rodent model [111]. Even though the majority of studies on Berberis plants have failed to pinpoint their mechanism of action, the plant has a long history of antipyretic effectiveness around the world.

14.2.3 *Cassia fistula* Linn. (Family Caesalpiniaceae)

Cassia fistula is grown throughout the world in the tropical regions, from the West Indies to India, and from South Africa to Egypt [126]. It is commonly known as Amaltas in Urdu, Indian laburnum in English, and Khiyar Shambar in Arabic. *C. fistula* is a fast-growing medium-sized plant that can reach a height of 10–20 m. The flowers are produced in pendulous racemes that are 20–40 cm long and have deciduous leaves that are 15–60 cm long. The fruit is a legume with a pungent odor and numerous seeds, about 30–60 cm in length and 1.5–2.5 cm in width. The tree produces strong, long-lasting wood that is used for site construction.

Studies show that the active agents in pods are fistulic acid, flavone-3-O-arabinopyranoside, 3-formyl-1-hydroxy-8-methoxyanthraquinone, and catechins [127, 128], while the bark contains active agents such as oxyanthraquinone, flavanol glycosides, dihydroxyanthraquinone, leucocyanidin, and fistucacidin [128, 129]. Other compounds such as lupeol, β-sitosterol, hexacosanol [130], kaempferol, rhein, fistulin, triterpenes [131], and 1,8-dihydroxy-3-anthraquinone derivative [132] have been reported many decades ago in the stem bark, flowers, and fruit pulp, respectively.

The plant has several medicinal uses. It is used as an effective antibacterial to aid in wound healing and to help the body control hormonal imbalances. According to a report from Nigeria, *C. fistula* has laxative properties and is regarded safe for stimulating bowel evacuation and defecation as well as relieving constipation [133]. There have been studies into hepatoprotective, anti-inflammatory, antitussive, antifungal, antipyretic, and other activities [134]. Of most importance are the antipyretic activities of ethanolic extract of *C. fistula* that were studied in experimental rats [135]. The findings show that extract at 250 and 500 mg/kg BW lowered TAB (typhoid) vaccine-induced pyrexia after one hour, while extract at 750 mg/kg BW diminished the body temperature after half an hour.

The antipyretic properties of a methanol extract of *C. fistula* buds were also tested on rats in two experimental groups for their effects on body temperature and pyrexia produced by yeast. The extract, at a dose of 200–400 mg/kg BW, was given to the two groups. Their findings are consistent with those obtained with paracetamol, a common antipyretic drug with dose-dependent antipyretic effects [136]. In a separate report, Singh et al. [137] investigated the antipyretic activities of *C. fistula* pods. The antipyretic activities produced using a methanol extract were much higher than those of the untreated groups, according to their findings. Despite evidence that *C. fistula* has antipyretic properties, there has been little renewed interest in the plant species in recent times, and the compounds are rarely characterized in the old literature.

14.2.4 *Crataeva magma* (Family Capparidaceae)

Crataeva magma is synonymous with *Nurvala crataeva* Buch. Ham., which is popularly called the three-leaved caper. *C. magna* is also known as *C. roxburghii or C. religiosa*. The name Crataeva was given in honor of Crataevus, a Greek botanist who lived around Hippocrates' time, while the epithet "religiosa" refers to its growth near religious sites [138]. *C. magna* is a medium-sized deciduous tree with a densely branching crown (about 30 m tall). Despite its origins in India, China, Bangladesh, and Burma, it is now grown widely in many countries around the world, including those in Africa, South America, and South Asia. The leaflets are ovate, whole, lanceolate/obovate, acute/or acuminate, and reticulately veined, with petioles (3.8–7.6 cm) on three-foliated leaves. Flowers can range in shade from light cream to pale yellow to reddish-yellow. Reniform seeds are implanted in the yellow and succulent pulp of the fruit in multiples [100].

Alkaloids, anthracene derivatives, flavonoids, phenolics, phytosterols, saponins, tannins, triterpenoids, and volatile oils have all been discovered by exhaustive phytochemical studies [119, 139]. The most pharmacologically active phytoconstituents isolated from *C. magna* are lupeol, quercetin, catechin, β-sitosterol, betulinic acid, diosgenin, kaempferol, and friedelin, among others [100]. Meanwhile, diverse metabolite compounds can be found in the leaves, barks (stem and root), as well as the fruits of *C. magna*. Isoquercetin, quercetin-3-O-D-glucoside, rutin, kaempferol-O-D-glucoside, methyl pentacosanoate, L-stachydrine, and dodecanoic anhydride have been discovered in the leaves [140, 141], as well as glucocapparin, friedelin triacontanol, β-sitosterol, triacontane, octanamide, pentadecane, and 12-tricosanone in the fruits (reviewed in [100]). The active compounds found in the stem bark include alkaloids such as cadabicine and its esters (cadabicine diacetate and cadabicine mether), as well as betulinic acid, cetyl alcohol, diosgenin, friedelin, gluco-capparin, lupeol, and β-sitosterol. Also, the root contains β-sitosterol, lupeol, rutin, quercetin, and other compounds including ϑ-taraxasterol and varunol (reviewed by [100]).

To test for antipyretic efficacy, Chidambaram et al. [142] induced rabbits with TAB vaccine-induced pyrexia followed by alcoholic extracts of the bark of *C. magna,* without isolating or characterizing the phytocompounds. When compared to paracetamol (100 mg/kg orally) at 200 and 400 mg/kg, there was a drop in temperature. The active molecules responsible for these pharmacologic effects were not established; they were thought to be terpenoids and flavonoids, involving betulinic acid, lupeol, 4-taraxasterol, rutin, and quercetin, which are known to be the main chemical components of *C. magna*. Given that prostaglandin is a body temperature regulator, this result appears to back up the theory that the plant has some influence on prostaglandin biosynthesis. On the other hand, Geetha and Varalakshmi [143] used oral administration (25 mg and 50 mg/kg, respectively) to investigate the antipyretic activity of two pentacyclic triterpenes, lupeol and its ester lupeol linoleate, even though lupeol is known to be the major active agent. According to the findings, lupeol and lupeol linoleate have no antipyretic effect individually.

Until now, the bulk of investigations on the plant genus have failed to distinguish the exclusive impact of each phytochemical compound from its combinatorial effects, suggesting that their potency may not always account for the reported efficacy individually rather than in combination. Furthermore, there are insufficient data to determine the mechanistic effects of the established compounds.

14.2.5 *Echinops kebericho* M. (Family Asteraceae)

Echinops kebericho M. is one of the most popular traditional herbs in Ethiopian folk medicine due to its antipyretic properties and potential benefit in treating a variety of ailments, including infectious and non-infectious diseases [95]. According to the literature, the aqueous root extract of *E. kebericho* contains phenols, tannins, flavonoids, terpenoids, and alkaloids [144]. In previous studies, flavonoids, alkaloids, tannins, resins, saponins, and terpenoids were all discovered [145–147].

The presence of several phytoconstituents from the methanol root extract of *E. kebericho* was determined using the qualitative test for secondary metabolites. Saponins, alkaloids, phenols, tannins, flavonoids, glycosides, and steroids were confirmed, which gives insight into the possible mechanisms [148]. The authors observed that the crude extract and aqueous fraction significantly lower the temperature of pyretic mice dose dependently, as evidenced by the finding that the average percentage of antipyretic activity increased with the concentration of the extracts and their solvent fractions (400 mg/kg) against the controls. The presence of active phytoconstituents such as flavonoids suggests the antipyretic actions are by suppressing various pyrogens such as proinflammatory cytokines (IL-1β and IL-6, IFN-α and TNF-α), and ultimately inhibiting the release of PGs including PGE2 and PGI2 [148].

The authors show in their subsequent report that the aqueous fraction is most efficacious as the antipyretic effect, implying that the extract's active phytoconstituents responsible for temperature reduction are mostly hydrophilic and soluble in more polar solvents such as distilled water. The antipyretic activity of the ethyl acetate fraction, on the other hand, was much lower than that of the aqueous fraction and significantly higher than that of the chloroform fraction [95]. Despite growing interest in the study of plant species, the literature suggests that different extraction procedures produce distinct secondary metabolites, which could explain why many studies have varying degrees of efficacy. As a result, more research is required to determine the best method of phytochemical extraction for a specific species and its parts.

14.2.6 *Enicostema littorale* (Family Gentianaceae)

Enicostemma littorale Blume is a (Gentian) perennial herb with several local names such as Nagajivha in Bengali and Chota chirayata in Hindi, depending on its origin. Generally, the English name is white head and it is predominantly found in South America, Africa, and Asia. *E. littorale* thrives in a wide range of environments, including savannas, grasslands, woodlands, and coastlines, and can survive in both wet and dry conditions. The plant is a perennial erect herb that grows to a height of 5–30 cm and is simple or branched at the base. The stems are glabrous and cylindric, with a decurrent ridge below the leaf [149].

Several secondary metabolites in this genus have been discovered in the past, but no clinical trials are currently being conducted to investigate their effects. Extensive research on *Enicostemma* spp. over the years reveals that various compounds from this genus have been isolated using different species and various extraction methods. Catechins, saponins, steroids, sapogenin, triterpenoids, flavonoids, xanthones, and a novel flavone C-glucoside obtained from the ethyl acetate soluble fraction of a distinct species, *E. verticillatum*, have all been isolated earlier [150]. Also, using an alcoholic extract, the swertiamarin compound was identified from *E. littorale* [151]. In 1966, Desai et al. [152] discovered six different phenolic acids, including vanillic acid, syringic acid, p-hydroxybenzoic acid, protocatechuic acid, p-coumaric acid, and ferulic acid.

More importantly, *E. littorale* has been used as a traditional medicine for its anti-inflammatory, hypoglycemic, antimalarial, and anticancer properties, among many others [153, 154]. Interestingly, an aqueous extract of *E. littorale* (aerial parts) displays an antipyretic effect, which could be attributed to the presence of alkaloids, steroids, and flavonoids linked to the suppression of PGE2 synthesis [155]. Although there are several reports on the medicinal uses of *E. littorale* including antioxidant, anti-inflammatory, antimalarial, anthelminthic, hepatoprotective, hypoglycemic, hypolipidemic, and anticancer properties, reports on its antipyretic use are still relatively sparse.

14.2.7 *Piper nigrum* (Family Piperaceae)

Piper nigrum L., often known as black pepper, has a wide range of applications. *P. nigrum* is grown for the fruit, which is known as a peppercorn and is dried and used as a spice and condiment. The fruit is a drupe (stonefruit) with a stone that encloses a single pepper seed and is around 5 mm in diameter when fresh and completely ripe [77]. Black pepper is believed to cure a variety of ailments, including digestive problems, chronic fatigue, oral abscesses, sunburn, and toothaches, due to its rich composition of various bioactive compounds.

Bioactive compounds discovered in *P. nigrum* include betaine, achyranthine, 10-tricosanone, 10-octacosanone, 4-tritriacontanone, pellitorine, piperidine, piperine, sylvamide, cepharadione A, and paprazine [156]. Piperine is the main active ingredient in *P. nigrum* [157]; an earlier study found that piperine accounts for approximately 98% of alkaloids in black pepper [158]. Other compounds (and isomers) such as amides, piperidine, pyrrolidines, and trace levels of safrole are among the other phytochemicals reported by Corinna et al. [159].

The function and chemistry of phytoconstituents in *P. nigrum* were recently detailed by Embuscado [160]. The use of black pepper reduced the levels of COX and nuclear factor kappa B (NFκB), as well as the amount of lipid peroxidation [161]. As a result, the compounds are suggested to have an antipyretic effect on PGE2 synthesis. The antipyretic action of *P. nigrum* compounds in animal models has been demonstrated in many investigations, including wet lab and computer simulations. Recently, the pharmacologic potential of *P. nigrum* methanol extract was investigated using molecular docking analysis of several secondary metabolites (bioactive phytoconstituents) in connection with pyrexia development and their druglike properties [77]. The docking score is ranked in the following manner: cepharadione A > paprazine > piperine > sylvamide > paracetamol > pellitorine > piperolactam D > 10-tricosanone, implying that cepharadione A, paprazine, and piperine would be more effective than paracetamol. According to these computational approaches, *P. nigrum* has promising biologic activities as an antipyretic, among other clinical applications [77].

Because research is scarce through clinical trials, little is known about the detrimental effects and drug–drug interaction. As a result, adequate knowledge of the constituents is required in various therapies. Concurrent administration of piperazine and chlorpromazine during an antipsychotic intervention, meanwhile, may worsen the effects and increase the risk of seizures. In animal models, piperazine plus chlorpromazine administration resulted in enhanced pharmacologic action of the latter provoking acute toxicity [162].

14.2.8 *Viola betonicifolia* (Family Violaceace)

Viola betonicifolia is found in Malaysia, Pakistan, China, India, Nepal, Sri Lanka, China, Malaysia, and Australia. Antipyretic, anticancer, purgative, astringent, and diaphoretic are all traditional properties of the herb. Also, it is useful for the treatment of epilepsy and sleeplessness, among other neurologic problems [163], sinusitis, skin and blood disorders, and pharyngitis [164]. Many different parts of the plant are used for various purposes. For instance, while the roots are used for kidney diseases, pneumonia, and bronchitis, the flowers are recommended for the treatment of asthma, coughs, and colds, while the leaves are useful for the treatment of boils [165]. The crude methanolic extract of *V. betonicifolia*, as well as the subsequent solvent fraction, were previously investigated for a variety of pharmacologic properties [166, 167]. These bioactive components are thought to be responsible for their antipyretic action.

The methanolic extract contains alkaloids, saponins, flavonoids, tannins, proteins, and phenolic compounds, based on phytochemical analysis [167]. In an animal model, the extract was shown to significantly reduce hyperthermia induced by yeast. The inhibition, according to the authors, was dose dependent. It remained significant up to three hours after extract administration, with 300 mg/kg producing the greatest antipyretic effect. The fever was reduced by 78.23% compared to the 90% reduction seen with paracetamol. However, because the acute toxicity test revealed that the plant is safe above 2000 mg/kg, a complete reversal of body temperature may be possible if the dose is upscaled.

Moreover, the inability to identify the exact biologic compound that is antipyretic and the specific locus with which it impedes the fever pathway was a shortcoming of the investigation. As a result, mechanistic research is required to isolate each of the pharmacologically active components and determine which of them might be used as a viable fever control drug.

14.3 Conclusion and Future Perspectives

Historically, plant-derived secondary metabolites have demonstrated potential therapeutic activities against many illnesses [168–171]. Several reports involving antipyretic plant metabolites on experimental models have been published over the years. However, whether these preliminary studies will advance through clinical trials to pharmaceutical market remains an open question. Careful clinical studies on various plant-derived antipyretic agents, that have undergone preclinical testing, are recommended. In addition, plants with analgesic and anti-inflammatory potential have also been suggested to have high antipyretic activity from crude extracts and their various solvent fractions. While folk medicine may not be able to uncover several antipyretics of plant origin at this time, further studies on phytochemicals with analgesic and anti-inflammatory properties could lead to the discovery of many more novel antipyretic agents.

One of the current challenges is that the majority of antipyretic studies do not identify active secondary metabolites. Moreso, among the existing plant species, there are only a few studies that demonstrate the mechanisms of action of their isolated compounds, and there is little renewed interest in some of these plant species compared to those investigated for other health issues such as anticancer, antidiabetic, hepatoprotective, and antinociceptive agents, to name a few. Meanwhile, extensive biologic and pharmacologic activities point to increased safety, reduced costs, and proven efficacy for a number of plant phytochemical compounds, making it logical to continue researching antipyretics derived from plants.

Currently, there is little research distinguishing individual roles of various isolated bioactive components from their combined activities in a variety of plant materials. Additional phytochemical and pharmacodynamic studies to determine the active substances responsible for their activities and the succinct mechanisms of their action will facilitate further exploration, which could lead to intriguing discoveries and progress to clinical trials.

More importantly, although little is known about adverse effects and drug–drug interactions in humans from many secondary plant metabolites due to a lack of research through clinical trials, antipyretics of plant origin are generally believed to be safer than synthetic drugs. The majority of investigations are conducted on animal models with little information about toxicity. Paracelsus, the father of toxicology, noted that dosage magnitude (i.e. the number of molecules) influences the likelihood of harmful effects. Thus, toxicity cannot be confined to an intrinsic biochemical property; all substances are toxic at a specific dosage. Until now, the toxicity level of several identified bioactive compounds suggested for pyrexia treatment has been rarely investigated, even though many individuals in developing countries ingest concoctions of multiherbal plant products in high quantities without any recommended dosage. The importance of toxicity cannot be overstated. Examining the relative toxicities of the commonly used bioactive plant substances in herbal dietary supplements, a recent study found about 47 toxic compounds to have adverse effects on the hepatic, cardiovascular, central nervous system, and digestive systems in 55 species from 46 plant groups curated from National Laboratory of Medicine journal and SciFinder databases [172]. By implication, there is a need for further studies to investigate the toxicity profile of those metabolites that have passed preliminary preclinical studies.

References

1 Ogoina, D. (2011). Fever, fever patterns and diseases called "fever". *J. Infect. Public Health* 4: 108–124.

2 Mabey D, Doherty T. The febrile patient. In: Parry E, Godfrey R, Mabey D, Gill G, editors. Principles of Medicine in Africa. 3 Cambridge: Cambridge University Press; 2004. p. 191–197.

3 Dinarello, C.A. and Gelfand, J.A. (2005). Fever and hyperthermia. *Harrisons Principles Internal Med.* 16 (1): 104.

4 Mackowiak, P.A. (2005). Temperature regulation and pathogenesis of fever. In: *Mandell, Douglas and Bennett's Principles and Practise of Infectious Disease*, 6e, vol. 1 (ed. J.E. Bennett, R. Dolin, and M.J. Blaser), 703–718. St. Louis, MO: Elsevier Churchill Livingstone.

5 Commission for Thermal Physiology of the International Union of Physiological Sciences (2001). Glossary of terms for thermal physiology, third edition. *Jpn. J. Physiol.* 51 (2): 245–280.

6 Bindu, S., Mazumder, S., and Bandyopadhyay, U. (2020). Non-steroidal anti-inflammatory drugs (NSAIDs) and organ damage: a current perspective. *Biochem. Pharmacol.* 180: 114147.

7 James, D.S. (1999). The multisystem adverse effects of NSAID therapy. *J. Osteopathic Med.* 99: 1–7.

8 Dreischulte, T., Morales, D.R., Bell, S., and Guthrie, B. (2015). Combined use of nonsteroidal anti-inflammatory drugs with diuretics and/or renin-angiotensin system inhibitors in the community increases the risk of acute kidney injury. *Kidney Int.* 88 (2): 396–403.

9 Arfe, A., Scotti, L., Varas-Lorenzo, C. et al. (2016). Non-steroidal anti-inflammatory F drugs and risk of heart failure in four European countries: nested case-control study. *Br. Med. J.* 354: 4857.

10 O'Grady, N.P., Barie, P.S., Bartlett, J.G. et al. (2008). Guidelines for evaluation of new fever in critically ill adult patients: 2008 update from the American College of Critical Care Medicine and the Infectious Diseases Society of America. *Crit. Care Med.* 36 (4): 1330–1349.

11 Conti, B., Tabarean, I., Andrei, C., and Bartfai, T. (2004). Cytokines and fever. *Front. Biosci.* 9: 1433–1449.

12 Aronoff, D.M. and Neilson, E.G. (2001). Antipyretics: mechanisms of action and clinical use in fever suppression. *Am. J. Med.* 111: 304–315.

13 Roth, J. and de Souza, G.E.P. (2001). Fever induction pathways: evidence from responses to systemic or local cytokine formation. *Braz. J. Med. Biol. Res.* 34 (3): 301–314.

14 Banks, W.A., Plotkin, S.R., and Kastin, A.J. (1995). Permeability of the blood-brain barrier to soluble cytokine receptors. *Neuroimmunomodulation* 2: 161–165.

15 Rummel, C., Barth, S.W., Voss, T. et al. (2005). Localised vs systemic inflammation in Guinea pigs: a role of prostaglandins at distinct points of the fever induction pathways? *Am. J. Physiol. Regul. Integr. Comp. Physiol.* 289: 340–347.

16 Ootsuka, Y., Blessing, W.W., Steiner, A.A., and Romanovsky, A.A. (2008). Fever response to intravenous prostaglandin E2 is mediated by the brain but does not require afferent vagal signalling. *Am. J. Physiol. Regul. Integr. Comp. Physiol.* 294: 1294–1303.

17 Romanovsky, A.A., Almeida, M.C., Aronoff, D.M. et al. (2005). Fever and hypothermia in systemic inflammation: recent discoveries and revisions. *Front. Biosci.* 10: 2193–2216.

18 Walter, E.J. and Carraretto, M. (2016). The neurological and cognitive consequences of hyperthermia. *Crit. Care* 20: 199.

19 Kiyatkin, E.A. (2007). Brain temperature fluctuations during physiological and pathological conditions. *Eur. J. Appl. Physiol.* 101: 3–17.

20 Dewhirst, M.W., Viglianti, B.L., Lora-Michiels, M. et al. (2003). Basic principles of thermal dosimetry and thermal thresholds for tissue damage from hyperthermia. *Int. J. Hyperthermia* 19: 267–294.

21 Earn, D.J., Andrews, P.W., and Bolker, B.M. (2014). Population-level effects of suppressing fever. *Proc. Biol. Sci.* 281 (1778): 20132570.

22 Aronoff, D.M. and Neilson, E.G. (2001). Antipyretics: mechanisms of action and clinical use in fever suppression. *Am. J. Med.* 111 (4): 304–315.

23 Etminan, M., Sodhi, M., and Ganjizadeh-Zavareh, S. (2021). Should antipyretics be used to relieve acute adverse events related to COVID-19 vaccines? *Chest* 159 (6): 2171–2172.

24 Safari, V.Z., Kamau, J.K., Nthiga, P.M. et al. (2016). Antipyretic, antiinflammatory and antinociceptive activities of aqueous bark extract of *Acacia Nilotica* (L.) Delile in albino mice. *J. Pain Manage. Med.* 2: 113.

25 Veronica, S.A., Cheruiyot, K.S., Bosibori, M.J. et al. (2017). Antiinflammatory, analgesic and antipyretic effects of dichloromethane stem bark extract of *Acacia mellifera*. *J. Phytopharmacol.* 6 (4): 239–246.

26 Abotsi, W.K.M., Lamptey, S.B., Afrane, S. et al. (2016). An evaluation of the anti-inflammatory, antipyretic and analgesic effects of hydroethanol leaf extract of *Albizia zygia* in animal models. *Pharm. Biol.* 55 (1): 338–348.

27 Olajide, O.A., Awe, S.O., Makinde, M.J. et al. (2000). Studies of anti-inflammatory, antipyretic and analgesic properties of *Alstonia boonei* stem bark. *J. Ethnopharmacol.* 7: 179–186.

28 Kumar, B.S.A., Lakshman, K., Jayaveera, K.N. et al. (2010). Antioxidant and antipyretic properties of methanolic extract of *Amaranthus spinosus* leaves. *Asian Pac. J. Trop. Med.* 3 (9): 702–706.

29 Sharma, V.C., Kaushik, A., Dey, Y.N. et al. (2020). Analgesic, anti-inflammatory and antipyretic activities of ethanolic extract of stem bark of *Anogeissus latifolia* Roxb. *Clin. Phytosci.* 6: 22.

30 Megwas, A.U., Akuodor, G.C., Chukwu, L.C. et al. (2020). Analgesic, anti-inflammatory and antipyretic activities of ethanol extract of *Annona senegalensis* leaves in experimental animal models. *Int. J. Basic Clin. Pharmacol.* 9 (10): 1477.

31 Jaffal, S.M., Oran, S.A., and Alsalem, M. (2020). Anti-nociceptive effect of *Arbutus andrachne* methanolic leaf extract mediated by CB1, TRPV1 and PPARs in mouse pain models. *Inflammopharmacology* 37: 189–195.

32 Sourabie, T.S., Nikiema, J.B., and Nacoulma, O.J. (2014). Monographycal survey of pharmacological potentials of *Argemone Mexicana* L. (Papaveraceae), a plant of Burkina Faso Pharmacopeia: determination of antihepatotoxic and antipyretic activities. *Int. J. Res. Pharmacy Biosci.* 1 (2): 1–10.

33 Begum, T.N., Muhammad Ilyas, M.H., and Anand, A.V. (2011). Antipyretic activity of azima tetracantha in experimental animals. *Int. J. Curr. Pharm. Res.* 1 (2): 41–44.

34 Channa, S., Dar, A., Anjum, S. et al. (2006). Anti-inflammatory activity of *Bacopa monniera* in rodents. *J. Ethnopharmacol.* 104 (1–2): 286–289.

35 Gupta, M., Mazumder, U.K.R., Sambath Kumar, R. et al. (2005). Anti-inflammatory, analgesic and antipyretic effects of methanol extract from *Bauhinia racemosa* stem bark in animal models. *J. Ethnopharmacol.* 98 (3): 267–273.

36 Zakaria, Z.A., Wen, L.Y., Abdul Rahman, N.I. et al. (2007). Anti-nociceptive, antiinflammatory and antipyretic properties of the aqueous extract of *Bauhinia purpurea* leaves in experimental animals. *Med. Princ. Pract.* 16: 443–449.

37 Mythreyi, D.R., Sasikala, E., and Geetha, A. (2012). Antipyretic activity of leaves of *Cadaba trifoliata*. *Int. J. Phytother. Res.* 2 (2): 2278–5701.

38 Chitme, H.R., Chandra, R., and Kaushik, S. (2005). Evaluation of antipyretic activity of *Calotropis gigantea* (Asclepiadaceae) in experimental animals. *Phytother. Res.* 19: 454–456.

39 Padhan, R., Agrahari, A.K., and Meher, A. (2010). A study on antipyretic activity of *Capparis zeylanica* Linn. Plant methanolic extract. *Int. J. Pharmaceut. Sci. Rev. Res.* 1 (3): 169–171.

40 Ghule, B.V., Murugananthan, G., and Yeole, P.G. (2007). Analgesic and antipyretic effects of *Capparis zeylanica* leaves. *Fitoterapia* 78: 365–369.

41 Maina, G.S., Maina, M.B., Muriithi, N.J. et al. (2015). Antipyretic properties of dichloromethane: methanolic leaf and root bark extracts of *Carissa edulis* in rats. *Asian J. Biomed. Pharmaceut. Sci.* 5 (43): 12–20.

42 Hasan, T., Jahan, E., Ahmed, K.S. et al. (2022). Rutin hydrate and extract from *Castanopsis tribuloides* reduces pyrexia via inhibiting microsomal prostaglandin E synthase-1. *Biomed. Pharmacother.* 148: 112774.

43 Dutta, T., Paulc, A., Majumderc, M. et al. (2020). Pharmacological evidence for the use of *Cissus assamica* as a medicinal plant in the management of pain and pyrexia. *Biochem. Biophys. Rep.* 21: 100715.

44 Bose, A., Mondal, S., Gupta, J.K. et al. (2007). Analgesic, anti-inflammatory and antipyretic activities of the ethanol extract and its fraction of *Cleome rutidosperma*. *Fitoterapia* 78: 515–520.

45 Narayanan, N., Thirugnanasambantham, P., Viswanathan, S. et al. (1999). Antinociceptive, anti-inflammatory and antipyretic effects of ethanol extract of *Clerodendron serratum* roots in experimental animals. *J. Ethnopharmacol.* 65 (3): 237–241.

46 Parimaladevi, B., Boominathan, R., and Subhash, C.M. (2004). Evaluation of antipyretic potential of *Clitoria ternatea* L. extract in rats. *Phytomedicine* 11 (4): 323–326.

47 Kanedi, M., Setiawan, W.A., Handayani, K., and Febriyanti, V. (2021). Antipyretic effect of ethanol extract of plant leaves of garden croton (*Codiaeum variegatum*) in male mice. *GSC Biol. Pharm. Sci.* 14 (3): 202–206.

48 Kushwah, C., Singh, N., Goswami, R., and Jain, N. (2021). Antipyretic activity of hydroalcoholic extract of leaves of *Colocasia esculenta*. *J. Drug Deliv. Ther.* 11 (1): 65–67.

49 Zakaria, Z.A., Kumar, G.H., Raden Mohd, R.N.S. et al. (2009). Antinociceptive, anti-inflammatory and antipyretic properties of an aqueous extract of *Corchorus capsularis* leaves in experimental animal models. *Pharm. Biol.* 47 (2): 104–110.

50 Binny, K., Kumar, S., and Thomas, D. (2010). Anti-inflammatory and antipyretic properties of the rhizome of Costus Speciosus (Koen.) sm. *J. Basic Clin. Pharmacy* 1 (3): 177–181.

51 Chidambaram, K., Albert, J., Karpagam, K., and Sivasubramanian, N. (2011). Antipyretic activity of *Crataeva magna* bark on Tab-Vaccine induced pyrexia. *Int. J. Pharm. Sci. Res.* 2: 856–859.

52 Azam, G., Noman, S., and Al-Amin, M. (2014). Phytochemical screening and antipyretic effect of *Curcuma zedoaria* Rosc. (Zingiberaceae) rhizome. *Br. J. Pharm. Res.* 4 (5): 569–575.

53 Zakaria, Z.A., Ghani, Z.D., Nor, R.N. et al. (2008). Antinociceptive, anti-inflammatory, and antipyretic properties of an aqueous extract of *Dicranopteris linearis* leaves in experimental animal models. *J. Nat. Med.* 62 (2): 179–187.

54 Rauf, A., Uddin, G., Siddiqui, B.S. et al. (2014). Antipyretic and antinociceptive activity of *Diospyros lotus* L. in animals. *Asian Pac. J. Trop. Biomed.* 4: S382–S386.

55 Perianayagam, J.B., Sharma, S.K., Joseph, A., and Christina, A.J.M. (2004). Evaluation of anti-pyretic and analgesic activity of *Emblica officinalis* Gaertn. *J. Ethnopharmacol.* 95: 83–85.

56 Ahamed, S.M., Swamy, S.K., Jayaverra, K.N. et al. (2008). Anti-inflammatory, antipyretic and analgesic activity of methanolic extract of Feronia Limonia fruit pulp. *Pharmacologyonline* 3: 852–857.

57 Singh, G., Singh, B., and Mehta, P. (2018). Antipyretic activity investigations on various extracts and fractions of Jwarnashak Panch Kashya (polyherbal formulation). *J. Pharmacogn. Phytochem.* 7 (3): 3017–3020.

58 Zhang, X., Wang, Y., Li, S. et al. (2019). The potential antipyretic mechanism of *Gardeniae fructus* and its heat-processed products with plasma metabolomics using rats with yeast-induced fever. *Front. Pharmacol.* 10: 491.

59 Parhi, P.K., Pattanayak, P., Mishra, P., and Pani, M.K. (2011). An in-vivo study on analgesic and antipyretic activity of bark extracts of *Gmelina arborea*. *Int. J. Pharm. Sci. Rev. Res.* 10 (2): 78–81.

60 Arora, R. and Mazumder, A. (2017). A study on antipyretic activity of methanolic extract of rhizomes of *Hedychium spicatum* plant. *J. Pharmacogn. Phytochem.* 6 (5): 1503–1506.

61 Zhang, B., Peng, H., Deng, Z., and Tsao, R. (2018). Phytochemicals of lentil (*Lens culinaris*) and their antioxidant and anti-inflammatory effects. *Food Bioactives* 1: 93–103.

62 Besra, S.E., Sharma, R.M., and Gomes, A. (1996). Antiiflammatory effect of petroleum ether extract of leaves of *Litchi chinensis* Gaertin. (Sapindasceae). *J. Ethnopharmacol.* 54: 1–6.

63 Alam, M.M., Emon, N.U., Alam, S. et al. (2021). Assessment of pharmacological activities of *Lygodium microphyllum* Cav. leaves in the management of pain, inflammation, pyrexia, diarrhea, and helminths: in vivo, in vitro and in silico approaches. *Biomed. Pharmacother.* 139: 111644.

64 Shekhawat, N. and Vijayvergia, R. (2010). Investigation of anti-inflammatory, analgesic and antipyretic properties of *madhuca indica* gmel. *Eur. J. Inflamm.* 8 (3): 165–171.

65 Gitonga, G.M., Ngeranwa, J., Machocho, A.K., and Kiruki, S. (2020). Qualitative phytochemical screening and antipyretic activity of methanolic stem bark extract of *Mangifera indica* in wistar rats. *J. Phytopharmacol.* 9 (4): 218–223.

66 Khan, M.D. (2011). Evaluation of analgesic and antipyretic activity of *Marsilea trifolia* Blanco. *Int. J. Sci. Eng. Res.* 2 (8): 2229–5518.

67 Zakaria, Z.A., Raden Mohd Nor, R.N.S., Kumar, G.H. et al. (2006). Antinociceptive, anti-inflammatory and antipyretic properties of *Melastoma malabathricum* leaves aqueous extract in experimental animals. *Can. J. Physiol. Pharmacol.* 84 (12): 1291–1299.

68 Aziz, U., Akther, R., Shahriar, M., and Bhuiyan, M.A. (2014). Invivo pharmacological investigation of *Mimosa pudica* L. *Int. J. Pharmacy Pharm. Sci.* 16 (2): 66–69.

69 Sahu, S.K., Das, D., Tripathy, N.K. et al. (2011). Anti-inflammatory, analgesic and antipyretic effects of *Mollugo Pentaphylla* L. *Rasayan J. Chem.* 4 (3): 533–538.

70 Nwobodo, N., Igwe, S.A., and Agbata, C.A. (2015). Antinociceptive, anti-inflammatory and antipyretic effects of aqueous extracts of *Morinda lucida* leaves in experimental animals. *Biomed. Pharmacol. J.* 4 (1): 61–67.

71 Zakaria, Z.A., Mohd Nor Hazalin, N.A., Mohd Zaid, S.N.H. et al. (2007). Antinociceptive, anti-inflammatory and antipyretic effects of *Muntingia calabura* aqueous extract in animal models. *J. Nat. Med.* 61: 443–448.

72 Deepa, P.K., Usha, P.T.A., Chandrasekharan Nair, A.M., and Prasannakumari, K.T. (2009). Antipyretic activity of seeds from red and white type of lotus (*Nelumbo nucifera*) in Albino rat. *Vet. World* 6: 213–214.

73 Makonnen, E., Debella, A., Zerihun, L. et al. (2003). Antipyretic properties of the aqueous and ethanol extracts of the leaves of *Ocimum suave* and *Ocimum lamiifolium* in mice. *J. Ethnopharmacol.* 88: 85–91.

74 Abidoye, E.O. and Aplakah, I. (2017). Antinociceptive and antipyretic properties of ethanol extract of *Oryza bathii* (Poaceae) in wistar rats. *Sokoto J. Vet. Sci.* 15 (2): 43–48.

75 Daud, A., Habib, N., and S'anchez Riera A. (2006). Anti-inflammatory, anti-nociceptive and antipyretic effects of extracts of *Phrygilanthus acutifolius* flowers. *J. Ethnopharmacol.* 108: 198–203.

76 Sireeratawong, S., Itharat, A., Lerdvuthisopon, N. et al. (2012). Anti-inflammatory, analgesic, and antipyretic activities of the ethanol extract of *Piper interruptum* Opiz. and *Piper chaba* Linn. *ISRN Pharmacol.* 2012: 480265.

77 Emon, N.U., Alam, S., Rudra, S. et al. (2021). Antidepressant, anxiolytic, antipyretic, and thrombolytic profiling of methanol extract of the aerial part of *Piper nigrum*: in vivo, in vitro, and in silico approaches. *Food Sci. Nutr.* 9 (2): 833–846.

78 Misra, V., Uddin, S.M., Srivastava, V., and Sharma, U. (2012). Antipyretic activity of plumeria rubra leaves extract. *Int. J. Pharm.* 2 (2): 330–332.

79 Ahmad, M., Jabeen, Q., Wajid, M. et al. (2013). Time and dose dependent antipyretic investigations of ethanolic leaves and fruits extracts of *Prosopis cineraria* L. (Druce). *J. Pharmacy Altern. Med.* 2 (2): 14–19.

80 Akuodor, G.C., Essien, A.D., Essiet, G.A. et al. (2013). Evaluation of antipyretic potential of *Pseudocedrela kotschyi* Schweint. Harms (Meliaceae). *Eur. J. Med. Plants* 3 (1): 105–113.

81 Vasudevan, S.C.N., Kariyil, B.J., and Neerakkal, I. (2019). Antipyretic activity of aqueous extract of heart wood of *Pterocarpus santalinus* L. in yeast induced pyrexia. *J. Pharmacogn. Phytochem.* 8 (3): 244–246.

82 Reanmongkol, W., Noppapan, T., and Subhadhirasakul, S. (2009). Antinociceptive, antipyretic, and anti-inflammatory activities of *Putranjiva roxburghii* wall. Leaf extract in experimental animals. *J. Nat. Med.* 63 (3): 290–296.

83 Subedi, N.K., Rahman, S.M., and Akbar, M.A. (2016). Analgesic and antipyretic activities of methanol extract and its fraction from the root of *Schoenoplectus grossus. Evid. Based Complement. Alternat. Med.* 2016: 3820704.

84 Mwonjoria, J.K., Kariuki, H.N., and Waweru, F.N. (2011). The antinociceptive antipyretic effects of *Solanum incanum* (Linneaus) in animal models. *Int. J. Phytopharmacol.* 2229: 7472.

85 Zakaria, Z.A., Sulaiman, M.R., Morsid, N.A. et al. (2009). Antinociceptive, anti-inflammatory and antipyretic effects of *Solanum nigrum* aqueous extract in animal models. *Methods Find. Exp. Clin. Pharmacol.* 31 (2): 81.

86 Vilela, F.C., Bitencourt, A.D., Cabral, L.D. et al. (2010). Anti-inflammatory and antipyretic effects of *Sonchus oleraceus* in rats. *J. Ethnopharmacol.* 127 (3): 737–741.

87 Nisar, M., Khan, I., Simjee, S.U. et al. (2008). Anticonvulsant, analgesic and antipyretic activities of *Taxus wallichiana* Zucc. *J. Ethnopharmacol.* 116 (3): 490–494.

88 Mascolo, N., Pinto, A., and Capasso, F. (1988). Antipyretic and analgesicstudies of the etanalic extract of *Teclea nobilis* Delile. *Phytother. Res.* 2 (3): 154–0156.

89 Al-Rehaily, A.J., Ahmad, M.S., Muhammad, I. et al. (2003). Furoquinoline alkaloids from *Teclea nobilis. Phytochemistry* 64: 1405–1411.

90 Saini, N.K. and Singha, M. (2012). Anti-inflammatory, analgesic and antipyretic activity of methanolic *Tecomaria capensis* leaves extract. *Asian Pac. J. Trop. Biomed.* 11: 870–874.

91 Priyanka, S., Pooja, S.K.C., and Rathore, K.S. (2011). Antipyretic activity of methanolic extract of root of *Tectona grandis* on albino rats. *J. Pharmacol. Toxicol.* l 2: 28–33.

92 Gupta, M., Mazumder, U.K., Manikandan, L. et al. (2003). Evaluation of antipyretic potential of *Vernonia cinerea* extract in rats. *Phytother. Res.* 17: 804–806.

93 Iwalewa, E.O., Iwalewa, O.J., and Adeboye, J.O. (2003). Analgesic, antipyretic, anti-inflammatory effects of methanol, chloroform and ether extracts of *Vernonia cinerea* less leaf. *J. Ethnopharmacol.* 86: 229–234.

94 Muhammad, N., Saeed, M., and Khan, H. (2012). Antipyretic, analgesic and anti-inflammatory activity of *Viola betonicifolia* whole plant. *BMC Complement. Altern. Med.* 12: 59.

95 Yimer, T., Emiru, Y.K., Kifle, Z.D. et al. (2021). Pharmacological evaluation of antipyretic and antioxidant activities of 80% methanol root extract and derived solvent fraction of *Echinops kebericho* M.(Asteraceae) in mice model. *Biomed. Res. Int.* 2021: 6670984.

96 Gaichu, D.M., Mawia, A.M., Gitonga, G.M. et al. (2017). Phytochemical screening and antipyretic activities of dichloroethane-methanolic leaf and stem bark extracts of *Ximenia americana* in rat models. *J. Herbmed. Pharmacol.* 6 (3): 107–113.

97 Soni, H. and Malik, J.K. (2021). Phyto-pharmacological potential of *Zizyphus jujube*: a review. *Scholars Int. J. Biochem.* 4 (2): 1–5.

98 Rauf A, Ali J, Khan H, Mubarak MS, Patel S. Emerging CAM *Ziziphus nummularia* with in vivo sedativehypnotic, antipyretic and analgesic attributes. *Biotech* 2016; 6(1):11

99 Alam, F., Din, K.M., Rasheed, R. et al. (2020). Phytochemical investigation, anti-inflammatory, antipyretic and antinociceptive activities of *Zanthoxylum armatum* DC extracts-in vivo and in vitro experiments. *Heliyon* 6 (11): e05571.

100 Kumar, D., Sharma, S., and Kumar, S. (2020). Botanical description, phytochemistry, traditional uses, and pharmacology of *Crataeva nurvala* Buch. Ham.: an updated review. *Fut. J. Pharm. Sci.* 6: 113.

101 Shu, Y.Z. (1998). Recent natural products-based drug development: a pharmaceutical industry perspective. *J. Nat. Prod.* 61: 1053–1071.

102 Graul, A.I. (2016). The year's new drugs. *Drug News Perspect.* 14: 12–31.

103 Kumar, D. (2016). Chemical and biological screening of selected medicinal plants. PhD dissertation, Kurukshetra University, Kurukshetra.

104 Intahphuak, S., Khonsung, P., and Panthong, A. (2009). Anti-inflammatory analgesic and antipyretic activities of virgin coconut oil. *Pharm. Biol.* 48 (2): 151–157.

105 Fawad, K., Ul-Islam, N., Subhan, F. et al. (2018). Novel hydroquinone derivatives alleviate algesia inflammation and pyrexia in the absence of gastric ulcerogenicity. *Trop. J. Pharm. Res.* 17 (1): 53.

106 Zhou, B., Tan, M., Lu, J. et al. (2012). Simultaneous determination of five active compounds in *Chimonanthus nitens* by double-development HPTLC and scanning densitometry. *Chem. Cent. J.* 6 (1): 46.

107 Afsar, T., Khan, M.R., Razak, S. et al. (2015). Antipyretic antiinflammatory and analgesic activity of *Acacia hydaspica* R. Parker and its phytochemical analysis. *BMC Complement. Altern. Med.* 15 (1): 136–147.

108 Ullah, I., Khan, J.A., Shahid, M. et al. (2016). Pharmacological screening of *Monotheca buxifolia* (Falc) A. DC. For antinociceptive anti-inflammatory and antipyretic activities. *BMC Complement. Altern. Med.* 16 (1): 1–8.

109 Dwivedi, J., Gupta, A., Verma, S. et al. (2019). Validated simultaneous high-performance thin-layer chromatographic analysis of ursolic acid, β-sitosterol, lupeol and quercetin in the methanolic fraction of *Ichnocarpus frutescens*. *J. Phys. Chem.* 32: 103–108.

110 Vasudevan, C.N., Kariyil, B.J., Nair, D.A., and Neerakkal, I. (2021). Screening of phytocompounds, molecular docking studies, and in vivo anti-inflammatory activity of heartwood aqueous extract of *Pterocarpus santalinus* L.f. *Asian Pac. J. Trop. Biomed.* 11 (2): 59–65.

111 Küpeli, E., Koşar, M., Yeşilada, E. et al. (2002). A comparative study on the anti-inflammatory, antinociceptive and antipyretic effects of isoquinoline alkaloids from the roots of Turkish Berberis species. *Life Sci.* 72 (6): 645–657.

112 Hämäläinen, M., Nieminen, R., Asmawi, M. et al. (2011). Effects of flavonoids on prostaglandin E-2 production and on COX-2 and mPGES-1 expressions in activated macrophages. *Planta Med.* 77: 1504–1511.

113 Xu, Y., Wang, C., Wu, L. et al. (2012). Comparison of inhibitory effects of nine flavonoids on prostaglandin E2 production and COX-2 expression in LPS-stimulated RAW264.7 macrophages. *Int. Conf. Biomed. Eng. Biotechnol.* 1: 981–984.

114 Sakamoto, W., Fujie, K., Nishihira, J. et al. (1991). Inhibition of PGE2 production in macrophages from vitamin E-treated rats. *Prostaglandins Leukot. Essent. Fatty Acids* 44 (2): 89–92.

115 Ferrario, A., Luna, M., Rucker, N. et al. (2011). Pro-apoptotic and anti-inflammatory properties of the green tea constituent epigallocatechin gallate increase photodynamic therapy responsiveness. *Lasers Surg. Med.* 43 (7): 644–650.

116 Oran, S.A. (2014). The status of medicinal plants in Jordan. *J. Agric. Sci. Technol.* 4: 461–467.

117 Oran, S.A. (2015). Selected wild aromatic plants in Jordan. *Int. J. Med. Plants* 108: 686–699.

118 Jaffal, S.M., Oran, S.A., and Alsalem, M.I. (2021). Anti-inflammatory and antipyretic potential of *Arbutus andrachne* L. methanolic leaf extract in rats. *Asian Pac. J. Trop. Biomed.* 11 (11): 491–499.

119 Canbolat, O., Ozkan, C.O., and Kamalak, A. (2007). Effects of NaOH treatment on condensed tannin contents and gas production kinetics of tree leaves. *Anim. Feed Sci. Technol.* 138 (2): 189–194.

120 Kumar, D.G., Deepa, P., Rathi, M.A. et al. (2012). Modulatory effects of *Crataeva nurvala* bark against testosterone and N-methyl-N-nitrosourea-induced oxidative damage in prostate of male albino rats. *Pharmacogn. Mag.* 8: 285–291.

121 Alsalem, M., Wong, A., Millns, P. et al. (2013). The contribution of the endogenous TRPV1 ligands 9-HODE and 13-HODE to nociceptive processing and their role in peripheral inflammatory pain mechanisms. *Br. J. Pharmacol.* 168 (8): 1961–1974.

122 Fraga, D., Zanoni, C.I.S., Rae, G.A. et al. (2009). Endogenous cannabinoids induce fever through the activation of CB1 receptors. *Br. J. Pharmacol.* 157 (8): 1494–1501.

123 Mokhber-Dezfuli, N., Saeidnia, S., Gohari, A.R., and Kurepaz-Mahmoodabadi, M. (2014). Phytochemistry and pharmacology of berberis species. *Pharmacogn. Rev.* 8 (15): 8–15.

124 Shahid, M., Rahim, T., Shahzad, A.N. et al. (2009). Ethnobotanical studies on *Berberis aristata* DC. Root extracts. *Afr. J. Biotechnol.* 8: 556–563.

125 Alamgeer, N.H., Rasool, S., Raza, S.A. et al. (2016). Anti-inflammatory, analgesic and antipyretic activities of the aqueous methanolic extract of *Berberis calliobotrys* in albino mice. *Acta Pol. Pharm.* 73 (3): 717–723.

126 Trease, G.E. and Evans, W.C. (1985). *Pharmacognosy*, 12e. London: Bailliere Tindall.

127 Bahorun, T., Neergheen, V.S., and Aruoma, O.I. (2005). Phytochemical constituents of *Cassia fistula*. *Afr. J. Biotechnol.* 4 (13): 1530–1540.

128 Martínez-Avila, G.C.G., Castro-Lopez, C., Rojas, R., and Martínez, G.C.G. (2018). Science screening of the *Cassia fistula* phytochemical constituents. *Annu. Rev. Nutr.* 2 (2): 1–3.

129 Chandra, P., Pandey, R., Kumar, B. et al. (2015). Quantification of multianalyte by UPLC-QqQLIT-MS/MS and in-vitro anti-proliferative screening in Cassia species. *Ind. Crop Prod.* 76: 1133–1141.

130 Sen, A.B. and Shukia, Y.N. (1968). Chemical examination of *Cassia fistula*. *J. Indian Chem. Soc.* 45: 744.

131 Kumar, A., Pande, C.S., and Kaul, R.K. (1966). Chemical examination of *Cassia fistula* flowers. *Indian J. Chem.* 4: 460.

132 Modi, M.R. and Khorana, M.L. (1952). A study of *Cassia fistula* pulp. *Indian J. Pharm.* 4: 61–63.

133 Akanmu, M.A., Iwalewa, E.O., Elujoba, A.A., and Adelusola, K.A. (2004). Toxicity potentials of *Cassia fistula* fruits as laxative with reference to senna. *Afr. J. Biomed. Res.* 7: 23–26.

134 Rahmani, A.H. (2015). *Cassia fistula* Linn: potential candidate in the health management. *Pharm. Res.* 7 (3): 217–224.

135 Gobianand, K., Vivekababdan, P., Pradeep, K. et al. (2010). Anti-inflammatory and antipyretic activities of Indian medicinal plant *Cassia fistula* Linn. (Golden shower) in Wister albino rats. *Int. J. Pharm.* 6: 719–725.

136 Kainsa, S. and Kumar, P. (2012). Pharmacological potentials of *Cassia auriculata* and *Cassia fistula* plants: a review. *Pak. J. Biol. Sci.* 15 (9): 408–417.

137 Singh, A., Singh, M.P., Alam, G. et al. (2012). Antipyretic activity of *Cassia fistula* Linn. *Pods. J. Nat. Prod. Plant Resour.* 2 (3): 385–388.

138 Parker, R.N. (1999). *Common Indian Trees and How to Know Them*. Jaipur: Pointer Publication.

139 Gakuya, D.W., Okumu, M.O., Kiama, S.G. et al. (2020). Traditional medicine in Kenya: past and current status, challenges and the way forward. *Sci. Afr.* 8: e00360.

140 Farnsworth, N.R., Akerele, O., and Bingel, A.S. (1985). Medicinal plants in therapy. *Bull. World Health Organ.* 63: 965–981.

141 Walia, N., Kaur, A., and Babbar, S.B. (2007). An efficient, in vitro cyclic production of shoots from adult trees of *Crataeva nurvala* Buch. *Ham. Plant Cell Report* 26: 277–284.

142 Chidambaram, K., Jaswanth, A., and Karpagam, K. (2011). Antipyretic activity of *Crateva magna* bark on Tab-Vaccine induced pyrexia. *Int. J. Pharm. Sci.* 2 (4): 856–859.

143 Geetha, T. and Varalakshmi, P. (2001). Anti-inflammatory activity of lupeol and lupeol linoleate in rats. *J. Ethnopharmacol.* 76 (1): 77–80.

144 Mekonnen, G., Negash, A., Gashu, M., and Belachew, N. (2023). *Echinops kebericho* aqueous root extract assisted green synthesis of zinc oxide nanoparticles for photocatalytic degradation of methylene blue. *Lett. Appl. NanoBioSci.* 12 (3): 72.

145 Ameya, G., Gure, A., and Dessalegn, E. (2016). Antimicrobial activity of *Echinops kebericho* against human pathogenic bacteria and fungi. *Afr. J. Tradit. Complement. Altern. Med.* 13: 199–203.

146 Bitew, H. and Hymete, A. (2019). The genus Echinops: phytochemistry and biological activities: a review. *Front. Pharmacol.* 10: https://doi.org/10.3389/fphar.2019.01234.

147 Deyno, S., Mtewa, A.G., Hope, D. et al. (2021). Antibacterial activities of Mesfin tuber extracts and isolation of the most active compound, dehydrocostus lactone. *Front. Pharmacol.* 11: 1–13.

148 Yimer, T., Birru, E.M., Adugna, M. et al. (2020). Evaluation of analgesic and anti-inflammatory activities of 80% methanol root extract of *Echinops kebericho* M. (Asteraceae). *J. Inflamm. Res.* 13: 647–658.

149 Saranya, R., Thirumalai, T., Hemalatha, M. et al. (2013). Pharmacognosy of *Enicostemma littorale*: a review. *Asian Pac. J. Trop. Biomed.* 3 (1): 79–84.

150 Jahan, E., Perveen, S., and Malik, A. (2009). Verticilliside, a new flavone C-glucoside from *Enicostemma verticillatum*. *J. Asian Nat. Prod. Res.* 11 (3): 257–260.

151 Leelaprakash, G. and Mohan, D.S. (2012). Antimicrobial activity and phytochemical screening of methanol extract of *Enicostemma axillare*. *Int. J. Pharmacy Pharm. Sci.* 4 (1): 342–348.

152 Desai, P.D., Ganguly, A.K., Govindachari, T.R. et al. (1966). Chemical investigation of some Indian medicinal plants: part II. *Indian J. Chem.* 4: 457–459.

153 Okokon, J.E., Nwafor, P.A., Abia, G.O., and Bankhede, H.K. (2012). Antipyretic and antimalarial activities of crude leaf extract and fractions of *Enicostema littorale*. *Asian Pac. J. Trop. Dis.* 2 (6): 442–447.

154 Patel, R., Patel, Y., Kunjadia, P., and Kunjadia, A. (2015). *Int. J. Curr. Microbiol. App. Sci.* 4 (1): 773–780.

155 Garad, M., Upadhya, M., Kokare, D., and Itankar, P. (2012). Aerial parts of *Enicostemma littorale* Blume serve as antipyretic and antacid: in vivo and in vitro evaluations. *Pharmacogn. Commun.* 2 (3): 42–45.

156 Ee, G., Lim, C., Lim, C. et al. (2009). Alkaloids from *Piper sarmentosum* and *Piper nigrum*. *Nat. Prod. Res.* 23 (15): 1416–1423.

157 Gulati, K., Verma, P., Rai, N., and Ray, A. (2021). Role of nutraceuticals in respiratory and allied diseases. In: *Nutraceuticals*, 2 (ed. R.C. Gupta, R. Lall, and A. Srivastava). Cambridge, MA: Academic Press, 101–115.

158 Hirasa, K. and Takemasa, M. (1998). *Spice Science and Technology*. Boca Raton, FL: CRC Press.

159 Dawid, C., Henze, A., Frank, O. et al. (2012). Structural and sensory characterization of key pungent and tingling compounds from black pepper (*Piper nigrum* L.). *J. Focused Food Agric. Chem.* 60 (11): 2884–2895.

160 Embuscado, M.E. (2018). Bioactives from spices and herbs. In: *Encyclopedia of Food Chemistry* (ed. L. Melton, F. Shahidi, and P. Varelis), 497–514. Cambridge, MA: Academic Press.

161 Shukla, S.K. and Kumar, V. (2013). Bioactive foods and supplements for protection against liver diseases. In: *Bioactive Food as Dietary Interventions for Liver and Gastrointestinal Disease* (ed. R. Watson and V. Preedy), 557–567. Cambridge, MA: Academic Press.

162 Sturman, G. (1974). Interaction between piperazine and chlorpromazine. *Br. J. Pharmacol.* 50 (1): 153–155.

163 Hamayun, M. (2005). Ethnobotanical profile of Utror and Gabral valleys, district Swat. *Pak. Ethnobot. Leaflets* 1: 9.

164 Bhatt, V. and Negi, G. (2006). Ethnomedicinal plant resources of Jaunsari tribe of Garhwal Himalaya, Uttaranchal. *Indian J. Trad. Knowl.* 5 (3): 331–335.

165 Husain, S.Z., Malik, R.N., Javaid, M., and Bibi, S. (2008). Ethonobotanical properties and uses of medicinal plants of Morgah biodiversity park, Rawalpindi. *Pak. J. Bot.* 40 (5): 1897–1911.

166 Muhammad, N. and Saeed, M. (2011). Biological screening of *Viola betonicifolia* Smith whole plant. *Afr. J. Pharmacy Pharmacol.* 5 (20): 2323–2329.

167 Muhammad, N., Saeed, M., Khan, H., and Haq, I. (2013). Evaluation of n-spiepr DHYPhexane extract of Viola betonicifolia for its neuropharmacological properties *J. Nat. Med.* 67 (1): 1–8.

168 Adedosu, O.T., Adeleke, G.E., Imodoye, S.O., Ajiboye, A.E. and Akinto, A.O. (2016). Aqueous extract of Hibiscus sabdariffa calyx showed antioxidative and ameliorate acetaminophen (paracetamol)-induced hepatotoxicity and nephrotoxicity in Rats. *Nat. Sci.* 14 (8): 28–37. doi.10.7537/marsnsj14081606169.

169 Busari, A.O., Salau, F.A., Bashir, S., Bolaji, M., Balogun, M.E., and Ibrahim, M. et al. (2021). Al-Hikmah *J. Health Sci.* 1 (1);7–13.

170 Adedokun, K.A., Imodoye, S.O., Bello, I.O., Lanihun, A., and Bello, I.O. (2023). Therapeutic potentials of medicinal plants for the treatment of cancer and significance of computational tools in anticancer drug discovery. In: Egbuna, C., Rudrapal, M., Tijjani, H. (Eds). Phytochemistry, computational tools, and databases in drug discovery. 1st ed. Elsevier, Amsterdam, 393–455. ISBN: 9780323905930.

171 Olusegun, O.H, Ademola, A.K., Olushola, S.M., Awe, E.O., and Oyenike, E.O. (2017). *J. Medical Sci. Clin.* 5 (7): 24808–24815. doi:https://dx.doi.org/10.18535/jmscr/v5i7.85.

172 Hudson, A. Lopez, E., AlMalki, A.J., Roe, A.L., and Calderon, A.I. (2018). A review of the toxicity of compounds found in herbal dietary supplements. *Planta Med.* 84 (9–10): 613–626.

15

Medicinal Herbs against Central Nervous System Disorders

Ahmed Olatunde[1], Neelma Munir[2], Godwin Anywar[3], Maria Hanif[2], Huma Waqif[2], Habibu Tijjani[4], Barbara Sawicka[5], and Akram Muhammad[6]

[1] *Department of Medical Biochemistry, Abubakar Tafawa Balewa University, Bauchi, Bauchi State, Nigeria*
[2] *Department of Biotechnology, Lahore College for Women University, Lahore, Punjab, Pakistan*
[3] *Department of Plant Sciences, Microbiology & Biotechnology, College of Natural Sciences, Makerere University, Kampala, Uganda*
[4] *Department of Biochemistry, Bauchi State University, Gadau, Bauchi State, Nigeria*
[5] *University of Life Sciences in Lublin, Lublin, Poland*
[6] *Department of Eastern Medicine, Government College University Faisalabad, Faisalabad, Punjab, Pakistan*

The central nervous system (CNS) controls major functions of the human body. Different senses like sight, hearing, taste, and smell are controlled by the CNS. The voluntary and involuntary functions of the body are also controlled by this system. The two main components of the CNS are the brain and spinal cord. The brain performs many functions such as managing thoughts and controlling the movements of different body parts, and it also predicts the surrounding environment and situation. The transmission of messages from the brain to different body parts is performed by the spinal cord. Neurons are present in the CNS. These neurons are connected to each other and function as a link in order to send and receive messages. These messages are delivered to the brain and spinal cord. The indications of CNS disorders show which part of the CNS is causing the problem. Nervous system disorders may be degenerative. These disorders may be acute and can lead to life-threatening problems. The symptoms of CNS disorders may be mild or severe.

Many plant extracts have been used for the treatment of different diseases in humans. These plant extracts can be used in a crude or purified form [1]. By using the different medicinally important plants and their extracts, traditional medicines have been helpful in the cure of diseases. In South America, South Asia, and South Africa, approximately 60% out of 300 000 seed plants are used for therapeutic interventions. Different diseases such as diabetes, obesity, malaria, and rheumatoid arthritis have been treated for a long time by using various plant extracts. These plant extracts are also widely used for the therapy of mental health disorders that influence the CNS. The most common chronic brain disorder is epilepsy, which affect 10 million people in Africa [2] and approximately 70 million people globally, 90% of whom are from developing countries [3, 4].

The detailed analysis or properties of plant extracts can be determined by genomic and proteomic research developments. These research developments help in the identification of

Phytochemical Drug Discovery for Central Nervous System Disorders: Biochemistry and Therapeutic Effects, First Edition. Edited by Chukwuebuka Egbuna and Mithun Rudrapal.

the purpose of discrete plant components and their effects on human health [5]. With respect to accentuation of the association between restorative plant genomes and the environment, the relationship between hereditary polymorphism and metabolic assorted variety has been attended to. In order to produce medicinal compounds in plant factories, different genomic technologies need to be applied to medicinal plants. For the forecast of chemodiversity and bioprospecting, many other approaches can be used, such as molecular phylogeny and phylogenomics [6]. Ethnomedicine and ethnopharmacology can bring promising outcomes equipped to increase the value of the rich natural resources of a nation. For the therapy of neurologic disorders, research in ethnopharmacy is required for the analysis of medicinal plant properties. These medicinal plants can also help people suffering from nervous disorders [7]. Plants contain various medicinal components that play a key role in the therapy of neurologic ailments. Such compounds include alkaloids, essential oils, mucilages, glucosides, fatty oils gums, and tannins. These constituents are helpful for the therapy of CNS disorders [7]. The use of plants and their extracts for treating neurologic disorders has provided new opportunities for research work [5].

Neurodegenerative diseases cause various changes, such as protein degradation, retarded protein accumulation in neurons, and mitochondrial defects. The crucial factor causing neurodegenerative disease is aging [8]. A few decades ago, many technologies were discovered to transfer numerous bioactive compounds and secondary metabolic particles to mitochondria in cells. These innovations have assumed a vital role in the betterment of medicines' pharmacokinetic profile, intracellular perforation, dispersal of medicine at the objective site, and improvement of the pharmacologic impacts [9]. In ancient times, herbs had great medicinal importance and they were used to treat many chronic diseases. Herbs are now regaining their popularity throughout the world. The traditionally known CNS-active herbal remedies can be easily analyzed with modern techniques to ascertain their efficacy. The currently accessible psychoactive medicines are not based on natural products or their bioactive constituents. Out of the 520 drugs endorsed by the US Food and Drug Administration (FDA) in 1997, 157 were of natural origin [10]. It has also been proven that there is greater success when efforts are made to find organic drug molecules for clinical use. About 60% of anticancer agents or drugs that were affirmed were organic products or their subsidiaries. Without focused projects for characteristic items, there will be no achievement.

15.1 Medicinal Plants as Interventions for Central Nervous System Disorders

There are various CNS receptors that are able to bind with different plant extracts and their parts. These parts influence the pharmacology and likewise assume a key role in the therapy of human ailments. Different studies indicate the interaction of plant extracts with the CNS both *in vivo* and *in vitro* [11, 12]. *Panax ginseng* (ginseng), *Withania somnifera* (ashwagandha), *Baccopa monnieri*, *Ginkgo biloba*, and *Centella asiatica* are the herbal medicines that have attracted a lot of enthusiasm for their remedial potential, as well as various other compounds like flavonoids, celastrol, trehalose, lycopene, sesamol, and resveratrol. However, their use remains a challenge due to extraction, manufacturing techniques, administration route, and toxicity [13]. Table 15.1 presents plants used against CNS disorders.

Table 15.1 Plants used in the management of some central nervous system (CNS) disorders.

Plant	Family	CNS disorder used against	Regions where plant is used	References
Abrus precatorius L.	Fabaceae	Epilepsy	Tanzania	[14]
Albizia coriaria Oliv	Fabaceae	Mental illness; nervous system disorders and epilepsy	Uganda; West Africa	[15, 16]
Cannabis sativa L.	Cannabaceae	Huntington's disease		[17]
Carissa spinarum L.	Apocynaceae	Epilepsy and mental disorders		[18]
Centella asiatica (L.) Urb.	Apiaceae	Huntington's disease; depression		[19, 20]
Clausena anisata (Willd.) Hook. f. ex Benth	Rutaceae	Epilepsy	Tanzania	[14]
Croton macrostachyus Hochst. ex Delile	Euphorbiaceae	Epilepsy	Cameroon	[21]
Curcuma longa L.	Zingiberaceae	Epilepsy; Huntington's disease; depression	Tanzania	[14, 19]
Cyperus articulates L.	Cyperaceae	Epilepsy		[22]
Erythrina abyssinica DC	Fabaceae	Mental illness	Uganda	[16]
Garcinia kola Heckel	*Clusiaceae*	Huntington's disease		[23]
Hoslundia opposita Vahl	Lamiaceae	Epilepsy	Tanzania	[14]
Hypericum perforatum L.	Hypericaceae	Mild to moderate depression		[24]
Melissa officinalis L.	Lamiaceae	Depression and nervous tension		[25]
Mondia whitei	Apocynaceae	Epilepsy		[26]
Mimosa pudica L.	Fabaceae	Nervous system disorders		[27]
Olea europaea L.	Oleaceae	Huntington's disease		[28]
Passiflora edulis Sims (synonym *Passiflora incarnata*)	Passifloraceae	Anxiety and depression		[29]
Persea americana Mill	Lauraceae	Childhood convulsions and epilepsy		[30]
Margaritaria discoidea (Baill.) G.L. Webster (synonym *Phyllanthus discoideus*)	Phyllanthaceae	Convulsion, epilepsy		[31]
Securidaca longipedunculata Fresen	Polygalaceae	Epilepsy	Northern Nigeria and Cameroon	[32]
Sesamum indicum L.	Pedaliaceae	Huntington's disease		[19]
Sutherlandia frutescens (L.) R. Br.	Fabaceae	Childhood seizures and epilepsy	South Africa	[33]

(*Continued*)

Table 15.1 (Continued)

Plant	Family	CNS disorder used against	Regions where plant is used	References
Withania somnifera L.	Solanaceae	Huntington's disease		[19]
Zanthoxylum chalybeum Engl.	Rutaceae	Mental illness	Uganda	[16]
Zanthoxylum capense (Thunb.) Harv.	Rutaceae	Epilepsy	South Africa	[34]
Avena sativa L.	Poaceae	Nervous disorder	Eurasia, Africa	[7]
Acorus calamus L.	Acoraceae	Neuroprotective	Central Asia, Russia, Siberia	[7]
Datura metel L.	Solanaceae	Induces relaxation	North America	[35]
Evolvulus alsinoides L.	Convolvulaceae	Psychotropic	East Asia	[36]
Ferula asafetida L.	Apiaceae	Epilepsy	Iran	[37]
Valeriana officinalis L.	Caprifoliaceae	Sedative	Europe	[38]
Annona squamosa L.	Annonaceae	Depression	America	[39]
Bacopa monnieri L.	Plantaginaceae	Mental health disorders	Eastern India	[40]
Annona muricata L.	Annonaceae	Sedative	Cuba	[41]
Biophytum sensitivum L.	Oxalidaceae	Insomnia	Nepal	[7]
Cassia occidentalis L.	Caesalpinaceae	Hysteria	Southeast Asia	[42]
Cassia fistula L.	Caesalpinaceae	Epilepsy	Southeast Asia	[43]
Papaver somniferum L.	Papaveraceae	Narcotic	Europe	[44]
Strychnos nux-vomica L.	Loganiaceae	Paralysis	Southeast Asia	[45]
Hyoscyamus niger L.	Solanaceae	Hypnotic	Europe	[46]
Panax ginseng Oken	Araliaceae	Stimulant	Russia	[47]
Aconitum napellus L.	Ranunculaceae	Neuralgia	Europe	[7]
Loranthus longifera Jacq.	Loranthaceae	Increases brain power	Africa	[48]

15.2 Some Medicinal Plants with Neuroprotective Action on Central Nervous System Disorders

15.2.1 *Desmodium adscendens, Cleome rutidosperma, Withania somnifera,* and *Panax ginseng*

Desmodium adscendens (Sw.) DC (family Fabaceae) is used as a medicinal plant in African traditional medicine. The ethanol extract of the leaves of *D. adscendens* is used to treat CNS disorders [49]. This plant is considered safe for consumption [50].

Cleome rutidosperma DC, commonly known as fringed spider or purple cleome, is a flowering plant that belongs to the family Cleomaceae. The plant is found in tropical Africa, but also seen in wet tropical areas of Asia and Oceania as a typical garden weed. Crude methanol, chloroform, and oil ether extracts of *C. rutidosperma* showed significant pain-relieving actions [51].

Withania somnifera (L.) Dunal (winter cherry), commonly known as ashwagandha, has been used in Ayurvedic medicine for stress management, protection of neurons, and boosting memory, among other applications. The extract of *W. somnifera* was reported to inhibit lipid peroxidation CNS diseases. Organic compounds such as withanolides found in the plant prevent the proliferation of cells in the CNS (Figure 15.1). The compound also prevents the release of corticosterone and inhibits the action of choline acetyltransferase. The plant promotes cell-to-cell activities such as presynapses and postsynapses and this action was linked to the presence of compounds such as withanolide A, withanolide IV, and withanolide VI. Withanolide A was first isolated from *W. somnifera* among other constituents and plays a vital role in the regeneration of dendrites and neuronal axons. The roots of extracts of *W. somnifera* exhibit neuroprotective properties [52].

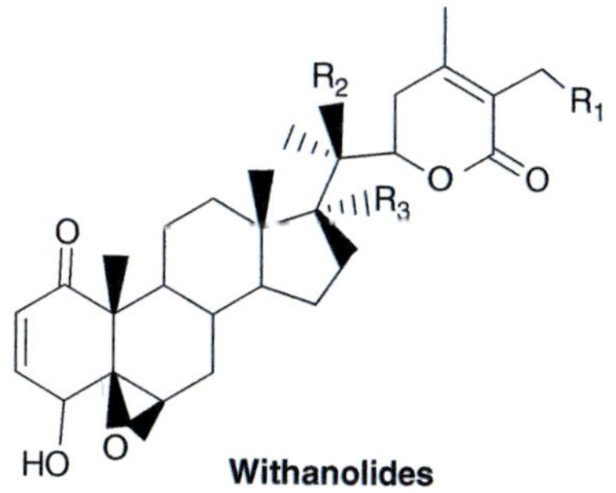

Figure 15.1 General structure for withanolides.

Ginseng is a medicinal herb found in China, Korea, and other parts of the world. Ginseng is used for the treatment of various ailments such as neurodegenerative disorder, diabetes, cancer, and hypertension. This herb was proved to be beneficial for boosting the immune system and prevention of illness. Nah et al. [53] showed that the plant can restrain voltage subordinate Ca^{2+} channels through a receptor connected to G protein that is sensitive to poison. Ginseng contains ginsenosides (Rb1 and Rg3) that are neuroprotective and this action makes the plant ideal for treating neurodegenerative illness. Also, the neuroprotective action of the plant is linked to its effect on dopaminergic neurons by hindering the level of nigral iron and possibly expanding the outflow of ferroportin (FP1) in Parkinson's disease. Chen et al. [54] stated that Rg1 inhibits the reactive oxygen species (ROS) synthesized by dopamine, transport of cytochrome c into cytosol, abrogation of caspase 3 action, and down-regulation of nitric oxide (NO) and inducible NO synthase [55].

15.2.2 *Baccopa monnieri* and *Rauvolfia serpentina*

Baccopa monnieri (L.) Wettst. is called Indian pennywort. It is known for its remarkable medicinal properties. The plant is commonly seen in India and Australia and it has beneficial actions on memory improvement as well as nerve cell regenerations. *B. monnieri* contains two saponins called bacoside A and bacoside B (Figure 15.2). These saponins are made up of sapogenins, bacogenins, and betulinic acid (Figure 15.3) and the plant contains various alkaloids. Bacoside A improves the memory [56]. *B. monnieri* has many other biologic properties such as being antioxidant, antistress, anti-inflammatory, and antimicrobial and is also used as a smooth muscle relaxant. Shinomol et al. [57] recommend the use of *B. monnieri* in Huntington's disease (HD) treatment. Mishra et al. [58] stated that the action of glutathione and glutathione reductase is enhanced in the presence of *B. monnieri* to inhibit oxidative stress.

Rauvolfia serpentina (L.) Benth. ex Kurz (Indian snakeroot) is commonly used to treat amnesia and dementia in humans and is mostly found in the Eastern and Western Ghats in India. It is used for the treatment of other types of nervous system disorders such as insomnia, mental diseases, and combative behaviors and it reduces anxiety and irritability.

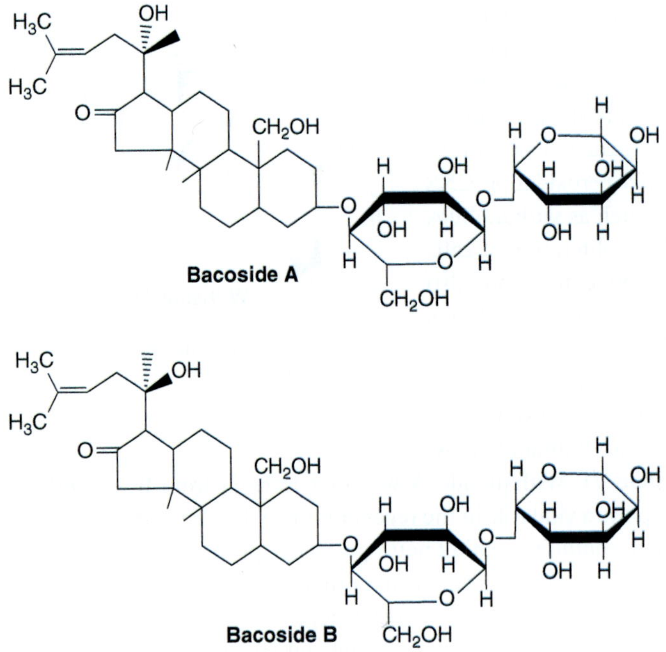

Figure 15.2 Structures of bacoside A and B.

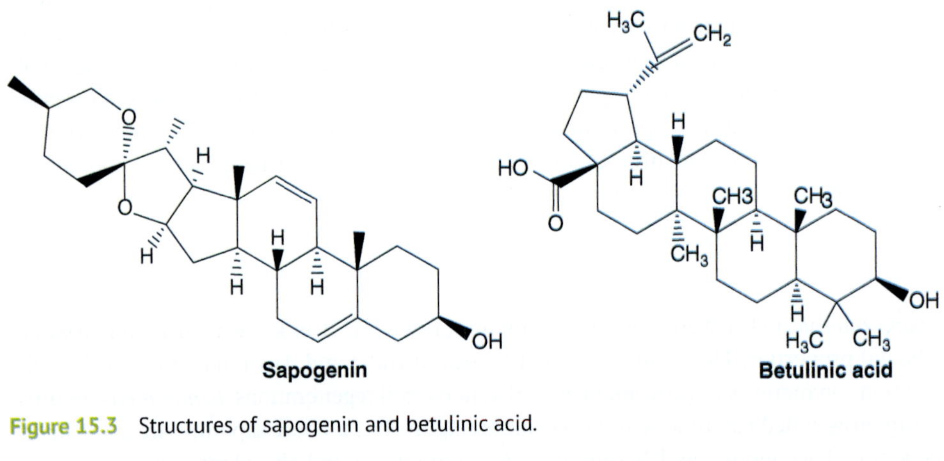

Figure 15.3 Structures of sapogenin and betulinic acid.

Many nervous system disorders such as schizophrenia, epilepsy, and psychosis can be treated with *R. serpentina*. Desperpidine, indobine, reserpine, and serpentine are the important compounds present in the plant and they are linked to the therapeutic action of *R. serpentina* against these disorders [59].

15.2.3 *Avena sativa* and *Annona squamosa*

Avena sativa L. (oat) is very effective for treating impaired nerves of the genitourinary system. *A. sativa* is found abundantly in Europe and contains secondary metabolites such as alkaloids, amino acids, and ascorbic acid, among others [7].

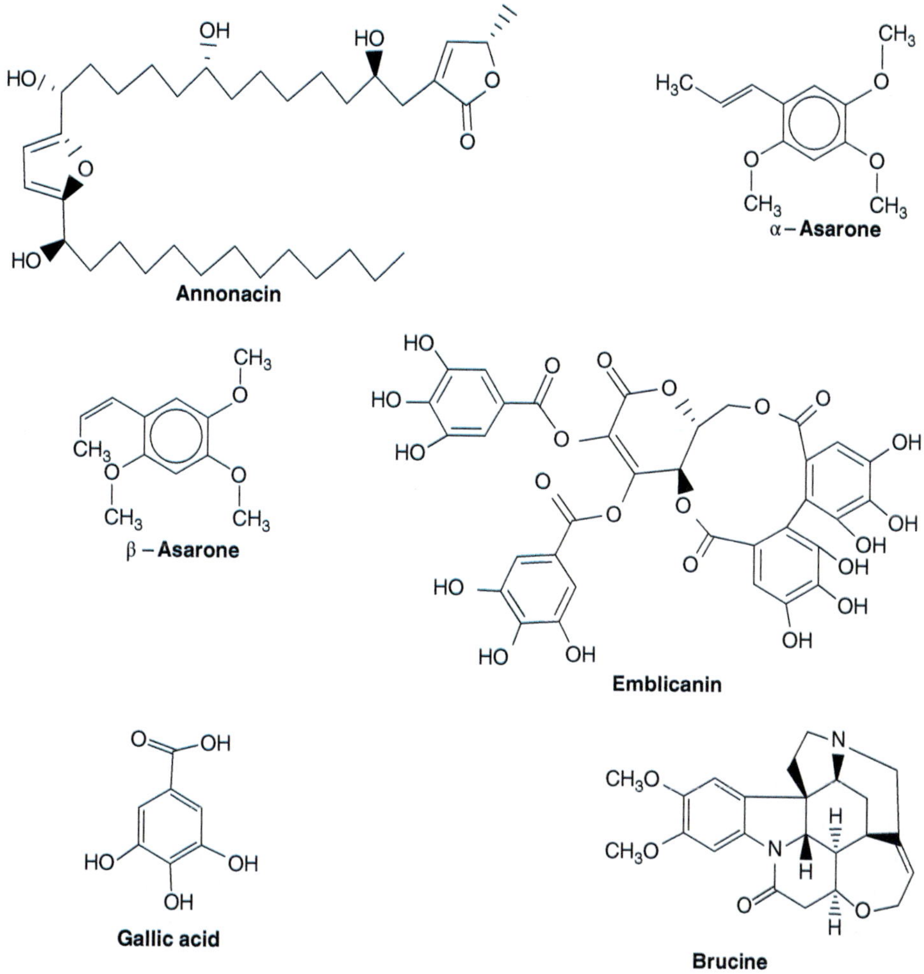

Figure 15.4 Structures of annonacin, α-asarone, β-asarone, emblicanin, gallic acid, and brucine.

The ethanol extracts of *Annona squamosa* L. (sugar apple) leaves possess neuropharmacologic properties in experimental mice subjected to the swimming test. The plant is commonly found in Cuba and other regions in the world. Annonacin (Figure 15.4) is an active constituent of *A. squamosa* and the ethanol extract of the plant leaves was reported to display antidepressant activity [60].

15.2.4 *Acorus calamus*, *Emblica officinalis*, and *Strychnos nux-vomica*

Another important medicinal plant used for managing CNS disorders is *Acorus calamus* L. (sweet flag), a plant reported to contain alpha-asarone and beta-asarone (Figure 15.4) as compounds responsible for its activity. Traditionally, chewing the rootstock of *A. calamus* can cause visual hallucinations. Furthermore, the neuroprotective effect of the plant was documented to be effective against stroke and in experimental rats chemically induced with neurodegeneration. The roots and rhizomes of *A. calamus* were reported to be effective for managing impaired functions of the CNS in India. In addition to this, pharmacologic activities like sedative, behavior modifying, CNS depressant, anticonvulsant, acetylcholinesterase inhibitory, and memory enhancing were reported [7].

Emblica officinalis Gaertn. (amla) is another beneficial plant for treating mental disorders such as memory loss, mental weakness, nervousness, nerve-cerebral pain, headache with pulsing and throbbing agony, memory loss, and others. *E. officinalis* possess active compounds such as emblicanin, ascorbic acid, polyphenols, and gallic acid (Figure 15.4), and these are linked to its neuroprotective efficacy against disorders of the CNS [7].

Strychnos nux-vomica L. (poison nut or Quaker buttons) is a plant that is abundantly found in Southeast Asia. The seed of the plant is used to treat nerve impairments and depression and it contains alkaloids such as strychnine, one of its main compounds. Also, the bark of this plant is used as an intervention for CNS disorders and contains an active component called brucine (Figure 15.4), which plays a vital role in its actions [7].

15.3 Some Central Nervous System Disorders and Medicinal Plant Interventions

15.3.1 Depression

The global burden caused by depression is on the increase [61]. Depression is an illness affecting more than 264 million people [62] and is characterized by sadness, loss of interest or pleasure, feelings of guilt or low self-esteem, loss of sleep or appetite, feelings of tiredness, and poor concentration. This condition can result in impairment of an individual's ability to function at work or school or cope with daily life, and sometimes leads to suicide [63]. With the advent of modern medicine, different effective treatments have been developed for managing mental disorders. However, between 76% and 85% of people in low- and middle-income countries (LMICs) do not have access to these treatments [64]. As a result of this, attention is now shifting to the use of medicinal plants as an alternative intervention for managing CNS disorders such as depression. Martins and Brijesh [65] conducted a detailed review of 110 medicinal plants that have been reported to have antidepressant activity. The review included the mode of action of plant species as antidepressants, the plant parts studied, their phytochemical composition, particularly the compounds with antidepressant activity, and the various antidepressant screening models used [65]. Some of the well-known medicinal plants with antidepressant activities are described here.

15.3.1.1 Family Amaryllidaceae – *Allium cepa*

Allium cepa L. (onion) exerted antidepressant-like activity in a behavioral model through the hypothalamic–pituitary–adrenal system. An aqueous bulb extract of *A. cepa* prevented an increase in metabolite/neurotransmitter ratio, suggesting its inhibition of monoamine oxidase (MAO). This inhibition was attributed to the presence of quercetin (flavonoids) in the plant (Figure 15.5) [66]. In addition, the administration of *A. cepa* reduced immobilization stress-induced behavioral deficits in mice measured by the forced swim test (FST). The activities of antioxidant enzymes such as superoxide dismutase, glutathione peroxidase, and catalase were also improved [67].

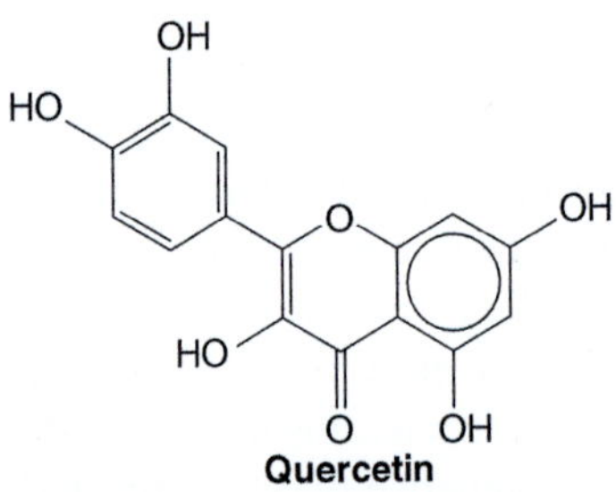

Figure 15.5 Structure of quercetin.

15.3.1.2 Family Plantaginaceae – *Bacopa monnieri*

Bacopa monnieri (L.) Wettst is a plant found in India, where it is used in traditional Ayurvedic medicine for treating epilepsy, insomnia, anxiety, and various other neuropharmacologic disorders. The plant contains several compounds such as bacoside A, bacoside A3, bacopaside I, bacopaside II, bacosaponin C (Figure 15.6), and bacosine [56, 68–70]. One of the compounds, bacoside A, isolated from the methanol extract of *B. monniera* was found to exert significant antidepressant activity in rodent models with induced depression in comparison to imipramine, a reference antidepressant drug [71]. The antidepressant action of *B. monniera* is mediated through the serotonergic and noradrenergic nervous systems [72]. It is reported that a mixture of bacopaside I and bacoside A inhibited the activity of MAO-A and MAO-B enzymes [70].

In another study, *B. monnieri* extract significantly attenuated depressive-like behaviors, normalized the levels of adrenocorticotropic hormone and corticosterone, and upregulated the expression of brain-derived neurotropic factor, neuronal marker doublecortin, and 5-bromo-2-deoxyuridine/neuronal nuclei (BrdU/NeuN) in chronic unpredictable stress

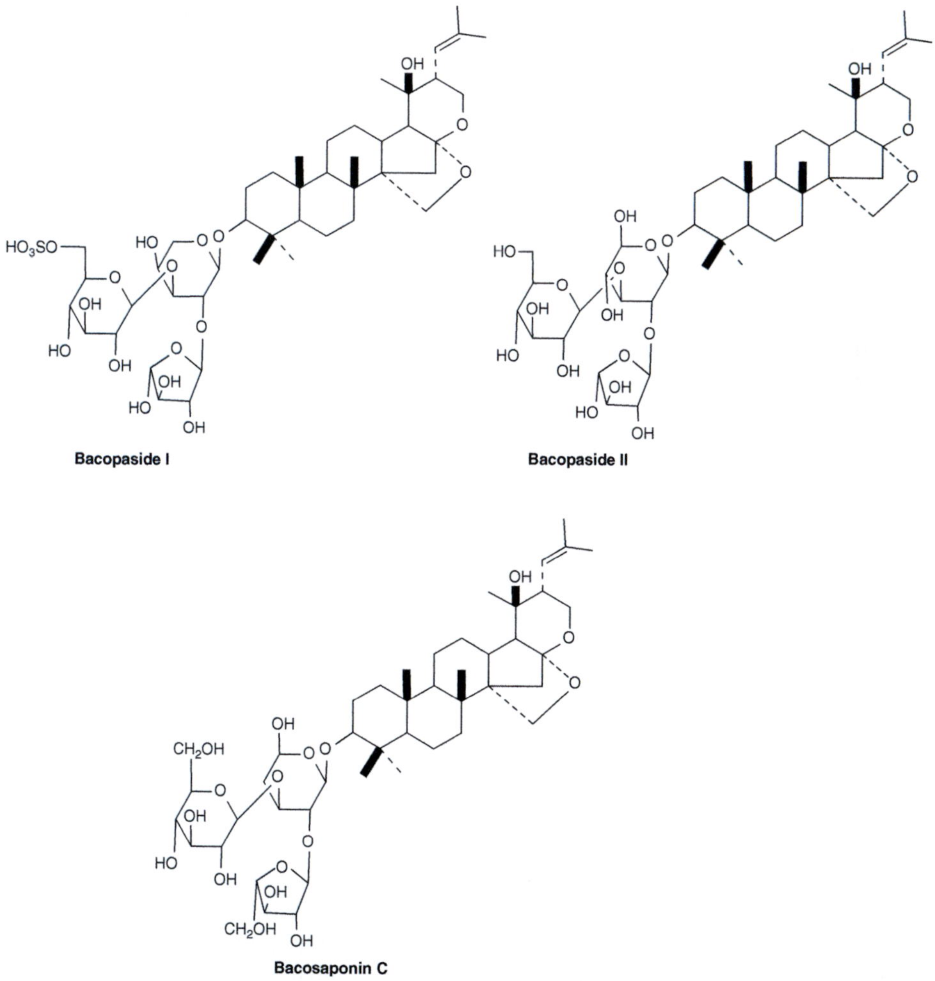

Figure 15.6 Structures of bacopaside I, bacopaside II, and bacosaponin C.

(CUS)-induced behavioral depression in rats [73]. Therefore, *B. monnieri* extract ameliorates CUS-induced behavioral depression in rats and can be used as a powerful therapeutic agent in treating depressive-like behavior [74].

In a randomized, double-blind, placebo-controlled clinical trial of up to 12 weeks of treatment in elderly individuals, a standardized *B. monnieri* extract was administered to investigate its effects on depression using the Center for Epidemiologic Studies Depression scale-10. It was observed that depression scores on this scale decreased over time for the group administered *B. monnieri*, but increased for the placebo group [75]. Furthermore, *B. monnieri* supplements have been shown to offer significant levels of protection against neurotoxicant-induced oxidative damage in the brain [69]. Choudhary et al. [19] have suggested that the strong antioxidant and protective effects against stress-mediated neuronal dysfunctions of *B. monnieri* could be useful in treating HD.

15.3.1.3 Family Fabaceae – *Glycyrrhiza glabra*

Dhingra and Sharma [76] investigated the antidepressant-like activity of *Glycyrrhiza glabra* L. in mouse models. They found out that the aqueous root extract powder of *G. glabra* (whose main constituent is glycyrrhizin) significantly attenuated the depressant-like effect in a tail suspension test (TST) after oral administration to Swiss male albino mice for one week. The antidepressant-like effect of *G. glabra* extract is mediated by increasing the level of brain norepinephrine and dopamine, but not by increasing the level of serotonin. Also, the plant has an MAO-inhibitory effect that may be playing a vital role in its antidepressant-like activity.

G. glabra has been reported to serve as a good potential adjuvant in treating depression. In support of this, an extract of *G. glabra* significantly increased the antidepressant activity of imipramine (15 mg/kg) and fluoxetine (20 mg/kg) when administered at doses of 30 and 100 mg/kg intraperitoneally (IP). The extract of *G. glabra* also significantly reduced the duration of immobility in mice in a TST, which is a widely used animal model of depression. At 30 mg/kg, the extract of *G. glabra* inhibited an eticlopride-induced increase in the duration of immobility in a TST, indicating dopamine D2/D3 receptor stimulation [77].

15.3.1.4 Family Lamiaceae – *Rosmarinus officinalis*

Rosmarinus officinalis L. (rosemary) is used for treating several ailments, including depression [78]. The hydroalcoholic extract of the stems and leaves of *R. officinalis* exerted an antidepressant-like effect in mice via interaction with the monoaminergic system. Additionally, the plant caused a significant reduction in the immobility time in an FST at 100 mg/kg and a TST at 10–100 mg/kg, compared to a control group [78].

In a related study, Machado et al. [79] showed that ursolic acid (Figure 15.7), a triterpenoid compound from *R. officinalis*, reduced the immobility time in a TST (0.01 and 0.1 mg/kg orally [PO]) and in an FST (10 mg/kg PO), similar to fluoxetine (10 mg/kg PO), imipramine (a standard drug for depression; 1 mg/kg PO), and bupropion (10 mg/kg PO). They concluded that the antidepressant-like effect of ursolic acid in the TST is likely mediated by interaction with the dopaminergic system, through the activation of dopamine D1 and D2 receptors. Also, carnosol and betulinic acid (Figure 15.7) from *R. officinalis* produced an antidepressant-like effect in a TST in Swiss mice [80].

In a similar study, Al Mukhtar and co-researchers found that an aqueous extract of *R. officinalis* (15 and 30 mg/kg) significantly reduced the immobility time and increased the FST in comparison to the untreated group, but with comparable antidepressant effects to imipramine in mice [81].

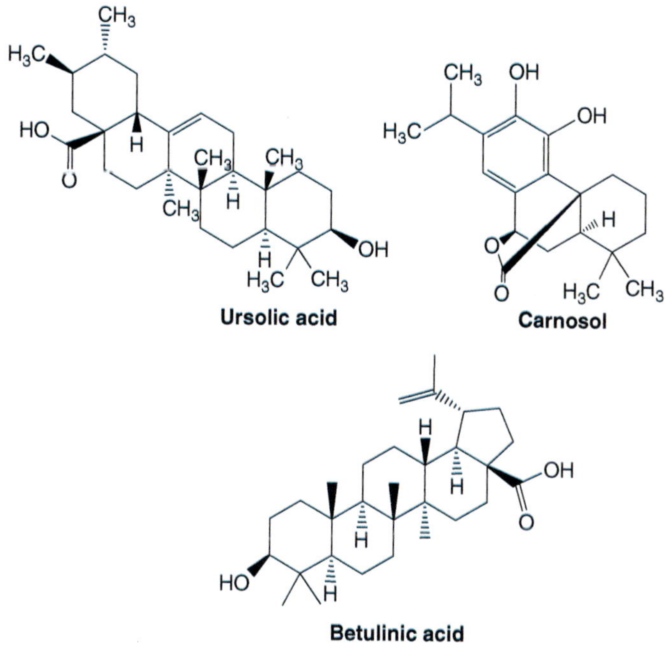

Figure 15.7 Structures of ursolic acid, carnosol, and betulinic acid.

15.3.1.5 Family Zingiberaceae – *Zingiber officinale*

A clinical trial was conducted to evaluate the efficacy of a mixture of *Boswellia carterii*, *Zingiber officinale*, and *Achillea millefolium* in severity of symptoms, anxiety, and depression on 60 patients with mild to moderate symptoms of irritable bowel syndrome (IBS). Using the Hospital Anxiety and Depression Scale, the mixture of *B. carterii*, *Z. officinale*, and *A. millefolium* was found to be effective in eliminating IBS symptoms and the related depression [82].

Ten terpenes purified from the hexane extract of *Z. officinale* were assessed for their in vitro MAO-A inhibition using the luminescence assay. The terpenes tested were 1,8-cineole, citronellal, geraniol, γ-terpinene, geranyl acetate, isobornyl acetate, terpinen-4-ol, (E, E)-α-farnesene, α-zingiberene, and β-sesquiphellandrene (Figure 15.8). The terpenes showed moderate inhibitory properties against the MAO-A enzyme in vitro with geraniol and (R)-terpinen-4-ol showing the strongest enzyme inhibitors with inhibition activity at 44.1% and 42.5%, respectively, at a concentration of 125 μg/ml. These findings highlight the potential of *Z. officinale* extracts in treating depression [83]. *In silico* analysis using docking studies using Argus lab 4.0.1 analysis intimated that gingerol and shogaol from *Z. officinale* can act as potent antidepressants, with the serotonin 5HT1A receptor protein as a potential target for treatment of depression [84].

15.3.2 Epilepsy

Epilepsy is a chronic non-communicable disease of the brain characterized by recurrent seizures due to excessive electrical discharges in a group of brain cells. The seizures vary in frequency and duration depending on which part of the brain the disturbance originates in,

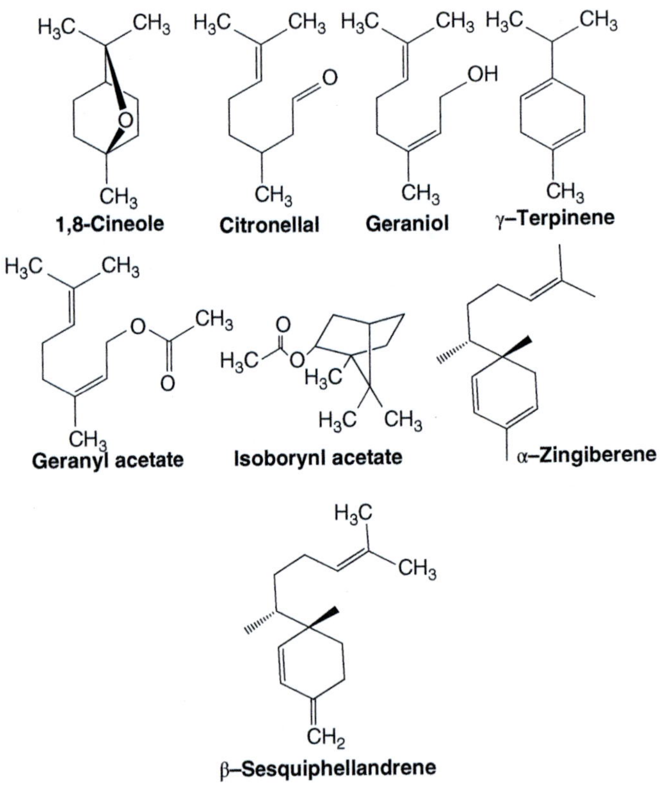

Figure 15.8 Structures of 1,8-cineole, citronellal, geraniol, γ-terpinene, geranyl acetate, isobornyl acetate, α-zingiberene, and β-sesquiphellandrene.

and how far it spreads. Some transient symptoms associated with the seizures include loss of consciousness, and disturbances of movement, sensation (including vision, hearing, and taste), mood, or other cognitive functions [85]. People with epilepsy often suffer from stigma in Africa [86]. Some of the important causes, especially in Africa, include infectious diseases such as meningitis, malaria, and cysticercosis, pathologic conditions, or diseases related to the perinatal period, febrile convulsions, and head trauma [85, 86]. There are about 50 million people living with epilepsy globally. Approximately 5 million people are diagnosed with epilepsy annually. Close to 80% of people with epilepsy live in LMICs. In high-income countries, there are about 49 per 100 000 people diagnosed with epilepsy annually compared to 139 per 100 000 in LMICs. Additionally, close to 80% of people with epilepsy live in LMICs [85], with the prevalence being highest in sub-Saharan Africa [87].

15.3.2.1 Family Hypoxidaceae – *Hypoxis hemerocallidea*

The corm of *Hypoxis hemerocallidea* Fisch, popularly known as "African potato," is widely used in South African traditional medicines for the treatment of various ailments, including childhood convulsions and epilepsy [88]. The aqueous extract of *H. hemerocallidea* exhibited anticonvulsant activity against pentylenetetrazole-, picrotoxin-, and bicuculline-induced seizures in male Balb C mice at 50–800 mg/kg when administered IP. *H. hemerocallidea* aqueous extract, significantly delayed ($p < 0.05–0.001$) the onset of, and antagonized, pentylenetetrazole-induced seizures. The extracts also antagonized picrotoxin-induced seizures, but only weakly antagonized bicuculline-induced seizures in a similar manner to the reference antiseizure standard drugs phenobarbitone and diazepam. *H. hemerocallidea* is thought to exert its antiseizure effect by enhancing GABAergic neurotransmission and/or action in the brain [89].

15.3.2.2 Family Piperaceae – *Piper methysticum*

Piper methysticum G. Forst (kava) is a plant native to the South Pacific islands. A traditional psychoactive beverage was prepared from the rhizomes, which was used for treating several ailments including insomnia and for creating a feeling of relaxation [90, 91]. *P. methysticum* contains starch and a resin from which arise several lactones (kavalactones/kava pyrones) such as kavain, dihydrokavain, yangonin, desmethoxyyangonin, methysticin, dihydromethysticin (Figure 15.9) [92]. The root also contains flavonoids [93].

P. methysticum extracts exhibit anticonvulsant, local anesthetic, and skeletal muscle relaxant properties in animal models. Also, *P. methysticum* is believed to exert antiseizure/anxiolytic effects by enabling γ-aminobutyric acid (GABA) transmission and inhibition of voltage-gated Na^+ channels in the brain, which leads to reduced neuronal excitability. Kavain is believed to be the main active ingredient responsible for this activity. Kavalactones facilitate GABA transmission, enhancing ligand binding to the GABAA receptor through a non-benzodiazepine (BDZ) receptor site [93, 94]. Kavalactones have also been shown to inhibit MAO-B in human platelets [95]. However, it is not clear to what extent tolerance occurs in humans and whether a rebound hyperexcitability can occur on sudden discontinuation [96].

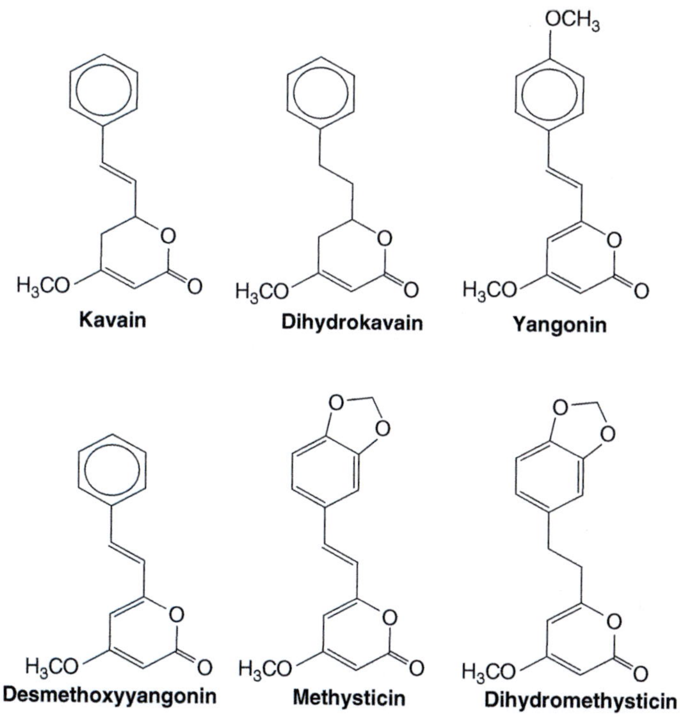

Figure 15.9 Structures of kavain, dihydrokavain, yangonin, desmethoxyyangonin, methysticin, and dihydromethysticin.

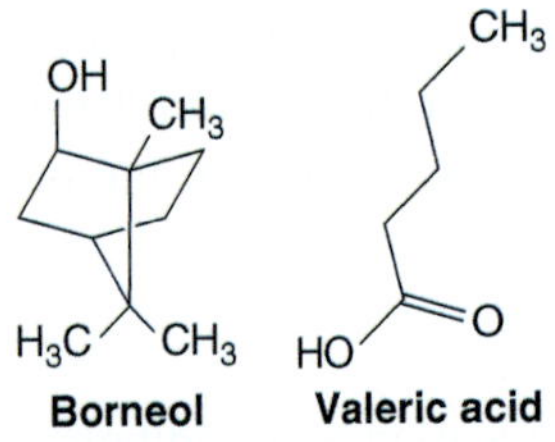

Figure 15.10 Structures of borneol and valeric acid.

15.3.2.3 Family Caprifoliaceae – *Valeriana officinalis*

Valeriana officinalis L. (valerian) is a flowering herb native to Europe and Asia, but now grown in most parts of the world. The use of valerian extends back at least 1000 years, and it gained a reputation in sixteenth-century Europe as a treatment for epilepsy [97]. The active compounds in its volatile oils are composed of monoterpenes, such as borneol, and sesquiterpenes, such as valeric acid (Figure 15.10) [98]. The aqueous extract of *V. officinalis* had anticonvulsant effects in male Sprague-Dawley rats in the animal model of temporal lobe epilepsy [99], whereas valerenic acid and the ethanol and aqueous extracts of *V. officinalis* delayed the onset of pentylenetetrazole-induced seizures in adult zebrafish [100]. The aqueous extract of *V. officinalis* was administered to mice at up to 800 mg/kg body weight IP. The anticonvulsive effect of *V. officinalis* was achieved via activation of the adenosine system [99]. Extracts from *V. officinalis* were shown to contain GABA, although its bioavailability in the brain when administered orally is not known. Additionally, aqueous extracts of valerian displaced GABA from receptors in rat brain cortex tissue [101]. *V. officinalis* extracts also have anxiolytic and antidepressant effects [102]. Constituents of *V. officinalis* inhibit enzymatic breakdown of GABA and enhance BDZ binding [103].

15.3.2.4 Family Phyllanthaceae – *Bridelia micrantha*

Bridelia micrantha (Hochst.) Baill is traditionally used for treating epilepsy in Cameroon [104]. *B. micrantha* extract was able to protect mice against maximal electroshock-, strychnine-, pentylenetetrazol-, picrotoxin-, and isonicotinic hydrazide acid-induced seizures by up to 80%. *B. micrantha* extract might show efficacy against secondarily generalized tonic–clonic seizures and primary generalized seizures in humans [105].

15.3.2.5 Family Rubiaceae – *Sarcocephalus latifolius*

A decoction from the root bark of *Nauclea latifolia* (Rubiaceae), a synonym for *Sarcocephalus latifolius* (Sm.) E. A. Bruce, protected mice against maximal electroshock-, pentylenetetrazol-, and strychnine-induced seizures. *N. latifolia* contains alkaloids, anthraquinones, flavonoids, glycosides, phenols, saponins, and tannins [31]. *N. latifolia* is also thought to have sedative psychoactive properties [106].

15.3.3 Huntington's Disease

HD is a rare genetic brain disorder that is caused by an autosomal dominant mutation of the Huntingtin (*HTT*) gene, which codes for a protein called Huntington. It results in progressive damage and loss of neurons [107–109]. The symptoms of HD can begin at any age from infancy to old age [109] and are cognitive, behavioral, and motor [110]. They include uncontrolled movement of the arms, legs, and head, and severe decline in cognition, irritability, depression, and other mood changes [111]. There is no cure to HD, but therapies that delay progression are available [107, 112]. There are significant differences in the prevalence of HD across different regions of the world, with Africa and Asia recording the lowest rates and the West having the highest rates [113].

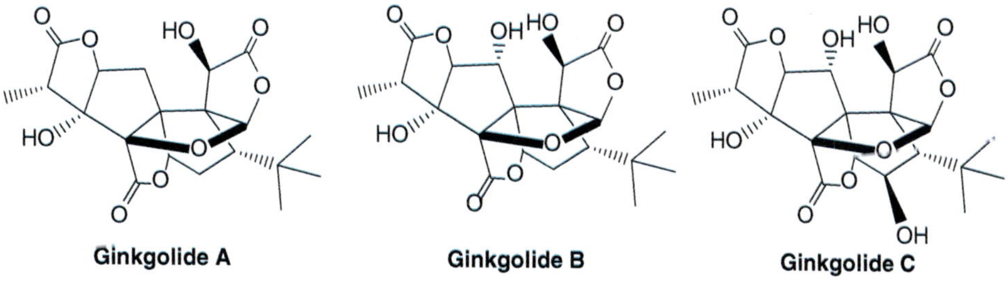

Figure 15.11 Structures of ginkgolide A, ginkgolide B, and ginkgolide C.

15.3.3.1 Family Ginkgoaceae – *Ginkgo biloba*

Ginkgo biloba L. is a dioecious tree with a long history of use in traditional Chinese medicine [114]. *G. biloba* extracts contain several compounds, including trilactonic diterpenes such as ginkgolide A, ginkgolide B, ginkgolide C (Figure 15.11); a trilactonic sesquiterpene, bilobalide; flavonoids such as quercetin, kaempferol, and isorhamnetin; biflavonoids such as ginkgetin, isoginkgetin, and sciadopitysin; and proanthocyanidins [114–116].

The 3-nitropropionic acid (3-NP) model is the animal model available for HD. Administration of *G. biloba* extract (100 mg/kg IP for 15 days) improved 3-NP-induced neurobehavioral deficits and also decreased the level of striatal malondialdehyde (MDA) [117]. Standardized *G. biloba* extract (EGb 761) also caused down- and upregulation of striatal glyceraldehyde-3-phosphate dehydrogenase and B-cell lymphoma-extra-large (Bcl-xl) expression levels, respectively [118], suggesting a neuroprotective role of EGb 761 in HD [118].

15.3.3.2 Family Araliaceae – *Panax ginseng*

Panax ginseng is the source of ginseng, which is a widely used herbal medicine. Ginsenosides are the main active components of ginseng, responsible for many of its beneficial actions. Nanomolar ginsenoside Rb1 and Rc effectively protected YAC128 medium spiny neurons (MSNs) from glutamate-induced apoptosis and Rg5 was protective at micro-molar concentration *in vitro*. These assays were performed using the HD assay with primary striatal MSN cultures from the YAC128 HD mouse model. The neuroprotective effects of Rb1, Rc, and Rg5 could correlate with their ability to inhibit glutamate-induced Ca^{2+} responses in cultured MSN and thus make them potential therapeutic agents for the treatment of HD [119].

In a separate study, the systemic administration of ginseng saponins produced significant protection against systemic 3-NP- and intrastriatal malonate-induced lesions in rat striatum in a dose-dependent manner. Ginseng significantly improved 3-NP-caused behavioral impairment and extended survival; caused a restoration of 3-NP-caused mitochondrial transmembrane potential reduction in cultured rat striatal neurons; and prevented 3-NP-induced striatal neuronal cell death in a dose-dependent manner. The neuroprotective effects might be achieved through the in vitro inhibition of 3-NP-induced intracellular Ca^{2+} elevations and cytotoxicity of striatal neurons [120].

15.3.3.3 Family Asteraceae – *Calendula officinalis*

Calendula officinalis L. (marigold) flower extract significantly attenuated behavioral alterations, oxidative damage, and striatal neuronal loss in 3-NP-treated rats and thus protected against 3-NP-induced neurotoxicity in rats. *C. officinalis* has anti-inflammatory and

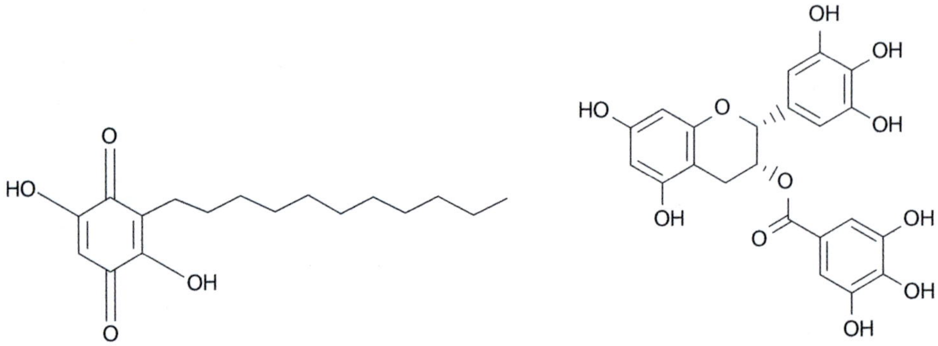

Embelin **Epigallocatechin-3-gallate**

Figure 15.12 Structures of embelin and epigallocatechin-3-gallate.

estrogenic properties that are thought to be responsible for its neuroprotective properties [121]. *C. officinalis* flowers have previously been shown to protect against monosodium glutamate-induced oxidative stress and excitotoxic brain damage in rats [122].

15.3.3.4 Family Primulaceae – *Embelia ribes*

In animals 3-NP induces severe neurotoxicity, which characterizes HD in humans [123]. Administration of 3-NP significantly altered the behavioral and neuronal antioxidant status and caused significant neuronal damage in the striatal region [124]. Embelin (Figure 15.12) from *Embelia ribes* Burm. F caused a significant reversal of behavioral and antioxidant status alterations and reversed the striatal neuronal damage induced by 3-NP at 10 and 20 mg/kg PO. Embelin also normalized the altered behavioral, biochemical, and histopathologic parameters in 3-NP-treated rats, suggesting its neuroprotective efficacy against HD, which might be attributed to its antioxidant properties [124].

15.3.3.5 Family Theaceae – *Camellia sinensis*

The polyphenol (2)-epigallocatechin-3-gallate from green tea, *Camellia sinensis* (L.) Kuntze, was shown to potently inhibit the aggregation of mutant HTT exon 1 protein in a dose-dependent manner. Epigallocatechin-3-gallate (Figure 15.12) modulates misfolding and oligomerization of mutant HTT exon 1 protein in vitro, indicating that it interfered with very early events in the aggregation process and significantly reduced polyQ-mediated HTT protein aggregation and cytotoxicity in a yeast model of HD. Transgenic HD flies overexpressing a pathogenic HTT exon 1 protein showed photoreceptor degeneration and motor function improvement when fed on epigallocatechin-3-gallate. Thus, modulators of HTT exon 1 misfolding and oligomerization like epigallocatechin-3-gallate are likely to reduce polyQ-mediated toxicity in vivo and provide a potential treatment for HD [125].

15.4 Some Mechanistic Actions of Medicinal Herbs against Central Nervous System Disorders

Medicinal plants play an important role in targeting the CNS. G protein-coupled receptors and ion channel receptors play a significant part in CNS disorders. They induce signal transduction actions in the CNS, which are activated by neurotransmitters and some other molecules, and inhibition results in impairment of CNS actions. Hence, they serve as targets for natural compounds in the management of CNS disorders [126]. There are many different studies showing that plant extracts and their metabolites have great effects on the ion channels and G protein-coupled receptors in the CNS.

15.4.1 In Vitro Studies

In vitro studies are a common approach used for analysis of the pharmacologic effect of CNS receptors of plant extracts. In this framework, binding analyses on serotonin, muscarinic, and dopamine receptors, along with GABAA and BDZ receptors, were conducted with water samples of 10 Chinese therapeutic drugs [127]. The plant extract showed the highest binding affinity with many different receptors. The GABAA receptors showed a strong binding affinity with *Gentianae macrophyllae*, *Angelicae sinensis*, and *Rhizoma chuanxiong*.

Different plant extracts showed greater affinity to the serotonin 5-HT1A and [76Br]4-bromodexetimide (BDEX) receptors, respectively. Usually, every plant extract shows interaction with two or more receptors, demonstrating the huge spectrum of accomplishment of plant extracts in the CNS. A study showed that mouse and rat neural membranes were related to the binding tests on BDZ and GABAA receptors. An interesting idea has been concentrating homogenates of human cerebral cortical cell membranes to attachment analyses on muscarinic and nicotinic receptors in the context of *Artemisia absinthium*, *Melissa officinalis*, and three Salvia species. The nicotine ligand choline has been detected in all the analyzed plant isolates. Isolates from *M. officinalis* demonstrated the strongest attraction to the nicotinic receptor, whereas *Salvia elegans* showed the strongest attraction to the muscarinic receptor.

Tanacetum partenium is used in Danish folk medicine and recently for the prophylactic medication of migraine issues. In addition, methanol and aqueous isolates of *Valeriana adscendens*, a psychoactive plant commonly used for effective medication in the Peruvian Andes, were screened for contact with CNS receptors [128]. The aqueous isolates displayed low attraction to the 5-HT1A receptor serotonin, however both isolates indicated attraction to the D1 dopamine receptor. In addition, the serotonin 5-HT2A and 5-HT2C, α1and α2 nor-adrenergic, and dopamine D2 receptors were not found to have any attraction.

15.4.2 In Vivo Studies

Plant isolates were tested in a variety of in vivo experiments, primarily in mice and rat species. For the treatment of epilepsy, SuHeXiang Wan (*Storax pili*) oil was used in Chinese medicine. Preinhalation with *S. pili* oil has been reported to postpone the onset of convulsions caused by pentylenetetrazole, suggesting GABAergic neuromodulation. It has also increased the sleeping time induced by pentobarbital. Medicinal herbal flavonoids have shown intermediate binding attraction to the GABAA receptor at the BDZ receptor region, most often in the type of partial in vivo agonists. Approaches based on the structure–activity relationship have electronegative groups integrating on the flavone backbone into the C6 and C30 sites to significantly improve binding affinity. In addition, 20-hydroxyl was formed as a key flavonoid moiety that stimulates the engineering of natural flavonoids with enhanced binding attraction and in vivo efficacy for medicinal formulations of CNS disorders. An interesting point indicated that treatment with saline-diluted potato juice in mice produced antiepileptic behavior in the bicuculline-induced seizure threshold test [129].

In the in vitro studies, betulin has been shown to bind and interact with GABAA receptor regions in the mouse brain, with in vivo seizures caused by the GABAA receptor antagonist bicuculline.

15.5 Conclusion

Medicinal plants show significant efficacy in the management of CNS disorders. Some of the roles of these plants in ameliorating CNS diseases involve inhibition of MAO activity, antioxidant actions, anti-inflammatory effects, inhibition of acetylcholinesterase activity, GABA stimulation, and others. These compounds found in these plants are responsible for their positive actions against CNS diseases. Hence, medicinal plants and other natural constituents could serve as an efficacious intervention in treating CNS diseases.

References

1 Carlini, E. (2003). Plants and the central nervous system. *Pharmacol. Biochem. Behav.* 75 (3): 501–512.

2 Farahani, M.S., Bahramsoltani, R., Farzaei, M.H. et al. (2015). Plant-derived natural medicines for the management of depression: an overview of mechanisms of action. *Rev. Neurosci.* 26 (3): 305–321.

3 Singh, A. and Trevick, S. (2016). The epidemiology of global epilepsy. *Neurol. Clin.* 34: 837–847.

4 Sahoo, N., Manchikanti, P., and Dey, S.H. (2011). Herbal drug patenting in India: IP potential. *J. Ethnopharmacol.* 137 (7): 289–297.

5 Hao, D.C. and Xiao, P.G. (2015). Genomics and evolution in traditional medicinal plants: road to a healthier life. *Evol. Bioinforma.* 11: 197–212.

6 Lundstrom, K., Pham, H.T., and Dinh, L.D. (2017). Interaction of plant extracts with central nervous system receptors. *Medicine* 4 (1): 1–12.

7 Sahoo, S. (2018). Medicinal uses of plants for nervous disorders. *Adv. Complement Altern. Med.* 3: 1–6.

8 Ratheesh, G., Tian, L., Venugopal, J.R. et al. (2017). Role of medicinal plants in neurodegenerative diseases. *Biomanufacturing Rev.* 2 (1): 2–19.

9 Couvreur, P. and Vauthier, C. (2006). Nanotechnology: intelligent design to treat complex disease. *Pharm. Res.* 23 (7): 1417–1450.

10 Cragg, G.M., Newman, D.J., and Snader, K.M. (1997). Natural products in drug discovery and development. *J. Nat. Prod.* 60 (1): 52–60.

11 Simmen, U., Higelin, J., Berger-Büter, K. et al. (2001). Neurochemical studies with St. John's wort in vitro. *Pharmacopsychiatry* 34 (Suppl. 1): 137–142.

12 Butterweck, V., Nahrstedt, A., Evans, J. et al. (2002). in vitro receptor screening of pure constituents of St. John's wort reveals novel interactions with a number of GPCRs. *Psychopharmacology* 162: 193–202.

13 Summerlin, N., Soo, E., Thakur, S. et al. (2015). Resveratrol nanoformulations: challenges and opportunities. *Int. J. Pharm.* 479 (2): 282–290.

14 Moshi, M.J., Kagashe, G.A.B., and Mbwambo, Z.H. (2005). Plants used to treat epilepsy by Tanzanian traditional healers. *J. Ethnopharmacol.* 97 (2): 327–336.

15 Adebesin, I.F., Akindele, A.J., and Adeyemi, O.O. (2015). Evaluation of neuropharmacological effects of aqueous leaf extract of *Albizia glaberrima* (Leguminosae) in mice. *J. Ethnopharmacol.* 160: 101–108.

16 Anywar, G., Kakudidi, E., Byamukama, R. et al. (2020). Indigenous traditional knowledge of medicinal plants used by herbalists in treating opportunistic infections among people living with HIV/AIDS in Uganda. *J. Ethnopharmacol.* 246: 112205.

17 Sagredo, O., Pazos, M.R., Satta, V. et al. (2011). Neuroprotective effects of phytocannabinoids-based medicines in experimental models of Huntington's disease. *J. Neurosci. Res.* 89 (9): 1509–1518.

18 Ya'u, J., Malami, S., Bako, Z. et al. (2017). Antipsychotic and sedative effects of residual aqueous fraction of ethanol root bark extract of *Carissa edulis* (Vahl) in mice. *J. Pharm. Bioresour.* 14 (2): 113–120.

19 Choudhary, S., Kumar, P., and Malik, J. (2013). Plants and phytochemicals for Huntington's disease. *Pharmacogn. Rev.* 7 (14): 81.

20 Hagemann, R.C., Burnham, T.H., Granick, B., and Neubauer, D. (1996). Gotu kola. In: *The Lawrence Review of Natural Products: Facts and Comparisons*, 41–42. St. Louis, MO: JB Lippincott.

21 Kamanyi, A., Mbiantcha, M., Nguelefack, T.B. et al. (2009). Anti-nociceptive and anti-inflammatory activities of extracts from the stem bark of *Croton macrostachyus* (Euphorbiaceae) in mice and rats. *J. Complement. Integr. Med.* 6 (1): 1255.

22 Bum, E.N., Gwa, C., Ntchapda, F. et al. (2002). Effect of the decoction of rhizomes of Cyperus articulatus on bicuculline-, N-methyl-D-aspartate-and strychnine-induced behavioural excitation and convulsions in mice. *J. Cameroon. Acad. Sci.* 2 (2): 91–95.

23 Ajayi, S.A., Ofusori, D.A., Ojo, G.B. et al. (2011). The microstructural effects of aqueous extract of *Garcinia kola* (Linn) on the hippocampus and cerebellum of malnourished mice. *Asian Pac. J. Trop. Biomed.* 1 (4): 261–265.

24 Bilia, A.R., Gallori, S., and Vincieri, F.F. (2002). St. John's wort and depression: efficacy, safety and tolerability-an update. *Life Sci.* 70 (26): 3077–3096.

25 Guginski, G., Luiz, A.P., Silva, M.D. et al. (2009). Mechanisms involved in the antinociception caused by ethanolic extract obtained from the leaves of *Melissa officinalis* (lemon balm) in mice. *Pharmacol. Biochem. Behav.* 93 (1): 10–16.

26 Adediwura, F.-J.A. and Ogunjobi, O.F. (2013). Antiepileptic activities of the extract and fractions of *Mondia whitei* (hook f.) skeel leaves. *Pharmacogn. J.* 5 (6): 256–258.

27 Ngo Bum, E., Dawack, D.L., Schmutz, M. et al. (2004). Anticonvulsant activity of Mimosa pudica decoction. *Fitoterapia* 75 (3): 309–314.

28 Tasset, I., Pontes, A.J., Hinojosa, A.J. et al. (2011). Olive oil reduces oxidative damage in a 3-nitropropionic acid-induced Huntington's disease-like rat model. *Nutr. Neurosci.* 14 (3): 106–111.

29 Jafarpoor, N., Abbasi-Maleki, S., Asadi-Samani, M., and Khayatnouri, M.H. (2014). Evaluation of antidepressant-like effect of hydroalcoholic extract of *Passiflora incarnata* in animal models of depression in male mice. *J. HerbMed. Pharmacol.* 3: 41–45.

30 Ojewole, J.A.O. and Amabeoku, G.J. (2006). Anticonvulsant effect of Persea americana mill (Lauraceae) (avocado) leaf aqueous extract in mice. *Phyther. Res.* 20 (8): 696–700.

31 Ngo Bum, E., Taiwe, G.S., Moto, F.C.O. et al. (2009). Anticonvulsant, anxiolytic, and sedative properties of the roots of Nauclea latifolia smith in mice. *Epilepsy Behav.* 15 (4): 434–440.

32 Muazu, J. and Kaita, A.H. (2008). A review of traditional plants used in the treatment of epilepsy amongst the Hausa/Fulani tribes of northern Nigeria. *Afr. J. Tradit. Complement. Altern. Med.* 5 (4): 387–390.

33 Van Wyk, B.E. and Albrecht, C. (2008). A review of the taxonomy, ethnobotany, chemistry and pharmacology of *Sutherlandia frutescens* (Fabaceae). *J. Ethnopharmacol.* 1199 (3): 620–629.

34 Amabeoku, G.J. and Kinyua, C.G. (2010). Evaluation of the anticonvulsant activity of *Zanthoxylum capense* (Thunb.) Harv. (Rutaceae) in mice. Int. *J. Pharmacol.* 6 (6): 844–853.

35 Soni, P., Siddiqui, A.A., Dwivedi, J., and Soni, V. (2012). Pharmacological properties of Datura stramonium l. as a potential medicinal tree: an overview. Asian Pac. *J. Trop. Biomed.* 2 (12): 1002–1008.

36 Nahata, A., Patil, U., and Dixit, V. (2009). Anxiolytic activity of Evolvulus alsinoides and Convulvulus pluricaulis in rodents. *Pharm. Biol.* 47 (5): 444–451.

37 Mahendra, P. and Bisht, S. (2012). Ferula asafoetida: traditional uses and pharmacological activity. *Pharmacogn. Rev.* 6 (12): 141–150.

38 Zabłocka-Słowińska, K., Jawna, K., and Biernat, J. (2013). Interactions between synthetic drugs used in treatment of selected central nervous system disorders and dietary supplements and herbal drugs. *Psychiatr. Pol.* 47 (3): 487–498.

39 Jamkhande, P.G. and Wattamwar, A.S. (2015). Annona reticulata linn. (bullock's heart): plant profile, phytochemistry and pharmacological properties. *J. Tradit. Complement. Med.* 5 (3): 144–152.

40 Kumar, V. (2016). Potential medicinal plants for CNS disorders: an overview. *Phyther. Res.* 20 (12): 1023–1035.

41 Spencer, P.S. and Palmer, V.S. Food plant chemicals linked with neurological and neurodegenerative disease. In: *Environmental Factors in Neurodegenerative Diseases* (ed. M. Aschner and L.G. Costa), 247–278. St. Louis, MO: Elsevier.

42 Vijayalakshmi, S., Ranjitha, J., Devi Rajeswari, V., and Bhagiyalakshmi, M. (2013). Pharmacological profile of *Cassia occidentalis* l.: a review. Int. *J. Pharm. Pharm. Sci.* 5 (3): 29–33.

43 Sharma, J., Gairola, S., Gaur, R. et al. (2013). Ethnomedicinal plants used for treating epilepsy by indigenous communities of sub-himalayan region of Uttarakhand, India. *J. Ethnopharmacol.* 150 (1): 353–370.

44 Prisinzano, T.E. (2009). Natural products as tools for neuroscience: discovery and development of novel agents to treat drug abuse. *J. Nat. Prod.* 72 (3): 581–587.

45 Patel, K., Laloo, D., Singh, G.K. et al. (2017). A review on medicinal uses, analytical techniques and pharmacological activities of strychnos nux-vomica linn.: a concise report. *Chin. J. Integr. Med.* https://doi.org/10.1007/s11655-016-2514-1.

46 Al-Snafi, A.E., Talab, T.A., and Majid, W.J. (2019). Medicinal plants with central nervous activity-an overview (part 1). *IOSR J. Pharm.* 9 (3): 52–102.

47 Rokot, N.T., Kairupan, T.S., Cheng, K.C. et al. (2016). A role of ginseng and its constituents in the treatment of central nervous system disorders. *Evid.-Based Complement Altern Med.* 2016: 12–20.

48 Ameer Omar, Z., Salman Ibrahim, M., Jin, Q.K., and AM. (2015). Loranthus ferrugineus: a mistletoe from traditional uses to laboratory bench. *J. Pharm.* 18 (1): 7–18.

49 N'gouemo, P., Baldy-Moulinier, B., and Nguemby-Bina, C. (1996). Effects of an ethanolic extract of Desmodium adscendens on central nervous system in rodents. *J. Ethnopharmacol.* 52 (2): 77–83.

50 Quaye, O., Cramer, P., Ofosuhene, M. et al. (2017). Acute and subchronic toxicity studies of aqueous extract of Desmodium adscendens (Sw) DC. *J. Evid.-Based Complement Altern Med.* 22 (4): 753–759.

51 Archi, F.F., Islam, S., Babu, M.A.H.K. et al. (2016). Potential evaluation of central nervous system anti-depressant activity of *Cleome rutidosperma* in mice. *Biomed. Res. Ther.* 3 (10): 889–901.

52 Yenesetti, S.C., Manjunath, M.J., and Muralidhara, C. (2016). Neuropharmacological properties of Withania somnifera – Indian ginseng: an overview on experimental evidence with emphasis on clinical trials and patents. *Recent Pat. CNS Drug Discov.* 10: 204–215.

53 Nah, S.Y., Park, H.J., and McCleskey, E.W. (1995). A trace component of ginseng that inhibits Ca2+ channels through a pertussis toxin-sensitive g protein. *Proc. Natl. Acad. Sci.* 92 (19): 8739–8743.

54 Chen, X.C., Zhu, Y.G., Zhu, L.A. et al. (2003). Ginsenoside Rg1 attenuates dopamine-induced apoptosis in pc12 cells by suppressing oxidative stress. *Eur. J. Pharmacol.* 473 (1): 1–7.

55 Liu, Q., Kou, J.P., and Yu, B.Y. (2011). Ginsenoside Rg1 protects against hydrogen peroxide-induced cell death in pc12 cells via inhibiting NF-kb activation. *Neurochem. Int.* 58 (1): 119–125.

56 Russo, A. and Borrelli, F. (2005). *Bacopa monniera*, a reputed nootropic plant: an overview. *Phytomedicine* 12 (4): 305–317.

57 Shinomol, G.K., Mythri, R.B., and Bharath, M.S. (2012). *Bacopa monnieri* extract offsets rotenone-induced cytotoxicity in dopaminergic cells and oxidative impairments in mice brain. *Cell. Mol. Neurobiol.* 32 (3): 455–465.

58 Mishra, S., Srivastava, S., Tripathi, R. et al. (2006). Phytochelatin synthesis and response of antioxidants during cadmium stress in *Bacopa monnieri* l. *Plant Physiol. Biochem.* 44 (1): 25–37.

59 Mukherjee, P.K., Kumar, V., Kumar, N.S., and Heinrich, M. (2008). The ayurvedic medicine clitoria ternatea-from traditional use to scientific assessment. *J. Ethnopharmacol.* 120 (3): 291–301.

60 Pattanaik, J., Kumar, Y., and Khatri, R.S. (2013). *Acorus calamus* linn. A herbal tonic for central nervous system. *J. Sci. Innov. Res.* 2 (5): 950–954.

61 World Health Organization (2021). Depression. https://www.who.int/news-room/fact-sheets/detail/depression (accessed 9 March 2023).

62 James, S.L., Abate, D., Abate, K.H. et al. (2018). Global, regional, and national incidence, prevalence, and years lived with disability for 354 diseases and injuries for 195 countries and territories, 1990–2017: a systematic analysis for the global burden of disease study 2017. *Lancet* 392 (10159): 1789–1858.

63 World Health Organization (2017). Depression and other common mental disorders: global health estimates. https://apps.who.int/iris/bitstream/handle/10665/254610/WHO-MSD-MER-2017.2-eng.pdf (accessed 9 March 2023).

64 Wang, P.S., Aguilar-Gaxiola, S., Alonso, J. et al. (2007). Use of mental health services for anxiety, mood, and substance disorders in 17 countries in the WHO world mental health surveys. *Lancet* 370 (9590): 841–850.

65 Martins, J. and Brijesh, S. (2018). Phytochemistry and pharmacology of anti-depressant medicinal plants: a review. *Biomed. Pharmacother.* 104: 343–365.

66 Sakakibara, H., Yoshino, S., Kawai, Y., and Terao, J. (2008). Antidepressant-like effect of onion (*Allium cepa* L.) powder in a rat behavioural model of depression. *Biosci. Biotechnol. Biochem.* 72 (1): 94–100.

67 Samad, N. and Saleem, A. (2018). Administration of *Allium cepa* L. bulb attenuates stress-produced anxiety and depression and improves memory in male mice. *Metab. Brain Dis.* 33 (1): 271–281.

68 Deepak, M. and Amit, A. (2004). The need for establishing identities of 'bacoside A and B', the putative major bioactive saponins of Indian medicinal plant *Bacopa monnieri*. *Phytomedicine* 11 (2/3): 264.

69 Shinomol, G.K. and Muralidhara. (2011). *Bacopa monnieri* modulates endogenous cytoplasmic and mitochondrial oxidative markers in prepubertal mice brain. *Phytomedicine* 18 (4): 317–326.

70 Singh, R., Ramakrishna, R., Bhateria, M., and Bhatta, R.S. (2014). in vitro evaluation of Bacopa monniera extract and individual constituents on human recombinant monoamine oxidase enzymes. *Phyther. Res.* 28 (9): 1419–1422.

71 Sairam, K., Dorababu, M., Goel, R.K., and Bhattacharya, S.K. (2002). Antidepressant activity of standardized extract of Bacopa monniera in experimental models of depression in rats. *Phytomedicine* 9 (3): 207–211.

72 Girish, C., Oommen, S., and Vishnu, R. (2016). Evidence for the involvement of the monoaminergic system in the antidepressant-like activity of methanolic extract of *Bacopa monnieri* in albino mice. *Int. J. Basic Clin. Pharmacol.* 5 (3): 914–922.

73 Kumar, S. and Mondal, A.C. (2016). Neuroprotective, neurotrophic and anti-oxidative role of *Bacopa monnieri* on CUS induced model of depression in rat. *Neurochem. Res.* 41 (11): 3083–3094.

74 Hazra, S., Kumar, S., Saha, G.K., and Mondal, A.C. (2017). Reversion of BDNF, Akt and CREB in Hippocampus of chronic unpredictable stress induced rats: effects of phytochemical, *Bacopa monnieri*. *Psychiatry Investig.* 14 (1): 74–80.

75 Calabrese, C., Gregory, W.L., Leo, M. et al. (2008). Effects of a standardized *Bacopa monnieri* extract on cognitive performance, anxiety, and depression in the elderly: a randomized, double-blind, placebo-controlled trial. *J. Altern Complement Med.* 14 (6): 707–713.

76 Dhingra, D. and Sharma, A. (2006). Antidepressant-like activity of *Glycyrrhiza glabra* L. in mouse models of immobility tests. *Prog Neuro-Psychopharmacology Biol. Psychiatry* 30 (3): 449–454.

77 Kasture, S.B., Barhate, S.A., and Mohan, M. (2018). Glycyrrhiza glabra as an adjuvant in treatment of parkinsonism and depression. *Planta Med.* 79 (9): PA310.

78 Machado, D.G., Bettio, L.E.B., Cunha, M.P. et al. (2009). Antidepressant-like effect of the extract of *Rosmarinus officinalis* in mice: involvement of the monoaminergic system. *Prog. Neuro-Psychopharmacology Biol. Psychiatry.* 33 (4): 642–650.

79 Machado, D.G., Neis, V.B., Balen, G.O. et al. (2012). Antidepressant-like effect of ursolic acid isolated from *Rosmarinus officinalis* L. in mice: evidence for the involvement of the dopaminergic system. *Pharmacol. Biochem. Behav.* 103 (2): 204–211.

80 Machado, D.G., Cunha, M.P., Neis, V.B. et al. (2013). Antidepressant-like effects of fractions, essential oil, carnosol and betulinic acid isolated from *Rosmarinus officinalis* L. *Food Chem.* 136 (2): 999–1005.

81 Al Mukhtar, E.J., Selman, S.M., Sahib, Z.H., and Naji, H. (2013). Antidepressant-like effect of *Rosmarinus officinalis* extract in male mice. *Med. J. Babylon.* 10 (4): 803–808.

82 Kazemian, A., Toghiani, A., Shafiei, K. et al. (2017). Evaluating the efficacy of mixture of *Boswellia carterii*, *Zingiber officinale*, and *Achillea millefolium* on severity of symptoms, anxiety, and depression in irritable bowel syndrome patients. *J. Res. Med. Sci.* 22: 120.

83 Kukula-Koch, W., Koch, W., Czernicka, L. et al. (2018). MAO-A inhibitory potential of terpene constituents from ginger rhizomes-a bioactivity guided fractionation. *Molecules* 23 (6): 1301.

84 Ittiyavirah, S.P. and Meera, P. (2013). In silico docking analysis of constituents of Zingiber officinale as antidepressant. *J. Pharmacogn. Phyther.* 5 (6): 101–105.

85 World Health Organization (2019). Epilepsy. https://www.who.int/news-room/fact-sheets/detail/epilepsy (accessed 9 March 2023).

86 Diop, A.G., Hesdorffer, D.C., Logroscino, G., and Hauser, W.A. (2005). Epilepsy and mortality in Africa: a review of the literature. *Epilepsia* 46 Supplem(1): 33–35.

87 Whiteford, H.A., Ferrari, A.J., Degenhardt, L. et al. (2015). Global burden of mental, neurological, and substance use disorders: an analysis from the global burden of disease study 2010. *Ment. Neurol. Subst. Use Disord.* 10 (2): e0116820.

88 Wyk, B.E., Van Oudtshoorn, B., and Gericke, N. (2002). *Medicinal Plants of South Africa*, 2e. Pretoria: Briza Publications.

89 Ojewole, J.A.O. (2008). Anticonvulsant activity of Hypoxis hemerocallidea Fisch. and C. A. Mey. (Hypoxidaceae) corm ('African potato') aqueous extract in mice. *Phyther. Res.* 22 (1): 91–96.

90 Piscopo, G. (1997). Kava kava: gift of the islands. *Altern. Med. Rev.* 2 (5): 355–364.

91 Stevinson, C., Huntley, A., and Ernst, E. (2002). A systematic review of the safety of kava extract in the treatment of anxiety. *Drug Saf.* 25 (4): 251–261.

92 Lebot, V., Merlin, M., and Lindstrom, L. (1997). *Kava: The Pacific Elixir: The Definitive Guide to its Ethnobotany, History, and Chemistry*. Rochester, VT: Inner Traditions/ Bear & Co.

93 Capasso, F., Gaginella, T.S., Grandolini, G., and Izzo, A.A. (2003). Plants and the nervous system. In: *Phytotherapy: A Quick Reference to Herbal Medicine* (ed. F. Capasso, T.S. Gaginella, G. Grandolini, and A.A. Izzo), 89–108. Berlin: Springer.

94 Jussofie, A., Schmiz, A., and Hiemke, C. (1994). Kavapyrone enriched extract from Piper methysticum as modulator of the GABA binding site in different regions of rat brain. *Psychopharmacology* 116 (4): 469–474.

95 Uebelhack, R., Franke, L., and Schewe, H.J. (1998). Inhibition of platelet MAO-B by kava pyrone-enriched extract from Piper methysticum Forster (kava-kava). *Pharmacopsychiatry* 31 (5): 187–192.

96 Spinella, M. (2001). Herbal medicines and epilepsy: the potential for benefit and adverse effects. *Epilepsy Behav.* 2 (6): 524–532.

97 Temkin, O. (1994). *The Falling Sickness: A History of Epilepsy from the Greeks to the Beginnings of Modern Neurology*, vol. 4. Baltimore, MD: Johns Hopkins University Press.

98 Houghton, P.J. (1999). The scientific basis for the reputed activity of valerian. *J. Pharm. Pharmacol.* 51 (5): 505–512.

99 Rezvani, M.E., Roohbakhsh, A., Allahtavakoli, M., and Shamsizadeh, A. (2010). Anticonvulsant effect of aqueous extract of *Valeriana officinalis* in amygdala-kindled rats: possible involvement of adenosine. *J. Ethnopharmacol.* 127 (2): 313–318.

100 Torres-Hernández, B.A., Del Valle-Mojica, L.M., and Ortíz, J.G. (2015). Valerenic acid and *Valeriana officinalis* extracts delay onset of Pentylenetetrazole (PTZ)-induced seizures in adult Danio rerio (zebrafish). *BMC Complement. Altern. Med.* 15: 228.

101 Santos, M.S., Ferreira, F., Faro, C. et al. (1994). The amount of GABA present in aqueous extracts of valerian is sufficient to account for [3H] GABA release in synaptosomes. *Planta Med.* 60 (5): 475–476.

102 Hattesohl, M., Feistel, B., Sievers, H. et al. (2008). Extracts of *Valeriana officinalis* L. s.l. show anxiolytic and antidepressant effects but neither sedative nor myorelaxant properties. *Phytomedicine* 15 (1): 2–15.

103 Riedel, E., Hänsel, R., and Ehrke, G. (1982). Inhibition of gamma-aminobutyric acid catabolism by valerenic acid derivatives. *Planta Med.* 46 (4): 219–220.

104 Nole, T., Lionel, T.D.W., Cedrix, T.F.S., and Gabriel, A.A. (2016). Ethnomedical and ethnopharmacological study of plants used for potential treatments of diabetes and arterial hypertension by indigenous people in three phytogeographic regions of Cameroon. *Diabetes Case Rep.* 1 (110): 2.

105 Ngo Bum, E., Ngah, E., Ngo Mune, R.M. et al. (2012). Decoctions of *Bridelia micrantha* and *Croton macrostachyus* may have anticonvulsant and sedative effects. *Epilepsy Behav.* 24 (3): 319–323.

106 Amos, S., Abbah, J., Chindo, B. et al. (2005). Neuropharmacological effects of the aqueous extract of Nauclea latifolia root bark in rats and mice. *J. Ethnopharmacol.* 97 (1): 53–57.

107 Ohlmeier, C., Saum, K.U., Galetzka, W. et al. (2019). Epidemiology and health care utilization of patients suffering from Huntington's disease in Germany: real world evidence based on German claims data. *BMC Neurol.* 19 (1): 318.

108 Reiner, A., Dragatsis, I., and Dietrich, P. (2011). Genetics and neuropathology of Huntington's disease. *Int. Rev. Neurobiol.* 98: 325–372.

109 Vonsattel, J.P.G. and DiFiglia, M. (1998). Huntington disease. *J. Neuropathol. Exp. Neurol.* 57 (5): 369.

110 McColgan, P. and Tabrizi, S.J. (2018). Huntington's disease: a clinical review. *Eur. J. Neurol.* 25 (1): 24–34.

111 Baig, S.S., Strong, M., and Quarrell, O.W.J. (2016). The global prevalence of Huntington's disease: a systematic review and discussion. *Neurodegener Dis. Manag.* 6 (4): 331–343.

112 Kavitha, R.V., Kumar, J.R., Egbuna, C., and Ifemeje, J.C. (2020). Phytochemicals as therapeutic interventions in neurodegenerative diseases. In: *Phytochemicals as Lead Compounds for New Drug Discovery* (ed. C. Egbuna, S. Kumar, J.C. Ifemeje, et al.), 161–178. St. Louis: MO: Elsevier.

113 Rawlins, M.D., Wexler, N.S., Wexler, A.R. et al. (2016). The prevalence of Huntington's disease. *Neuroepidemiology* 46 (2): 144–153.

114 McKenna, D.J., Jones, K., and Hughes, K. (2001). Efficacy, safety, and use of Ginkgo biloba in clinical and preclinical applications. *Altern. Ther. Health Med.* 7 (5): 70–90.

115 DeFeudis, F.V. and Drieu, K. (2000). Ginkgo biloba extract (EGb 761) and CNS functions: basic studies and clinical applications. *Curr. Drug Targets* 1 (1): 25–58.

116 Nakanishi, K. (2005). Terpene trilactones from Gingko biloba: from ancient times to the 21st century. *Bioorg. Med. Chem.* 13 (17): 4987–5000.

117 Mahdy, H.M., Tadros, M.G., Mohamed, M.R. et al. (2011). The effect of Ginkgo biloba extract on 3-nitropropionic acid-induced neurotoxicity in rats. *Neurochem. Int.* 59 (6): 770–778.

118 Birks, J., Grimley, E.V., and Van Dongen, M. (2002). Ginkgo biloba for cognitive impairment and dementia. *Cochrane Database Syst. Rev.* 4: CD003120.

119 Wu, J., Jeong, H.K., Bulin, S.E. et al. (2009). Ginsenosides protect striatal neurons in a cellular model of Huntington's disease. *J. Neurosci. Res.* 87 (8): 1904–1912.

120 Kim, J.H., Kim, S., Yoon, I.S. et al. (2005). Protective effects of ginseng saponins on 3-nitropropionic acid-induced striatal degeneration in rats. *Neuropharmacology* 48 (5): 743–756.

121 Shivasharan, B.D., Nagakannan, P., Thippeswamy, B.S. et al. (2013). Protective effect of *Calendula officinalis* Linn. Flowers against 3-nitropropionic acid induced experimental Huntington's disease in rats. *Drug Chem. Toxicol.* 36 (4): 466–473.

122 Nagakannan, P., Shivasharan, B.D., Thippeswamy, B.S. et al. (2012). Protective effect of hydroalcoholic extract of Mimusops elengi Linn. Flowers against middle cerebral artery occlusion induced brain injury in rats. *J. Ethnopharmacol.* 140 (2): 247–254.

123 Tariq, M., Khan, H.A., Elfaki, I. et al. (2005). Neuroprotective effect of nicotine against 3-nitropropionic acid (3-NP)-induced experimental Huntington's disease in rats. *Brain Res. Bull.* 67 (1–2): 161–168.

124 Dhadde, S.B., Nagakannan, P., Roopesh, M. et al. (2016). Effect of embelin against 3-nitropropionic acid-induced Huntington's disease in rats. *Biomed. Pharmacother.* 77: 52–58.

125 Ehrnhoefer, D.E., Duennwald, M., Markovic, P. et al. (2006). Green tea (−)-epigallocatechin-gallate modulates early events in huntingtin misfolding and reduces toxicity in Huntington's disease models. *Hum. Mol. Genet.* 15 (18): 2743–2751.

126 Nazıroglu, M. and Demirdas, A. (2015). Psychiatric disorders and TRP channels: focus on psychotropic drugs. *Curr. Neuropharmacol.* 13 (2): 248–257.

127 Liao, J., Jan, Y., Huang, S. et al. (1995). Evaluation with receptor binding assay on the water extracts of ten CNS-active chinese herbal drugs. *Proc. Natl. Sci. Counc. Repub. China B* 19 (3): 151–158.

128 Nencini, C., Cavallo, F., Capasso, A. et al. (2006). Binding studies for serotoninergic, dopaminergic and noradrenergic receptors of Valeriana adscendens trel. Extracts. *J. Ethnopharmacol.* 108 (2): 185–187.

129 Muceniece, R., Saleniece, K., Krigere, L. et al. (2008). Potato (*Solanum tuberosum*) juice exerts an anticonvulsant effect in mice through binding to GABA receptors. *Planta* 74 (5): 491–496.

16

Important Antihistaminic Plants and Their Potential Role in Health

Salwa Bouabdallah[1,2,3], *Hagar A. Sobhy*[4], *Babatunde O. Adetuyi*[4], *Omayma A. Eldahshan*[4], *and Chukwuebuka Egbuna*[5,6]

[1] *Environmental Biomonitoring Laboratory LBE (LR01/ES14), Faculty of Sciences Bizerta, Carthage University, Zarzouna, Tunisia*
[2] *Faculty of Sciences Tunis, University of Tunis El-Manar, Tunis, Tunisia*
[3] *Pharmacognosy Department, Faculty of Pharmacy, Ain Shams University, Cairo, Egypt*
[4] *Department of Natural Sciences, Faculty of Pure and Applied Sciences, Precious Cornerstone University, Ibadan, Oyo State, Nigeria*
[5] *Africa Centre of Excellence for Public Health and Toxicological Research (ACE-PUTOR), University of Port Harcourt, Choba, Rivers State, Nigeria*
[6] *Department of Biochemistry, Chukwuemeka Odumegwu Ojukwu University, Uli, Anambra State, Nigeria*

Allergy is an immune reaction in which the immune system responds to an external stimulus from the environment (pollen, pet, dander, foods) and thus produces several hypersensitivity reactions [1]. It is a disorder of the immune system, which no longer tolerates substances that are harmless to the body: allergens. The term "allergy," derived from the Greek *allos* meaning other and *ergon* meaning reaction, was defined by Hungarian-born Bela Schick and Clemens von Pirquet in 1906 [2]. An allergen is a substance that causes an allergic reaction, a biologic aggressor. It is an antigen that causes an allergy by hypersensitivity in sensitive people who have already been in contact with it. Such a chemical substance causes a violent reaction of the immune system after contact with the skin, ingestion, or inhalation [3].

The spread of allergic conditions has increased strongly in the last decade [1, 4]. Drug allergy triggers the same hypersensitivity reactions that occur in response to medication. β-Lactam antibiotics such as cephalosporins and penicillin, sulfonamides, angiotensin-converting enzyme (ACE) inhibitors, local anesthetics, and chemotherapeutic drugs are the most common drugs that have been reported as inducing undesirable hypersensitivity reactions in some patients [5].

Histamine remains the principal conducting mediator and the discovery of its physiologic action was demonstrated by Dale et al., who isolated histamine from the mold ergot [6]. Histamine shows its potential via the binding of histamine receptors, especially H1. Histamine receptors are a group of the G-protein-coupled receptor family that differ in their location and their histamine-binding characteristics [7].

Depending on the locality, histamine can produce sneezing, itching, and rhinorrhea in the nose, edema, erythema, and itching in the skin, and bronchoconstriction in the

Phytochemical Drug Discovery for Central Nervous System Disorders: Biochemistry and Therapeutic Effects, First Edition. Edited by Chukwuebuka Egbuna and Mithun Rudrapal.

respiratory system [8]. The mechanism involved in allergic reactions is very complex, but in general it is expressed by the binding of immunoglobulin E (IgE) antibodies to an allergen, then to mast cells, leading to the diagnosis of allergy based on medical history as well as blood and/or skin tests [9, 10]. Allergy treatment strategies consist of prescribing antiallergic drugs to hypersensitive patients. Daniel Bovet and his student discovered the first antihistamines. First-generation H1 antihistamines exert an inverse agonist effect [6]. They have been reported to show adverse effects, including anticholinergic effects [11]. Adverse effects of H1 antihistamines such as chlorpheniramine [7] manifest as dizziness, sedation, and lack of concentration [12]. Second-generation H1 antihistamines, such as loratadine and cetirizine, are more specific, effective with fewer sedative side effects compared to the first generation [12]. Corticosteroids are used to control allergic conditions, although special precautions must be taken for their safety [13]. Mast cell stabilizers such as cromoglicic acid are clinically recommended to prevent allergic reactions by inhibiting the release of allergic mediators from mast cells [14]. Leukotriene inhibitors such as zafirlukast and montelukast are effective for the treatment of asthma and allergic rhinitis. Allergen immunotherapy, which consists of repeated exposure to allergens, can achieve allergen tolerance and may be effective in controlling mild allergic conditions [15].

The current antihistaminic drugs have some limitations in terms of efficacy and safety. Most allergic conditions are chronic and long-term therapy is basic. Therefore, most allergic patients look for better alternatives to synthetic antihistaminic agents. Several plants and their derived natural products have been reported to be safe, effective, and inexpensive antiallergic agents. In this chapter we highlight the relevant use of plants from folk medicine or scientific-based evidence for natural antiallergic drugs. Special emphasis is given to natural compounds with reported antiallergic activity.

Currently, several plant-derived formulas have reached clinical trials as safe antiallergics, with similar mechanisms to synthetic drugs.

16.1 Antihistaminic Plants

16.1.1 Family Acanthaceae

Asystasia gangetica T. Adams is used for the management of asthma. The hexane, ethyl acetate, and methanol extracts of the leaves of *A. gangetica* were evaluated for antiasthmatic activity (guinea pig trachea; rat stomach strip; guinea pig ileal preparation; and egg albumin-induced acute inflammation). The results indicated that the extracts did not exhibit relaxant or contractile activity in isolated tissue preparations, but they inhibited the contraction evoked by spasmogens [15].

16.1.2 Family Amaranthaceae

Aerva lanata L. is a prostrate herbaceous plant that is recognized by its white axillary bunches of small wooly flowers. It is abundant in the warmer parts of India. Ethanol extract of aerial parts of *A. lanata* at 100 g/ml has been studied in the isolated goat tracheal chain preparation model, and 30 and 60 mg/kg doses orally in clonidine-induced catalepsy and mast cells [12].

16.1.3 Family Amaryllidaceae

Crinum glaucum is popular among the Yoruba of southwest Nigeria. It is a traditional effective remedy for convulsions, cough, and asthma. The aqueous extract of *C. glaucum* possesses antiallergic activity (doses 100–400 mg/kg) by reduction in area of dye leak in passive cutaneous anaphylactic reaction, protecting histamine-induced bronchoconstriction in the guinea pig and degranulation of mast cells [16].

Curculigo orchioides Gaertn. is a small herb, widely dispersed in China, Malaysia, Japan, and India. Alcoholic extract of rhizomes (doses 100–400 mg/kg) shows mast cell stabilizing and antihistaminic activity on compound 48/80-induced mast cell degranulation and systemic anaphylaxis [17]. The rhizomes extract inhibited histamine-induced contraction in goat trachea, bronchoconstriction, guinea pig ileum, and guinea pigs, as well as egg albumin-induced passive paw anaphylaxis in rats [18].

16.1.4 Family Asclepiadaceae

Hemidesmus indicus R. is a twining shrub commonly found in India. Bhujbal et al. reported antiasthmatic activity of ethanol extract of *H. indicus roots* at doses of 25, 50, and 100 mg/kg using isolated goat tracheal chain preparation, passive paw anaphylaxis in rats, and clonidine-induced catalepsy in mice [19, 20].

16.1.5 Family Asteraceae

Ageratum conyzoides is an erect, annual herbaceous plant, native to tropical America, but with a distribution range in tropical and subtropical areas around the world. The hydroalcoholic extract of leaves of *A. conyzoides* at doses of 250, 500, and 1000 mg/kg shows antihistaminic activity by inhibiting clonidine-induced catalepsy in mice [21].

Sphaeranthus indicus Kurz is a medicinal plant used in folk medicine. The ethanol and ethyl acetate extracts at doses of 100, 150, and 300 mg/kg are slightly more effective against sheep serum and compound 48/80-induced mast cell degranulation than the standard drug ketotifen [22].

16.1.6 Family Caesalpiniaceae

Cassia sophera is used in traditional medicine for treatment of asthma and bronchitis. Chloroform, ethyl acetate, and ethanol fractions of leaves of *C. sophera* provided significant antiasthmatic activity in carrageenan-induced paw edema, histamine-induced bronchoconstriction, clonidine- and haloperidol-induced catalepsy, milk-induced leukocytosis, and eosinophilia and passive paw anaphylaxis animal models at doses of 250, 500, and 750 mg/kg. The activity may be due to flavonoid compounds [23].

16.1.7 Family Casuarinaceae

Casuarina equisetifolia L. is a green tree (height up to 50 m) cultivated in coastal regions from Gujarat to Orissa, some parts of West Bengal, and in Andamans. The methanol extract of wood and bark shows antihistaminic activity by inhibiting the histamine-induced contraction of trachea (10–80 μc/ml), clonidine-induced catalepsy, and mast cell degranulation at 100 mg/kg [24].

16.1.8 Family Cruciferae

Lepidium sativum L., commonly known as Asaliyo, is an annual glabrous herb, cultivated as a salad plant in India. The ethyl acetate, methanol, ethanol extracts, and n-butanol fractions inhibit bronchospasm induced by acetylcholine and histamine [25].

16.1.9 Family Cucurbitacea

Momordica dioica is a climbing creeper plant. Its fruits and leaves are traditionally used to treat asthma, bronchitis, and tridosha. Methanol and aqueous extract of pulp show antihistaminic activity by inhibiting clonidine-induced catalepsy in mice (50 mg/kg) [26].

16.1.10 Family Euphorbiaceae

Euphorbia hirta is a wild herbaceous plant growing in the hotter parts of India. Popularly it is known as asthma weed. Ethanol extract of the whole aerial part of the plant (doses 100–1000 mg/kg) shows antihistaminic and antiallergic activities by inhibiting passive cutaneous anaphylaxis and paw anaphylaxis reaction, and protection of mast cell from degranulation [27].

16.1.11 Family Fabaceae

Amburana cearensis is a medicinal plant common to the northeast of Brazil, known as caatinga (savannah), and popularly used in respiratory path diseases and asthma. The flavonoid isokaempferide isolated from the bark of *A. cearensis* shows significant relaxation of KCl-induced contraction on guinea pig trachea [28].

16.1.12 Family Lamiaceae

Mentha spicata L., the ethyl acetate fractions of *M. spicata* leaves, manifested antihistaminic activity by hindering antigen-stimulated rat basophile [29].

16.1.13 Family Moraceae

Ficus bengalensis L. is a very tall tree (height of 30 m) and sends up many aerial roots from the branches. Ethyl acetate, aqueous, and ethanol extracts possess antihistaminic activity by inhibiting clonidine-induced catalepsy in mice (dose 50 mg/kg) [30, 31].

16.1.14 Family Myricaceae

Myrica esculenta Buch. Ham, commonly known as Kaiphal, is used for treatment of bronchitis and asthma in Ayurveda. The antiallergic and anti-inflammatory activities of ethanol extract of the aerial parts were reported using allergic pleurisy and acetic acid–induced vascular permeability in mice (75 and 150 mg/kg) [32]. The stem bark of this plant shows bronchodilatation and antianaphylactic activities by inhibiting acetylcholine-induced bronchospasm in guinea pigs and egg albumin–induced anaphylaxis in guinea pigs (75 mg/kg), and by relaxing acetylcholine- and histamine-induced guinea pig trachea and ileum [33].

16.1.15 Family Myrtaceae

Eugenia jambolana Lamk. has been reported to be useful as an astringent in diarrhea as well as dysentery. The bark ethanolic extract showed anti-inflammatory activity in carrageenan and formaldehyde paw edema. The same extract was also shown to inhibit histamine-, serotonin (5-HT)-, and prostaglandin 2-induced paw edema [2, 4].

Syziygium cumini is among the most important genera of the Myrtaceae family. It is widely used in folk medicine. The antiallergic properties of an aqueous leaf extract of *Syzygium cumini* (L.) Skeels was investigated. Oral administration of *S. cumini* (25–100 mg/kg) in Swiss mice inhibited paw edema induced by compound 48/80. *S. cumini* treatment also inhibited the edema induced by histamine and 5-HT. It prevented mast cell degranulation and the consequent histamine release in Wistar rat peritoneal mast cells induced by compound 48/80. Pretreatment of BALB/c mice with 100 mg/kg of the extract significantly inhibited eosinophil accumulation in allergic pleurisy. This effect was related to the inhibition of interleukin (IL)-5 and cytokine C-C motif chemokine 11 (CCL11)/eotaxin levels in pleural lavage fluid. The plant exhibited an antiallergic effect of *S. cumini*. The anti-edematogenic effect is due to the inhibition of mast cell degranulation and of histamine and serotonin effects, whereas the inhibition of eosinophil accumulation in the allergic pleurisy model is probably due to an impairment of CCL11/eotaxin and IL-5 production [34].

16.1.16 Family Olacaceae

Nyctanthes arbortristis is traditionally used in the treatment of asthma. The petroleum ether extract shows antihistaminic activity by inhibiting clonidine-induced catalepsy in mice at a dose of 50 mg/kg [35].

Olea europea is a small green tree (3–6 m high) with hoary, rigid branches, and a grayish bark. Aqueous extract of ripe olive is able to inhibit clonidine-induced peritoneal mast cell degranulation in rats and catalepsy in mice at doses of 4 and 8 mg/kg. The extract protects histamine-induced contraction of goat trachea and guinea pig ileum at a concentration of 100 g/ml [36].

16.1.17 Family Piperaceae

Piper betel L. is traditionally used to cure cold, cough, pruritus, rheumatism, and asthma. Ethanol and aqueous extracts of the leaves at doses of 100 and 200 mg/kg exhibited antiasthmatic activity on histamine-induced bronchoconstriction in guinea pigs and on contraction of guinea pig tracheal chain and isolated guinea pig ileum preparation [37].

16.1.18 Family Poaceae

Cynodon dactylon, one of the most commonly occurring perennial weeds everywhere in India, is commonly known as Dhub. The petroleum ether, chloroform, and methanol extracts of the whole plant showed antianaphylactic activity. Fractions isolated from chloroform extract seem to be more effective at doses of 10, 25, 50, and 100 mg/kg induced using compound 48/80 [38].

16.1.19 Family Polygalaceae

Phymatodes scolopendria (Burm.) is a creeping fern growing in the sandy areas of the east coast of Madagascar. The ethanol extract of its aerial parts exhibited bronchodilator activity by inhibiting histamine or carbachol precontracted guinea pig trachea [39].

Securidaca longipedunculata Tresen, a popular herbal drug, was investigated using guinea pig tracheal chain; guinea pig ileum preparation; on the rat stomach strip; and egg albumin-induced hind paw edema. Results indicate that the extract blocked the effects of isoprenaline and histamine on the guinea pig tracheal chain. It shows inhibition contraction of isolated guinea pig ileum and rat stomach strip, caused by histamine and 5-HT [26].

16.1.20 Family Scrophulariaceae

Petroleum ether, chloroform, methanol, and water extracts of the leaves of *Bacopa monnieri* L. at a dose of 10 g/ml were used for mast cell stabilizing activity in rats. The results of the observed investigation indicated that all the extracts largely inhibited mast cell degranulation [40, 41].

Striga orobanchioides Benth. is a parasitic plant that lives on the roots of various plants. The ethanol and aqueous extracts of the whole plant showed mast cell stabilizing activity by inhibiting histamine-induced contractions of the guinea pig ileum at concentrations of 2.5–25 g/ml in a dose-related manner [42–44].

16.1.21 Family Verbenaceae

Clerodendrum serratum L., known as bharangi in Ayurveda, is useful in treating pain, respiratory diseases, inflammation, rheumatism, and malarial fever. The ethanol extract of the roots of *C. serratum* showed antiasthmatic activity using isolated goat tracheal chain preparation, clonidine-induced catalepsy, milk-induced leukocytosis, and eosinophilia in mice at doses of 50 100, and 200 mg/kg [45].

16.1.22 Family Zygophylaceae

Tribulus terrestris is an annual herbaceous plant, commonly used in Ayurveda, that is widely dispersed in the world. *T. terrestris* showed significantly decreased scratching or rubbing around the nose of the rat model. *T. terrestris* has an inhibitory effect on the process of allergic rhinitis and it can be used in relieving symptoms of allergic rhinitis [46].

16.2 Bioactive Compounds with Antihistaminic Activities

There has been considerable interest in the use of natural products for the treatment of various disorders. Histamine is a potent mediator of immediate allergic reactions. The synthetic antihistamines such as diphenhydramine, chlorpheniramine, promethazine, mepyramine, and loratadine have various adverse effects such as drowsiness, dry mouth, confusion, nervousness, restlessness, and irritability. Accordingly, isolation of natural phytoconstituents that possess antihistaminic activities significantly assists in overcoming the problems of synthetic antihistamines. Table 16.1 shows the isolated phytoconstituents that have been proven to possess prominent antihistaminic activities (the numbers relate to the structures in Figure 16.1).

Table 16.1 Summary of isolated phytoconstituents with reported antihistaminic activities.

Phytoconstituent	Plant species	Main effects on histamine	References
Bidenlignaside A	*Bidens parviflora* WILLD	Inhibitory effect on histamine (IC50 = 25.1 μg) compared to indomethacin (IC50 = 24.3 μM)	[41]
Bidenlignaside B	*Bidens parviflora* WILLD	Inhibitory effect on histamine (IC50 = 37.2 μg) compared to indomethacin (IC50 = 24.3 μM)	
Ergosterol	*Grifola frondosa*	Suppression of β-hexosaminidase and histamine release at concentrations of 10, 20, and 50 μg/ml, in a dose-dependent manner	[42]
Isoliquiritigenin	*Glycyrrhiza uralensis*	Inhibitory effect on histamine (IC50 = 2.3 ± 0.1 μM) compared to ranitidine (IC50 = 1.0 ± 0.7 μM)	[43]
Hypophyllanthin	*Phyllanthus amarus*	Antihistaminic activity through binding with the H1 receptor and interfering with its activation	[44]
Ortho-vanillic acid	*Amomum xanthioides*	Suppression of histamine release	[45]
Geniposide	*Gardenia jasminoides*	Inhibition of the release of histamine from mast cells, lowering serum histamine levels (% of inhibition = 70%)	[46]
2-Methyl, 1-hexadecanol	*Ceasalpinia bonducella*	Relaxation of guinea pig ileum in histamine-induced contraction (% of relaxation = 100, 92.77, 87.32, and 81.51%) in a dose-dependent manner	[47]
Evolitrine	*Acronychia pedunculata*	Inhibition of histamine (% inhibition = 35%) compared to chlorpheniramine (% inhibition = 55%)	[48]
1,2,4,5-Tetramethoxybenzene	*Amomum xanthioides*	Dose-dependently attenuated histamine release	[49]
Nigellone	*Nigella sativa*	Inhibition of histamine release (minimum inhibitory concentration [MIC] = 11 μg/ml)	[50]

(*Continued*)

Table 16.1 (Continued)

Phytoconstituent	Plant species	Main effects on histamine	References
Paeoniflorin	*Paeonia lactiflora* Pall		[51]
5-Hydroxy-1-methylpiperidin-2-one	*Tragia involucrata*	Protection against histamine-induced triple response (% protection = 49.05 ± 2.26%) compared to chlorpheniramine (50.86 ± 2.54%)	[52]
Resveratrol	*Cissus sicyoides*	Inhibition of histamine (% inhibition = 82.4 ± 0.1%) compared to sodium cromoglicate (% inhibition = 38.2 ± 1.4%)	[53]
Protoanemonin	*Anemone raddeana* Regel	Resists the contraction of isolated ileum from guinea pig induced by histamine	[54]
Rosmarinic acid	*Ehretia philippinensis*	Antihistamine release activity (IC50 = 18 μM)	[55]
Methyl rosmarinate	*Ehretia philippinensis*	Antihistamine release activity (activity 66% at 107 μM)	
Thymonin	*Mentha spicata*	Antihistaminic (inhibitory activity of exocytosis of rat basophils with IC50 = 6.4 μM)	[29]
5,6-Dihydroxy-7,8,3′,4′-tetramethoxyflavone	*Mentha spicata*	Antihistaminic (inhibitory activity of exocytosis of rat basophils with IC50 = 56 μM)	
Palmitamide	*Polyscias guilfoylei*	Inhibition of histamine (IC50 = 38.65 ± 1.21 μg/ml) compared to diclofenac (IC50 = 9.00 ± 1.25)	[56]
Tamarixetin 3,7-di-O-α-L-rhamnopyranoside	*P. guilfoylei*	Inhibition of histamine (IC50 = 36.96 ± 1.27 μg/ml) compared to diclofenac (IC50 = 9.00 ± 1.25)	
Geniposide	*G. jasminoides*		[46]
Quercetin	*Allium cepa, Malus domestica, Camellia sinensis,* and *Fagopyrum esculentum*	Concentration-dependent inhibition of immunoglobulin E-mediated histamine release	[57]

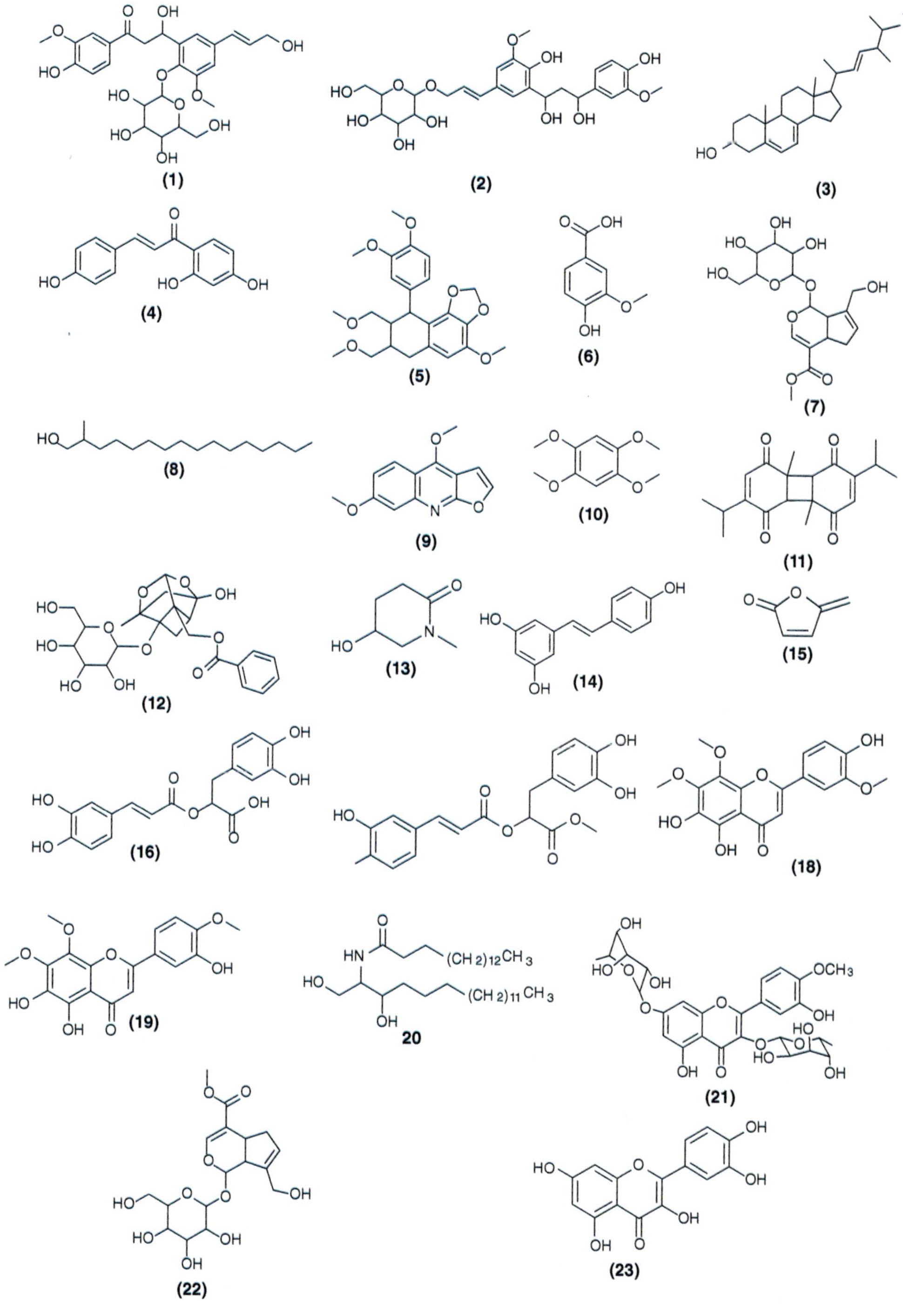

Figure 16.1 Chemical structures of isolated phytoconstituents with reported antihistaminic activities.

16.3 Conclusion

The chapter has highlighted the efficacy of many plants and plant extracts as antihistaminics. Furthermore, many phytoconstituents have also proved their effectiveness as antihistaminics, such as quercetin, catechin, flavones, and saponin. Therefore, hard work should be carried out on the integration of traditional medicine into national healthcare systems, including regulatory guidelines for the clinical trial of herbal drugs (toxicity and safety in many indications).

References

1 Holgate, S.T. and Polosa, R. (2008). Treatment strategies for allergy and asthma. *Nat. Rev. Immunol.* 8: 218–230.

2 Donno, D.M., Bittesnich, D., Chetta, A. et al. (2000). The effect of inflammation on mucociliary clearance in asthma. *Chest* 118 (4): 1142–1149.

3 Tattersfield, A.E., Knox, A.J., Britton, J.R., and Hall, I.P. (2002). Asthma. *Lancet* 360: 1313–1322.

4 Masoli, M., Fabian, D., Holt, S., and Beasley, R. (2004). The global burden of asthma: executive summary of the GINA dissemination committee report. *Allergy* 59: 469–478.

5 Bousquet, J., Bousquet, P.J., Godard, P., and Daures, J.P. (2005). The public health implications of asthma. *Bull. World Health Organ.* 83: 548–554.

6 Braman, S.S. (2006). The global burden of asthma. *Chest* 130: 4S–12S.

7 Greenberger, P.A. (2003). Therapy in management of rhinitis asthma complex. *Allergy Asthma Proc.* 24: 403–407.

8 Lyengar, M.A., Jambaiah, K.M., and Rao, G.M. (1994). Studies on an antiasthma kada: a proprietary herbal combination part-I clinical study. *Indian Drug* 31 (5): 183–186.

9 Henricks, P.A. and Nijkamp, F.P. (2001). Reactive oxygen species as mediators in asthma. *Pulm. Pharmacol. Ther.* 14: 409–420.

10 Slader, C.A., Reddel, H.K., Jenkins, C.R. et al. (2006). Complementary and alternative medicine use in asthma: who is using what? *Respirology* 11: 373–387.

11 Tripathi, K.D. (2001). *Essentials of Medical Pharmacology*, 4e. New Delhi: Jaypee Brothers Medical Publishers.

12 Kumar, D., Prasad, D.N., Parkash, J. et al. (2009). Antiasthmatic activity of ethanolic extract of *Aerva lanata* Linn. *Pharmacologyonline* 2: 1075–1081.

13 Tote, M.V., Mahire, N.B., Jain, A.P. et al. (2009). Effect of *Ageratum conyzoides* Linn on clonidine and haloperidol induced catalepsy in mice. *Pharmacologyonline* 2: 186–194.

14 Bhalke, R.D. and Gosavi, S.A. (2009). Antistress and antiallergic effect of Argemone mexicana stem in asthma. *Arch. Pharm Sci Res* 1 (1): 127–129.

15 Akah, P.A., Ezike, A.C., Nwafor, S.V. et al. (2003). Evaluation of the anti-asthmatic property of *Asystasia gangetica* leaf extracts. *J. Ethnopharmacol.* 89: 25–36.

16 Okpo, S.O. and Adeyemi, O.O. (2002). The anti-allergic effects of *Crinum glaucum* aqueous extract. *Phytomedicine* 9: 438–441.

17 Venkatesh, P., Mukherjee, P.K., Satheesh Kumar, N. et al. (2009). Mast cell stabilization and antihistaminic potentials of *Curculigo orchioides* rhizomes. *J. Ethnopharmacol.* 126: 434–436.

18 Pandit, P., Singh, A., Bafna, A.R. et al. (2008). Evaluation of antiasthmatic activity of *Curculigo orchioides* Gaertn. Rhizomes. *Indian J Pharm Sci* 70 (4): 440–444.

19 Akah, P.A., Gamaniel, K.S., Samson, A., and Wambebe, C.O. (1997). Evaluation of Nigerian traditional medicine: effects of Gakani, a herbal antiasthmatic drug. *J. Ethnopharmacol.* 55: 87–92.

20 Bhujbal, S.S., Kumar, D., Deoda, R.S. et al. (2009). Antiasthmatic activity of roots of *Hemidesmus indicus* R. Br. *Pharmacologyonline* 1: 209–216.

21 Taur, D.J. and Patil, R.Y. (2011). Some medicinal plants with antiasthmatic potential: a current status. *Asian Pac. J. Trop. Biomed.* 1 (5): 413–418.

22 Mathew, J.E., Srinivasan, K.K., Dinakaran, V., and Joseph, A. (2009). Mast cell stabilizing effects of Sphaeranthus indicus. *J. Ethnopharmacol.* 122: 394–396.

23 Nagore, D.H., Ghosh, V.K., and Patil, M.J. (2009). Evaluation of antiasthmatic activity of *Cassia sophera* Linn. *Phcog Mag* 5 (19): 109–118.

24 Aher, A.N., Pal, S.C., Patil, U.K. et al. (2009). Evaluation of anthistaminic activity of *Casuarina equisetifolia* frost (Casuarinaceae). *Pharmacologyonline* 1: 1144–1149.

25 Mali, R.G., Mahajan, S.G., and Mehta, A.A. (2008). Studies on bronchodilatory effect of Lepidium sativum against allergen induced bronchospasm in guinea pigs. *Phcog Mag* 4 (15): 189–192.

26 Rakh, M.S., Raut, D.N., Chavan, M.J., and Chaudhari, S.R. (2010). Effect of various extracts of Momordica dioica pulp on clonidine and haloperidolinduced catalepsy in mice. *Pharmacologyonline* 1: 1–11.

27 Youssouf, M.S., Kaiser, P., Tahir, M. et al. (2007). Anti-anaphylactic effect of *Euphorbia hirta*. *Fitoterapia* 78: 535–539.

28 Leal, L.K.A.M., Costa, M.F., Pitombeira, M. et al. (2006). Mechanisms underlying the relaxation induced by isokaempferide from *Amburana cearensis* in the guinea-pig isolated trachea. *Life Sci.* 79: 98–104.

29 Yamamura, S., Ozawa, K., Ohtani, K. et al. (1998). Antihistaminic flavones and aliphatic glycosides from *Mentha spicata*. *Phytochemistry* 48 (1): 131–136.

30 Taur, D.J., Nirmal, S.A., and Patil, R.Y. (2007). Effect of various extracts of Ficus bengalensis bark on clonidine and haloperidol-induced catalepsy in mice. *Pharmacologyonline* 3: 470–477.

31 Taur, D.J. and Patil, R.Y. (2009). Effect of bio-fractions isolated from Ficus bengalensis bark on clonidine induced catalepsy. *J. Pharm. Res.* 2 (11): 1676–1677.

32 Patel, K.G., Rao, N.J., Gajera, V.G. et al. (2010). Antiallergic activity of stem bark of Myrica esculenta Buch. Ham. (Myricaceae). *J. Young Pharm.* 2 (1): 74–78.

33 Patel, K.G., Bhalodia, P.N., Patel, A.D. et al. (2008). Evaluation of bronchodilator and anti-anaphylactic activity of Myrica sapida. *Iran. Biomed. J.* 12 (3): 191–196.

34 Brito, F.A., Lima, L.A., Ramos, M.F.S. et al. (2007). Pharmacological study of anti-allergic activity of *Syzygium cumini* (L.) skeels pharmacology. *Braz. J. Med. Biol. Res.* 40 (1): 105–115.

35 Nirmal, S.A., Pal, S.C., and Mandal, S.C. (2009). Antihistaminic activity of Nyctanthes arbortristis bark. *Pharmacologyonline* 3: 924–928.

36 Chandak, R., Devdhe, S., and Changediya, V. (2009). Evaluation of antihistaminic activity of aqueous extract of ripe olives of Olea europea. *J. Pharm. Res.* 2 (3): 416–420.

37 Jawale, N.M., Shewale, A.B., Nerkar, G.S., and Patil, V.R. (2009). Evalution of antihistaminic activity of leaves of *Piper betel* Linn. *Pharmacologyonline* 3: 966–977.

38 Savali, A.S., Biradar, P.R., and Jirankali, M.C. (2010). Antianaphylactic and mast cell stabilization activity of Cynodon dactylon. *Int J Pharm Pharm Sci* 2 (2): 69–73.

39 Ramanitrahasimbola, D., Rakotondramanana, D.A., Rasoanaivo, P. et al. (2005). Bronchodilator activity of Phymatodes scolopendria (Burm.) Ching and its bioactive constituent. *J. Ethnopharmacol.* 102: 400–407.

40 Samiulla, D.S., Prashanth, D., and Amit, A. (2001). Mast cell stabilising activity of *Bacopa monnieri*. *Fitoterapia* 72: 284–285.

41 Wang, N.L., Wang, J., Yao, X.S., and Kitanaka, S. (2006). Two Neolignan glucosides and antihistamine release activities from *Bidens parviflora* W ILLD. *Chem. Pharm. Bull.* 54 (8): 1190–1192.

42 Kawai, J., Mori, K., and Hirasawa, N. (2019). Grifola frondosa extract and ergosterol reduce allergic reactions in an allergy mouse model by suppressing the degranulation of mast cells. *Biosci. Biotechnol. Biochem.* 83 (12): 2280–2287.

43 Kim, D.-C., Choi, S.-Y., Kim, S.-H. et al. (2006). Isoliquiritigenin selectively inhibits H_2 histamine receptor signaling. *Mol. Pharmacol.* 70 (2): 493–500.

44 Abd Rani, N.Z., Lam, K.W., Jalil, J. et al. (2021). Mechanistic studies of the antiallergic activity of Phyllanthus amarus Schum. & Thonn. And its compounds. *Molecules* 26 (3): 695.

45 Kim, Y.-Y., Je, I.-G., Kim, M.J. et al. (2017). 2-Hydroxy-3-methoxybenzoic acid attenuates mast cell-mediated allergic reaction in mice via modulation of the FcεRI signaling pathway. *Acta Pharmacol. Sin.* 38 (1): 90–99.

46 Sung, Y.-Y., Lee, A.Y., and Kim, H.K. (2014). The Gardenia jasminoides extract and its constituent, geniposide, elicit anti-allergic effects on atopic dermatitis by inhibiting histamine in vitro and in vivo. *J. Ethnopharmacol.* 156: 33–40.

47 Vikhe, S. and Nirmal, S. (2018). Antiallergic and antihistaminic actions of Ceasalpinia bonducella seeds: possible role in treatment of asthma. *J. Ethnopharmacol.* 216: 251–258.

48 Ratnayake, W., Suresh, T., Abeysekera, A. et al. (2019). Acute anti-inflammatory and anti-nociceptive activities of crude extracts, alkaloid fraction and evolitrine from *Acronychia pedunculata* leaves. *J. Ethnopharmacol.* 238: 111827.

49 Je, I.-G., Choi, H.G., Kim, H.-H. et al. (2015). Inhibitory effect of 1, 2, 4, 5-tetramethoxybenzene on mast cell-mediated allergic inflammation through suppression of IκB kinase complex. *Toxicol. Appl. Pharmacol.* 287 (2): 119–127.

50 Koshak, A., Koshak, E., and Heinrich, M. (2017). Medicinal benefits of Nigella sativa in bronchial asthma: a literature review. *Saudi Pharma. J.* 25 (8): 1130–1136.

51 Wang, G. and Cheng, N. (2018). Paeoniflorin inhibits mast cell–mediated allergic inflammation in allergic rhinitis. *J. Cell. Biochem.* 119 (10): 8636–8642.

52 Yadav, S.A., Ramalingam, S., Raj, A.J., and Subban, R. (2015). Antihistamine from Tragia involucrata L. leaves. *J. Complement. Integr. Med.* 12 (3): 217–226.

53 Quilez, A., Saenz, M., Garcia, M., and De la Puerta, R. (2004). Phytochemical analysis and anti-allergic study of Agave intermixta Trel. and *Cissus sicyoides* L. *J. Pharm. Pharmacol.* 56 (9): 1185–1189.

54 Sun, Y.-X., Liu, J.-C., and Liu, D.-Y. (2011). Phytochemicals and bioactivities of *Anemone raddeana* regel: a review. *Die Pharmazie* 66 (11): 813–821.

55 Simpol, L.R., Otsuka, H., Ohtani, K. et al. (1994). Nitrile glucosides and rosmarinic acid, the histamine inhibitor from Ehretia philippinensis. *Phytochemistry* 36 (1): 91–95.

56 Ashmawy, N.S., Gad, H.A., Al-Musayeib, N. et al. (2019). Phytoconstituents from *Polyscias guilfoylei* leaves with histamine-release inhibition activity. *Z. Naturforsch. C* 74 (5–6): 145–150.

57 Kimata, M., Shichijo, M., Miura, T. et al. (2000). Effects of luteolin, quercetin and baicalein on immunoglobulin E-mediated mediator release from human cultured mast cells. *Clin. Exp. Allergy.* 30 (4): 501–508.

17

Effect of Plant-Based Anticonvulsant Products and Phytochemicals

Muhammad Akram[1], Sadia Zafar[2], Hassan Shah[1], Zerfishan Riaz[1], Khawaja S. Ahmad[3], Muhammad Riaz[4], Naveed Munir[5], Muhammad Jahangeer[5], Imtiaz M. Tahir[6], Michael P. Okoh[7], Muhammad A. Ishfaq[8], David Pérez-Jorge[9], Vanessa de Andrade Royo[10], Muhammad M. Aslam[11], Chukwuebuka Egbuna[12,13], and Chukwunonso O. Igboekwe[12]

[1] *Department of Eastern Medicine and Surgery, Government College University Faisalabad, Faisalabad, Punjab, Pakistan*
[2] *Department of Botany, Division of Science and Technology, University of Education Lahore, Lahore, Punjab, Pakistan*
[3] *Department of Botany, University of Poonch, Rawalakot, Azad Jammu & Kashmir, Pakistan*
[4] *Department of Allied Health Sciences, Government College University Faisalabad, Faisalabad, Punjab, Pakistan*
[5] *Department of Biochemistry, Government College University Faisalabad, Faisalabad, Punjab, Pakistan*
[6] *College of Allied Health Professionals, Government College University Faisalabad, Faisalabad, Punjab, Pakistan*
[7] *Department of Medical Biochemistry, Faculty of Basic Medical Sciences, College of Health Sciences, University of Abuja, Abuja, Federal Capital Territory, Nigeria*
[8] *Department of Chemistry, University of Poonch, Rawalakot, Azad Jammu & Kashmir, Pakistan*
[9] *Department of Didactics and Educational Research, University of La Laguna, San Cristóbal de La Laguna, Tenerife, Spain*
[10] *Department of General Biology, Laboratory of Natural Products, State, University of Montes Claros, Montes Claros, Minas Gerais, Brazil*
[11] *Nuclear Institute of Agriculture (NIA), Tando Jam, Sindh, Pakistan*
[12] *Africa Centre of Excellence in Public Health and Toxicological Research (ACE-PUTOR), University of Port-Harcourt, Choba, Rivers State, Nigeria*
[13] *Department of Biochemistry, Chukwuemeka Odumegwu Ojukwu University, Uli, Anambra State, Nigeria*

Hippocrates, in the fourth to fifth centuries BCE, considered epilepsy and melancholia as closely related, writing, "Melancholics ordinarily become epileptics, and epileptics, melancholics" [1]. Epileptic seizures occur because of a variety of causes, which include acquired, genetic, and developmental. Seizures can suddenly occur for a few seconds or minutes and stop by themselves [2]. Epilepsy is the most common neurologic symptom or condition that could occur or appear at any age; it is estimated that about 50 million people suffer from epilepsy disorder [3, 4].

Epilepsy in classical literature is termed "Saraa" or "Mirgi." It is a group of disorders characterized by unpredictable seizures. The disorders were assumed to have appeared because of localized, unnecessary, and unusually synchronized neuronal electrical discharges [5, 6]. Thus, epileptic seizure is the second leading neurologic episodic dysfunction, and is caused by unusual neuronal firing. Other symptoms are loss of awareness, disturbance of movement, sensation, mood, or other cognitive functions [6, 7]. One of the major side effects of antiepileptic drugs (AEDs) is that they cause systemic, cognitive impairment and dose-related neurotoxicity [8]. Since the currently available AEDs often have undesirable side

Phytochemical Drug Discovery for Central Nervous System Disorders: Biochemistry and Therapeutic Effects, First Edition. Edited by Chukwuebuka Egbuna and Mithun Rudrapal.

effects, new alternative therapeutic strategies based on medicinal plants have been proposed [9]. For testing anticonvulsive compounds, the pentylenetetrazole (PTZ) kindling model is generally used. It applies action mostly through the picrotox in the site of the γ-aminobutyric acid (GABA) receptor, which is inhibited/blocked by PTZ [10].

Though herbal medicine is widely used in antiepileptic treatment, there is a lack of robust evidence for the efficacy and toxicity of most herbs [11]. Drugs used for treating epilepsy have certain side effects, which are dependent on the type of medication, amount of dose, and duration of treatment. Mostly, a high dose of drugs causes side effects [12]. Globally, natural products and their active components constitute over 50% of all drugs. However, one of the major drawbacks of medicinal plants now is that cellular targets for their active components remain largely undefined [13], although they are used because of their wide range of biologic targets, including hnRNPA2 (a factor associated with numerous cellular malignancies and mRNA metabolism/splicing), compared to AEDs that are produced synthetically. There is a general belief that herbal-based drugs are less toxic compared to synthetic compounds when used at their optimum protective dose levels. Although the evidence supporting the use of complementary and alternative medicine for the treatment of epilepsy is extremely sparse, most physicians worldwide believe that herbal-based drugs should be integrated with the use of conventional antiseizure medications in some patients [14].

17.1 Types of Epileptic Seizures

Many different types of seizures can be identified based on their clinical phenomena [15]. There are two main types of seizures, generalized seizures and focal seizures, which are also called partial seizures. Partial (focal/local) seizures are those in which there is clinical or electrographic evidence to suggest that the attacks have a localized onset in the brain, usually in a portion of a hemisphere, while generalized seizures are those in which the evidence for a localized onset is missing [16]. Partial seizures are subdivided into simple partial seizures and complex partial seizures, while generalized seizures are categorized as absence (non-convulsive), atonic, myoclonic, clonic, tonic, tonic–clonic, and febrile seizures. In addition to classifying seizures that occur in patients with epilepsy, the patients themselves are classified into appropriate types of epilepsy or epileptic syndromes, characterized by different types of seizures, etiologies, ages of onset, and electroencephalographic (EEG) characteristics [17].

17.2 Basic Mechanisms of Epilepsy

Initiation of seizure is categorized by two conditions: high-frequency action potentials burst and neuronal hypersynchronization [18, 19]. γ-Aminobutyric acid aminotransferase (GABA-AT) is a pyridoxal 5′-phosphate (PLP)-dependent enzyme that degrades GABA in the brain. GABA is an important inhibitory neurotransmitter that plays important neurologic roles in the brain. Therefore, GABA-AT is an important drug target that regulates the GABA level. Novel and potent drug development to inhibit GABA-AT is still a very challenging task [20]. The bursting activity is the cause of prolonged depolarization due to extracellular Ca^{++} influx, which causes opening of voltage-dependent sodium channels, generation of action potentials, and Na^{+} influx. According to the type of cell, hyperpolarizing potential is facilitated by K+ efflux, Cl^- influx, and GABA receptors. GABA is the inhibitory neurotransmitter that stops the mind from sending messages (Figure 17.1a) [21].

Drugs that increase the inhibition of GABA are considered active anticonvulsants. It includes benzodiazepines, which recover GABA binding and cause enhanced frequency of Cl^- channel openings. Synthesis of a few GABA inhibitors can cause seizures, including 4-d eoxypyridinoline,l-allylglycine, thiosemicarbazide, and isoniazid [22]. Glutamate and potassium are the two major hypotheses suggested for the development of epilepsy. Glutamate is the major excitatory neurotransmitter and type of amino acid in the brain. In normal conditions, synapses release glutamate, which is picked up by astrocytes and by glutamine synthetase it is quickly changed into non-excitotoxic amino acid glutamine [23] and is a major precursor of GABA.

The significant recycling mechanism of GABA and glutamate in the brain is the glutamate–glutamine cycle. A deficiency of glutamine synthetase (GS) leads to increased extracellular glutamate levels as well as raising glutamate levels in astrocytes. Glutamate levels can be

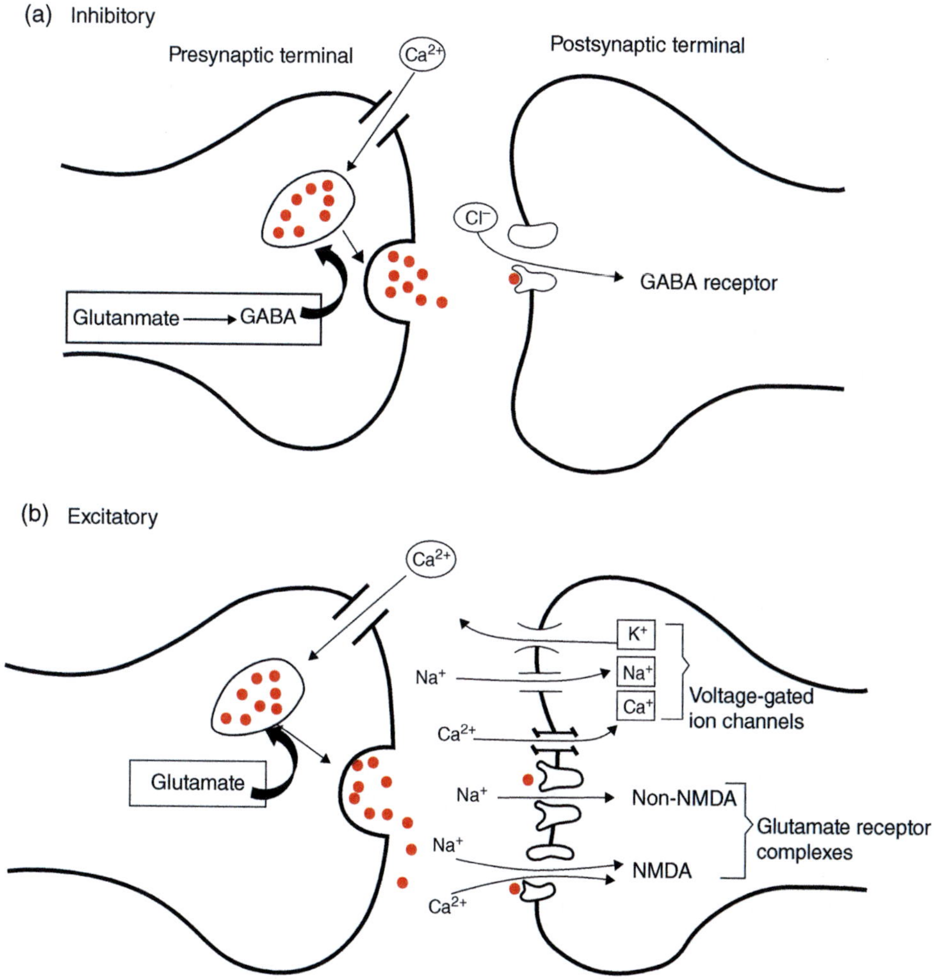

Figure 17.1 Mechanism of normal synaptic transmission: (a) inhibitory mechanism; (b) excitatory mechanism. GABA, γ-aminobutyric acid; NMDA, N-methyl-d-aspartate.

raised by impaired removal of glutamate in the extracellular space and increasing the release of glutamate from astrocytes and neurons [24]. The releasing process of glutamate is mainly performed by glutamate transporters – that is, sodium-dependent glutamate transporter 1 (GLT-1, also known as EAAT2) – that move glutamate and potassium across the plasma membrane, and either downregulation or misfunctioning of GLT-1 may raise the level of glutamate (Figure 17.1b) [25].

Two sets of benzenesulfonamide-based effective human carbonic anhydrase (hCA) inhibitors have been developed using the tail approach. These sulfonamides were found to be orally active anticonvulsants, and were non-toxic in neuronal cell lines and in animal models [26]. Raised levels of glutamate have been reported in human brain tissues and animal models of epilepsy, and it is known that glutamate-induced excitotoxicity causes neuronal death in epilepsy; that is, temporal lobe epilepsy (TLE) [27]. Stimulative glutamatergics are included in evoked, acute, and momentary seizures. Through N-methyl-D-aspartate (NMDA) and non-NMDA receptors, glutamate exerts its excitatory effects to enhance Na^+ and Ca^{++} conductance [28].

Two different modes of releasing glutamate present a new way of thinking about epilepsy in astrocytes: TREK-1, a two-pore potassium channel, engendered by Bestrophin-1 (Best1) activation, is an anion channel activated by calcium for slow glutamate release; and G-protein-coupled receptors (GPCRs) for fast release [29]. Best1 is expressed close to synapses while TREK-1 is preferentially present at cell bodies and processes of astrocytes. About a half-century ago the potassium hypothesis of epilepsy was proposed [30]. The involvement of TREK-1 depends on the activation and inactivation of the potassium channel, and downregulation or any type of mutation of potassium channels causes rodent and human epilepsies [31]. Stronger long-lasting and prolonged depolarization of glial cells and neurons by astrocytes will result in impaired buffering of potassium that is spatial in response to the release of potassium-dependent activity, and it may also contribute to the generation of seizures in this particular state of human TLE [32].

17.3 Epilepsy and Oxidative Stress

Oxidative stress causes cellular damage and dysfunction and later causes cell death through nucleotide, lipid, and protein oxidation (Figure 17.2) [33]. The biologic significance of oxidative stress caused by reactive oxygen species (ROS) is dependent on their quantities, chemical nature, subcellular and tissue localization/location, and the rates of their formation and degradation [34]. A seizure is related to an imbalance in homeostasis between oxidants and antioxidants [35]. Antioxidant treatments to decrease ROS have attracted more attention in curing the disorder of epilepsy. The mechanism of action of these plants was earlier postulated [36] to include the activation of metabolizing enzymes that detoxify carcinogens, the suppression of DNA adduct formation, inhibition of the production of ROS, the regulation of cell–cycle arrest, and the induction of apoptosis [37].

Both oxidative stress and hypoxia may lead to the epigenetic modification of DNA, which may result in the activation of cytokines and lead to initiation of the neuroinflammation process. Neuroinflammation is the primary factor that leads to dysfunction or inactivated mutation in the glutamatergic system and neurodegeneration, with the resultant dysregulation of GABA synthesis, which may lead to the development of epilepsy (Figure 17.2) [38]. The high production of free radicals in mitochondria causes mitochondrial dysfunction, and due to this dysfunction energy may be depleted in a living body and may result in cell death

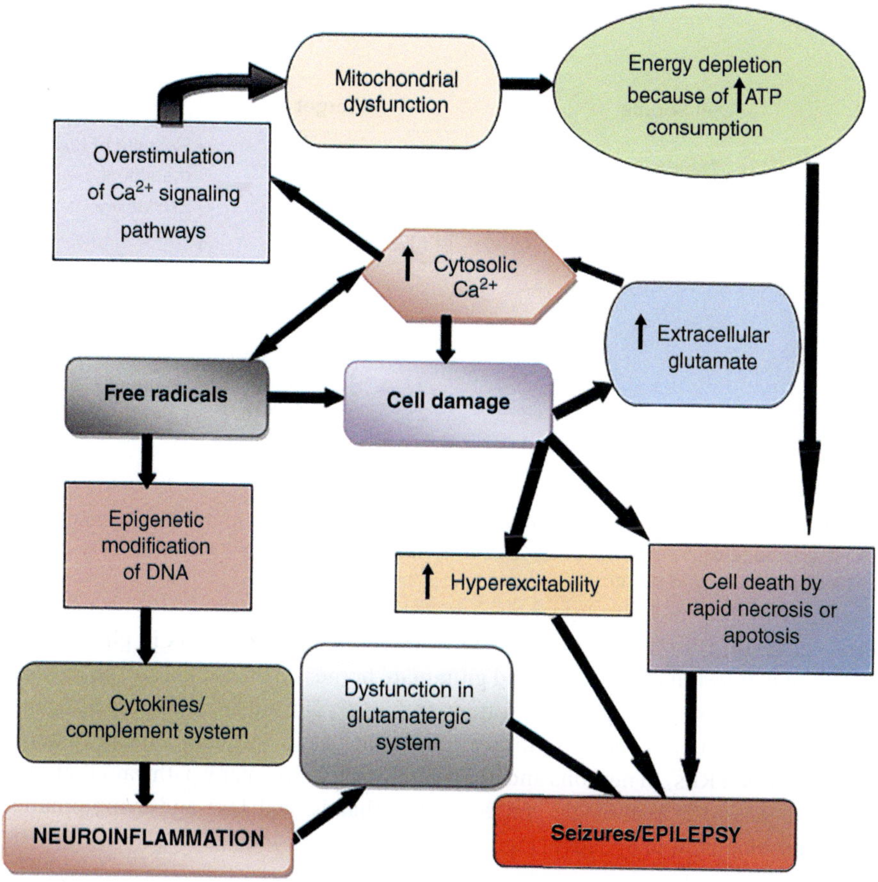

Figure 17.2 Role of free radicals in the development of epilepsy through various dysfunctions, i.e. neuroinflammation, mitochondrial dysfunction, cell damage, cell death, and ultimately seizure and epilepsy development. ATP, adenosine triphosphate.

by rapid necrosis or delayed apoptosis, which leads to the development of seizures. High production of free radicals causes cell damage, which increases the extracellular concentration of glutamate, and the resulting higher cytosolic calcium concentration increases hyperexcitability and activates many pathways that can cause long-term and acute pathologic consequences, including epilepsy, cell death, and many others (Figure 17.2).

17.4 Epilepsy and Inflammation

Experimental results showed that seizures enhance the inflammatory mediators that are involved in initiating epileptic action [39]. Direct anti-inflammatory drug treatment leads to a decrease in epileptic seizures. Glia triggered an inflammatory response by electrical stimulation [40]. Interleukin-6 (IL-6), which overexpresses cytokines, astrocytes, and tumor necrosis factor-α are described to appear in neurologic dysfunction (age dependent) as it reduced the frequency of spontaneous seizure and the seizure threshold [41].

The first event of inflammation involved in epilepsy is cell injury or inflammation in the brain, which leads to a rapid outpouring of damage-associated molecular patterns (DAMPs)

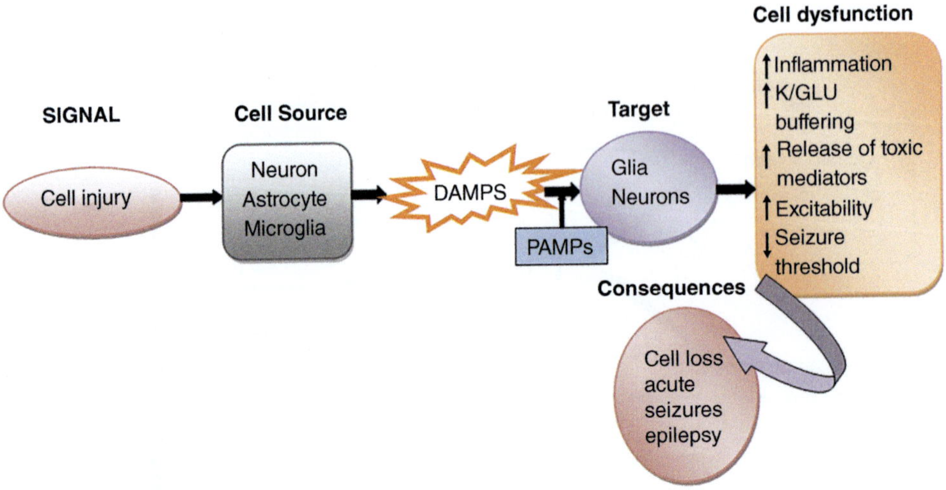

Figure 17.3 Role/mechanism of inflammation or injury in the brain leading to the development of epilepsy. DAMPs, damage-associated molecular patterns; K/GLU, potassium glutamate; PAMPs, pathogen-associated molecular patterns.

from injured cells. DAMPs increase the number of inflammatory mediators in glia, which causes the loss of extracellular water, K^+, and glutamate homeostasis and boosts the release of toxic intermediators such as gliotransmitters and ROS. During infections, pathogen-associated molecular patterns (PAMPs) diffuse the brain tissue and can activate pattern recognition receptors (PRRs) in neurons and glia, thereby activating many pathways that can cause acute and long-term pathologic consequences; that is, cell loss and symptomatic seizures or the development of epilepsy (Figure 17.3) [42].

17.5 Tests for Seizure Induction

The most commonly used models to devise novel anticonvulsant drugs are the pentylenetetrazole (PTZ) seizure test and the maximal electroshock seizure (MES) test [43]. The positive results that are obtained from the PTZ seizure test provide information on drugs that are effective against spike-wave seizures, and MES tests provide some insight into the ability of a given drug to penetrate the blood–brain barrier and exert a central nervous system (CNS) effect [44]. Indeed, both models are well suited for evaluating anticonvulsant activity and are non-selective concerning mechanism [15]. PTZ may cause seizures by inhibitory chloride ion channel associated with GABA receptors. Besides these tests, many others such as the pilocarpine (PILO) and kainate (KAI) tests are also provided. The sequential analysis of AEDs in PILO and KAI models provides a correlation between seizure behavior and pharmacologic effects and includes brain substrate [45].

17.6 Medicinal Plants Used to Treat Epilepsy

Until the second decade of the twentieth century, epilepsy was treated mainly by traditional remedies. Today, many medicinal plants used in different cultures and regions of the world have been identified. Most of the phytochemical components of these plants have been identified and, in some cases, their targets located. Therefore, it is possible that new, effective, and affordable anticonvulsant drugs can be obtained from a medicinal plant [26].

The major concerns with the use of AEDs are the limited availability, adverse side effects, high cost, and low efficacy [46]. Furthermore, pathologic changes that occurred during the epilepsy process are not reverted or cured with the use of most AEDs, so novel therapeutics are the requirement in the current era, with high potential to manage epilepsy with low side effects [47]. Moreover, drug resistance developed in about 30–40% of patients due to regularly using drugs, and intractable epilepsy develops in most patients [48]. Traditional healers are the first choice in many parts of the world due to several reasons, like the high cost of AEDs, low supply, low efficacy, and lack of tolerance of AEDs [49, 50].

Medicinal plants have always remained the primary choice of treatment for most populations worldwide, especially in Africa and Asia, due to easy availability, low cost, good tolerance, high efficacy, as well as few reported side effects [51]. Moreover, some herbs have revealed potential as safe and new therapeutic options [52, 53]. It was reported that medicinal plants are being used to treat epilepsy as sedative and antiepileptic agents; however, there is still a lack of controlled experimental reports on their therapeutic use [54]. Further, the anticonvulsant activities of various plants have been explored [55, 56]. The reports cover 23 medicinal plants used in Chinese traditional medicine for the treatment of epilepsy, but none of them has been developed into a standard medication for the treatment of seizures. The therapeutic potential of herbal plants and some of their bioactive compounds have been the subject of extensive research. Data on essential oils, methods, bioactive components, type of extract, mechanisms, reference, and mode of action associated with plants with anticonvulsant properties are summarized in Table 17.1.

For instance, *Cymbopogon citratus* (DC) Stapf. is consumed as an infusion in folk medicine due to its pharmacologic properties and action in the CNS. It has been tested for its effects on PTZ-induced seizures in zebrafish (*Danio rerio*) [9].

The anticonvulsant properties of several cannabinoids, namely Δ9-tetrahydrocannabinol and cannabidiol (CBD), have been demonstrated in both in vitro and in vivo studies.

17.7 Conclusion

Research has established that most synthetic drugs have varieties of side effects with respect to individual makeup, hence there is a need to explore orthodox and natural drugs that are a good source of bioactive compounds that may have no or fewer side effects for the treatment of epilepsy. These products have been proven to have expected pharmacologic actions that may require a long-term procedure, therefore administration of synthetic drugs could be withdrawn gradually over a period of six months. Natural products and medicinal plants have a significant role in curing neurologic disease, including degenerative diseases, reperfusion, Alzheimer's disease, and brain ischemia. In truth, existing attention to traditional medicine is leading to quick progress and examination of several remedies useful in numerous cultural clusters across the world. Science has shown that the complexity of most synthetic/chemical drugs and their metabolic process may be the main effect of drug resistance and drug–drug interactions. Therefore, in the interest of public health, researchers within the scientific community should explore alternatives like natural products with bioactive chemical compounds for the treatment of epilepsy globally.

Table 17.1 Antiepileptic properties of medicinal plants and mechanisms of action in treatment of epilepsy.

Compounds	Method of seizure induction	Mechanism of action	References
Aqueous extract of *Trichosanthes dioica* Roxb. fruits	Delivering electroshock, tonic–clonic convulsions induced by PTZ injection in mice	Action on generalized tonic–clonic and focal seizures	[57]
Methanol extract of *Ficus platyphylla*	In mice PTZ 13 convulsant injections, in two-way shuttle box, and performance of learning tested	Attraction for 3H-GABA binding assay and undifferentiated glutamate receptors reduces K^+-stimulated glutamate	[57]
Aqueous extract of *Feretia apodanthera*	Administered 30 mg/kg PTZ injection in mice, T-maze tests and elevated plus-maze included in behavioral tests	Improved GSH and MDH levels in brain, increased activity of AChE and BuChE in brain	[58]
Oil of *Nigella sativa*	Electroshock and PTZ injection in mice	Strong antioxidant activity	[59]
Ethanolic extract of leaves of *Psidium guajava* (guava)	Delivering 50 mA electroshock, tonic–clonic convulsions induced by PTZ 70 mg/kg injection in mice	Selective inhibition of NMDA receptor	[60]
Methanol extract of *Trachyspermum ammi*	4 mg/kg injection of strychnine-induced seizure in rats	Stimulates GABA responses by exciting GABAA receptors and enhancing opening of chloride ion channel	[61]
Hydroethanolic extract of rhizomes of *Zingiber officinale* (ginger)	Tail vein administration of PTZ	Hinders NO production, decrease NO synthase, antioxidant activity	[62]
Root of *Anacyclus pyrethrum* hydroalcoholic extract	Administered PTZ 60 mg/kg intraperitoneal injection, T-maze tests and elevated plus-maze are included in behavioral tests	Improved GSH and MDH levels of brain, increased activity of AChE and BuChE in brain	[63]
Aqueous root extract of *Valeriana officinalis*	Amygdala kindled rats	Activation of adenosine system, GABA receptor binding	[64]
Butanol extract of *Panax ginseng* root	By 40 mg/kg administration of pilocarpine hydrochloride	Enhanced hippocampal volume, enhanced T2 relaxation time in rat entorhinal and piriform cortex, thalamus, amygdala, and hippocampus	[65]
Essential oil of *Pimpinella anisum*	In rat administered 120 mg/kg PTZ	Inhibits dark neuron production and also inhibits initiation of long-term potentiation	[66]

Table 17.1 (Continued)

Compounds	Method of seizure induction	Mechanism of action	References
Essential oil of *Cyperus rotundus*	MES-induced convulsion	Inhibits Na^+ channels, blocking of glutamatergic excitation facilitated by NMDA receptor	[67]
Essential oil and methanolic extract of *Zhumeria majdae*	In mice 110 mg/kg PTZ and MES models	Inhibits Na^+ channels, blocking of glutamatergic excitation facilitated by NMDA receptor	[68]
Essential oil of *Angelica archangelica* roots	In mice 80 mg/kg PTZ and MES models	Blocks excitation of glutamatergic	[69]
Essential oils of *Cymbopogon winterianus* Jowitt	In mice pilocarpine-induced convulsions 350 mg/kg PTZ administration	GABAergic, degenerates autoregulation of glutamate release	[70]
Essential oils of *Rosmarinus officinalis*	In mice administered 80 mg/kg PTZ	In midbrain region raises GABA levels	[71]
Essential oils of *Ocimumbasilicum*	In mice administered 80 mg/kg PTZ	Regulates activation of glutamate expression, decrease in ACh, direct contact with NMDA receptor	[71]
Essential oils of *Mentha spicata*	In mice administered 80 mg/kg PTZ	GABAergic	[71]
Essential oils of *Lavandula angustifolia*	In mice administered 80 mg/kg PTZ	Regulates activation of glutamate expression, decrease in ACh, direct contact with NMDA receptor	[71]
Essential oils of *Mentha piperita*	In mice administered 80 mg/kg PTZ	GABAergic	[71]
Essential oils of *Origanum dictamnus*	In mice administered 80 mg/kg PTZ	GABAergic	[71]
Essential oils of *Origanum vulgare*	In mice administered 80 mg/kg PTZ	GABAergic	[71]
Essential oils of *Mentha pulegium*	In mice administered 80 mg/kg PTZ	GABAergic	[71]
Methanolic extract of turmeric	In larvae of zebrafish 40 mmol/l PTZ, in mice 50 and 100 mg/kg PTZ infusion in tail vein	Increased lipophilicity, crosses blood–brain barrier without difficulty, inhibits lipid peroxidation and oxidative DNA impairment	[44]

(*Continued*)

Table 17.1 (Continued)

Compounds	Method of seizure induction	Mechanism of action	References
α-Asarone	Administration of 320 and 350 mg/kg pilocarpine, administration of 85 mg/kg PTZ, in mice MES test	Pyrogallol, autooxidation inhibition, superoxide dismutase-like activity, glutathione oxidation inhibition (hydroperoxide dependent)	[72]
Aqueous extract of *Acorus calamus*	In mice 80 mg/kg PTZ and MES models	Blocks NMDA receptors	[73]
Methanolic extract and essential oil of *Bunium persicum*	In NMRI male mice convulsions induced by 110 mg/kg PTZ, MES models	GABAergic	[74]
Chlorophyll constituent phytol	In mice 400 mg/kg pilocarpine hydrochloride	In hippocampus variations in AChE activity, muscarinic activation	[75]
Soy extract	In ovariectomized rat 40 mg/kg PTZ is administered for 14 days or just a single high-dose 90 mg/kg injection of PTZ	Severity of seizure affected by phytoestrogens of soy	[76]
Hydroalcoholic extract of *Zizyphus jujuba*	In rat administered 60 mg/kg PTZ, MES	Increases BuChE, GSH, and AChE in brain, reduces brain MDA levels	[77]
Hydroalcoholic extract of *Emblica officinalis*	In rat 60 mg/kg PTZ administered	Increases GSH levels and decreases MDA levels in brain	[78]
Naringin (present in grapefruit, bioflavonoid)	In rat 60 mg/kg PTZ administered	Increases GSH levels and decreases MDA levels in brain, reduces tumor necrosis factor-α	[79]
Anisomeles malabarica leaves (flavonoid fraction)	In rat 50 mg/kg PTZ administered, MES	Reduces tonic hindlimb delayed phase	[80]
Rhizoma Curcumae (curcumol)	In culture hippocampal neurons	Facilitation of recombinant GABAA receptors in hippocampal neurons, enhances tonic and phasic GABAergic inhibition in hippocampus	[81]
Citrus flowers hydroalcoholic extract	In rat 90 mg/kg PTZ administered	Minimal clonic seizure	[82]
Vitexin (flavonoid)	Intracerebroventricularly administered 90 mg/kg PTZ	Ligand for GABAA benzodiazepine receptors and exerts anticonvulsant effects	[82]

Table 17.1 (Continued)

Compounds	Method of seizure induction	Mechanism of action	References
Quercetin (flavonoid)	Total 15 injections of 35 mg/kg PTZ administered every day	Increases sulfhydryl and decreases MDA levels in brain	[83]
Centella asiatica (gotu kola)	60 mg/kg PTZ injection	Raised ACh levels, increased AChE activity	[84]
Curcumin	In mice 35 mg/kg PTZ administered	Reduces AChE activity, increases norepinephrine, reduces nitrite levels	[85]
Oil of *Nigella sativa*	Single-dose injection of 380 mg/kg pilocarpine	K^+ ATPase activity, reduces raised NO levels, reduces hippocampal Na^+ decrease, raises angiotensin-converting enzyme	[86]

ACh, acetylcholine; AChE, acetylcholinesterase; ATP, adenosine triphosphate; BuCHe, butyrylcholinesterase; GABA, γ-aminobutyric acid; GSH, glutathione; MDA, malondialdehyde; MDH, malate dehydrogenase; MES, maximal electroshock seizure; NMDA, N-methyl D-aspartate; NO, nitric oxide; PTZ, pentylenetetrazole.

References

1 Lewis, A.J. (1934). Melancholia: a historical review: part I. *J. Ment. Sci.* 80: 1–42.

2 Yalçin, A.D., Kaymaz, A., and Forta, H. (1997). Childhood occipital epilepsy: seizure manifestations and electroencephalographic features. *Brain Dev.* 19 (6): 408–413.

3 Fisher, R.S., Boas, W.V.E., Blume, W. et al. (2005). Epileptic seizures and epilepsy: definitions proposed by the international league against epilepsy (ILAE) and the International Bureau for Epilepsy (IBE). *Epilepsia* 46 (4): 470–472.

4 World Health Organization (2019). Epilepsy. https://www.who.int/news-room/fact-sheets/detail/epilepsy (accessed 8 March 2023).

5 Nakagawa, E., Aimi, Y., Yasuhara, O. et al. (2000). Enhancement of progenitor cell division in the dentate gyrus triggered by initial limbic seizures in rat models of epilepsy. *Epilepsia* 41 (1): 10–18.

6 Mustafa, S. and Arif, Y. (2020). Epilepsy in pregnancy and role of Unani medicine: a review. *J. Integr. Community Health* 9: n2, 28–n31.

7 Scott, B.W., Wojtowicz, J.M., and Burnham, W.M. (2000). Neurogenesis in the dentate gyrus of the rat following electroconvulsive shock seizures. *Exp. Neurol.* 165 (2): 231–236.

8 Perucca, E., Gram, L., Avanzini, G., and Dulac, O. (1998). Antiepileptic drugs as a cause of worsening seizures. *Epilepsia* 39 (1): 5–17.

9 Hacke, A.C.M., Miyoshi, E., Marques, J.A., and Pereira, R.P. (2021). Cymbopogon citratus (DC.) Stapf, citral and geraniol exhibit anticonvulsant and neuroprotective effects in pentylenetetrazole-induced seizures in zebrafish. *J. Ethnopharmacol.* 275: 114142.

10 Velisek, L., Kubova, H., Pohl, M. et al. (1992). Pentylenetetrazol-induced seizures in rats: an ontogenetic study. *Naunyn Schmiedeberg's Arch. Pharmacol.* 346 (5): 588–591.

11 Liu, W., Ge, T., Pan, Z. et al. (2017). The effects of herbal medicine on epilepsy. *Oncotarget* 8 (29): 48385–48397.

12 Vermeulen, J. and Aldenkamp, A.P. (1995). Cognitive side-effects of chronic antiepileptic drug treatment: a review of 25 years of research. *Epilepsy Res.* 22 (2): 65–95.

13 Okoh, M.P., Alli, L.A., Tolvanen, M.E.E., and NWegbu, M.M. (2019). Herbal drug use in sickle cell disease management; trends and perspectives in sub-Saharan Africa - a systematic review. *Curr. Drug Discov. Technol.* 16 (4): 372–385.

14 Assadi-Pooya, A.A., Brigo, F., Lattanzi, S. et al. (2021). Complementary and alternative medicine in epilepsy: a global survey of physicians' opinions. *Eplepsy & Behavior.* 117: 107835.

15 Löscher, W. (1998). New visions in the pharmacology of anticonvulsion. *Eur. J. Pharmacol.* 342 (1): 1–13.

16 Da Silva, F.H.L., Blanes, W., Kalitzin, S.N. et al. (2003). Dynamical diseases of brain systems: different routes to epileptic seizures. *IEEE Trans. Biomed. Eng.* 50 (5): 540–548.

17 Commission on Classification and Terminology of the International League Against Epilepsy (2003). Proposal for revised classification of epilepsies and epileptic syndromes. *Epilepsia* 30: 389–399.

18 Richerson, G.B. and Wu, Y. (2004). Role of the GABA transporter in epilepsy. *Adv. Exp. Med. Biol.* 548: 76–91.

19 Marafiga, J.R., Pasquetti, M.V., Calcagnotto, M.E. et al. (2020). GABAergic interneurons in epilepsy: More than a simple change in inhibition. *Epilepsy Behav.* 121 (Pt B): 106935.

20 Al-Obaidi, A., Elmezayen, A.D., and Yelekçi, K. (2020). Homology modeling of human GABA-AT and devise some novel and potent inhibitors via computer-aided drug design techniques. *J. Biomol. Struct. Dyn.* 39 (11): 4100–4110.

21 Schuler, V., Lüscher, C., Blanchet, C. et al. (2001). Epilepsy, hyperalgesia, impaired memory, and loss of pre-and postsynaptic GABAB responses in mice lacking GABAB (1). *Neuron* 31 (1): 47–58.

22 Gale, K. (1992). GABA and epilepsy: basic concepts from preclinical research. *Epilepsia* 33: S3–S12.

23 During, M.J. and Spencer, D.D. (1993). Extracellular hippocampal glutamate and spontaneous seizure in the conscious human brain. *Lancet* 341 (8861): 1607–1610.

24 Had-Aissouni, L. (2012). Toward a new role for plasma membrane sodium-dependent glutamate transporters of astrocytes: maintenance of antioxidant defenses beyond extracellular glutamate clearance. *Amino Acids* 42 (1): 181–197.

25 Tanaka, K., Watase, K., Manabe, T. et al. (1997). Epilepsy and exacerbation of brain injury in mice lacking the glutamate transporter GLT-1. *Science* 276 (5319): 1699–1702.

26 Mishra, C.B., Kumari, S., Angeli, A. et al. (2021). Discovery of potent carbonic anhydrase inhibitors as effective anticonvulsant agents: drug design, synthesis, and *in vitro* and *in vivo* investigations. *J. Med. Chem.* 64 (6): 3100–3114.

27 Coulter, D.A. and Eid, T. (2012). Astrocytic regulation of glutamate homeostasis in epilepsy. *Glia* 60 (8): 1215–1226.

28 Doherty, J. and Dingledine, R. (2002). The roles of metabotropic glutamate receptors in seizures and epilepsy. *Curr. Drug Targets CNS Neurol. Disord.* 1 (3): 251–260.

29 Woo, D.H., Han, K.-S., Shim, J.W. et al. (2012). TREK-1 and Best1 channels mediate fast and slow glutamate release in astrocytes upon GPCR activation. *Cell* 151 (1): 25–40.

30 Fisher, R.S., Pedley, T.A., Moody, W.J., and Prince, D.A. (1976). The role of extracellular potassium in hippocampal epilepsy. *Arch. Neurol.* 33 (2): 76–83.

31 Benarroch, E.E. (2009). Potassium channels: brief overview and implications in epilepsy. *Neurology* 72 (7): 664–669.

32 Hinterkeuser, S., Schröder, W., Hager, G. et al. (2000). Astrocytes in the hippocampus of patients with temporal lobe epilepsy display changes in potassium conductances. *Eur. J. Neurosci.* 12 (6): 2087–2096.

33 Sudha, K., Rao, A.V., and Rao, A. (2001). Oxidative stress and antioxidants in epilepsy. *Clin. Chim. Acta* 303 (1–2): 19–24.

34 Okoh, M.P. (2019). Antioxidants an epigenetics regulator for the prevention of disease and aging process. *J. Genetics Engr. and Biotech. Res.* 1 (1): 1–5.

35 Gluck, M.R., Jayatilleke, E., Shaw, S. et al. (2000). CNS oxidative stress associated with the kainic acid rodent model of experimental epilepsy. *Epilepsy Res.* 39 (1): 63–71.

36 Blokhina, O., Virolainen, E., and Fagerstedt, K.V. (2003). Antioxidants, oxidative damage and oxygen deprivation stress: a review. *Ann. Bot.* 91: 179–194.

37 Souza, S.M.C., Aquino, L.C., Milach, A.C. Jr. et al. (2006). Antiinflammatory and antiulcer properties of tannins from MyracrodruonurundeuvaAllemão (Anacardiaceae) in rodents. *Phytother. Res.* 21: 220–225.

38 Kobylarek, D., Iwanowski, P., Lewandowska, Z. et al. (2019). Advances in the potential biomarkers of epilepsy. *Front. Neurol.* 10: https://doi.org/10.3389/fneur.2019.00685.

39 Marchi, N., Granata, T., and Janigro, D. (2014). Inflammatory pathways of seizure disorders. *Trends Neurosci.* 37 (2): 55–65.

40 Kwon, Y.S., Pineda, E., Auvin, S. et al. (2013). Neuroprotective and antiepileptogenic effects of combination of anti-inflammatory drugs in the immature brain. *J. Neuroinflammation* 10 (1): 798.

41 Virta, M., Hurme, M., and Helminen, M. (2002). Increased plasma levels of pro-and anti-inflammatory cytokines in patients with febrile seizures. *Epilepsia* 43 (8): 920–923.

42 Vezzani, A., Fujinami, R.S., White, H.S. et al. (2016). Infections, inflammation and epilepsy. *Acta Neuropathol.* 131 (2): 211–234.

43 Löscher, W. and Schmidt, D. (1988). Which animal models should be used in the search for new antiepileptic drugs? A proposal based on experimental and clinical considerations. *Epilepsy Res.* 2 (3): 145–181.

44 Orellana-Paucar, A.M., Serruys, A.-S.K., Afrikanova, T. et al. (2012). Anticonvulsant activity of bisabolene sesquiterpenoids of Curcuma longa in zebrafish and mouse seizure models. *Epilepsy Behav.* 24 (1): 14–22.

45 Leite, J.P., Garcia-Cairasco, N., and Cavalheiro, E. (2002). New insights from the use of pilocarpine and kainate models. *Epilepsy Res.* 50 (1–2): 93–103.

46 Manchishi, M.S. (2018). Recent advances in antiepileptic herbal medicine. *Curr. Neuropharmacol.* 16 (1): 79–83.

47 Sriranjini, S.J., Sandhya, K., and Mamta, V.S. (2015). Ayurveda and botanical drugs for epilepsy: current evidence and future prospects. *Epilepsy Behav.* 52: 290–296.

48 Shetty, A.K. and Upadhya, D. (2016). GABA-ergic cell therapy for epilepsy: advances, limitations and challenges. *Neurosci. Biobehav. Rev.* 62: 35–47.

49 Al Asmi, A., Al Maniri, A., Al-Farsi, Y.M. et al. (2013). Types and sociodemographic correlates of complementary and alternative medicine (CAM) use among people with epilepsy in Oman. *Epilepsy Behav.* 29 (2): 361–366.

50 Sucher, N.J. and Carles, M.C. (2015). A pharmacological basis of herbal medicines for epilepsy. *Epilepsy Behav.* 52: 308–318.

51 Barnes, P.M., Bloom, B., and Nahin, R.L. (2008). Complementary and alternative medicine use among adults and children; United States, 2007. *Natl. Health Stat. Report* 12: 1–23.

52 Kakooza-Mwesige, A. (2015). The importance of botanical treatments in traditional societies and challenges in developing countries. *Epilepsy Behav.* 52: 297–307.

53 Samuels, N., Finkelstein, Y., Singer, S.R., and Oberbaum, M. (2008). Herbal medicine and epilepsy: proconvulsive effects and interactions with antiepileptic drugs. *Epilepsia* 49 (3): 373–380.

54 Sahu, S., Dutta, G., Mandal, N. et al. (2012). Anticonvulsant effect of Marsilea quadrifolia Linn. On pentylenetetrazole induced seizure: a behavioral and EEG study in rats. *J. Ethnopharmacol.* 141 (1): 537–541.

55 Lian, X.Y., Zhang, Z.Z., and Stringer, J.L. (2005). Anticonvulsant activity of ginseng on seizures induced by chemical convulsants. *Epilepsia* 46 (1): 15–22.

56 Xiao, F., Yan, B., Chen, L., and Zhou, D. (2015). Review of the use of botanicals for epilepsy in complementary medical systems – traditional Chinese medicine. *Epilepsy Behav.* 52: 281–289.

57 Singh, P., Garg, V.K., Sharma, P.K., and Gupta, S. (2012). Antiepileptic activity of aqueous extract of Tricosanthes dioica Roxb. *Asian J. Plant Sci.* 2 (1): 45–47.

58 Taiwe, G., Moto, F., Ayissi, E. et al. (2015). Effects of a lyophilized aqueous extract of *Feretia apodanthera* Del.(Rubiaceae) on pentylenetetrazole-induced kindling, oxidative stress, and cognitive impairment in mice. *Epilepsy Behav.* 43: 100–108.

59 Jaykare, S.C., Motghare, V.M., Padwal, S.L. et al. (2013). Evaluation of anticonvulsant activity of the seed oil extract of *Nigella sativa*: an experimental study. *Hygeia. J. Drugs Med.* 5: 21–26.

60 Pushpa, V., Shetty, K.P., Sushma, N. et al. (2014). Evaluation of the anticonvulsant activity of ethanol extract of *Psidium guajava* (guava leaves) in albino mice. *Int. J. Pharm. Sci. Res.* 5 (10): 4288–4292.

61 Rajput, M.A., Khan, R.A., and Feroz, Z. (2013). Evaluation of antiepileptic activity of the methanol extract of *Trachyspermum ammi* (L.). *Arch. Biol. Sci.* 65 (3): 815–819.

62 Hosseini, A. and Mirazi, N. (2014). Acute administration of ginger (*Zingiber officinale* rhizomes) extract on timed intravenous pentylenetetrazol infusion seizure model in mice. *Epilepsy Res.* 108 (3): 411–419.

63 Pahuja, M., Mehla, J., Reeta, K. et al. (2012). Root extract of *Anacyclus pyrethrum* ameliorates seizures, seizure-induced oxidative stress and cognitive impairment in experimental animals. *Epilepsy Res.* 98 (2–3): 157–165.

64 Rezvani, M.E., Roohbakhsh, A., Allahtavakoli, M., and Shamsizadeh, A. (2010). Anticonvulsant effect of aqueous extract of *Valeriana officinalis* in amygdala-kindled rats: possible involvement of adenosine. *J. Ethnopharmacol.* 127 (2): 313–318.

65 Suleymanova, E., Gulyaev, M., and Chepurnova, N. (2014). Ginseng extract attenuates early MRI changes after status epilepticus and decreases subsequent reduction of hippocampal volume in the rat brain. *Epilepsy Res.* 108 (2): 223–231.

66 Karimzadeh, F., Hosseini, M., Mangeng, D. et al. (2012). Anticonvulsant and neuroprotective effects of *Pimpinella anisum* in rat brain. *BMC Complement. Altern. Med.* 12 (1): 76.

67 Biradar, S., Kangralkar, V., Mandavkar, Y. et al. (2010). Antiinflammatory, antiarthritic, analgesic and anticonvulsant activity of *Cyperus essential* oils. *Int J Pharm Pharm Sci* 2 (4): 112–115.

68 Mandegary, A., Sharififar, F., Abdar, M., and Arab-Nozari, M. (2012). Anticonvulsant activity and toxicity of essential oil and methanolic extract of *Zhumeria majdae* Rech, a unique Iranian plant in mice. *Neurochem. Res.* 37 (12): 2725–2730.

69 Pathak, S., Wanjari, M., Jain, S., and Tripathi, M. (2010). Evaluation of antiseizure activity of essential oil from roots of *Angelica archangelica* Linn. In mice. *Indian J. Pharm. Sci.* 72 (3): 371.

70 Silva, M.R., Ximenes, R.M., Costa, J.G.M.d. et al. (2010). Comparative anticonvulsant activities of the essential oils (EOs) from *Cymbopogon winterianus* Jowitt and *Cymbopogon citratus* (DC) Stapf. In mice. *Naunyn Schmiedeberg's Arch. Pharmacol.* 381 (5): 415–426.

71 Koutroumanidou, E., Kimbaris, A., Kortsaris, A. et al. (2013). Increased seizure latency and decreased severity of pentylenetetrazol-induced seizures in mice after essential oil administration. *Epilepsy Res. Treat.* 2013: 532657.

72 Pages, N., Maurois, P., Delplanque, B. et al. (2010). Activities of α-asarone in various animal seizure models and in biochemical assays might be essentially accounted for by antioxidant properties. *Neurosci. Res.* 68 (4): 337–344.

73 Gopalakrishna, H., Pemminati, S., Giri, S. et al. (2010). Effect of *Acorus calamus* on electrical and chemical induced seizures in mice. *Int. J. Appl. Biol. Pharm. Technol.* 1 (2): 465–473.

74 Mandegary, A., Arab-Nozari, M., Ramiar, H., and Sharififar, F. (2012). Anticonvulsant activity of the essential oil and methanolic extract of *Bunium persicum* (Boiss). B. Fedtsch. *J. Ethnopharmacol.* 140 (2): 447–451.

75 Costa, J., Ferreira, P., De Sousa, D. et al. (2012). Anticonvulsant effect of phytol in a pilocarpine model in mice. *Neurosci. Lett.* 523 (2): 115–118.

76 Ebrahimzadeh, A.B., Hosseini, M., Mohammadpour, T. et al. (2011). Effects of soy extract on pentylenetetrazol-induced seizures in ovariectomized rats. *Zhong Xi Yi Jie He Xue Bao.* 9 (6): 611–618.

77 Pahuja, M., Mehla, J., Reeta, K. et al. (2011). Hydroalcoholic extract of *Zizyphus jujuba* ameliorates seizures, oxidative stress, and cognitive impairment in experimental models of epilepsy in rats. *Epilepsy Behav.* 21 (4): 356–363.

78 Golechha, M., Bhatia, J., and Arya, D.S. (2010). Hydroalcoholic extract of *Emblica officinalis* Gaertn. affords protection against PTZ-induced seizures, oxidative stress and cognitive impairment in rats. *Indian J. Exp. Biol.* 48 (5): 474–478.

79 Golechha, M., Sarangal, V., Bhatia, J. et al. (2014). Naringin ameliorates pentylenetetrazol-induced seizures and associated oxidative stress, inflammation, and cognitive impairment in rats: possible mechanisms of neuroprotection. *Epilepsy Behav.* 41: 98–102.

80 Choudhary, N., Bijjem, K.R.V., and Kalia, A.N. (2011). Antiepileptic potential of flavonoids fraction from the leaves of *Anisomeles malabarica*. *J. Ethnopharmacol.* 135 (2): 238–242.

81 Ding, J., Wang, J.-J., Huang, C. et al. (2014). Curcumol from Rhizoma Curcumae suppresses epileptic seizure by facilitation of GABA (a) receptors. *Neuropharmacology* 81: 244–255.

82 Hosseini, M., Pkan, P., Rakhshandeh, H. et al. (2011). The effect of hydro-alcoholic extract of citrus flower on pentylenetetrazole and maximal electroshock-induced seizures in mice. *World Appl. Sci. J.* 15 (8): 1104–1109.

83 Nassiri-Asl, M., Moghbelinejad, S., Abbasi, E. et al. (2013). Effects of quercetin on oxidative stress and memory retrieval in kindled rats. *Epilepsy Behav.* 28 (2): 151–155.

84 Visweswari, G., Prasad, K.S., Chetan, P.S. et al. (2010). Evaluation of the anticonvulsant effect of *Centella asiatica* (gotu kola) in pentylenetetrazol-induced seizures with respect to cholinergic neurotransmission. *Epilepsy Behav.* 17 (3): 332–335.

85 Choudhary, K.M., Mishra, A., Poroikov, V.V., and Goel, R.K. (2013). Ameliorative effect of curcumin on seizure severity, depression like behavior, learning and memory deficit in post-pentylenetetrazole-kindled mice. *Eur. J. Pharmacol.* 704 (1–3): 33–40.

86 Ezz, H.S.A., Khadrawy, Y.A., and Noor, N.A. (2011). The neuroprotective effect of curcumin and *Nigella sativa* oil against oxidative stress in the pilocarpine model of epilepsy: a comparison with valproate. *Neurochem. Res.* 36 (11): 2195.

18

Application of Nanophytomedicine for the Treatment of Central Nervous System Disorders

Nilesh Rarokar[1], *Nilambari Gurav*[2], *and Shailendra Gurav*[3]

[1] *Department of Pharmaceutical Sciences, Rashtrasant Tukadoji Maharaj Nagpur University, Nagpur - 440 033, Maharashtra, India*
[2] *PES's Rajaram and Tarabai Bandekar College of Pharmacy, Ponda, Goa University, Panaji, Goa - 403 101, India*
[3] *Department of Pharmacognosy, Goa College of Pharmacy, Goa University, Panaji, Goa - 403 001, India*

The central nervous system (CNS) is a vital organ system of the body that coordinates and controls all the activities of the different body parts. Hence, medicine or treatment therapy for CNS-related disorders plays a crucial role in clinical practice. However, traditional clinical treatment and the therapeutic agents used in that treatment have not been successful due to the problems associated with the anatomy and physiologic hurdles like the blood–brain barrier (BBB). Phytomedicines can resolve the issues reported by various researchers [1–3]. Phytomedicines from the species *Asparagus racemosus, Centella asiatica, Stephania japonica, Aegle marmelos, Coccinia grandis, Tabernaemontana divaricata, Bacopa monnieri, Abroma augusta,* and *Scoparia dulcis* have frequently been used for the treatment of various CNS-related disorders [4]. These phytomedicines are safe with low or no side effects/toxicity, low cost, and ease of accessibility with higher efficacy. However, their poor solubility and permeability and lack of desired biopharmaceutical qualities like absorption, distribution, and bioavailability make them therapeutically ineffective.

Moreover, nanotechnological approaches could be needed to develop a better formulation for getting the desired therapeutic effect. This has led to the development of nanomedicine, in which herbal medicines are either modified or encapsulated in nanocarriers, including liposomes, niosomes, cubosomes, phytosomes, and dendrimer micelles [5]. Phytomedicine could also be conjugated with various polymers to improve its efficacy. Many researchers have used the solubility enhancement strategy for fine-tuning biopharmaceutical properties like bioavailability and permeability.

In this chapter we discuss the importance and applications of nanophytomedicine for the effective target delivery of drugs in treating CNS disorders. We also highlight the various nanotechnologic approaches for developing an ideal delivery system with higher efficacy.

Phytochemical Drug Discovery for Central Nervous System Disorders: Biochemistry and Therapeutic Effects, First Edition. Edited by Chukwuebuka Egbuna and Mithun Rudrapal.

18.1 Neurodegenerative Disease and the Blood–Brain Barrier

18.1.1 Problems Associated with Treatment of Central Nervous System Disorders

The CNS is crucial for human health, since the brain controls all physiologic functions. Infections have been a prominent source of significant CNS morbidity in the last two decades. Identifying the causative organism and clinical symptoms associated with infectious CNS disorders are critical components of treatment options for CNS diseases. Many promising medications have been identified to treat various CNS illnesses. Still, certain features such as the BBB and the blood–cerebrospinal fluid barrier (BCSFB) usually limit the therapeutic efficacy of these drugs. Because of the ineffectiveness and failure of recently accessible treatment options for CNS illnesses, better and more effective treatment strategies are needed to overcome these limitations [6].

CNS diseases with neurodegenerative characteristics, such as aging-related deterioration in cognitive function and neuronal loss, are the most difficult to treat. However, many different disorders contribute to the loss of neuronal activity in the CNS. Many CNS disorders, including Alzheimer's disease (AD), amyotrophic lateral sclerosis (ALS or Lou Gehrig's disease), Huntington's disease (HD), Parkinson's disease (PD), head trauma, epilepsy, and stroke, have neurodegeneration as a common problem. Unfortunately, there have only been a few successful therapeutic methods available for neurodegenerative illnesses.

One of the most challenging steps in treating neurodegeneration is getting medications into the CNS by crossing the BBB. Current neurodegenerative disease therapy is ineffective because it focuses on alleviating symptoms rather than affecting the underlying condition. Furthermore, long-term treatment necessitates high doses and amounts of medications, but the problem of severe side effects is still unresolved. To accomplish successful therapy, a range of treatment methods like symptomatic treatment and direct disease management may need to be utilized at this time. Because of the difficulty in administering small molecule and protein/nucleic acid therapies to the CNS, these treatments are believed to be off-limits [7]. CNS disorders receive significantly fewer new drug approvals than other drugs due to the time and cost required for development, a higher risk of clinical failure, and unresolved mechanisms and pathophysiology. Progress in this subject has been impeded by the intricacy of CNS disorders and the biologic logistics of drug distribution. Many diseases are more challenging to treat because of a lack of understanding of the root causes, making it more difficult to find effective remedies. However, there is no evident cause for most illnesses in most cases. There are several plausible causes, including genetic predisposition and environmental effects.

Furthermore, the disease's symptoms are only understood by trial and error in many situations. Schizophrenia and depression, for example, are successfully treated with receptor-altering medicines, but the lack of understanding of these receptors results in uneven therapeutic efficacy and considerable adverse effects. The absence of physiologically realistic models to research these disorders increases the difficulty [8]. Getting macromolecular medicines into the CNS has been challenging for biomedical researchers for many years. Due to the selective permeability of the BBB, drugs must first penetrate it to reach their intended destinations in the brain. This intricate physiologic gatekeeper separates circulating blood from brain extracellular fluid and prevents circulating chemicals from entering the blood, which freely diffuses into the brain [9].

18.1.2 Role of the Blood–Brain Barrier

The BBB is a concept used to describe how chemicals diffuse to most bodily tissues but cannot access the brain. It is a neurovascular unit (NVU) component that functions as a blood–brain interface facilitating an electrochemical communication between the CNS and the peripheral nervous system (PNS). The BBB also protects the brain by controlling the transport of serum factors and neurotoxins to the brain and maintaining a consistent brain environment. It provides an interface on luminal and albumin membranes for transport. It also acts as a secretory body and a metabolic barrier [10–12]. The BBB is critical for transporting biomolecules into and out of the neuronal system of the brain [13].

18.1.3 Blood–Brain Barrier Crossing Mechanism

The brain's microvascular system creates extracellular fluid, which it separates from the blood via the BBB membrane. The astrocyte endfeet, basement membrane, and tight connections between endothelial cells are the most basic parts of the BBB, which comprises five separate sections (pericytes, astrocytes, neurons, basement membrane, and junction complexes). Endothelial cells called pericytes, which make up 20% of the cell's outer surface, are descended from vascular smooth muscle cells and participate in contracting and relaxing the brain capillaries to control blood flow. Glial cells, also known as astrocytes, link the brain's capillaries and neurons and keep the BBB functional by feeding neurons and protecting it from oxidative stress and metal toxicity. The permeability can be improved by temporarily disrupting extracellular matrix proteins, which is the most convenient method for achieving desired drug transportation. There are three types of junctions: adherent, tight, and gap; the tight junction is the essential component with significant transendothelial resistance. Tight connections rather than considerable fenestration between endothelial cells limit drug transport across the BBB. By encapsulating capillaries with pericytes and astrocytes, the electrical resistance between endothelial cells rises to 1500–2000 cm^2. Passive transcellular diffusion allows only tiny molecules like water, gases, and lipid-soluble substances to move through the BBB. Specific proteins are required to transport large molecules like glucose, amino acids, and most drugs with a high electrical charge, polarity, and hydrophilicity. This means that distributing and delivering medications to the brain is challenging [13].

BBB regulates the entry of some drugs into the brain while removing harmful molecules from circulation. This effective permeable barrier permits the CNS milieu to maintain homeostasis and brain circuits to function correctly [14]. For proper neuronal function, the BBB is critical in creating and maintaining a healthy CNS [15]. Paracellular and transcellular pathways allow different chemicals to enter the CNS through the BBB. Under normal physiologic conditions, low molecular weight compounds can enter CNS parenchyma by crossing the BBB in three ways:

- Tiny hydrophilic compounds can diffuse paracellularly to the brain.
- Lipid-soluble compounds like steroidal hormones tend to passive transcellular diffusion across the BBB. On the other hand, this route is restricted to low molecular weight

(<500 Da), lipid-soluble, neutral (at physiologic pH) molecules, and has fewer than eight hydrogen bonds.

- Endogenous transporters (carriers) mediate most of the remaining nutrients' entry into the brain, driven by the concentration gradient of the nutrients and the help of appropriate transporters [16]. Passive diffusion, active efflux, carrier-mediated transport (CMT), receptor-mediated transport (RMT), major facilitator superfamily, and immune cell migration over the BBB are all modes of transport through the BBB [17].

The BBB protects the brain and maintains CNS homeostasis. However, comprehension of the BBB is incomplete, and breaching it remains a significant barrier to innovative brain therapeutics.

The following are modes of transportation:

- *Carrier-mediated transport* (CMT) – the BBB blocks the movement of critical polar chemicals like glucose and amino acids, essential for metabolism, from the brain to the rest of the body. As a result, the brain's essential nutrients must be delivered differently. As part of the solute carrier (SLC) transporter gene family, CMTs are included in the CMT subfamily. Many membrane-bound transporter genes have been found that encode membrane-bound proteins that transport a wide range of substrates [18]. A vast variety of substances, including carbohydrates, amino acids, fatty acids, mono- and polycarboxylic acids, hormones, nucleotides and organic anions, and vitamins and amines, can be transported across the cell membrane by SLC transporters.

- *Receptor-mediated transport* (RMT) – because of the binding of amino acids to peptides, bigger peptides and proteins are impeded from crossing the BBB. However, crossing the BBB with RMT systems using specific neuroactive peptides or regulatory proteins is possible. Endocytotic processes, such as transcytosis, allow these large molecular weight solutes to enter the CNS intact [17].

- *Immune cell movement across the BBB* – compared to other tissues, neutrophil infiltration into the CNS is highly minimal and tightly regulated, making the CNS an immune-privileged zone [19].

- *Passive diffusion* – although the BBB can allow a wide range of lipid-soluble compounds to enter the brain, nearly 98% of all tiny molecules cannot. Only 5% of the medications in the Comprehensive Medicinal Chemistry (CMC) database are used to treat the CNS [20].

- *Active efflux* – BBB luminal, blood-facing endothelial plasma membrane expresses several ABC proteins. They are adenosine triphosphate (ATP)-driven xenobiotic and endogenous metabolite efflux pumps that modulate drug permeability [20].

- *Major facilitator superfamily* – docosahexaenoic acid (DHA) and other essential omega-3 fatty acids are transported into the brain by an endothelial facilitator superfamily domain-containing protein (MFSD2a).

18.1.4 Phytomedicine/Neuroprotective Drugs Reported for Central Nervous System Disorders

Drugs studied in previous research on CNS-related disorders are outlined in Table 18.1.

Table 18.1 Previous studies on central nervous system (CNS)-related disorders.

Drug	Disease	Key findings	References
Memantine	Stroke, CNS trauma, Parkinson's disease, Alzheimer's disease, vascular dementia	Memantine evaluated for neuroprotection action in dementia, and long-term safety established	[21]
Cholinesterase inhibitors (ChEIs)	Alzheimer's disease	Safety and efficacy of ChEIs in Alzheimer's disease evaluated, and technology offers more advances in this area	[22]
Ginkgo biloba, St. John's wort, kava-kava, valerian, *Bacopa monniera,* and *Convolvulus pluricaulis*	CNS disorder (dementia of Alzheimer's type, anxiety, poor cognition, depression, nervous disorders)	Various herbs used for CNS disorders and herbal remedies as potential medicinal plants for CNS disorders	[23]
Intravenous recombinant tissue plasminogen activator	Acute ischemic stroke	Implementation of hypothermia best treatment for acute ischemic stroke	[24]
Ibogaine, hesperetin, ginsenoside Rg3, paeoniflorin, quercetin, ginsenoside Rg3, Curcumin	Epilepsy	Extracts in herbal medicine can show best pharmacologic action on epilepsy	[25]
ChEIs, memantine, anti-amyloid therapy, β-secretase enzyme, γ-secretase enzyme, immunization, monoclonal antibodies	Alzheimer's disease	Dementia is multifaceted disease that needs direct treatment. Patients need pharmacologic treatment plus biopsychosocial care. Potential drug treatments in Alzheimer's disease	[26]
α- and β-asarone	Depression, Parkinson's disease, psychosis, Alzheimer's disease, anxiety, epileptic diseases	Pharmacokinetic studies of drugs demonstrated that oral bioavailability of both α- and β-asarone is poor with short plasma half-lives, and well distributed to highly perfused organs, especially brain, due to high lipophilicity	[27]
Centella asiatica	Parkinsonism	Extract of *C. asiatica* ECa233 in rotenone-induced parkinsonism rats shows neuroprotective effect	[28]
Carotenoids	Neurodegenerative disease	Carotenoids may be used as drug to prevent and treat neurodegenerative disorder	[29]
Limonene and D-limonene	Alzheimer's disease, epilepsy, stroke, migraine	Use of natural limonene has neuroprotective potential [30]	[30]
Angelica sinensis	Ischemic stroke	Can be used for ischemic stroke treatment	[31]

18.2 Nano Approaches to Central Nervous System Drug Delivery

18.2.1 Types of Nanocarriers

18.2.1.1 Liposomes

Liposomes were first-generation nanocarriers that proved helpful in delivering drugs in the 1970s. In order to transfer drugs, proteins, and other bioactive molecules to the body, these microscopic vesicles have hydrophilic compartments in the center and phospholipid bilayers on either one or more sides. These vesicles resemble the cell membrane in their morphology. Large and small unilamellar (10–50 nm) and multilamellar (20–100 nm) bilayers are divided into three categories: small, medium, and large unilamellar. Hydrogen bonds and van der Walls forces make these structures reversible because of their non-covalent interactions. It means that ordinary liposomes have a brief time in the body, since reticuloendothelial system (RES) cells quickly remove them from circulation. As a result, numerous attempts have been made to develop liposomes that circulate for a long time and are specifically targeted.

Bilayers of lipids are used to make liposomal nanoparticles. They have a sphere-like form. Liposomes with diameters ranging from 50 to 500 nm are formed when an amphiphilic lipid is added to an aqueous or another hydrophilic liquid. Sonication and reverse-phase evaporation are the most common methods for creating liposomal nanoparticulated systems. Thermosensitive nanoparticles release the encapsulated drug, a liposome nanoparticle type, at a predetermined temperature. The high-intensity ultrasound, microwaves, or radio waves are used to stimulate targeted drug release [32].

18.2.1.2 Micelles

Polymeric and non-polymeric micelles are vesicles of amphiphilic copolymers and amphiphilic surfactants, respectively. Many research groups are currently investigating this as a viable method for administering drugs to the CNS. When it comes to biodistribution, polymeric micelles outperform non-polymeric micelles. Micelles are composed of an outside hydrophilic environment and an inner hydrophobic core (polycaprolactone, phospholipids, and fatty acids) that allows hydrophobic pharmaceuticals to be loaded. The hydrophilic environment is predominantly composed of poly(ethylene glycol) (PEG). In an aqueous environment, the hydrophilic coating on the exterior of micelles stabilizes them. In addition, it increases their circulation time in the bloodstream, shielding them from the RES and increasing their accumulation in leaky vascular regions [33].

18.2.1.3 Solid Lipid Nanoparticles

Solid lipid nanoparticles (SLNs) have proven to be a unique drug carrier for novel and tailored drug delivery systems. Aqueous colloidal nanocarriers consisting of physiologic lipids such as triglycerides, fatty acids, and steroids, which are dispersed in water or an aqueous surfactant solution, are known as hydrophobic nanocarriers. Once the SLNs have cooled, the final step is to solidify the material. By combining spatially varied or solid lipids with liquids, researchers have developed nanostructured lipocarriers, which have the potential to improve the drug-loading capacity and the long-term stability of SLNs. Researchers have found that PEG modification increases BBB penetration and drug delivery to the CNS when combined with anticancer drugs like camptothecin and doxorubicin [32].

18.2.1.4 Phytosomes

The polar phytochemicals are transformed into lipid-compatible molecular complexes known as phospholipid complexes, supramolecular complexes, or herbosomes. They are also known as phytosomes and help improve the solubility of standardized plant extracts by encapsulating phospholipids, which tends to increase bioavailability. Phytosomes have a better bioavailability than typical plant extracts because they can pass through lipid membranes and reach the systemic circulation. The term "phytosome" is made up of two parts: *phyto* means "plant," and *some* means "cell-like structure." Phytosomes are small, cell-like structures formed using a mixture of lecithin and plant bioactives/extract in a suitable solvent. They are phyto-phospholipid complexes composed of equimolar or stoichiometric ratios of phospholipids and natural bioactives such as polyphenols or plant extracts. The most common ratio utilized is 1 : 1. The most prevalent requirement for selecting an appropriate lipid is the ratio of phosphatidyl groups present in that lipid [33].

Phytosomes are also used to enhance the solubility of poorly soluble drugs. Among all solubility enhancement technologies, the phospholipid complexation technique is found to be superior. Many researchers have employed this technique successfully to enhance solubility, permeability, and oral bioavailability of many extracts and plant bioactives such as *B. monnieri* extract, *C. asiatica* extract, apigenin, kaempferol, and daidzein. This system has created more interest in research due to a number of benefits. These include easy preparation, high drug/extract loading capacity, long-term stability and controlled release, maximization of therapeutic activity, fewer side effects, and gastric protection from luminal enzymes.

18.2.1.5 Nanosponges

Nanosponges are nanoscale drug carriers that have a three-dimensional structure generated by crosslinking polymers and are used for conventional and targeted drug delivery. Nanosponges can be prepared in different shapes and sizes. They are biodegradable and biocompatible with minimum toxicity. They also show higher encapsulation efficiency with about 10–25 nm diameter [34]. Furthermore, the solubility of poorly water-soluble compounds can be improved by using nanosponges, which can encapsulate hydrophilic and lipophilic moieties. Nano-sized particles interact with bioactive molecules by two different mechanisms: these bioactives may capture the nanoparticles that entangle with them or form a coupling with nano-sized particles [35].

18.2.1.6 Nanoemulsions

A nanoemulsion is a colloidal particulate system ranging from 10 to 1000 nm and acts as a nanocarrier of drug molecules. Because they are smaller in size than the wavelength of visible light, nanoemulsions are transparent or transparent to milky-white solid spheres with a negative charge on the surface. A miniemulsion, either oil-in-water or water-in-oil dispersion stabilized by an interfacial film of a surfactant molecule, is also referred to as a nanoemulsion. By delivering a drug to the brain via the nasal route, nanoemulsions can avoid the issues associated with oral administration. The different dosage forms like liquids, creams, gels, foams, and sprays containing nanoemulsions can be administered by different routes, including oral, parenteral, ocular, and nasal. Both hydrophobic and lipophilic drugs can be administered by formulating nanoemulsions. Ultrasonication and high-pressure homogenization can be used to prepare nanoemulsions [36].

18.2.1.7 Dendrimers

Dendrimers are three-dimensional molecules containing an initial core, many internal layers, repeating units, and several active surface groups at the end. The number of dendrimer generations increases with the number of dendrimer branches and surface groups (represented by G). Dendrimers have a reputation for high performance and low dispersion. Because they have several surface groups and a hydrophobic core, they can carry a vast number of medications and serve various purposes. The most compatible molecules for drug delivery are the nanosize structures, transported across the cellular membranes and physiologic barriers by endocytosis and cellular internalization. Different dendrimers such as polyamidoamine (PAMAM) dendrimers, polyhydroxylamine dendrimers, and polypropylene amine have clinical applications in drug administration. They could also be used in gene therapy to deliver polycationic dendrimers that form polyplexes with nucleic acids [37].

18.2.1.8 Nanoparticles

Nanoparticles are solid colloidal particles ranging from 1 to 1000 nm (1 μm) composed of macromolecular materials in which drug or biologically active material is dissolved, entrapped, or encapsulated, or to which drug is adsorbed or attached [38, 39]. Nanoparticles have three layers: a surface layer, a shell layer, and the core where the drug cargo is entrapped. If we consider brain delivery, several macromolecules can be delivered in the form of nanoparticles. Drug transport through the BBB via nanoparticles can be facilitated by endocytosis (drug release from endothelial cells) and transcytosis into brain cells. The surface modification of nanoparticles could also be done for specific plasma protein adsorption and desired cellular uptake of the drug [40].

18.2.2 Techniques/Preparation/Methods

Methods of preparation of nanocarriers are illustrated in Figure 18.1.

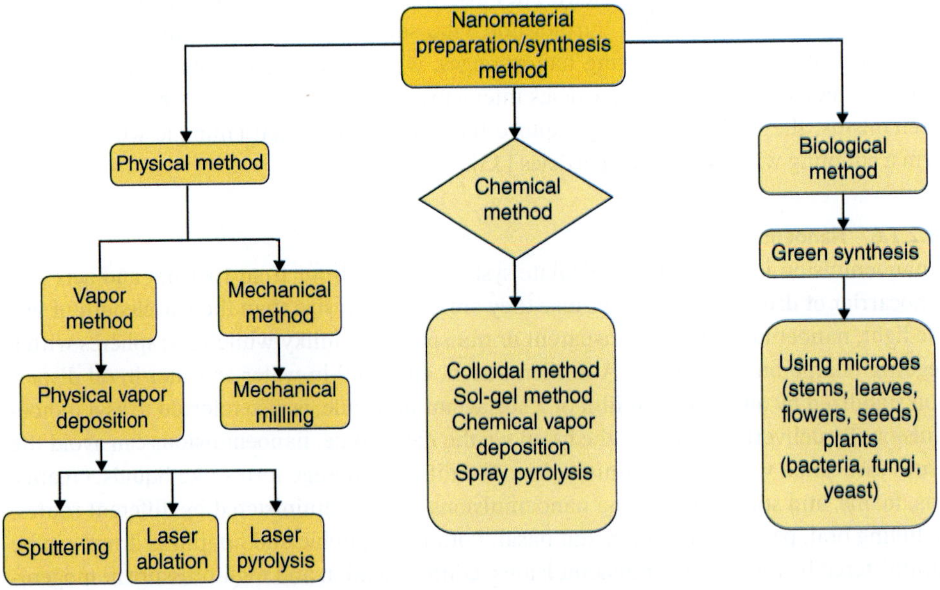

Figure 18.1 Methods of preparation of nanocarriers.

18.2.3 Mechanism of Action of Drug Release

Drug adsorption on brain capillary endothelial cells is dependent on hydrophilicity, surface charge, and targeting the ligand properties of drug-loaded nanocarriers. The brain's endothelial cells have a negative surface charge, which interacts with the positive charge on the surface of nanomaterials to form electrostatic interactions, augmenting and facilitating drug absorption. Additionally, the lipophilic nature of nanocarriers affects drug absorption. Nanomaterials reenter the bloodstream and release the drug payload by targeting low-density lipoprotein receptors found on microvessel brain capillary endothelial cells or normal endocytosis [41]. The drug release in the bloodstream after entry into the brain cells could be responsible for the improved bioavailability.

Therapeutic applications of different nanocarriers are outlined in Table 18.2.

Table 18.2 Applications of different nanocarriers.

Drug	Study title	Therapeutic applications	References
Paclitaxel	A hindsight reflection on clinical studies of poly(l-glutamic acid)-paclitaxel	Anticancer agent extracted from bark of Pacific yew tree	[42]
Ginkgo biloba	Role of *Ginkgo biloba* extract, against isoproterenol induced cardiac toxicity in rats	Has antioxidant activity and protects heart from free radical-mediated toxicity of isoproterenol	[43]
Terminalia arjuna	Antibacterial efficacy of silver nanoparticles synthesized employing *Terminalia arjuna* bark extract	Used as cardioprotective agent, antioxidant, antimicrobial	[44]
Curcumin, *Syzygium cumini*	Current advances in the use of nanophytomedicine therapies for human cardiovascular diseases	Antihypercholesterolemic and antihypertensive effects of curcumin nanoemulsion, antidiabetic and gastroprotective properties	[45]
Rabdosia rubescens (Donglingcao)	Inhibition of platelet aggregation and the release of P-selectin from platelets by cilostazol	Use of silver nanoparticles in inflammation and thrombosis	[46, 47]
Cherry	Cherry antioxidants: from farm to table	Used in cancers, heart disease, and diabetes	[48]
Garlic (*Allium sativum* L.) oil	Nanophytomedicine for prevention of metabolic syndrome: a pharmacological and biopharmaceutical review	Antidiabetic, antihyperlipidemic, and antiatherogenic effects. Nanoemulsion	[49]
Gold bhasma	Nanophytomedicine and drug formulations	Ayurvedic nanoparticles used for cancer therapy	[50]
Tea	Recent trends in the development of nanophytobioactive compounds and delivery systems for their possible role in reducing oxidative stress in Parkinson's disease models	Rich in flavonoids, reduces risk of Parkinson's disease in human trials	[51]

18.3 Nanophytomedicine for Treatment of Central Nervous System Disorders

18.3.1 Alzheimer's Disease

AD is characterized by cognitive impairment triggered by amyloid aggregation, hyperphosphorylated neurofibrillary tangle formation, and cholinergic system malfunction. The neuronal channel's integrity can erode after the disease has begun. Numerous tailored nanomedicines have paved the way due to nanomedicines' ability to target and transport drugs across the BBB. In conjunction with phytomedicine, nanotechnology is a developing medical miracle that shows promising results in the treatment of AD [52].

Resveratrol (3,5,4′-trihydroxystilbene) is a polyphenolic flavonoid that occurs naturally in cis and trans isomers. The trans isomers show prevalent and desired biologic activity. Resveratrol's neuroprotective effects in neurologic illnesses are linked to protecting neurons from oxidative damage and toxicity and inhibiting apoptotic neuronal death. Several plants like mulberries, peanuts, rhubarb, and grape seeds and skins contain resveratrol. About 5–10% of resveratrol is present in the biomass, depending on the quality of the grapes. Moreover, red wine and purple grapes contain a variety of flavonoids and have a higher resveratrol content.

SLNs are composed of hydrophobic lipids and have sizes in the nanometer range. Hence this could be a promising approach for the brain delivery of various drugs. They also have the ability of hepatic bypass and are able to cross the BBB. Modified polymers are used to prepare the SLNs, which prolong the circulation time of SLN in the bloodstream and show a longer duration of action. Aside from improving blood circulation duration, PEG chains provide steric stability, allowing ligands to attach to BBB nutrition transport systems [53].

In AD patients, quercetin acts by preventing lipid peroxidation, protein oxidation, and neuronal cell death. Amyloid-β peptide (1–42) aggregation is inhibited by quercetin, which acts as a neuroprotectant. Among the various nanoparticles, lipid nanoparticles are ideal carriers for the delivery of quercetin to the brain. Pharmaceutically they are biocompatible and biodegradable, hence they are not harmful to the human body. They can be used as a targeting carrier if the conjugation approach has been employed [54].

18.3.2 Parkinson's Disease

PD is a chronic neurodegenerative disorder that affects patients worldwide. The decrease in the dopamine level in PD patients due to the deterioration of the substantia nigra leads to loss of movement, affecting their social and family life. Hence, it will become challenging to handle such patients. This condition arises in those aged 65–69 and significantly above 80 years [55]. The treatment of PD has some clinical difficulties due to the uncooperative behavior of patients. However, it can be treated with various natural bioactives. The use of phytoactives such as phenolic acids and flavonoids for treatment purposes could be the paradigm in clinical therapy. Various mechanisms could minimize oxidative stress and repair the CNS. Furthermore, by binding with metal ions, flavonoids reduce free radical generation and slow the progression of PD.

Another phytoconstituent, curcumin, has established its potential in treating PD. The use of nanotechnology in formulating a delivery system for curcumin enhances its bioavailability and therapeutic efficacy. Nanocurcumin helps reduce oxidative stress and apoptosis in the brain of flies and mice. The curcumin-loaded poly(lactic-co-glycolic acid) (PLGA) nanoparticles show improved bioavailability over the others [56].

Puerarin (PU) is a major active component found in *Pueraria lobata*. It helps to inhibit tyrosine hydroxylase (TH) + dopamine (DA) neuronal mortality. PU also can preserve DA neurons in animal models of 6-hydroxydopamine (6-OHDA)-mediated neurotoxicity, as reported in the literature. This protective effect was connected to activating the phosphatidylinositol 3-kinase/protein kinase B signaling pathway, which prevented p53 accumulation in the nucleus and caspase-3-mediated apoptosis. Moreover, PU is a Class IV drug in the Biopharmaceutics Classification System, including poorly water-soluble and permeable drugs. Therefore, various groups have developed different strategies to overcome these drug-related problems. PU can be encapsulated in micelles prepared from polyethylene glycol-phosphatidylethanolamine, PEGylated mesoporous silica nanoparticles, and RGD-modified and PEGylated lipid nanoparticles [57].

18.3.3 Epilepsy

Epilepsy is a set of illnesses characterized by a wide range of comorbidities (depression, anxiety, learning disabilities, attention-deficit hyperactivity disorder, intellectual disability, and autism). Yet no drugs have been proved to cure epilepsy. As a result, pharmacologic treatment focuses on symptom control via long-term administration of antiepileptic drugs (AEDs). Epilepsy patients can be treated in an emergency or using acute or chronic medication. In the first example, an immediate effect of the treatment on the brain is required, whereas in the second case, side effects are minimized and patient compliance is improved [58].

Catechin-rich fruits (black grapes, peaches, strawberries, and broad beans) and phytochemicals (epicatechin, catechins, catechin hydrate, epigallocatechin, epicatechin gallate, and epigallocatechin gallate) are used to treat epilepsy. Catechin hydrate, a flavonoid, is abundant in green tea and grape seed. In the past, catechin hydrate was found to have antioxidant and anti-inflammatory activities. Biodegradable and biocompatible nanoparticles that release catechin hydrate in a continuous and controlled manner are becoming increasingly popular in nasal gel compositions. Catechin hydrate was evaluated for its brain bioavailability by direct nose-to-CNS distribution using chitosan-coated PLGA nanoparticles [59].

Natural black pepper contains the alkaloid piperine, which has potent anticonvulsant effects. Hence it is widely used in epilepsy, but the solubility and bioavailability issues restrict its clinical use. Researchers have found several ways to improve the bioactivity and solubility of the piperine molecule, including biodegradable polymeric hydrogels, liposomes, and piperine-encapsulated nanoparticles. As previously documented, piperine encapsulated in copper oxide quantum dots/hyaluronic acid (CuQDs@HA)/PLGA nanovesicles has a more significant potential to enhance bioavailability and anticonvulsant effect than free piperine [60].

18.3.4 Stroke

Stroke is a leading cause of mortality and disability all over the world. The most common type of stroke is ischemic, which represents 87% of all cases. There are very few therapeutic options available for clinical therapy that target brain tissues. Tissue-type plasminogen activator (tPA) injections commonly break blood clots in occluded arteries. In stroke treatment, this is the only US Food and Drug Administration (FDA)-approved treatment option.

Sulfonylurea receptor 1 and transient receptor potential melastatin 4 (SUR1-TRPM4) cation channels have high affinity for glyburide, an antagonist, which is one of the possible compounds for stroke treatment.

According to a recent study, over 200 medicinal herbs may be beneficial in treating stroke. However, such plant-derived bioactives have limited CNS permeability; nanomedicines are a promising strategy for resolving permeation-related problems.

Betulinic acid or glyburide-loaded nanoparticles have been shown in some studies to be incredibly effective at reaching the brain and aiding functional recovery in stroke-bearing rats. The utilization of betulinic acid nanoparticles as a drug carrier for glyburide was also seen and validated by positron emission tomography/computed tomography (PET/CT) [61]. Resveratrol has proven antioxidant and anti-inflammatory activity and can be used in brain illnesses such as stroke, but its bioavailability problem leads to limited clinical use. Reported studies have demonstrated nanoformulations for improved bioavailability and efficacy of resveratrol. In another study, resveratrol-loaded nanostructured lipid carriers were 40–160 times more effective than free resveratrol in the treatment of middle cerebral artery occlusion [62].

18.3.5 Huntington's Disease

HD is a hereditary neurodegenerative illness. The cell death due to HD results in motor, cognitive, and mental impairments in adult HD patients. HD is a multifaceted condition with epigenetic variables and oxidative stress at its core. Its genesis and progression are also influenced by oxidative stress, as it damages cellular structures, impairs DNA repair, and causes mitochondrial malfunction. Increased oxidative damage has been found in the postmortem brains of HD patients. There is currently no viable treatment available to cure or slow the course of HD. Most HD therapeutics in clinical trials have targeted the *HTT* gene, with RNA interference (RNAi)-based therapy to deliver adeno-associated viral vectors (AAV), genome editing based on clustered regularly interspaced short palindromic repeats (CRISPR), and targeting of single-nucleotide polymorphisms (SNPs). These therapies have loopholes like gene interference for a more extended period and cleavage with safety risks. Brain function is dependent on the presence of selenium. On the other hand, selenium has been shown in observational epidemiologic studies to improve cognitive and neurologic function. Reduced brain function and neuronal death may be caused by low selenium levels in the brain or problems with selenium homeostasis. Among the other forms of selenium, the nanoparticles show improved bioavailability, higher efficacy, and lower toxicity [63].

Some research teams worked on nose-to-brain drug delivery for clinical therapy of HD. The anti-HTT small interfering RNA (siRNA)-loaded chitosan nanoparticles proved their higher efficacy when studied in a transgenic YAC128 mouse HD model. The nanoformulations also demonstrated their potential for targeted delivery of the anti-HTT siRNA. It shows a more than 50% decrease in HTT mRNA expression. Intranasal delivery of siRNA nanoparticles is a viable treatment option for safely and effectively decreasing mutant HTT expression [64]. Tetrabenazine is a potential molecule to treat HD, but the problems associated with biopharmaceutical properties restrict its use in clinical therapy. The nasal administration of tetrabenazine-loaded nanoemulsion overcame these issues of lower solubility and first-pass metabolism [65].

18.3.6 Multiple Necrosis

Apoptosis was coined in the early 1970s after researchers discovered novel patterns of cell death. During apoptosis, morphologic and biochemical events looked to be well ordered, whereas necrosis appeared to be a disordered (or unintended) form of cell death. Apoptosis was later an example of preprogrammed cell death (PCD). According to the theory, cell death induced by pathologic stimuli is a unique example of the evolved conservative mechanism of cell elimination in animals and plants in response to morphogenetic and homeostatic signals. The discovery of genes responsible for biochemical pathways of suicidal cell death (in the nematode *Caenorhabditis elegans*) also contributed to the concept's success. Initially, it was considered that PCD and apoptosis were purely dependent on the expression of new ("death") genes. As a result, epigenetic cell death strategies have been largely dismissed; that is, ways that do not require de novo protein synthesis [66]. In China, Chinese herbal medicine has gained popularity as adjuvant therapy for femoral head osteonecrosis (ONFH) [67].

18.3.7 Tumors/Gliomas/Glioblastomas

Gliomas are brain tumors that resemble astrocytes, oligodendrocytes, and ependymal cells (astrocytomas, oligodendrogliomas). These belong to a group of malignancies for which there is no proper clinical therapy to cure or increase patient life. Gliomas are classified primarily by their morphologic similarity to types of glial cells, the architecture of cells, and their histologic structures with diffusion infiltration ability [68]. Curcumin, derived from the *Curcuma longa* rhizome, has shown anticancer potential with less toxicity and greater efficacy at a higher dose. Curcumin's anticancer properties have been utilized to treat glioma, and it has been proved by using different animal models to prevent tumor initiation, development, and metastasis. Several curcumin delivery approaches were reported, including curcumin-loaded glyceryl monooleate nanoparticles (GMO-NPs) made of GMO, and Pluronic™ F-68 (ThermoFisher Scientific, Waltham, MA, USA) was used as a surfactant. This could be a novel approach to treating glioblastoma [69].

18.4 Challenges in Nanophytomedicine

Nanotechnology's burgeoning involvement in therapeutics is not without its difficulties, and its outcomes are not always encouraging. It has also been linked to several hazards. Nanotechnology is currently being used to carry medications across the BBB [70, 71]. Nanomedicine is in its infancy in preclinical and clinical research, but sufficient data are still needed to support its superiority over traditional neurotherapeutics [72]. Oxidative stress, amino acid imbalance, and BBB disruption are possible consequences of nanocarrier composition and may lead to neurotoxicity [73]. Aside from ligand targeting, nanoparticles' tiny size and surface area may present issues such as particle aggregation, interparticular friction, limited drug loading, and a high clearance rate that may preclude their application in diagnostics or drug administration.

Toxic effects such as neuroinflammation, excitotoxicity, and DNA damage can be caused by the increased surface area of nanoparticles, which in turn increases ROS generation. Chemical and physical features and the route of administration all influence nanocarriers' toxicity profile [74]. For nano-drugs to be safe for human consumption, they must be biocompatible and biodegradable. Drug delivery via nanotechnology faces a variety of obstacles. Additionally, it is important to understand the frontiers. Anatomic restrictions and multimodal neural contact make nano-based drug delivery more difficult because nanomedicines must engage with neurons to display their response [75]. The ultimate goal of nanophytomedicine for CNS delivery systems is to engineer safe nanocarriers for prolonged usage without unwanted side effects and make them effective in bench-to-bedside clinical translation.

18.5 Conclusion

New drug carrier systems loaded with nanotechnology are commonly called nanomedicine. They can be utilized for various medical purposes, including therapy, diagnosis, and prevention. However, concerns about the prudent integration of nanotechnology concepts and methodologies with phytomedicines have recently emerged. There has been great interest in using these formulations in clinical settings because of their potential for improving absorption, bioavailability, and pharmacologic performance in loaded phytomedicines. These advantages and the fact that they dissolve quickly and begin working sooner have sparked numerous research projects. In addition, pharmaceutical companies are interested in providing better alternatives to conventional dosage forms to improve people's health through "herbal treatments" that incorporate technological technologies. However, nanotechnology-based nanophytomedicinal products have yet to find their way into clinical settings despite their enormous potential and benefits. This is due to several factors, including the need for complex and advanced machinery processes, scaling issues, high preparation costs, environmental and health risks associated with the generation of extremely small particles, and the short shelf life of such developed formulations. These factors are still considered major obstacles and must be overcome.

References

1 Kantati, Y.T., Kodjo, K.M., Dogbeavou, K.S. et al. (2016). Ethnopharmacological survey of plant species used in folk medicine against central nervous system disorders in Togo. *J. Ethnopharmacol.* 181: 214–220.

2 Kinda, P.T., Zerbo, P., Guenné, S. et al. (2017). Medicinal plants used for neuropsychiatric disorders treatment in the Hauts Bassins region of Burkina Faso. *Medicines* 4 (2): 32.

3 Kpemissi, M., Eklu-Gadegbeku, K., Veerapur, V.P. et al. (2019). Antioxidant and nephroprotection activities of *Combretum micranthum*: a phytochemical, in-vitro and ex-vivo studies. *Heliyon* 5 (3): e01365.

4 Uddin, M. and Zidorn, C. (2020). Traditional herbal medicines against CNS disorders from Bangladesh. *Natural Products Bioprospect.* 10 (6): 377–410.

5 Rodrigues, K., Gurav, S., Joshi, A. et al. (2020). Porous polymeric carrier system for modified drug release of boswellic acid. *Chem. Sci.* 11 (2): 1–2.

6 Jahangir, M.A., Anand, C., Muheem, A. et al. (2020). Nano phytomedicine based delivery system for CNS disease. *Curr. Drug Metab.* 21 (9): 661–673.

7 Ingle, A.P., Shende, S., Gupta, I. et al. (2018). Metal nanoparticles in management of diseases of the central nervous system. In: *The Microbiology of Central Nervous System Infections* (ed. K. Kon and M. Rai), 81–98. Cambridge, MA: Academic Press.

8 Barchet, T.M. and Amiji, M.M. (2009). Challenges and opportunities in CNS delivery of therapeutics for neurodegenerative diseases. *Expert Opin. Drug Deliv.* 6 (3): 211–225.

9 DiNunzio, J.C. and Williams, R.O. III (2008). CNS disorders – current treatment options and the prospects for advanced therapies. *Drug Dev. Ind. Pharm.* 34 (11): 1141–1167.

10 Reynolds, J.L. and Mahato, R.I. (2017). Nanomedicines for the treatment of CNS diseases. *J. Neuroimmune Pharmacol.* 12 (1): 1–5.

11 Cai, Z., Qiao, P.F., Wan, C.Q. et al. (2018). Role of blood-brain barrier in Alzheimer's disease. *J. Alzheimers Dis.* 63 (4): 1223–1234.

12 Rhea, E.M. and Banks, W.A. (2019). Role of the blood-brain barrier in central nervous system insulin resistance. *Front. Neurosci.* 13: 521.

13 Zhou, Y., Peng, Z., Seven, E.S., and Leblanc, R.M. (2018). Crossing the blood-brain barrier with nanoparticles. *J. Control. Release* 270: 290–303.

14 Khan, N.H., Mir, M., Ngowi, E.E. et al. (2021). Nanomedicine: a promising way to manage Alzheimer's disease. *Front. Bioeng. Biotechnol.* 9: 630055.

15 Haddad-Tóvolli, R., Dragano, N.R., Ramalho, A.F., and Velloso, L.A. (2017). Development and function of the blood-brain barrier in the context of metabolic control. *Front. Neurosci.* 11: 224.

16 Liebner, S., Dijkhuizen, R.M., Reiss, Y. et al. (2018). Functional morphology of the blood–brain barrier in health and disease. *Acta Neuropathol.* 135 (3): 311–336.

17 Li, G., Shao, K., and Umeshappa, C.S. (2019). Recent progress in blood–brain barrier transportation research. In: *Brain Targeted Drug Delivery Systems* (ed. H. Gao and X. Gao), 33–51. Cambridge, MA: Academic Press.

18 Kadry, H., Noorani, B., and Cucullo, L. (2020). A blood–brain barrier overview on structure, function, impairment, and biomarkers of integrity. *Fluids Barriers CNS* 17 (1): 1–24.

19 Saint-Pol, J., Gosselet, F., Duban-Deweer, S. et al. (2020). Targeting and crossing the blood-brain barrier with extracellular vesicles. *Cells* 9 (4): 851.

20 Wong, A.D., Ye, M., Levy, A.F. et al. (2013). The blood-brain barrier: an engineering perspective. *Front. Neuroeng.* 6: 7.

21 Jain, K.K. (2000). Evaluation of memantine for neuroprotection in dementia. *Expert Opin. Investig. Drugs* 9 (6): 1397–1406.

22 Francis, P.T., Nordberg, A., and Arnold, S.E. (2005). A preclinical view of cholinesterase inhibitors in neuroprotection: do they provide more than symptomatic benefits in Alzheimer's disease? *Trends Pharmacol. Sci.* 26 (2): 104–111.

23 Kumar, V. (2006). Potential medicinal plants for CNS disorders: an overview. *Phytother. Res.* 20 (12): 1023–1035.

24 Froehler, M.T. and Ovbiagele, B. (2010). Therapeutic hypothermia for acute ischemic stroke. *Expert Rev. Cardiovasc. Ther.* 8 (4): 593–603.

25 Sucher, N.J. and Carles, M.C. (2015). A pharmacological basis of herbal medicines for epilepsy. *Epilepsy Behav.* 52: 308–318.

26 Briggs, R., Kennelly, S.P., and O'Neill, D. (2016). Drug treatments in Alzheimer's disease. *Clin. Med.* 16 (3): 247.

27 Chellian, R., Pandy, V., and Mohamed, Z. (2017). Pharmacology and toxicology of α-and β-Asarone: a review of preclinical evidence. *Phytomedicine* 32: 41–58.

28 Teerapattarakan, N., Benya-Aphikul, H., Tansawat, R. et al. (2018). Neuroprotective effect of a standardized extract of *Centella asiatica* ECa233 in rotenone-induced parkinsonism rats. *Phytomedicine* 44: 65–73.

29 Manochkumar, J., Doss, C.G., El-Seedi, H.R. et al. (2021). The neuroprotective potential of carotenoids in vitro and in vivo. *Phytomedicine* 91: 153676.

30 Eddin, L.B., Jha, N.K., Meeran, M.N. et al. (2021). Neuroprotective potential of limonene and limonene containing natural products. *Molecules* 26 (15): 4535.

31 Han, Y., Chen, Y., Zhang, Q. et al. (2021). Overview of therapeutic potentiality of *Angelica sinensis* for ischemic stroke. *Phytomedicine* 90: 153652.

32 Jahangir, M.A., Khan, S., Singh, A.D. et al. (2022). Nanophytomedicine in clinical management: an introductory evidence-based review. *J. Pharm. Res. Sci. Technol.* 6 (1): 26–37.

33 Gurav, S., Usapkar, P., Gurav, N. et al. (2022). Preparation, characterization and evaluation (in-vitro, ex-vivo and in-vivo) of naturosomal carriers for enhanced delivery and therapeutic efficacy of hesperetin. *PLoS One* 17 (11): e0274916.

34 Rodrigues, K., Nadaf, S., Rarokar, N. et al. (2022). QBD approach for the development of hesperetin loaded colloidal nanosponges for sustained delivery: in-vitro, ex-vivo, and in-vivo assessment. *OpenNano* 7: 100045.

35 Yata, V.K., Ranjan, S., Dasgupta, N., and Lichtfouse, E. (ed.) (2020). *Nanopharmaceuticals: Principles and Applications*, vol. 3. Cham: Springer International.

36 Tiwari, K. and Bhattacharya, S. (2022). The ascension of nanosponges as a drug delivery carrier: preparation, characterization, and applications. *J. Mater. Sci. Mater. Med.* 33 (3): 1–21.

37 Bonferoni, M.C., Rossi, S., Sandri, G. et al. (2019). Nanoemulsions for “nose-to-brain” drug delivery. *Pharmaceutics* 11 (2): 84.

38 Dessai, S., Ayyanar, M., Amalraj, S. et al. (2022). Bioflavonoid mediated synthesis of TiO_2 nanoparticles: characterization and their biomedical applications. *Mater. Lett.* 311: 131639.

39 Saoji, S.D., Rarokar, N.R., Dhore, P.W. et al. (2022). Phospholipid based colloidal nanocarriers for enhanced solubility and therapeutic efficacy of withanolides. *J. Drug Deliv. Sci. Technol.* 70: 103251.

40 Saeedi, M., Eslamifar, M., Khezri, K., and Dizaj, S.M. (2019). Applications of nanotechnology in drug delivery to the central nervous system. *Biomed. Pharmacother.* 111: 666–675.

41 Kreuter, J. (2014). Drug delivery to the central nervous system by polymeric nanoparticles: what do we know? *Adv. Drug Deliv. Rev.* 71: 2–14.

42 Zhao, J., Koay, E.J., Li, T. et al. (2018). A hindsight reflection on the clinical studies of poly (l-glutamic acid)-paclitaxel. *Wiley Interdiscip. Rev. Nanomed. Nanobiotechnol.* 10 (3): e1497.

43 Badore, N.S., Das, P.K., Pillai, S., and Thakur, A. (2017). Role of Ginkgo biloba extract, against isoproterenol induced cardiac toxicity in rats. *Indian J. Pharm. Educ. Res.* 51 (4): 691–699.

44 Ahmed, Q., Gupta, N., Kumar, A., and Nimesh, S. (2017). Antibacterial efficacy of silver nanoparticles synthesized employing Terminalia arjuna bark extract. *Artif. Cells Nanomed. Biotechnol.* 45 (6): 1192–1200.

45 Hesari, M., Mohammadi, P., Khademi, F. et al. (2021). Current advances in the use of nanophytomedicine therapies for human cardiovascular diseases. *Int. J. Nanomedicine* 16: 3293.

46 Kariyazono, H., Nakamura, K., Shinkawa, T. et al. (2001). Inhibition of platelet aggregation and the release of P-selectin from platelets by cilostazol. *Thromb. Res.* 101 (6): 445–453.

47 Ferretti, G., Bacchetti, T., Belleggia, A., and Neri, D. (2010). Cherry antioxidants: from farm to table. *Molecules* 15 (10): 6993–7005.

48 Scalbert, A., Manach, C., Morand, C. et al. (2005). Dietary polyphenols and the prevention of diseases. *Crit. Rev. Food Sci. Nutr.* 45 (4): 287–306.

49 Nouri, Z., Hajialyani, M., Izadi, Z. et al. (2020). Nanophytomedicines for the prevention of metabolic syndrome: a pharmacological and biopharmaceutical review. *Front. Bioeng. Biotechnol.* 8: 425.

50 Othayoth, R.A., Kalivarapu, S.R., and Botlagunta, M.A. (2014). Nanophytomedicine and drug formulations. *Int. J. Nanotechnol. Appl.* 4: 1–8.

51 Gunasekaran, T., Haile, T., Nigusse, T., and Dhanaraju, M.D. (2014). Nanotechnology: an effective tool for enhancing bioavailability and bioactivity of phytomedicine. *Asian Pac. J. Trop. Biomed.* 4: S1–S7.

52 Ovais, M., Zia, N., Ahmad, I. et al. (2018). Phyto-therapeutic and nanomedicinal approaches to cure Alzheimer's disease: present status and future opportunities. *Front. Aging Neurosci.* 10: 284.

53 Loureiro, J.A., Andrade, S., Duarte, A. et al. (2017). Resveratrol and grape extract-loaded solid lipid nanoparticles for the treatment of Alzheimer's disease. *Molecules* 22 (2): 277.

54 Pinheiro, R.G., Granja, A., Loureiro, J.A. et al. (2020). Quercetin lipid nanoparticles functionalized with transferrin for Alzheimer's disease. *Eur. J. Pharm. Sci.* 148: 105314.

55 Sharma, S., Rabbani, S.A., Agarwal, T. et al. (2021). Nanotechnology driven approaches for the management of Parkinson's disease: current status and future perspectives. *Curr. Drug Metab.* 22 (4): 287–298.

56 Chen, T., Liu, W., Xiong, S. et al. (2019). Nanoparticles mediating the sustained puerarin release facilitate improved brain delivery to treat Parkinson's disease. *ACS Appl. Mater. Interfaces* 11 (48): 45276–45289.

57 Musumeci, T., Bonaccorso, A., and Puglisi, G. (2019). Epilepsy disease and nose-to-brain delivery of polymeric nanoparticles: an overview. *Pharmaceutics* 11 (3): 118.

58 Ahmad, N., Ahmad, R., Alrasheed, R.A. et al. (2020). Quantification and evaluations of catechin hydrate polymeric nanoparticles used in brain targeting for the treatment of epilepsy. *Pharmaceutics* 12 (3): 203.

59 Zhu, D., Zhang, W.G., Nie, X.D. et al. (2020). Rational design of ultra-small photoluminescent copper nano-dots loaded PLGA micro-vessels for targeted co-delivery of natural piperine molecules for the treatment for epilepsy. *J. Photochem. Photobiol. B Biol.* 205: 111805.

60 Deng, G., Ma, C., Zhao, H. et al. (2019). Anti-edema and antioxidant combination therapy for ischemic stroke via glyburide-loaded betulinic acid nanoparticles. *Theranostics* 9 (23): 6991.

61 Ashafaq, M., Alam, M.I., Khan, A. et al. (2021). Nanoparticles of resveratrol attenuates oxidative stress and inflammation after ischemic stroke in rats. *Int. Immunopharmacol.* 94: 107494.

62 Cong, W., Bai, R., Li, Y.F. et al. (2019). Selenium nanoparticles as an efficient nanomedicine for the therapy of Huntington's disease. *ACS Appl. Mater. Interfaces* 11 (38): 34725–34735.

63 Sava, V., Fihurka, O., Khvorova, A., and Sanchez-Ramos, J. (2020). Enriched chitosan nanoparticles loaded with siRNA are effective in lowering Huntington's disease gene expression following intranasal administration. *Nanomed. Nanotechnol. Biol. Med.* 24: 102119.

64 Arora, A., Kumar, S., Ali, J., and Baboota, S. (2020). Intranasal delivery of tetrabenazine nanoemulsion via olfactory region for better treatment of hyperkinetic movement associated with Huntington's disease: pharmacokinetic and brain delivery study. *Chem. Phys. Lipids* 230: 104917.

65 Proskuryakov, S.Y., Konoplyannikov, A.G., and Gabai, V.L. (2003). Necrosis: a specific form of programmed cell death? *Exp. Cell Res.* 283 (1): 1–6.

66 Zhang, Q., Yang, F., Chen, Y. et al. (2018). Chinese herbal medicine formulas as adjuvant therapy for osteonecrosis of the femoral head: a systematic review and meta-analysis of randomized controlled trials. *Medicine* 97 (36): e12196.

67 Glaser, T., Han, I., Wu, L., and Zeng, X. (2017). Targeted nanotechnology in glioblastoma multiforme. *Front. Pharmacol.* 8: 166.

68 Zanotto-Filho, A., Coradini, K., Braganhol, E. et al. (2013). Curcumin-loaded lipid-core nanocapsules as a strategy to improve pharmacological efficacy of curcumin in glioma treatment. *Eur. J. Pharm. Biopharm.* 83 (2): 156–167.

69 Dilnawaz, F. and Sahoo, S.K. (2013). Enhanced accumulation of curcumin and temozolomide loaded magnetic nanoparticles executes profound cytotoxic effect in glioblastoma spheroid model. *Eur. J. Pharm. Biopharm.* 85 (3): 452–462.

70 Wong, H.L., Wu, X.Y., and Bendayan, R. (2012). Nanotechnological advances for the delivery of CNS therapeutics. *Adv. Drug Deliv. Rev.* 64 (7): 686–700.

71 Naqvi, S., Panghal, A., and Flora, S.J. (2020). Nanotechnology: a promising approach for delivery of neuroprotective drugs. *Front. Neurosci.* 14: 494.

72 Dinda, C. and S, Pattnaik G. (2013). Nanobiotechnology-based drug delivery in brain targeting. *Curr. Pharm. Biotechnol.* 14 (15): 1264–1274.

73 Sharma, H.S. and Sharma, A. (2007). Nanoparticles aggravate heat stress induced cognitive deficits, blood–brain barrier disruption, edema formation and brain pathology. *Prog. Brain Res.* 162: 245–273.

74 Vega-Villa, K.R., Takemoto, J.K., Yáñez, J.A. et al. (2008). Clinical toxicities of nanocarrier systems. *Adv. Drug Deliv. Rev.* 60 (8): 929–938.

75 Jain, K.K. (2007). Nanobiotechnology-based drug delivery to the central nervous system. *Neurodegen. Dis.* 4 (4): 287–291.

Index

a

Aβ deposition 61
abnormal involuntary movement scale (AIMS) 83
Acanthopanax 22
Acanthopanax senticosus roots and stems (ASRS) 22
accelerated approval 3
accelerate synuclein aggregation 17
acetylcholine (ACh) 55, 316
acetylcholine-induced bronchospasm 388
acetylcholinesterase (AChE) 54, 89, 267, 274, 318
acetylcholinesterase inhibitors (AChEI) 12
AChE *see* acetylcholinesterase (AChE)
AChEI *see* acetylcholinesterasc inhibitors (AChEI)
acid-sensing ion channels 138
Aconitum 185
Acorus calamus Linn. 88, 139, 365–366
acquired immune deficiency syndrome (AIDS) 235
active efflux 416
acute flaccid paralysis (AFP) 267
AD *see* Alzheimer's disease (AD)
adenosine triphosphate (ATP)-sensitive K+ channels 21, 239
adenosine triphosphate (ATP) synthase membrane 87
Aduhelm® 12
advanced glycation end products (AGE) 238
Aegle marmelos 274
AFP *see* acute flaccid paralysis (AFP)
Ageratum conyzoides 387
aggregated proteins 221
aggregation-competent intermediate 17
Alafia barteri oliver 185
alkaloids 184–185
allergic reactions 386
aloe vera 292
α-synuclein 19, 27
  phytochemicals targeting 27–28
Alpinia 22
Alstonia scholaris 139
Alzheimer's disease (AD) 29, 43, 222
  medicinal plants targeting
    amyloid beta 61–65
    cholinergic system 56–61
    tau-related pathways 67
  natural products, with anti-Alzheimer's activity 66, 68–69
  pathogenesis
    amyloid plaques 44
    cholinergic functions 45
    neuroinflammation 45
    NFT 45
    oxidative stress 45
    synaptic damage 45
  treatment strategies
    amyloid hypothesis 46
    cholinergic hypothesis 55
    and clinical trials 47–53
    dendritic hypothesis 54–55
    metabolic hypothesis 55–56
    strategies for, 5-HT_6 obstruction 55
    tau proteins hypothesis 46, 54
Amaryllidaceae 297
Amburana cearensis 388
α-amino-3-hydroxy-5-methyl-4-isoxazole propionic acid receptors (AMPAR) 319
Amondys 45® 12
amphetamine-based medications 4
amylin 56
β-amyloid protein (Aβ) 317
amyloid beta (Aβ) peptide 11
amyloid precursor protein (APP) 43

Phytochemical Drug Discovery for Central Nervous System Disorders: Biochemistry and Therapeutic Effects, First Edition. Edited by Chukwuebuka Egbuna and Mithun Rudrapal.

amyotrophic lateral sclerosis (ALS) 222, 226–227, 414
angiogenesis targeting 162
angiopoietin 1 (Ang-1) expression 164
angiotensin-converting enzyme (ACE) inhibitors 385
Annona squamosa 365
anti-Alzheimer's agent 317
anti-dengue flavivirus herbal agents 264
antidepressants 8
antiepileptic drugs (AEDs) 56, 397
antihistaminic plants, family
  Acanthaceae 386
  Amaranthaceae 386
  Amaryllidaceae 387
  Asclepiadaceae 387
  Asteraceae 387
  Caesalpiniaceae 387
  Casuarinaceae 387
  Cruciferae 388
  Cucurbitacea 388
  Euphorbiaceae 388
  Fabaceae 388
  Moraceae 388
  Myricaceae 388
  Myrtaceae 389
  Olacaceae 389
  Piperaceae 389
  Poaceae 389
  Polygalaceae 390
  Scrophulariaceae 390
  Verbenaceae 390
  Zygophylaceae 390
anti-inflammatory agents
  bioactive compounds
    curcumin 238
    epigallocatechin 3-gallate 239–240
    eugenol 238–239
    guggulsterone 240
    resveratrol 240–241
    thymoquinone 241
  medicinal plants
    Allium sativum 232
    Azadirachta indica 233
    Cassia tora 233
    Euphorbia hirta 233–234
    Garcinia mangostana 234
    Ricinus communis 235–236
    Scutellaria baicalensis 236–237
    Solanum melongena 237
  neuroinflammation role, in neurodegenerative diseases 220–222
  neuroinflammatory drugs
    for ALS 226–227
    for MS 227–230
    for NMO 225–226
    for Parkinson's disease 230–231
    for stroke 223
    for traumatic brain injury 224
anti-inflammatory drugs 182
anti-inflammatory isoflavone 20
antioxidant-responsive element (ARE) activation 237
antioxidative enzymes 172
antipyretic agents, of plant origin
  Arbutus andrachne (Ericaceae) 341–343
  Berberis spp. (Berberidaceae) 343
  Cassia fistula Linn. (Caesalpiniaceae) 343–344
  Crataeva magma (Capparidaceae) 344–345
  Echinops kebericho M. (Asteraceae) 345–346
  Enicostema littorale (Gentianaceae) 346
  Piper nigrum (Piperaceae) 346–347
  Viola betonicifolia (Violaceace) 347–348
anxious distress 206
apigenin (4',5,7-trihydroxyflavone) 319
apolipoprotein E gene (APOE) 43
apoptosis targeting 162
arachidonic acid (AA) pathway 239
Arbutus andrachne (Ericaceae) 341–343
Artemisia dracunculus 140
arthropod-borne viruses (Arboviruses) 268
ashwagandha extract 65
asiatic acid 279
Asparagus racemosus 211
Asteraceae 95
Astragalus membranaceus 21
Atropa belladonna 27
attention-deficit/hyperactivity disorder (ADHD) 4
Avena sativa 364
avian infectious bronchitis (IB) 252

b

baby hamster kidney (BHK)-21 cell-derived vaccines 263
Bacopa monnieri 89, 211, 363
bacoside A and B 364
BBB *see* blood-brain barrier (BBB)
BCSFB *see* blood-cerebrospinal fluid barrier (BCSFB)
benign prostatic hyperplasia (BPH) 211
Berberidaceae (*Berberis* spp.) 343
berberine 21
β-blockers 9
bhilavanol A and B 279
biflavonoid 266
bioactive compounds 342
  with antihistaminic activities 390–393
blackberries 275–276

blocking cholinesterase enzymes 12
blood-brain barrier (BBB) 3, 165, 221, 222, 413
blood-cerebrospinal fluid barrier (BCSFB) 414
blood cleansing agents 292
blood purifiers 292
blueberries 276
bovine mammary epithelial cells (BME-UV1)
reduced proinflammatory cytokines 235
brain
conditions 302–303
functions of 300
homeostasis 301
learning and memory 302
motivation 301
motor control 300–301
physiology
electrical activity 300
metabolism 300
neurotransmitters 299
receptors 299
brain-derived neurotrophic factor (BDNF) 18, 45, 163, 283
brief psychiatric rating scale (BPRS) 83
5-bromo-2-deoxyuridine/neuronal nuclei (BrdU/NeuN) 367
Buddleja globose 185
Butea monosperma 139
butyrylcholinesterase (BChE) 12

C

Caenorhabditis elegans 275
caffeine (1,3,7-trimethylpurine-2,6-dione) 318
calcitonin gene-related peptide (CGRP) 8
calcium channel blockers 13
Calendula officinalis L. 173
Camellia 22
cannabidiol (CBD) 403
cannabinoid receptor 1 (CB1) 331
carbonic anhydrase inhibitors 13
carbon nanotubes (CNT) 173
carrier-mediated transport (CMT) 416
caspase-3-mediated cell death 25
Cassia spp. 23
C. fistula Linn. (Caesalpiniaceae) 343–344
C. obtusifolia L. 23
C. sophera 387
C. tora L. 23
Casuarina equisetifolia L. 387
catalase (CAT) 89, 154, 239
catechins 318
catechol O-methyltransferase (COMT) inhibitors 10
celastrol 21
Celastrus panicutalus 90
cellular nucleophiles 19
Centella asiatica 279
Centers for Disease Control and Prevention (CDC) 181
central nervous system (CNS) disorder 1, 88, 161, 252, 359
BBB
immune cell movement 416
mechanism 415–416
role of 415
controlling vectors, of viral diseases 268–269
drugs approved, by FDA
ADHD 4–8
Alzheimer's disease 11–12
epilepsy/seizure 12–13
migraine 8–9
multiple sclerosis 11
muscular dystrophy 12
Parkinson's disease 9–11
etiology 2
future perspective 269
incidence and prevalence, neurologic disorders 2
mechanistic actions, of medicinal herbs
in vitro studies 375
in vivo studies 375–376
nano approaches
mechanism of action, drug release 421
nanocarriers, types of 418–420
techniques/preparation/methods 420
nanophytomedicine, treatment of
AD 422
challenges in 425–426
epilepsy 423
HD 424
multiple necrosis 425
Parkinson's disease 422–423
stroke 423–424
pathogenesis 3
phytomedicine/neuroprotective drugs 416–417
plant and phytochemicals
HIV 267–268
HSV 264–265
JEV 262–263
poliomyelitis virus 266–267
rabies virus 265–266
SARS Cov-2 virus 252, 262
TBE virus 264
against viral infections 253–261
VZV 266
WNV 263–264
problems, with CNS disorder treatment 414–415
viral infections 252

cerebral cortex 298
cerebral edema 302
cerebrovascular changes (CVC) 43
chemotherapy-induced peripheral neuropathy (CIPN) 139, 153–154
cherries 276–277
choline acetyl-transferase (ChAT) 55
chronic constriction injury (CCI) 140
chronic mild stress (CMS) 320
chronic unpredictable stress (CUS)-induced behavioral depression 367
Chrysanthemi Flos 23
Chrysanthemum spp. 23
  C. indicum L. 23
  C. morifolium R. 23
Cistanche spp. 23
  C. deserticola 23
  C. tubulosa 23
Citrullus spp.
  C. colocynthis 140
  C. lanatus 274
Citrus grandis 274
c-Jun N-terminal kinase (JNK) phosphorylation 23
Cleome rutidosperma 362
Clerodendrum serratum L. 390
clinically isolated syndrome (CIS) 11
clustered regularly interspaced short palindromic repeats (CRISPR) 424
CMS *see* chronic mild stress (CMS)
CNT *see* carbon nanotubes (CNT)
cobalamin (vitamin B12) 281
collagen-induced arthritis (CIA) 238
combination antiretroviral therapy (cART) 267
conventional healing strategies 291
Coptis xhinensis 21
Coriandrum sativum, medicinal properties of 304
coronavirus disease 2019 (COVID-19) 252
Crataeva magma (Capparidaceae) 344–345
C-reactive protein (CRP) 236
Crinum glaucum 387
critical neurotransmitter 17
Crocus sativus, medicinal properties of 303
CRP *see* C-reactive protein (CRP)
Cuphea aequipetala (CAE) 239
Curculigo orchioides 387
Curcuma longa 140, 296
  medicinal properties of 305
curcumin-loaded glyceryl monooleate nanoparticles (GMO-NPs) 425
CXC chemokine receptor 4 (CXCR4) expression 164
cyclic adenosine monophosphate (cAMP) 83
cyclic adenosine monophosphate (cAMP)-response element-binding protein (CREB) 163, 320
cyclooxygenase 162
cyclooxygenase (COX)-2 enzyme 26, 184, 328
Cymbopogon citratus 403
Cynodon dactylon 389
Cyperus Rotundus Linn. 92
cytokine C-C motif chemokine 11 (CCL11)/ eotaxin 389

d

Dale's principle 299
damage-associated molecular patterns (DAMPs) 401
DDS *see* diallyldisulfide (DDS)
Deflazacort 12
6-dehydrogingerdione 297
depression
  causes of
    environmental factors 204
    genetic factors 204
    risk factors 204
  diagnosis 205
  medicinal plants
    Allium cepa (Amaryllidaceae) 366
    Bacopa monnieri (Plantaginaceae) 367–368
    Glycyrrhiza glabra (Fabaceae) 368
    Rosmarinus officinalis (Lamiaceae) 368
    Zingiber officinale (Zingiberaceae) 369
  phytochemicals 208
  symptoms of
    in children 205
    in men 204
    in women 205
  treatment
    medicinal plants, as antidepressants 208–213
    natural treatment 208
  types of
    depression, in pregnancy 207
    depression, with psychosis 206–207
    major depressive disorder 206
    PDD 206
Desmodium adscendens 362
dexmethylphenidate hydrochloride 8
diabetes-linked complications 43
diallyldisulfide (DDS) 296
diallylmonosulfide (DMS) 296
diallyltetrasulfide (DTTS) 296
diallyltrisulfide (DTS) 296
diroximel fumarate 11
disease-modifying agents for MS (DMAMS) 11
DMS *see* diallylmonosulfide (DMS)

docosahexaenoic acid (DHA) 416
dopamine D2 receptors (DRD2) 82, 88
dopamine metabolism 19
dopamine transporter (DAT) 82
Drosophila melanogaster 22
D9-tetrahydrocannabinol 403
DTTS *see* diallyltetrasulfide (DTTS)
Duchenne muscular dystrophy (DMD) 4

e

echinacoside 23
Echinops kebericho M. (Asteraceae) 345–346
egg albumin-induced anaphylaxis 388
electroencephalography (EEG) 300, 398
electron transport chain (ETC) 18, 327
Emblica officinalis 275, 366
Emergency use authorizations (EUA) 3
endogenous transporters 416
endothelial cells 415
Enicostema littorale (Gentianaceae) 346
enteroviruses 266
enzyme-linked immunosorbent assay (ELISA) 137
epicatechin 155
(–)-epicatechin (EC) 317
(–)-epicatechin gallate (ECG) 317
epidermal growth factor (EGF) 18
(–)-epigallocatechin gallate (EGCG) 317
epigallocatechin 3-gallate 239–240
epigallocatechin gallate (EGCG) 154
epilepsy
  antiepileptic properties, of medicinal plants 404–407
  basic mechanisms 398–400
  epileptic seizures, types of 398
  and inflammation 401–402
  medicinal plants
    Bridelia micrantha (Phyllanthaceae) 372
    Hypoxis hemerocallidea (Hypoxidaceae) 370–371
    Piper methysticum (Piperaceae) 371
    Sarcocephalus latifolius (Rubiaceae) 372
    treating 402–403
    Valeriana officinalis (Caprifoliaceae) 372
  and oxidative stress 400–401
  seizure induction tests 402
Eremobium aegyptiacum 61
ergot alkaloids 8
Ericaceae (*Arbutus andrachne*) 341–343
erythrocyte sedimentation rate (ESR) 238
ESR *see* erythrocyte sedimentation rate (ESR)
ethanol extract from *Clerodendrum cyrthophyllum* Turcz (EE-CC) 239
ethanolic extract from *Vatairea macrocarpa* (EEVM) 239
Eucalyptus citriodora 268
Eugenia jambolana 389
eugenol (4-allyl-2-methoxyphenol) 238
Euphorbia hirta 388
European plants 27
excitatory-inhibitory synaptic imbalance 81
excitotoxicity targeting 162
extracellular signal-regulated kinase (ERK) activation 163

f

F-box protein (FBXO7) 18
femoral head osteonecrosis (ONFH) 425
ferroportin (FP1) 363
Ferula assafoetida, medicinal properties of 304–305
ferulic acid (FA) 173
fever development, mechanism
  humoral pathway 328–330
  neural pathway 330
fibroblast growth factor-2 (FGF2)-Akt 25
Ficus spp.
  F. bengalensis L. 388
  F. religiosa 92–93
Flaviviridae flavivirus *see* Japanese encephalitis virus
flavone-3-O-arabinopyranoside 343
flavonoids 184, 298
foci forming unit reduction assay (FFURA) 263
Foeniculum vulgare L. 56
folate (vitamin B9) 281
folic acids 283
forced swim test (FST) 211, 366
Fourier transform infrared spectroscopy (FITR)-spectral analysis 343
Fragaria ananassa 276
fragile X mental retardation protein (FMRP) 87
front-facing projection 301
fumarate therapy 11
Fyn kinase inhibitor 54

g

GABA *see* γ-aminobutyric acid (GABA)
$GABA_A$ receptor (GABAAR) 319
GAD-immunoreactivity assays (GAD-ir) 87
galanthamine-like alkaloids 267
γ-aminobutyric acid (GABA) 9, 81, 299, 316
γ-aminobutyric acid aminotransferase (GABA-AT) 398
γ-aminobutyric acid type A (GABAA) receptors 154
gastric caeca 269

Gastrodia spp. 23–24
G. elata 23
generalized anxiety disorder (GAD) 209
gensinoside Rd-mediated neuroprotection 21
gingerenone-A 297
gingerols 297
Ginkgo spp. 24
G. biloba 140, 213
ginsenosides 21
glioblastomas 425
gliomas 425
glutamate 316
blockers 13
levels 274
metabotropic receptors 138
glutamate dehydrogenase (GDH) 319
glutamic acid decarboxylase (GAD) 86
glutamyl peptide 295
glutathione (GSH) 90, 163
depletion 19
glutathione peroxidase 1 (GPx1) 164, 239
glutathione peroxidase (GPx) 21, 89, 172
glutathione S-transferase (GST) 239
glycogen synthase kinase-3 beta (p-GSK-3β) 163
glycosaminoglycan (GAG) 264
glycosides 185
Glycyrrhiza glabra 93
G-protein-coupled receptors (GPCRs) 400
grape seed proanthocyanidin extract (GSPE) 318
guanine nucleotide-binding protein subunit beta-like protein 1 gene (GNB1L) 87
Gynostemma spp. 24
G. pentaphyllum 24

h

Harpagophytum procumbens 185
H-current modulators 13
heat shock protein 72 (HSP72) antibody 165
heme oxygenase 1 (HO-1) 164
Hemidesmus indicus R. 387
herbalism 291
herbal medications 331
herbal medicine 293
herpes simplex virus (HSV) 236, 252, 264–265
heterotypic immunity 267
high-fat diet 139
histamine 385
histamine-induced bronchoconstriction 389
H_2O_2-induced PC12 cell damage 275
5-HT1 receptor modulators 162
human alpha herpesvirus 1 (HHV-1) 264
human alpha herpesvirus 2 (HHV-2) 264
human brain 161
human carbonic anhydrase (hCA) inhibitors 400
human immunodeficiency virus (HIV) 252, 267–268
HIV-1 222
Huntington's disease (HD) 317, 424
medicinal plants 372
Calendula officinalis (Asteraceae) 373–374
Camellia sinensis (Theaceae) 374
Embelia ribes (Primulaceae) 374
Ginkgo biloba (Ginkgoaceae) 373
Panax ginseng (Araliaceae) 373
15-hydroperoxy-eicasotetraenoyl-phosphatidylethanolamines (15-HpETE-PE) 18
hydroxyapatite (HA) 233
6-hydroxydopamine (6-OHDA)-mediated neurotoxicity 423
12(S)-hydroxyeicosatetraenoic acid 172
8-hydroxyguanosine (8-OHG) 164
hydroxymethylfurfural (HMF) 234
5-hydroxytryptamine (5-HT) 83
Hyoscyamus niger L. 27
hyperexcitability 138
hyperphosphorylated tau 65
hypoxia inducible factor-1 alpha (HIF-1α)/VEGF pathway 164
Hypoxis hemerocallidea 268

i

immune-mediated inflammatory disease 11
immunomodulators 11
Indian Ayurvedic medical system 297
Indian ginseng 26
induced pluripotent stem cells (iPSC)-derived neurons 18
inducible nitric oxide synthase (iNOS) 154, 184, 232
inducible NO synthase 363
inflammation 219, 220
inflammatory cytokines (tumor necrosis factor (TNF))-α 18
inflammatory process 219
iNOS, 154 *see* inducible nitric oxide synthase (iNOS)
interleukin-6 (IL-6) 89, 401
interleukin (IL)-5 389
intracerebral hemorrhage 302
irritable bowel syndrome (IBS) 369
ischemic stroke, treatment strategies
neuroprotective strategy 162
neurorestorative strategy 162–163

j

Japanese encephalitis virus (JEV) 252, 262–263
Japanese kampo medicine 65

JEV *see* Japanese encephalitis virus (JEV)
jyotishmati 90

k

kaempferol 3-O-rutinoside 97
kainate receptors (KAR) 319
kainate (KAI) tests 402
KAR *see* kainate receptors (KAR)
kinase 2 (JAK2) phosphorylation 163
Korean Seonghyangjeongkisan 65

l

lactate dehydrogenase (LD) 21
lactic acid-like excretory secretions 269
Lasmiditan Reyvow 9
Lavandula officinalis 209–210
learned helplessness test (LH) 211
Lepidium sativum L. 388
leukotriene B4 (LTB4) 241
leukotrienes (LTs) 219
lipid peroxidation 164
lipid peroxide (LPO) 19
lipopolysaccharide (LPS) 220, 330
  LPS-induced neurotoxicity 20
liposomes circulation time 165
5-lipoxygenase (5-LOX) 154, 241
12/15-lipoxygenase overexpression 172
lipoxygenase (LOX) pathways 185
liquid chromatography-mass spectrometry (LC-MS) 342
liquiritigenin ((2S)-4',7-dihydroxyflavan-4-one) 318
long-term stockpiling 297
Loranthus micranthus 185
low-and middle-income countries (LMICs) 366
low molecular weight (LMW) 19
LPO *see* lipid peroxide (LPO)
LPS-induced inflammation 26
LPS-induced toll-like receptor (TLR)-4-mediated neuroinflammation 277
L-theanine 212
luteolin (3',4',5,7-tetrahydroxyflavone) 318

m

magnetoencephalography (MEG) 300
major depressive disorder 205
malondialdehyde (MDA) 24, 90, 140, 163, 279, 303
malpighian tubules 269
Matricaria spp.
  M. chamomilla 153
  M. recutita 213
matrix metalloproteinase (MMP)-2 241
matrix metalloproteinase (MMP)-9 activation 140
maximal electroshock seizure (MES) test 402
mechanical allodynia 137
medhya rasayana plant 90
medicinal plants 291
  antidepressants, for depression
    Asparagus racemosus 211
    Bacopa monniera 211
    Ginkgo biloba 213
    herbal plants 209
    Kaempferia parviflora 210–211
    Lavandula officinalis 209–210
    Matricaria recutita 213
    Momordica charantia 212
    Passiflora foetida 211
    Valeriana officinalis 212
  for CNS disorders 360–362
  components 29–37
  interventions and CNS disorders
    depression 366–369
    epilepsy 369–372
    HD 372–374
  with neuroprotective action 362–366
  phytochemical constituents
    in garlic 297–298
    in ginger 297
    in onions 295–296
    in turmeric 296–297
  phytochemicals 293, 295
  protective effects, on brain
    Coriandrum sativum 304
    Crocus sativus 303
    Curcuma longa 305
    Ferula assafoetida 304–305
    Nigella sativa 304
    Thymus vulgaris 305
  therapeutic value and importance of 292–293
  types and uses of 293, 294
mediodorsal (MD) 87
medium spiny neurons (MSNs) 373
Melissa officinalis L. 56
Mentha spicata L. 388
methanolic extract of roots (MAR) 211
methiin 296
6-methoxyflavone 154
methylphenidate-based medications 4
methylphenidate hydrochloride 4
1-methyl-4-phenylpyridinium (MPP+) 18
1-methyl-4-phenyl-1,2,3,6-tetrahydropyridine (MPTP) 18
microglial stimulation 138
microscopic organisms
  gram-negative 298
  gram-positive 298
microtubule stabilizers 54
Middle East respiratory syndrome (MERS) 252

mitochondrial-mediated cascade 23
mitochondrial membrane potential (MMP) stabilization 21
mitochondrial oxidative stress (MOS) 327
mitogen-activated protein kinase 3 (MAPK3) 84, 165, 235
Momordica spp.
  M. charantia 140, 212
  M. dioica 388
monoamine oxidase (MAO) 280, 366
monoamine oxidase B (MAO-B) 10, 26
monoclonal antibodies 11
monocyte chemoattractant protein-1 (MCP-1) 232
mood disorders 283
Morris water maze 164
MOS *see* mitochondrial oxidative stress (MOS)
motor cortex 301
Moutan Cortex Radicis 24
Mucuna pruriens (MP) 26–27
multiple sclerosis (MS) 227–230
muscarinic metabotropic receptor (mAChR) 317
muscle-mind signals 299
muscular dystrophy 12
myeloperoxidase (MPO) 239
Myrica esculenta 388

n

N-acetylcysteine (NAC) 88
Namzaric® 12
nanocarriers
  applications 421
  types
    dendrimers 420
    liposomes 418
    micelles 418
    nanoemulsions 419
    nanoparticles 420
    nanosponges 419
    phytosomes 419
    SLN 418
nanomicelles 173
nanoparticles (NP) 173
Nauclea latifolia 372
nerve growth factor (NGF) 153
neural crest stem cells (NSC) 173
neuroactive ligand-receptor interaction 88
neurodegenerative diseases 43, 44, 273, 303, 360
neurofibrillary tangles (NFT) 45
  intracellular hyperphosphorylated NFT 65
neurogenic locus notch homolog protein 4 88
neuroinflammation 220
neuroinflammatory variations 154
neuromyelitis optica (NMO) 225–226
neuron 316
neuronal impairment 139
neuronal potassium channel 13
neuropathic pain 155
  animal models, for studying 139
  chronic 137
  mechanisms of 138
  medicinal plants and phytochemicals 139–144, 146–152
  role of, plants and phytochemicals
    CIPN 153–154
    diabetic neuropathy 140, 145, 153
    neuropathic pain signaling pathways 154–155
    sciatic nerve chronic constriction injury 154
neuroplasticity 138
neuroprotective strategy 162
neurorestorative strategy 162–163
neurotoxic 6-hydroxydopamine (6-OHDA) 25
neurotransmitters 84
  cholinergic signaling, modulation of 317–318
  classifications 311–315
  and functions 316–317
  phytochemicals effect on
    AChE 318
    GABAergic signaling 318–319
    glutamatergic signaling 319–320
    serotonergic and dopaminergic signaling 320–321
  sources and structures 311–315
N-formyl-methionyl-leucyl-phenylalanine (fMLP) 238
NGF *see* nerve growth factor (NGF)
nicotinamide adenine dinucleotide phosphate (NADPH) oxidase-1 (NOX-1) 154
nicotinic ionotropic receptor (nAChR) 317
Nigella sativa, medicinal properties of 304
nitric oxide (NO) 20, 162, 163, 219, 363
nitric oxide (NO)-cyclic guanosine monophosphate (cGMP)-potassium (K) channels pathway 140
nitrogenated phytochemicals
  alkaloids 21
  saponins
    astragaloside IV 21
    ginseng 21
3-nitropropionic acid (3-NP) model 373
N-methyl-D-aspartate receptor (NMDAR) 12, 54, 91, 138, 162, 281, 319
nobiletin 320
nociceptive neurotransmitters 182
nociceptive transmission 182
non-benzodiazepine (BDZ) receptor 371

non-neuronal cells 138
non-pharmacologic therapies 331
non-polymeric micelles 418
non-steroidal anti-inflammatory drugs (NSAIDs) 8, 182, 220, 327
non-stimulant medications 8
non-structural protein (NS5RNA)-dependent RNA polymerase (RdRp) 263
non-toxic botanical extract *see* PMI-5011
norepinephrine (NE) 82
normalized glycemia 172
NP *see* nanoparticles (NP)
3-NP-induced neurobehavioral deficits 373
Nrf2 expression 172
nuclear factor erythroid 2-related factor 2 (Nrf2) 22, 236
nuclear factor kappa B (NF-κB) activation 184
nutraceuticals 273
 as adjunctive therapy 284
 in Alzheimer's disease
  alkaloids 281
  apigenin 280
  carnosic acid and rosmarinic acid 281
  carotenoids 280
  crocin 281
  curcumin 280
  flavonoids 279
  genistein, daidzein, glycitin 280
  resveratrol 280
 in depression 283
 fruits, for cognition and brain health 274–278
 in improving neurodegeneration
  asiatic acid 279
  kaempferol 278
  quercetin 278
  withanine 278–279
 in Parkinson's disease 281–283
 in psychotic disorders 283
Nyctanthes arbortristis 389

o

occludin 163
Ocimum sanctum 139
ofatumumab 11
Olea europea 389
omega-3 fatty acids 283
ononitol monohydrate (OM) 233
opioids 182
opium latex 185
oral polio vaccine (OPV) 267
oranges 277
organum vasculosum laminae terminalis (OVLT) 328
ovalbumin (OVA) 238
oxaliplatin 153
oxaliplatin-induced mechanical allodynia 154
oxidative stress 400
 targeting 162
oxygen-glucose deprivation/reoxygenation [OGD/R]-induced stroke 163

p

Paeonia 24–25
Paeonia spp. 24
 P. suffruticosa Andrews 24, 65
paeoniflorin 24
pain therapy 181
Panax spp.
 P. ginseng 184
 P. notoginseng 25
Papaver somniferum 185
Parkinson's disease (PD) 9–11, 17, 222, 317
 chinese herbal medications 22–26
 herbal medicines, from India 26–27
 medicinal plants used 38–39
  components 29–37
 natural dopaminergic neuroprotective compounds 19–21
 nitrogenated phytochemicals
  alkaloids 21
  saponins 21
 pathogenesis of
  metal ion imbalance, in brain 18
  misfolding and aggregation, of proteins 17
  mitochondrial dysfunction 18
  neuroinflammation 18
  oxidative stress 19
  protease-mediated degradation inhibition 18
partial (focal/local) seizures 398
Passiflora foetida 211
passive diffusion 416
pathogen-associated molecular patterns 221
PD *see* Parkinson's disease (PD)
PDD *see* persistent depressive disorder (PDD)
pentylenetetrazole (PTZ)
 kindling model 398
 seizure test 402
pericytes 415
peripheral nervous system (PNS) 316
peroxisome proliferator-activated receptor (PPAR) signaling pathways 236
persistent depressive disorder (PDD) 205
phenolic constituents 297
phenylethanoid glycoside compound 23
pheochromocytoma (PC12) cell 61
Phoenix sylvestris 274

phorbol 12-myristate 13-acetate (PMA)-activated HepG2 cells 234
phosphoactivation of extracellular signal regulated kinase (pERK) 140
phospholipase A 162
phospholipase A2 (PLA2) 184, 219
phosphorylated Akt (p-Akt) 163
photofrin-mediated photodynamic therapy 341
Phyllanthus emblica 274–275
Phymatodes scolopendria 390
pilocarpine (PILO) tests 402
Piper spp.
  P. betel L. 389
  P. nigrum (Piperaceae) 346–347
plant-based analgesics
  drugs and mechanisms of action 182–184
  future development 194
  limitations 194
  medicinal plants found, in nigeria 185–193
  plant-derived lead compounds
    alkaloids 184–185
    flavonoids 184
    glycosides 185
    saponins 184
    terpenoids 184
plant-derived secondary metabolites 348
plant flavone 319
plums 277
PMI-5011 172
PNS *see* peripheral nervous system (PNS)
poliomyelitis virus 252
poly(ethylene glycol) (PEG) 418
poly(D,L-lactide-co-glycolide) acid (PLGA) 173
poly-ADP ribose polymerase (PARP) 23
polyamidoamine (PAMAM) dendrimers 173, 420
polyethylene glycol (PEG) 165
Polygala spp. 25
  P. tenuifolia Radix Polygalae 25
Polygonum spp. 25–26
  P. aviculare 25
  P. cuspidatum 320
  P. multiflorum Thunb 25
polyhydroxylamine dendrimers 420
polymeric micelles 418
poly(lactic-co-glycolic acid) (PLGA) nanoparticles 423
polyphenol (2)-epigallocatechin-3-gallate 374
polyphenols
  acacetin 20
  baicalein 20
  biochanin A 20
  curcumin 19–20
  fustin 20
  resveratrol 19
  sesamin 19
  6-shogaol 20
polysaccharide capsule 3
pomegranates 277
  peel extract 235
positive and negative syndrome scale (PANSS) 83
positron emission tomography/computed tomography (PET/CT) 424
posttraumatic neuralgias 138
preformed pyrogenic factors (PFPF) 330
premenstrual syndrome (PMS) 206
preoptic area (POA) 330
preprogrammed cell death (PCD) 425
presenilin 1 (PSEN1) 43
presenilin 2 (PSEN 2) 43
prophylactic medications 9
proprioception 300
prostaglandin E2 (PGE2) 154, 185, 328
prostaglandins 219
protease-activated receptor 2 (PAR2) 237
proteasome 18
protein kinase C α gene (PRKCA) 87
protein kinase C β gene (PRKCB) 87
protein phosphatase 2A (PP2A) expression 163
prunes 277
Prunus domestica 277
Pseudopterogorgia elisabethae 185
psychiatric disorders 209
psychotic disorders 283
Pterodon pubescens 140
Punica granatum 235
pyrexia 328
  development
    fever development, mechanism 328–330
    plant metabolites, as antipyretics 331–340
pyridoxal 5'-phosphate (PLP)-dependent enzyme 398
pyridoxin (vitamin B6) 281

q

quercetin 297, 366

r

rabies virus 252, 265–266
Radix Polygalae 25
rare, endangered, and threatened (RET) 94
raspberries 276
Rauvolfia serpentina 94–95, 363–364
R. communis extract (RCE) 236
reactive oxygen species (ROS) 17, 162, 172, 236, 273, 363
receptor activator of nuclear factor kappa-B ligand (RANKL) 236
receptor-mediated transport (RMT) 416
red grapes 277

relapsing-remitting MS (RRMS) 11
reserpine 95
resveratrol 240–241
reticuloendothelial system (RES) cells 418
reticulon 4 gene (RTN4) 87
reward/punishment mechanism 301
rheumatoid arthritis (RA) 304
Rhinacanthus nasutus 65
Rhizoma corydalis 21
Rhodiola rosea 212
Rhododendron hortense 268
Rhus verniciflua 20
ROS *see* reactive oxygen species (ROS)
rosemary (*Rosmarinus officinalis*) 318
rutin 155

S
sage (*Salvia pachyphylla*) 318
(+)-S-alk(en)yl-L-cysteine sulfoxide 295
S-allyl cysteine 296
S-allyl mercaptocysteine 296
Salvia triloba L. 56
saponins 184, 298
schizophrenia
  ayurvedic plants, treatment for
    Acorus calamus 88
    Allium cepa 88
    Bacopa monnieri 89
    Carum carvi 89–90
    Celastrus panicutalus 90
    Centella asiatica 90–91
    Convolvulus pluricaulis 91
    Coriandrum sativum 91
    Cuminum cyminum L. 91–92
    Eclipta alba 92
    Ficus religiosa 92–93
    Glycyrrhiza glabra 93
    Moringa oleifera 93
    Nigella sativa 93–94
    Piper longum 94
    Rauwolfia serpentina 94–95
    Sesbania grandiflora 95
    Sphaeranthus indicus 95–96
    Tinospora cordifolia 96
    Valeriana wallichii 96
    Withania somnifera 97
    Ziziphus mauritiana 97
  mechanisms of action
    dopamine receptors/transporters 82–83
    genes 87–88
    glutamate transporters 84
    phytochemicals and neurologic 98–121
    serotonin receptors/transporters 83–84
    VMAT 82
sciatic nerve
  chronic constriction injury 154
  injury 139
Scutellaria baicalensis 20, 319
secretory vesicles 82
Securidaca longipedunculata 390
selective norepinephrine reuptake inhibitors (SNRIs) 8, 33
selenium (Se) nanoparticles 173
Semecarpus anacardium 279
Senna septemtrionalis (SSE) 239
sensory/nervous system 298
serotonin-1A receptors (5-HT$_{1A}$) 84
serotonin-1B receptors (5-HT$_{1B}$) 84
serotonin 5-HT1F agonists 8
serotonin receptor 55
Sesbania grandiflora 95
severe acute respiratory syndrome (SARS) 252
shogaols 297
signal transducer and activator of transcription 3 (STAT3) 163
silexan 209
single-nucleotide polymorphisms (SNP) 18, 424
S-methyl-L-cysteine sulfoxide 296
Snedrella nodiflora 153
Solanum torvum 275
solid lipid nanoparticles (SLN) 418
solute carrier family 6 member 4 (SLC6A4) 84
solute carrier (SLC) transporter 416
Spathodea campanulate 185
Sphaeranthus indicus Kurz 387
spinal serotonergic system (5-HT$_7$) 140
S-propenyl-L-cysteines 296
S-propyl cysteine sulfoxide 295
steroids 298
strawberries 276
streptozocin (STZ)-induced diabetic rat 172
streptozotocin-induced diabetes 139
streptozotocin-induced neuropathic pain 140
Striga orobanchioides 390
stroke
  active compounds, structure of 172
  treatment of
    medicinal plants for 163–164
    nanomedicine for 165–174
    natural products for 165
  in vitro and in vivo studies on 166–171
stroke-homing peptide (SHp) 165
stromal cell-derived factor 1 (SDF-1) 164
Strychnos nux-vomica 366
subdural hematoma 302
substance-induced mood disorder (SIMD) 206
subventricular zone (SVZ) 163
superior temporal gyrus (STG) 87

superoxide dismutase (SOD) 21, 140, 154, 163, 172, 239
Sutherlandia frutescens 268
synaptic vesicle monoamine transporter gene (SVMT) 82
Syzygium guineense 185

t

tachykinin-3(TAC3) 87
tail suspension test (TST) 368
Tanacetum spp.
  T. cinerariifolium 268
  T. coccineum 268
  T. partenium 375
  T. parthenium 184
tau protein 66
T-cell signaling pathway cytokine (interleukin (IL)-2) 18
temporal lobe epilepsy (TLE) 400
Terminalia chebula 275
terpenoids 184, 298
tetrahydro berberine 21
2,3,5,4'-tetrahydroxystilbene-2-O-D-glucoside (TSG) 25
Teucrium polium L. 56
therapeutic plant 291–293
thermal hyperalgesia 137
thiobarbituric acid-reacting substances (TBARS) 90
Thymus vulgaris, medicinal properties of 305
tick-borne encephalitis (TBE) virus 252, 264
Tinospora cordifolia (TC) 26, 96
Tinospora malabarica 96
tissue polypeptide antigen (TPA) test 239
tissue-type plasminogen activator (tPA) injections 423
toll-like receptor-4 (TLR4)/NF-B 25, 328
traditional Chinese medicine (TCM) 232
transection-induced neuropathic pain 140
transforming growth factor (TGF)-β 241
transient ischemic attack (TIA) 1
transient receptor potential vanilloid 1 (TRPV1) 140, 331
transports dopamine 82
tree of life 90
Tribulus terrestris 390
Trifolium pratense 20
3,5,4'-trihydroxy-trans-stilbene 240 *see also* resveratrol
Tripterygium wilfordii 21
tumor necrosis factors (TNF) 89, 137, 184, 328
tyrosine hydroxylase-immunoreactive neuron survival 25
tyrosine kinase B (TrkB) receptors 18, 163
tyrosine 3-monooxygenase/tryptophan 5-monooxygenase activation protein zeta gene (YWHAZ) 87
tyrosine-protein kinase genes 87

u

ubiquitin proteasome system (UPS) 18
ultra-high performance liquid chromatography-tandem mass spectrometry (UHPLC-MS) 137
ultraviolet (UV) 264
US Food and Drug Administration (FDA) 208, 360

v

valerian 212
Valeriana spp.
  V. officinalis 212
  V. wallichii 96
varicella zoster virus (VZV) 252, 266
vascular endothelial cells 220
vascular endothelial growth factor (VEGF) 162
VEGF *see* vascular endothelial growth factor (VEGF)
vesicular glutamate transporter (VGTs) 82
vesicular monoamine transporters (VMATs) 82
Viola betonicifolia (Violaceace) 347–348
vitamin B6 282
von Willebrand factor (vWF) 164
VZV *see* varicella zoster virus (VZV)

w

Waltheria indica 184
water-soluble garlic polysaccharide (WSGP) 232
West Nile virus (WNV) 252, 263–264
winter cherry/poison gooseberry 26 *see also* *Withania somnifera* (WS)
Withania somnifera (WS) 26, 97, 212–213, 363
World Health Organization (WHO) 43, 182, 203, 251, 327

x

Xylopia aethiopica 153

z

zingerone 297
Zingiberaceae 297
Zingiber officinale Roscoe 297
Ziziphus spp.
  Z. jujube 318
  Z. mauritiana 97
zonula occludens-1 (ZO-1) 163

Printed and bound by CPI Group (UK) Ltd, Croydon, CR0 4YY
17/09/2026
14964258-0002